Analytic Function Theory of Several Variables

Junjiro Noguchi

Analytic Function Theory of Several Variables

Elements of Oka's Coherence

Springer

Junjiro Noguchi (Emeritus)
The University of Tokyo
Tokyo
Japan

and

Tokyo Institute of Technology
Tokyo
Japan

ISBN 978-981-10-9124-7 ISBN 978-981-10-0291-5 (eBook)
DOI 10.1007/978-981-10-0291-5

Mathematics Subject Classification (2010): 32-01, 32-03, 32Axx, 32Cxx, 32Dxx, 32Exx, 32Txx

Printed on acid-free paper

This Springer imprint is published by SpringerNature
The registered company is Springer Science+Business Media Singapore Pte Ltd.

The original version of the bookfrontmatter was revised: The copyright for the Japanese edition is included. The Erratum to the bookfrontmatter is available at
DOI 10.1007/978-981-10-0291-5_10

Preface

The title of this book was taken from the series of papers to which Dr. Kiyoshi Oka devoted his life:

"Sur les fonctions analytiques de plusieurs variables."

A term such as "complex function theory in several variables," "function theory in several complex variables," or "complex analysis in several variables" is used almost in the same sense as the present one. "Several variables" means not only the case where the independent variables are plural, but also where the dependent variables are plural, and the fundamental theory provided in this book is indispensable. The most fundamental part of the theory is the *Coherence Theorems* found and proved by K. Oka ([62], VII, VIII). These theorems together with the notion of coherence itself are indispensable, not only in the study of complex analysis, complex geometry or the theory of complex manifolds of general dimension, but also in a large area of modern Mathematics to which analytic function theory of several variables provides a foundation. For example, the theory of M. Sato's hyperfunctions is based on coherent sheaves and the Oka–Cartan Fundamental Theorem. The situation for complex function theory of one variable or the theory of Riemann surfaces is similar, when a little advanced content is involved.

The purpose of this book is to develop the theory of Oka's Coherence Theorems as a standard subject in a reasonable volume size for those students at the level of the first year of a graduate course in Mathematics, who have learned the elements of Mathematics such as the general theory of sets and topology, some algebra (groups, rings, modules, etc.), and complex function theory of one variable. It is an interesting question as to what kind of approach is the best to provide such contents in a course of Mathematics. It turns out that the best and the easiest is to begin with Oka's Coherence Theorem (Oka's First Coherence Theorem), opposite to the order in existing books, and then to deal with the Oka–Cartan Fundamental Theorem on holomorphically convex domains.

In view of the theory of Sato's hyperfunctions mentioned above, due to an introductory book by A. Kaneko ([34], p. 307) the Oka–Cartan Theorem on polynomially convex domains suffices for that purpose, and in the case of the present book it is included in the content up to Chap. 4, where the theorem is proved on holomorphically convex domains. Up to the proof of the Oka–Cartan Fundamental Theorem on holomorphically convex domains in Chap. 4, the notion of abstract manifolds will not appear. When the content at the end of Chap. 4 is presented, the definition of manifolds will have been taught in some other lectures. Then it is easy to introduce the notion of Stein manifolds, and the Oka–Cartan Fundamental Theorem on Stein manifolds.

We took account of the above considerations in organizing the materials of this book. It is intended to present the contents as comprehensively as possible for the readers who are starting to learn Mathematics. Citations from other books or sources are avoided or minimized, so that the readers just after finishing a standard textbook on complex function theory of one variable together with general topology and elementary algebra may be able to read the contents by themselves. In Chap. 2 very basic contents of algebra are cited from other books, but they may be already learned in class lectures or easily understood by referring to some textbooks. Although in Chap. 6 the existence of primitive elements in the finite field extension is cited, the facts from ring theory are proved.

The subjects taught in Mathematics major courses, such as general theory of sets and topology, complex analysis in one variable and algebra on groups, rings and modules are marvelously merged, so that such a far reaching result as the Oka–Cartan Fundamental Theorem is proved; therefore the contents of the present textbook may be suitable to be presented at the beginning of graduate courses in Mathematics. This book provides the complete self-contained proofs of the following:

- Oka's three Coherence Theorems ($\mathscr{O}_{\mathbf{C}^n}$, geometric ideal sheaves, and the normalization sheaves of complex spaces).
- The Oka–Cartan Fundamental Theorem.
- Oka's Theorem on Levi's Problem (Hartogs' Inverse Problem) for Riemann domains.

As seen in the list of references, there are already a number of excellent books on analytic function theory of several variables, each of which is specialized in its specific theme. But it is rather difficult to find a book dealing with all the above three themes in a self-contained manner at elementary level. The present textbook, for instance, should be read before reading Hörmander's book [33] on the theory of $\bar{\partial}$-equation based on the theory of Hilbert spaces, or Grauert and Remmert [27] or [28]. The present text shares a common part with those of S. Hitotsumatsu [31], R.C. Gunning and H. Rossi [29], or T. Nishino [49], but the overall structure is different, and may be easier than those for readers.

The theory presented in this textbook was established by the 1960s, and one can say without exaggeration that almost all the essential parts are due to K. Oka's ideas and his works; the central core is his coherence theorems. The standing viewpoint of this book is the one described in the introductions of Oka [62], VII and VIII. Being based on the coherence of analytic functions, one can see at a glance the path achieved in Oka [62], I–VI and can reach in a step to the forthcoming Levi's Problem of pseudoconvexity (Hartogs' Inverse Problem). When K. Oka was writing VII (*Oka's First Coherence Theorem*), he had in hand the proofs of the coherence of geometric ideal sheaves (*Oka's Second Coherence Theorem*, Oka VII, VIII) and the normalization sheaves of structure sheaves of complex spaces (*Oka's Third Coherence Theorem*, Oka VIII). In many references the coherence of geometric ideal sheaves is attributed to H. Cartan [10], but as H. Cartan pointed out in [10], K. Oka had already obtained its proof when he wrote VII. In fact, a key preparation of the proof of the coherence of geometric ideal sheaves was already discussed and proved in Oka VII (1948) (cf. Problème (K) in it), which was used by Cartan [10] (1950) and by Oka VIII (1951). In this way the works of Oka VII and VIII form one set of works: It might be the most plausible version of history that H. Cartan gave an independent proof of geometric ideal sheaves referring to Oka VII between Oka VII and VIII for his own aim of completing the program proposed in [9]. Therefore, we refer in this text to those three coherence theorems as:

- Oka's First Coherence Theorem (the sheaf of germs of holomorphic functions);
- Oka's Second Coherence Theorem (geometric ideal sheaves (ideal sheaves of analytic subsets);
- Oka's Third Coherence Theorem (normalization sheaves).

This will be discussed in more detail at Chap. 9.

In this textbook, we prove Oka's First Coherence Theorem first (Chap. 2) just after some necessary definitions and a preparation from one variable theory (Chap. 1). This is new, and different from the other existing standard monographs. In Chap. 3 we prepare the cohomology theory of sheaves. We prove the Oka–Cartan Fundamental Theorem on holomorphically convex domains in Chap. 4, at the end of which the definition of Stein manifolds is given and the Oka–Cartan Fundamental Theorem on them is proved.

In Chap. 5 we show the equivalence of domains of holomorphy and holomorphically convex domains. Then the solutions of Cousin Problems I and II and the Oka Principle are described.

In Chap. 6 we deal with the theory of analytic sets. We investigate the structure of analytic sets and prove Oka's Second Coherence Theorem claiming the coherence of geometric ideal sheaves. As a result, we see that the set of singular points of an analytic set is again analytic. Then we introduce the concept of complex spaces. After the definition of normality of structure sheaves, we prove Oka's Third Coherence Theorem on the normalization of the structure sheaf of a complex space.

In Chap. 7 we give a solution of Levi's Problem (Hartogs' Inverse Problem). K. Oka solved this in the two-dimensional case in Oka VI (1942), and then for Riemann domains (unramified covering domains) of general dimension in Oka IX

(1953). On the course we describe *plurisubharmonic functions* introduced by K. Oka VI (1942) in order to solve Levi's Problem (Hartogs' Inverse Problem).

As for Levi's Problem (Hartogs' Inverse Problem), there is an interesting comment on the reason why he dealt only with the two-dimensional case in VI, in footnote (3) of Oka VIII, Introduction: "Précisément dit, … pour le problème des convexités, nous l'avons expliqué pour les deux variables complexes, pour diminuer la répétition ultérieure inévitable". In the introduction of Oka VII (1948/1950) he had mentioned a possibility to apply his coherence theorems to this problem (but, that part was deleted by a modification by H. Cartan). Furthermore, in the first sentence of Oka VIII (1951), Oka was writing that the solution of Levi's Problem (Hartogs' Inverse Problem) for unramified covering domains over $\mathbf{C}^n$ had been written and sent in 1943 as a research report to Teiji Takagi, then professor at the University of Tokyo, which was written in Japanese. The manuscript was complete just before the translation into French. But, it was time for him to begin thinking of coherent sheaves, *idéal de domaines indéterminés* in his own terms; even the notion was not at all clear then. He probably preferred to use his time not to translate the manuscript but to concentrate on thinking of *idéal de domaines indéterminés*. Fortunately, his handwritten report to T. Takagi remains and can be seen on the website "K. Oka Library" [68] (Posthumous Papers, Vol. 1 §7, dated 12 December 1943). Up to Oka VIII, he had believed that it would be possible to solve Levi's Problem (Hartogs' Inverse Problem) even for ramified covering domains, and proved the series of the coherence theorems for that purpose. Oka, however, preferred to write his IX limited to the case of unramified covering, solving Levi's Problem (Hartogs' Inverse Problem). (Later, a counterexample for the ramified case was found, and his choice turned out to be correct.)

In Chap. 7 we deal first with domains in $\mathbf{C}^n$, and then with Riemann domains over $\mathbf{C}^n$. The method is due to Grauert's Theorem of the finite dimensionality of higher cohomologies of coherent sheaves on strongly pseudoconvex domains.

Finally, in Chap. 8 we describe the topology in the space of sections of coherent sheaves, and the convergence of holomorphic functions on a complex space in general. Then we prove the Cartan–Serre Theorem on the finite dimensionality of cohomologies of coherent sheaves over compact complex spaces, and establish the above-mentioned Grauert's Theorem on domains with strongly pseudoconvex boundary in a complex manifold. In the final section, we apply Grauert's Theorem to prove Kodaira's Embedding Theorem. It is very nice to see such a fundamental theorem, which gives a bridge of Kodaira–Hodge theory and of complex projective algebraic geometry, to be proved as an application of Grauert's Theorem, which shows a supple possibility of Oka's Coherence Theorems.

In Chap. 7, there are not many references to Chap. 6. Therefore it is possible to skip Chap. 6 to read it. On the other hand, for those readers who like to learn the basics of analytic sets and complex spaces, they may proceed with Chaps. 1–2, and then may go to Chap. 6.

This book is based on the lectures which the author has delivered intermittently for about ten years at the Department of Mathematics, the University of Tokyo. In the course of reading the notes and writing proofs from them, Professors Hideaki

Kazama and Shigeharu Takayama gave valuable suggestions. Professor Hiroshi Yamaguchi provided a great deal of advice and suggestions on the records of Professor Kiyoshi Oka. The author expresses sincere gratitude to those three professors. Writing this book, the discussions with the members of the Monday seminar at the University of Tokyo were very helpful, and some colleagues kindly provided a number of references that the author did not know. The author is grateful to all of them. In the last year the author had oppotuities to give an intensive course of the contents of this book at Kanazawa University, Kyushu University and Tokyo Institute of Technology; in particular, the lecture at Kyushu University which was arranged by Professor Joe Kamimoto was very helpful. The author thanks him deeply. Last but not least the author would like to express his deepest thanks to Mr. Hiroya Oka and Professor Akira Takeuchi. Mr. H. Oka kindly agreed with printing some pictures of Professor Kiyoshi Oka at the end of this book, which were taken from some photo albums made by Professor Akira Takeuchi.

Komaba, Tokyo
Fall 2012

Junjiro Noguchi

Added in the English Version

In the course of Grauert's proof of Oka's Theorem on Levi's Problem (Hartogs' Inverse Problem) L. Schwartz's finiteness theorem plays a key role (cf. Chap. 7), in the same way as in the Cartan–Serre Theorem (Chap. 8). The proof of L. Schwartz's finiteness theorem in the Japanese version is due to L. Bers [6], which is rather long and involved. Here in this English version, we give a very simple proof of L. Schwartz's finiteness theorem from J.-P. Demailly's notes [13].

During the preparation of the present English version, the author had the opporunity to give a series of lectures from March to May 2014 at the University of Roma II, "Tor Vergata" by kind invitation of Professor Filippo Bracci. Professor Joël Merker kindly invited the author to stay at University Paris Sud (Orsay) for a month from October to November 2014, where the author gave seminary talks on the contents of this book and had helpful discussions with him; he read through the manuscript with great care, and gave numerous useful remarks and comments. Translating Chap. 9, the author owes many suggestions and improvements of English expressions to Professor Alan Huckleberry. The author would like to express his sincere gratitude to Professors P. Bracci, J. Merker and A. Huckleberry.

Kamakura
Spring 2015

Junjiro Noguchi

Conventions

(i) The set of natural numbers (positive integers) is denoted by $\mathbf{N}$, the set of integers by $\mathbf{Z}$, the set of rational numbers by $\mathbf{Q}$, the set of real numbers by $\mathbf{R}$, the set of complex numbers by $\mathbf{C}$, and the imaginary unit by i, as usual. The set of non-negative integers (resp. numbers) is denoted by $\mathbf{Z}^+$ (resp. $\mathbf{R}^+$).

(ii) For a complex number $z = x + iy \in \mathbf{C}$ we set $\Re z = x$ and $\Im z = y$.

(iii) Theorems, equations, etc., are numbered consecutively. Here an equation is numbered as (1.1.1) with parentheses; the first 1 stands for the chapter number and the second 1 for the section number.

(iv) *Monotone increasing* and *monotone decreasing* are used in the sense including the case of equality: e.g., a sequence of functions $\{\varphi_\nu(x)\}_{\nu=1}^{\infty}$ is said to be monotone increasing if for every point x of the defining domain $\varphi_\nu(x) \leq \varphi_{\nu+1}(x)$ for all $\nu = 1, 2, \ldots$.

(v) A map $f : X \to Y$ between locally compact topological spaces is said to be *proper* if for every compact subset $K \subset Y$, the inverse image $f^{-1}K$ is also compact.

(vi) Manifolds are assumed to be connected, unless anything else is specified.

(vii) The symbol $\Subset$ stands for the relative compactness; e.g., $\Delta(a; r) \Subset U$ means that the topological closure $\overline{\Delta(a; r)}$ is compact in U.

(viii) The symbols $O(1), o(1)$, etc., follow after Landau's.

(ix) For a set S, $|S|$ denotes its cardinality.

(x) A map $f : X \to Y$ is said to be *injective* or an *injection* if $f(x_1) \neq f(x_2)$ for every distinct $x_1, x_2 \in X$, and to be *surjective* or a *surjection* if $f(X) = Y$. If f is injective and surjective, it is said to be bijective.

(xi) If a map $f : X \to Y$ between topological spaces X, Y is proper and the inverse image $f^{-1}\{y\}$ is always finite for all $y \in Y$, f is called a *finite map*. The restriction of f to a subset $E \subset X$ is denoted by $f|_E$.

(xii) A function f defined on an open subset $U \subset \mathbf{R}^m$ is said to be of C^k-class if f is k-times continuously differentiable. $C^k(U)$ denotes the set of all functions of C^k-class on U. $C_0^k(U)$ stands for the set of all $f \in C^k(U)$ with compact support.

(xiii) In general, for a differential form α we write $\alpha^k = \alpha \wedge \cdots \wedge \alpha$ (k-times).
(xiv) A polynomial in one variable with coefficients in a ring with $1(\neq 0)$ whose leading coefficient is 1 is called a monic polynomial.
(xv) A neighborhood is always assumed to be open, unless otherwise mentioned.
(xvi) A ring is commutative and contains $1 \neq 0$.

Contents

Chapter 1
Holomorphic Functions

We recall some basics from complex function theory in one variable, and then define holomorphic functions in several variables. We explain Hartogs' phenomenon, which is a special property in several variables caused by the increase in the number of variables from a single variable. We will see that the concept of "holomorphic convexity" arises naturally. In the last section, the notion of a sheaf will be introduced.

1.1 Holomorphic Functions of One Variable

Let $z = x + iy$ $(x, y \in \mathbf{R})$ denote the natural complex coordinate of the complex plane $\mathbf{C}$. We call $\Re z := x$ (resp. $\Im z := y$) the real part (resp. imaginary part) of z. In this section U denotes an open subset of $\mathbf{C}$; if it is connected, it is called a *domain*. We write $\Omega \Subset U$ if Ω is a relatively compact subset of U.

Functions are complex-valued unless otherwise mentioned.

Definition 1.1.1 A function $f : U \to \mathbf{C}$ is said to be complex differentiable at a point $a \in U$ if the following limit exists:

$$f'(a) = \lim_{h \to 0} \frac{f(a+h) - f(a)}{h}.$$

The limit $f'(a)$ is called the derivative of $f(z)$ at a. If $f(z)$ is complex differentiable at every point of U, $f(z)$ is called a *holomorphic function* on U.

By $\mathscr{O}(U)$ we denote the set of all holomorphic functions on U, which forms a ring with $1 (\neq 0)$. A composite of holomorphic functions is holomorphic.

Theorem 1.1.2 (Cauchy's Theorem) *Let $f : U \to \mathbf{C}$ be a holomorphic function. Let $\Omega \Subset U$ be an open subset such that the boundary $\partial\Omega$ consists of finitely many positively oriented (as the boundary of Ω) piecewise C^1 curves. Then we have*

J. Noguchi, *Analytic Function Theory of Several Variables*,
DOI 10.1007/978-981-10-0291-5_1

$$\int_{\partial\Omega} f(\zeta)d\zeta = 0,$$

$$f(z) = \frac{1}{2\pi i}\int_{\partial\Omega} \frac{f(\zeta)}{\zeta - z} d\zeta, \quad z \in \Omega.$$

By the integral expression above we immediately see that $f(z)$ is of C^∞-class, and the derivative $f'(z)$ as a function in $z \in \Omega$ is again holomorphic.

Definition 1.1.3 (*Normal convergence*) Let $\sum_{\nu=0}^{\infty} f_\nu$ be a series of functions on U. We say that $\sum_{\nu=0}^{\infty} f_\nu$ *normally converges* if the series of absolute values, $\sum_{\nu=0}^{\infty} |f_\nu(z)|$ converges uniformly on every compact subset of U.

In the above definition, if moreover, all f_j are holomorphic in U, then so is the limit $f(z) = \sum_{\nu=0}^{\infty} f_\nu(z)$.

We denote the open disk of radius $r > 0$ with center $a \in \mathbf{C}$ by

$$\Delta(a; r) = \{z \in \mathbf{C}; |z - a| < r\},$$

and we call the topological closure $\overline{\Delta(a; r)}$ a closed disk. We set $\Delta^*(a; r) = \Delta(a; r) \setminus \{a\}$, which is called a punctured disk.

A holomorphic function $f \in \mathcal{O}(\Delta(a; r))$ is expanded as a power series,

$$f(z) = \sum_{\nu=0}^{\infty} c_\nu (z - a)^\nu, \quad z \in \Delta(a; r). \tag{1.1.4}$$

The right-hand side normally converges. Because of this property, holomorphic functions are also called *analytic functions*.

For a C^1 function $f(z)$ we define the holomorphic partial differential and the anti-holomorphic partial differential, respectively by

$$\frac{\partial f}{\partial z} = \frac{1}{2}\left(\frac{\partial f}{\partial x} + \frac{1}{i}\frac{\partial f}{\partial y}\right), \quad \frac{\partial f}{\partial \bar{z}} = \frac{1}{2}\left(\frac{\partial f}{\partial x} - \frac{1}{i}\frac{\partial f}{\partial y}\right).$$

Remark 1.1.5 The so-called Cauchy–Riemann equations for the complex-valued function $f(z) = f(x, y) = u(x, y) + iv(x, y)$ in variable $z = x + iy$ is equivalent to $\partial f/\partial \bar{z} = 0$. If $f(z)$ is complex differentiable at a, then $f'(a) = \frac{\partial f}{\partial z}(a)$.

Proposition 1.1.6 *A function f of C^1-class on U satisfies $\partial f/\partial \bar{z} = 0$ if and only if $f \in \mathcal{O}(U)$.*

Theorem 1.1.7 (Identity theorem) *Let U be a domain of* $\mathbf{C}$ *and let $f \in \mathcal{O}(U)$.*

(i) *If there is a subset $E \subset U$ with an accumulation point in U, and $f(z) = 0$ for all $z \in E$, then $f(z) \equiv 0$ in U.*

(ii) *If there is a point* $a \in U$ *at which all derivatives* $f^{(\nu)}(a) = 0$ $(\nu = 0, 1, \ldots)$*, then* $f(z) \equiv 0$ *in* U.

We recall:

Definition 1.1.8 (*Analytic continuation*) Let U, V be open subsets of $\mathbf{C}$ with $V \not\subset U$, let V be connected, and let $f \in \mathscr{O}(U)$, $g \in \mathscr{O}(V)$. Assume that $U \cap V \neq \emptyset$, and let W be a connected component of $U \cap V$. If $f|_W = g|_W$ holds, then an analytic function h over $U \cup V$, which may be multi-valued in general, is defined by

$$h(z) = \begin{cases} f(z), & z \in U, \\ g(z), & z \in V. \end{cases}$$

We call h (sometimes, also g) an *analytic continuation* of f, and say that f is *analytically continued over* V *through* W.

By the Identity Theorem 1.1.7, h is uniquely determined if it exists (*uniqueness of analytic continuation*), but it is not necessarily one-valued.

Remark 1.1.9 Let U, V and W be as above. Then, $\partial U \cap \partial W \cap V \neq \emptyset$.

Lemma 1.1.10 *Let* U *be a domain of* $\mathbf{C}$. *Then there is a discrete sequence* $\{b_\nu\}_{\nu=1}^{\infty}$ *of distinct points of* U *without accumulation point in* U, *satisfying the following property: For every point* $c \in \partial U \cap \partial W \cap V$ *with* V *and* W *as given in Definition 1.1.8, there is a subsequence* $\{b_{\nu_\mu}\}_{\mu=1}^{\infty}$ *contained in* W, *which converges to* c.

Proof Let A be the set of all rational points (i.e., points with rational real and imaginary parts) in U. Then A is dense in U and countable. Set $A = \{a_\nu\}_{\nu=1}^{\infty}$. For $z \in U$ we set

$$\delta(z) = \sup\{r > 0;\ \Delta(z; r) \subset U\}.$$

Note that $\partial \Delta(z; \delta(z)) \cap \partial U \neq \emptyset$. For every a_ν we take a point $b_\nu \in \Delta(a_\nu; \delta(a_\nu))$ with $\delta(b_\nu) < 1/\nu$. It follows that $\{b_\nu\}_{\nu=1}^{\infty}$ is discrete and has no accumulation point in U. Here, we may take distinct b_ν, $\nu = 1, 2, \ldots$.

Let $c \in \partial U \cap \partial W \cap V$ be as given. Then there is a subsequence $\{a_{\nu_\mu}\}_{\mu=1}^{\infty}$ of $\{a_\nu\}_{\nu=1}^{\infty}$ such that $a_{\nu_\mu} \in W$ and $\lim_{\mu\to\infty} a_{\nu_\mu} = c$. It follows that $\delta(a_{\nu_\mu}) \to 0$ as $\mu \to \infty$. By the choices we see that for all sufficiently large $\mu \gg 1$, $b_{\nu_\mu} \in W$, and $\lim_{\mu\to\infty} b_{\nu_\mu} = c$. □

Theorem 1.1.11 (Riemann's Extension Theorem) *If a holomorphic function* $f \in \mathscr{O}(\Delta^*(a; r))$ *on the punctured disk is bounded around the center* a, *then there exists* $\tilde{f} \in \mathscr{O}(\Delta(a; r))$ *such that* $\tilde{f}|_{\Delta^*(a;r)} = f$; *i.e.,* f *is analytically continued over* $\Delta(a; r)$ *through* $\Delta^*(a; r)$.

Let $U_i \subset \mathbf{C}$, $i = 1, 2$, be two open sets. Assume that a holomorphic function f on U_1 takes values in U_2. If $f : U_1 \to U_2$ as a map is bijective, the inverse $f^{-1} : U_2 \to U_1$ is holomorphic. In this case we call $f : U_1 \to U_2$ a *biholomorphic map*.

As the items after the residue theorem in the course of complex function theory of one variable we may list the following three important theorems of (i) Riemann, (ii) Mittag-Leffler, and (iii) Weierstrass (cf., e.g., [52]).[1] Weierstrass's $\wp$-function (doubly periodic meromorphic function) is a special case of (ii), and the infinite product expression of $\sin z$ or $\cos z$ is a special case of (iii).

We recall them for convenience:

Theorem 1.1.12 (Riemann's Mapping Theorem) *Let $\Omega \subset \mathbf{C}$ be a simply connected domain. If $\Omega \neq \mathbf{C}$, then there exists a biholomorphic mapping $\varphi : \Omega \to \Delta(0; 1)$.*

Theorem 1.1.13 (Mittag-Leffler's Theorem) *Let $U \subset \mathbf{C}$ be a domain and let $\{a_\nu\}_{\nu=1}^{\infty}$ be a discrete subset of U without accumulation point in U. For given poles (so called, the main parts) at a_ν*

$$Q_\nu(z) = \frac{c_{\nu k_\nu}}{(z-a_\nu)^{k_\nu}} + \frac{c_{\nu k_\nu - 1}}{(z-a_\nu)^{k_\nu - 1}} + \cdots + \frac{c_{\nu 1}}{z-a_\nu}, \quad k_\nu \in \mathbf{N}, \quad c_{\nu j} \in \mathbf{C},$$

there is a meromorphic function f on U such that $f(z) - Q_\nu(z)$ is holomorphic about every a_ν.

Theorem 1.1.14 (Weierstrass' Theorem) *Let $U \subset \mathbf{C}$ be a domain, let $\{a_\nu\}_{\nu=1}^{\infty}$ be a discrete subset of U without accumulation point in U, and let $m_\nu \in \mathbf{N}$ be any given natural numbers for $\nu \in \mathbf{N}$. Then there is a holomorphic function f on U such that f has zeros of order m_ν at all a_ν and has no zeros other than $\{a_\nu\}_{\nu=1}^{\infty}$.*

In the course of the proofs of Theorems 1.1.13 and 1.1.14, Runge's Approximation Theorem plays an important role. When the domain is simply connected, it is stated as follows.

Theorem 1.1.15 (Runge's Approximation Theorem) *Let $\Omega \subset \mathbf{C}$ be a simply connected domain and let $f \in \mathscr{O}(\Omega)$. Then f is uniformly approximated by polynomials on every compact subset of Ω.*

The next important fact follows from Theorem 1.1.14:

Theorem 1.1.16 *On every domain $U \subset \mathbf{C}$ there exists a holomorphic function $f \in \mathscr{O}(U)$ whose domain of existence is U; i.e., f cannot be analytically extended over any neighborhood of every boundary point $a \in \partial U$ of U (in Riemann's sphere).*

Proof Take a discrete sequence $\{b_\nu\}_{\nu=1}^{\infty}$ of U obtained by Lemma 1.1.10. By Theorem 1.1.14 there is a function $f \in \mathscr{O}(U)$ such that f has a zero of order 1 at every b_ν and has no zero outside $\{b_\nu\}$.

Suppose that there is a connected open set V with $V \not\subset U$ and a connected component W of $U \cap V$, and that f is analytically continued over V through W. Then f would vanish on a sequence of points with an accumulation point in V. By the Identity Theorem 1.1.7, $f(z) \equiv 0$ (identically), and this is a contradiction. Therefore, f has U as a domain of existence. □

[1] As a consequence of the treatment of this book, Mittag-Leffler's and Weierstrass' theorems will be proved.

In the theory of holomorphic or meromorphic functions of one variable, Theorems 1.1.13 and 1.1.14 imply various important consequences. In general, a problem to obtain a function which has prescribed values at prescribed points is called an *interpolation problem.*

Theorem 1.1.17 (Interpolation theorem) *Let $U \subset \mathbf{C}$ be a domain and take a discrete subset $\{\mathrm{P}_\nu\}_{\nu=1}^{\infty}$ of U without accumulation point in U. Let $\{A_\nu\}_{\nu=1}^{\infty}$ be arbitrarily given numbers. Then there is a function $f \in \mathscr{O}(U)$ satisfying*

$$f(\mathrm{P}_\nu) = A_\nu, \ {}^\forall \nu = 1, 2, \ldots .$$

Proof Let $g(z) \in \mathscr{O}(U)$ be a holomorphic function which vanishes on $\{\mathrm{P}_\nu\}_\nu$ with order 1 and has no zero elsewhere (Weierstrass' Theorem). With $\alpha_\nu = g'(\mathrm{P}_\nu)(\neq 0)$, we consider the following meromorphic function (main part) about each P_ν:

$$h_\nu(z) = \frac{A_\nu}{\alpha_\nu(z - \mathrm{P}_\nu)}, \quad \nu = 1, 2, \ldots .$$

By Mittag-Leffler's Theorem there is a meromorphic function h on U having these as main parts at P_ν. Set $f(z) = g(z)h(z)$. By construction, $f \in \mathscr{O}(U)$ and f is expanded about the removable singularity P_ν as follows:

$$\begin{aligned} f(z) &= \left(\frac{A_\nu}{\alpha_\nu(z - \mathrm{P}_\nu)} + O(1) \right) \cdot (z - \mathrm{P}_\nu)\,(\alpha_\nu + O(z - \mathrm{P}_\nu)) \\ &= A_\nu + O(z - \mathrm{P}_\nu). \end{aligned}$$

Therefore, $f(\mathrm{P}_\nu) = A_\nu$, $\nu = 1, 2, \ldots$. □

By this theorem we may construct rather freely holomorphic functions on U with prescribed properties. Therefore, this guarantees a significance to investigate the general properties of holomorphic functions.

In the case of several variables, Theorem 1.1.16 does not hold anymore. This leads to a notion of a domain of holomorphy, and to notions of holomorphic convexity and of pseudoconvexity (Inverse Problem of Hartogs and the Levi Problem). Then there arises a problem whether Mittag-Leffler's Theorem 1.1.13 and Weierstrass' Theorem 1.1.14 remain valid on a domain of holomorphy (Cousin's Problems I, II); in a crucial step we need Runge's Approximation Theorem (in an extended form) as in the case of one variable.

The same arises in Interpolation Theorem 1.1.17; in the case of several variables, it depends on the shape of a domain if we may freely construct holomorphic functions on it (cf. Corollary 4.4.21 (i)).

Kiyoshi Oka solved all of these problems (see K. Oka [62], [65]). They are the contents which we will describe henceforth in this book.

1.2 Holomorphic Functions of Several Variables

1.2.1 Definitions

We write $z = (z_1, \dots, z_n)$ for the complex coordinate system of n-dimensional complex vector space $\mathbf{C}^n$. When tensor calculus appears, it is convenient to use superscripts, but in this section we use subscripts. We write z_j by the real part and the imaginary part as

$$z_j = x_j + iy_j, \quad 1 \le j \le n.$$

We define a norm $\|z\|$ of $z = (z_1, \dots, z_n)$ by

$$\|z\| = \sqrt{\sum\nolimits_{j=1}^{n} |z_j|^2} = \sqrt{\sum\nolimits_{j=1}^{n} \left(|x_j|^2 + |y_j|^2\right)}.$$

We denote a *ball* (or an open ball) of radius $r > 0$ with center $a \in \mathbf{C}^n$ by

$$B(a; r) = \{z \in \mathbf{C}^n; \|z - a\| < r\}.$$

Its boundary $\partial B(a; r) = \{z \in \mathbf{C}^n; \|z - a\| = r\}$ is called a *sphere*.

Let φ be a differentiable function and set vector fields and differential forms as follows:

$$(1.2.1)\qquad \begin{aligned} &\frac{\partial\varphi}{\partial z_j} = \frac{1}{2}\left(\frac{\partial\varphi}{\partial x_j} + \frac{1}{i}\frac{\partial\varphi}{\partial y_j}\right), && \frac{\partial\varphi}{\partial \bar z_j} = \frac{1}{2}\left(\frac{\partial\varphi}{\partial x_j} - \frac{1}{i}\frac{\partial\varphi}{\partial y_j}\right), \\ &dz_j = dx_j + idy_j, && d\bar z_j = dx_j - idy_j, \\ &\partial\varphi = \sum_{j=1}^{n} \frac{\partial\varphi}{\partial z_j} dz_j, && \bar\partial\varphi = \sum_{j=1}^{n} \frac{\partial\varphi}{\partial \bar z_j} d\bar z_j, \\ &d\varphi = \sum_{j=1}^{n} \left(\frac{\partial\varphi}{\partial x_j} dx_j + \frac{\partial\varphi}{\partial y_j} dy_j\right), && d^c\varphi = \frac{i}{4\pi}(\bar\partial\varphi - \partial\varphi). \end{aligned}$$

Here, dx_j, dy_j etc. are symbols of linearly independent vectors, duals of vector fields $\frac{\partial}{\partial x_j}$, $\frac{\partial}{\partial y_j}$, etc., and $d\varphi$ is called an exterior differential (a more general treatment will be given in Sect. 3.5). With the above notation we have

$$d\varphi = \partial\varphi + \bar\partial\varphi.$$

These symbols will be used throughout this book.

If the variables z_j are functions $z_j(\xi) = z_j(\xi_1, \dots, \xi_m)$ of class C^1 in $\xi = (\xi_1, \dots, \xi_m)$, the following hold:

(1.2.2)
$$\frac{\partial \varphi(z(\xi))}{\partial \xi_k} = \sum_{j=1}^{n} \left(\frac{\partial \varphi}{\partial z_j}(z(\xi)) \cdot \frac{\partial z_j}{\partial \xi_k}(\xi) + \frac{\partial \varphi}{\partial \bar{z}_j}(z(\xi)) \cdot \frac{\partial \bar{z}_j}{\partial \xi_k}(\xi) \right),$$
$$\frac{\partial \varphi(z(\xi))}{\partial \bar{\xi}_k} = \sum_{j=1}^{n} \left(\frac{\partial \varphi}{\partial z_j}(z(\xi)) \cdot \frac{\partial z_j}{\partial \bar{\xi}_k}(\xi) + \frac{\partial \varphi}{\partial \bar{z}_j}(z(\xi)) \cdot \frac{\partial \bar{z}_j}{\partial \bar{\xi}_k}(\xi) \right).$$

This is checked by Leibnitz's formula for a composed function in real variables.

Definition 1.2.3 A function $f : U \to \mathbf{C}$ defined on an open set $U \subset \mathbf{C}^n$ is said to be *holomorphic* if f is of C^1-class and satisfies

$$\bar{\partial} f = 0 \quad \text{on } U;$$

that is, $\partial f / \partial \bar{z}_j = 0$ on U for all j.

We denote by $\mathcal{O}(U)$ the set of all holomorphic functions on U. Then $\mathcal{O}(U)$ is a ring with unit $1 \neq 0$. For a closed set $F \subset \mathbf{C}^n$ we set

$$\mathcal{O}(F) = \{f \in \mathcal{O}(V);\ V \text{ is a neighborhood of } F\}.$$

For two open sets $\Omega_1 \subset \mathbf{C}^n$ and $\Omega_2 \subset \mathbf{C}^m$, a map

$$\varphi : \Omega_1 \to \Omega_2$$

whose components are holomorphic functions is called a *holomorphic map*. Then, if $g \in \mathcal{O}(\Omega_2)$, the composed function $f = g \circ \varphi$ is again holomorphic by (1.2.2).

If φ carries a holomorphic inverse map $\varphi^{-1} : \Omega_2 \to \Omega_1$, φ is called a *biholomorphic map*. In this case, as seen in Theorem 1.2.41 of the next section, $n = m$ follows, and Ω_1 and Ω_2 are said to be *holomorphically isomorphic*.

As in the case of one variable, a connected open subset of $\mathbf{C}^n$ is called a *domain*. Let $\Omega_1 \subset \mathbf{C}$ be a domain, and let $\Omega'_1 \Subset \Omega_1$ be a subdomain such that the boundary $C_1 = \partial \Omega'_1$ consists of oriented piecewise C^1 curves. Let $U_{n-1} \subset \mathbf{C}^{n-1}$ be an open set and set $U = \Omega_1 \times U_{n-1}$. Let $f \in \mathcal{O}(U)$, and let $z_1 \in \Omega'_1$, $z' = (z_2, \ldots, z_n) \in U_{n-1}$. Applying Theorem 1.1.2 for the variable z_1, we obtain

(1.2.4)
$$f(z_1, z') = \frac{1}{2\pi i} \int_{C_1} \frac{f(\zeta_1, z')}{\zeta_1 - z_1} d\zeta_1.$$

This holds not only for z_1, but also for other variables z_j in the same way.

Definition 1.2.5 Let $\Omega_j \subset \mathbf{C}$ $(1 \leq j \leq n)$ be n domains. Then the product $\Omega = \Omega_1 \times \cdots \times \Omega_n \subset \mathbf{C}^n$ is called a *cylinder domain*. In particular, when Ω is (affine) convex, equivalently when each domain Ω_j is (affine) convex, we call Ω a *convex cylinder domain*.

For $a = (a_1, \ldots, a_n) \in \mathbf{C}^n$ and $r = (r_1, \ldots, r_n)$, $r_j > 0$ (which we call *polyradius*) we define a *polydisk* by

$$\mathrm{P}\Delta(a; r) = \prod_{j=1}^{n} \Delta(a_j; r_j).$$

Let $f \in \mathscr{O}\left(\overline{\mathrm{P}\Delta(a; r)}\right)$. By making use of (1.2.4) for each variable successively we get

$$f(z) = \left(\frac{1}{2\pi i}\right)^n \int_{|\zeta_1 - a_1| = r_1} d\zeta_1 \cdots \int_{|\zeta_n - a_n| = r_n} d\zeta_n \frac{f(\zeta_1, \ldots, \zeta_n)}{\prod_{j=1}^n (\zeta_j - z_j)}. \tag{1.2.6}$$

In this integral expression we see that it is possible to change the order of integration and differentiation, so that

Proposition 1.2.7 *A holomorphic function is of C^∞-class.*

For a multi-index $\alpha = (\alpha_1, \ldots, \alpha_n) \in (\mathbf{Z}^+)^n$ we set

$$z^\alpha = z_1^{\alpha_1} \cdots z_n^{\alpha_n}, \ |\alpha| = \alpha_1 + \cdots + \alpha_n, \ \alpha! = \alpha_1! \cdots \alpha_n!,$$

$$\partial_j = \frac{\partial}{\partial z_j}, \qquad \partial^\alpha = \partial_1^{\alpha_1} \cdots \partial_n^{\alpha_n}.$$

Lemma 1.2.8 *Let* $0 < \theta < 1$ *be fixed. For every* $f \in \mathscr{O}(\mathrm{P}\Delta(a; r))$ *and every* $z \in \overline{\mathrm{P}\Delta(a; \theta r)}$ *we have*

$$|\partial^\alpha f(z)| \le \frac{\alpha!}{(1-\theta)^{|\alpha|+n} r^\alpha} \cdot \sup_{\mathrm{P}\Delta(a;r)} |f|.$$

Proof Choose $\theta r_j < r'_j < r_j$, $1 \le j \le n$. For $z \in \overline{\mathrm{P}\Delta(a; \theta r)}$, (1.2.6) yields

$$\partial^\alpha f(z) = \alpha! \left(\frac{1}{2\pi i}\right)^n \int_{|\zeta_1 - a_1| = r'_1} d\zeta_1 \cdots \int_{|\zeta_n - a_n| = r'_n} d\zeta_n \frac{f(\zeta_1, \ldots, \zeta_n)}{\prod_{j=1}^n (\zeta_j - z_j)^{\alpha_j + 1}}.$$

Therefore,

$$|\partial^\alpha f(z)| \le \frac{\alpha! \prod_{j=1}^n r'_j}{\prod_{j=1}^n |r'_j - \theta r_j|^{\alpha_j + 1}} \cdot \sup_{\mathrm{P}\Delta(a;r)} |f|.$$

Letting $r'_j \to r_j$, we get the desired formula. □

Theorem 1.2.9 *Let* $U \subset \mathbf{C}^n$ *be an open set, and let* $K \Subset U$ *be a compact subset. For any multi-index* $\alpha \in (\mathbf{Z}^+)^n$ *there exists a positive constant* $C_{K,\alpha}$ *such that*

$$|\partial^\alpha f(z)| \le C_{K,\alpha} \sup_U |f|, \quad {}^\forall f \in \mathscr{O}(U), \ {}^\forall z \in K.$$

Proof For any point $a \in K$ we take a polydisk $\mathrm{P}\Delta(a; r) \subset U$. As

$$K \subset \bigcup_{a \in K} \mathrm{P}\Delta\left(a; \tfrac{1}{2}r\right)$$

and K is compact, there are finitely many points $a_\nu \in K, r_\nu, 1 \le \nu \le N$ (Heine–Borel Theorem) such that

$$K \subset \bigcup_{\nu=1}^{N} \mathrm{P}\Delta\left(a_\nu; \tfrac{1}{2}r_\nu\right).$$

Lemma 1.2.8 with $\theta = \frac{1}{2}$ implies

$$|\partial^\alpha f(z)| \le \left(\max_{1 \le \nu \le N} \frac{2^{|\alpha|+n}\alpha!}{r_\nu^\alpha} \right) \cdot \sup_U |f|.$$

Thus it suffices to set $C_{K,\alpha} = \max_{1 \le \nu \le N} \frac{2^{|\alpha|+n}\alpha!}{r_\nu^\alpha}$. □

Theorem 1.2.10 *Let $\{f_\nu\}_{\nu=1}^\infty$ be a sequence of holomorphic functions on an open set U of $\mathbf{C}^n$. If $\{f_\nu\}_{\nu=1}^\infty$ converges uniformly on every compact subset of U, then the limit function is also holomorphic on U.*

Proof We know that the limit function $f = \lim_{\nu\to\infty} f_\nu$ is continuous in U. For any point $a \in U$ we take a closed polydisk $\overline{\mathrm{P}\Delta(a; r)} \Subset U$. Using (1.2.6) for f_ν, we get

$$f_\nu(z) = \left(\frac{1}{2\pi i}\right)^n \int_{|\zeta_1 - a_1| = r_1} d\zeta_1 \cdots \int_{|\zeta_n - a_n| = r_n} d\zeta_n \frac{f_\nu(\zeta_1, \ldots, \zeta_n)}{\prod_{j=1}^n (\zeta_j - z_j)},$$
$$z \in \mathrm{P}\Delta(a; r).$$

Since the convergence $f_\nu \to f$ $(\nu \to \infty)$ is uniform on $\overline{\mathrm{P}\Delta(a, r)}$, the continuous function f satisfies

$$f(z) = \left(\frac{1}{2\pi i}\right)^n \int_{|\zeta_1 - a_1| = r_1} d\zeta_1 \cdots \int_{|\zeta_n - a_n| = r_n} d\zeta_n \frac{f(\zeta_1, \ldots, \zeta_n)}{\prod_{j=1}^n (\zeta_j - z_j)},$$
$$z \in \mathrm{P}\Delta(a; r).$$

As the order of integration and partial differentiation can be changed, the right-hand side of the integral expression above implies that f is of C^∞-class, and moreover

$$\bar\partial f(z) = 0, \quad z \in \mathrm{P}\Delta(a; r).$$

Therefore, $f(z)$ is holomorphic in U. □

The definition of the *normal convergence* in n variables is the same as in Definition 1.1.3.

We consider the following power series expansion of the kernel in (1.2.6):

$$\frac{1}{\prod_{j=1}^{n}(\zeta_j - z_j)} = \prod_{j=1}^{n} \frac{1}{(\zeta_j - a_j)\left(1 - \frac{z_j - a_j}{\zeta_j - a_j}\right)}$$
$$= \prod_{j=1}^{n} \sum_{\alpha_j=0}^{\infty} \frac{(z_j - a_j)^{\alpha_j}}{(\zeta_j - a_j)^{\alpha_j+1}}, \quad z \in \mathrm{P}\Delta(a, r), \quad |\zeta_j - a_j| = r_j.$$

As the product of this right-hand side is expanded to a series in any order, it is normally convergent, and in particular absolutely convergent, so that the limit is independent from the order. From this we get a power series expansion of $f(z)$:

$$f(z) = \sum_{\alpha} c_\alpha (z - a)^\alpha. \tag{1.2.11}$$

Because of this expansion, a holomorphic function is also called an *analytic function*. The coefficients in (1.2.11) satisfy

$$\partial^\alpha f(a) = \alpha! \cdot c_\alpha, \qquad c_\alpha = \frac{1}{\alpha!}\partial^\alpha f(a). \tag{1.2.12}$$

Setting homogeneous polynomials $P_\nu(z - a) = \sum_{|\alpha|=\nu} c_\alpha (z - a)^\alpha$ of degree $\nu = 0, 1, \ldots$, we have

$$f(z) = \sum_{\nu=0}^{\infty} P_\nu(z - a), \tag{1.2.13}$$

which is called the *homogeneous polynomial expansion* of f.

When we write $f = 0$ (on U) as a function, it means that f is a constant function with value 0 (on U).

Theorem 1.2.14 (Identity theorem) *Let U be a domain, and let $f \in \mathscr{O}(U)$.*

(i) *If there exists a non-empty open subset $U' \subset U$ such that the restriction $f|_{U'} = 0$, then $f = 0$ on U.*
(ii) *If there exists a point $a \in U$ such that for every $\alpha \in (\mathbf{Z}^+)^n$, $\partial^\alpha f(a) = 0$, then $f = 0$.*

Proof We show (i) and (ii) together. Let Ω be the set of all points $z \in U$ such that there is a neighborhood $V \ni z$ with $f|_V = 0$. Then, by definition, Ω is an open subset. If $a \in \Omega$, then, of course, $\partial^\alpha f(a) = 0$ for every $\alpha \in (\mathbf{Z}^+)^n$; if this holds, then in a polydisk neighborhood $\mathrm{P}\Delta(a; r)$ where formula (1.2.11) holds, $f|_{\mathrm{P}\Delta(a;r)} = 0$, so that $a \in \Omega$.

By assumption $\Omega \neq \emptyset$. Take an accumulation point z_0 of Ω in U. Then we take a point $a \in \Omega$ close enough to z_0 such that $U \Supset \mathrm{P}\Delta(a; r) \ni z_0$. Then, $f|_{\mathrm{P}\Delta(a;r)}$ is expanded to a power series as in (1.2.11). Since $a \in \Omega$ is the center of $P\Delta(a; r)$, all coefficients $c_\alpha = 0$. Therefore, $f|_{P\Delta(a;r)} = 0$, and $z_0 \in \Omega$. We see that Ω is closed. Since U is connected, $\Omega = U$. □

Corollary 1.2.15 (i) *Let U be a domain and let $f \in \mathscr{O}(U)$ such that $f \neq 0$. Then the set $V(f) = \{z \in U;\ f(z) = 0\}$ contains no interior point.*
(ii) *If U is a domain, the ring $\mathscr{O}(U)$ is an integral domain.*

Proof (i) This immediately follows from Theorem 1.2.14 (i).
(ii) Let $f_j \in \mathscr{O}(U)$, $f_j \neq 0, j = 1, 2$. Put $W = \{z \in U;\ f_1(z) \neq 0\}$. Then W is a non-empty open subset. From (i) above follows $f_2|_W(z) \not\equiv 0$. Therefore, there is a point $a \in W$ such that $f_2(a) \neq 0$, so that $f_1(a) f_2(a) \neq 0$; i.e., $f_1 f_2 \neq 0$. □

Let a holomorphic function f be expanded to a power series in $P\Delta(a; r)$ as in (1.2.11). Taking positive numbers $s_j < r_j$, $1 \leq j \leq n$, we set $z_j = a_j + s_j e^{i\theta_j}$, $0 \leq \theta_j \leq 2\pi$, and consider the integration of $|f(a_1 + s_1 e^{i\theta_1}, \ldots, a_n + s_n e^{i\theta_n})|^2$. Noting that $\int_0^{2\pi} e^{i(\mu-\nu)\theta} d\theta = 2\pi \delta_{\mu,\nu}$ (Kronecker's symbol), we have

$$(1.2.16) \qquad \left(\frac{1}{2\pi}\right)^n \int_0^{2\pi} d\theta_1 \cdots \int_0^{2\pi} d\theta_n |f(a_1 + s_1 e^{i\theta_1}, \ldots, a_n + s_n e^{i\theta_n})|^2$$
$$= \sum_{|\alpha| \geq 0} |c_\alpha|^2 s_1^{2\alpha_1} \cdots s_n^{2\alpha_n} \geq |c_0|^2 = |f(a)|^2.$$

Theorem 1.2.17 (Maximum principle) *Let U be a domain, and let $f \in \mathscr{O}(U)$. If $|f(z)|$ attains a maximal value at a point $a \in U$, then f is a constant function.*

Proof Suppose that $|f(a)|$ $(a \in U)$ is maximal. Then there is a polydisk neighborhood $P\Delta(a; r)$ such that

$$(1.2.18) \qquad |f(z)|^2 \leq |f(a)|^2, \quad {}^\forall z \in P\Delta(a; r).$$

In $P\Delta(a; r)$ we expand $f(z) = \sum c_\alpha (z - a)^\alpha$ to a power series. It follows from (1.2.18) and (1.2.16) that

$$c_\alpha = 0, \quad {}^\forall |\alpha| > 0.$$

By the Identity Theorem 1.2.14, f is constant. □

1.2.2 Montel's Theorem

Theorem 1.2.19 *Let $\{f_\nu\}$ be a sequence of holomorphic functions on a domain Ω. Assume that $\{f_\nu\}$ is uniformly bounded on Ω. Then there is a subsequence of $\{f_\nu\}$ which converges uniformly on compact subsets of Ω, and the limit function is holomorphic in Ω.*

Proof Let M be a number such that $|f_\nu(z)| \le M$ for all ν and $z \in \Omega$. Take an arbitrary compact subset $K \Subset \Omega$. Then we take an open subset U such that $K \Subset U \Subset \Omega$. By Theorem 1.2.9 applied for $\bar{U}$, there is a positive constant C such that

$$(1.2.20) \qquad |\partial_j f_\nu(z)| \le CM, \quad z \in \bar{U}, \quad 1 \le j \le n, \quad \nu = 1, 2, \ldots.$$

We take a number $\delta > 0$ so that for every two points $z, w \in K$ with $\|z - w\| < \delta$

$$tz + (1-t)w \in U, \quad 0 \le {}^\forall t \le 1.$$

In this case it follows from (1.2.20) and Schwarz's inequality that

$$\begin{aligned} |f_\nu(z) - f_\nu(w)| &\le \left| \int_0^1 \left(\sum_{j=1}^n \partial_j f_\nu(tz + (1-t)w) \cdot (z_j - w_j) \right) dt \right| \\ &\le \sqrt{n} CM \|z - w\|, \quad z, w \in K. \end{aligned}$$

Thus, the restriction $\{f_\nu|_K\}$ to K is uniformly bounded and equicontinuous on K. The Ascoli–Arzelà Theorem implies an existence of a uniform convergent subsequence of $\{f_\nu|_K\}$.

Let $K_\mu \Subset \Omega$, $\mu = 1, 2, \ldots$, be an increasing sequence of compact subsets such that

$$K_\mu \subset K^\circ_{\mu+1}, \quad \Omega = \bigcup_{\mu=1}^\infty K^\circ_\mu,$$

where K°_μ denotes the set of its interior points. We choose a subsequence $\{f^{(1)}_{\nu(\lambda)}\}_\lambda$ of $\{f_\nu\}_\nu$ which converges uniformly on K_1. Then we choose a subsequence $\{f^{(2)}_{\nu(\lambda)}\}_\lambda$ of $\{f^{(1)}_{\nu(\lambda)}\}_\lambda$ which converges uniformly on K_2. In this way we inductively choose a subsequence $\{f^{(\mu)}_{\nu(\lambda)}\}_\lambda$ which converges uniformly on K_μ. Then the diagonal subsequence $\{f^{(\lambda)}_{\nu(\lambda)}\}_\lambda$ converges uniformly on each K_μ (Cantor's diagonal argument). That is, the subsequence $\{f^{(\lambda)}_{\nu(\lambda)}\}_\lambda$ converges uniformly on every compact subset of Ω. The limit function $f = \lim_{\lambda\to\infty} f^{(\lambda)}_{\nu(\lambda)}$ is holomorphic by Theorem 1.2.10. □

1.2.3 Approximation Theorem

This is a prototype of Runge's Theorem in several complex variables.

Theorem 1.2.21 *Let $\Omega = \prod_j \Omega_j$ be a cylinder domain such that all $\Omega_j \subset \mathbf{C}$ are simply connected domains. Let $f \in \mathscr{O}(\Omega)$. Then f can be approximated uniformly on each compact subset $K \Subset \Omega$ by polynomials; that is, for every $\varepsilon > 0$ there is a polynomial $P_\varepsilon(z_1, \ldots, z_n)$ such that*

$$|f(z) - P_\varepsilon(z)| < \varepsilon, \quad z \in K.$$

Proof By Riemann's Mapping Theorem 1.1.12 there is a biholomorphic map $\psi_j : \Omega_j \to \Delta(0; 1) = U_j$ or $\Omega_j = \mathbf{C} = U_j$ for each j. Setting a cylinder domain $U = \prod U_j$, we have a biholomorphic map $\psi : \Omega \to U$. By (1.2.11), $f \circ \psi^{-1}(\zeta_1, \ldots, \zeta_n)$ is expanded to a power series:

$$f \circ \psi^{-1}(\zeta_1, \ldots, \zeta_n) = \sum_\alpha c_\alpha \zeta^\alpha.$$

For every $\varepsilon > 0$ there is a sufficiently large $N \in \mathbf{N}$ such that

$$\left| f \circ \psi^{-1}(\zeta_1, \ldots, \zeta_n) - \sum_{|\alpha| \le N} c_\alpha \zeta^\alpha \right| < \varepsilon, \quad \zeta \in \psi(K).$$

Substituting $\zeta_j = \psi_j(z_j)$, we get

$$\left| f(z_1, \ldots, z_n) - \sum_{|\alpha| \le N} c_\alpha \psi_1(z_1)^{\alpha_1} \cdots \psi_n(z_n)^{\alpha_n} \right| < \varepsilon, \quad z \in K.$$

Let K_j denote the projection of K to the z_j-coordinate. Then $K_j \Subset \Omega_j$. By Runge's Theorem 1.1.15, $\psi_j(z_j)$ can be approximated uniformly on K_j by a polynomial $Q_j(z_j)$ in z_j. Set $Q(z) = (Q_1(z_1), \ldots, Q_n(z_n))$. Then we have

$$\left| f(z_1, \ldots, z_n) - \sum_{|\alpha| \le N} c_\alpha Q(z)^\alpha \right| < 2\varepsilon, \quad z \in K.$$

Thus it suffices to set $P_\varepsilon(z) = \sum_{|\alpha| \le N} c_\alpha Q(z)^\alpha$. □

Corollary 1.2.22 *Let $\Omega \subset \mathbf{C}^n$ be a convex cylinder domain, and let $f \in \mathscr{O}(\Omega)$. Then f can be approximated uniformly on each compact subset $K \Subset \Omega$ by polynomials.*

This is a special case of Theorem 1.2.21.

1.2.4 Analytic Continuation

The analytic continuation in several variables is defined in the same way as in the case of one variable: See Definition 1.1.8. By the Identity Theorem 1.2.14, the analytic continuation is unique, provided that it exists (*the uniqueness of analytic continuation*), but it is not necessarily one-valued.

Suppose $n = 1$, and consider any domain $U \subsetneq \mathbf{C}$. For every boundary point $a \in \partial U$ the function $f(z) = \frac{1}{z-a}$ is holomorphic in U, and cannot be analytically continued over any domain containing the point a.

In several variables ($n \geq 2$), according to the shape of a domain $U \subset \mathbf{C}^n$, there is a much larger domain $\tilde{U} \supsetneq U$ such that every $f \in \mathscr{O}(U)$ is analytically continued to $\tilde{f} \in \mathscr{O}(\tilde{U})$. This phenomenon does not occur in one variable. Furthermore, $\tilde{U}$ may not be univalent; there is an example of $\tilde{U}$ such that it covers some part of $\mathbf{C}^n$ infinitely many times. This problem will be discussed in detail in Chap. 5, but since the phenomenon is an important point that gave a motivation for the development of analytic function theory in several variables, we explain it by giving an example. It will be helpful to comprehend why the development of the theory described after Chap. 2 took such a path.

Now, we give an example of several variables, for which Theorem 1.1.16 in one variable no longer holds. Assume $n \geq 2$. Let $a = (a_1, \ldots, a_n) \in \mathbf{C}^n$ and define $\Omega_{\mathrm{H}}(a; \gamma) \subset \mathbf{C}^n$ as follows: Take n pairs of positive numbers, $\gamma = (\gamma_j)_{1 \leq j \leq n}$ and $0 < \delta_j < \gamma_j$, $1 \leq j \leq n$, and set

$$
\begin{aligned}
(1.2.23) \qquad \Omega_1 &= \{z = (z_1, \ldots, z_n) \in \mathbf{C}^n;\ |z_1 - a_1| < \gamma_1,\\
&\qquad |z_j - a_j| < \delta_j,\ 2 \leq j \leq n\},\\
\Omega_2 &= \{z = (z_1, \ldots, z_n) \in \mathbf{C}^n;\ \delta_1 < |z_1 - a_1| < \gamma_1,\\
&\qquad |z_j - a_j| < \gamma_j,\ 2 \leq j \leq n\},\\
\Omega_{\mathrm{H}}(a; \gamma) &= \Omega_1 \cup \Omega_2 \quad \text{(Fig. 1.1)}.
\end{aligned}
$$

Take any $f \in \mathscr{O}(\Omega_{\mathrm{H}}(a; \gamma))$. Taking $\delta_1 < r_1 < \gamma_1$, we have the following integral expression for $z = (z_j) \in \Omega_1$ with $|z_1 - a_1| < r_1$,

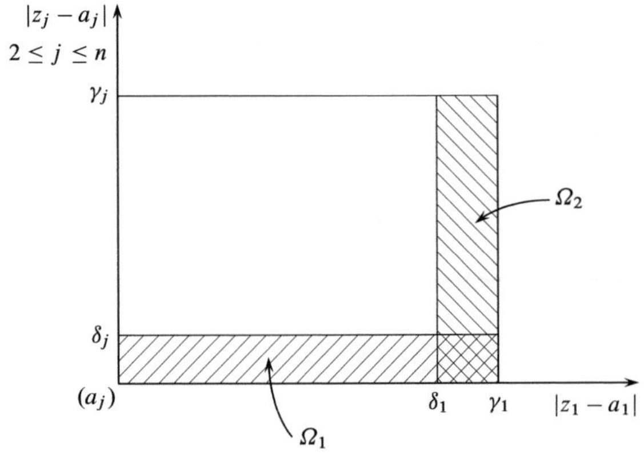

Fig. 1.1 Hartogs' domain $\Omega_{\mathrm{H}}(a; \gamma)$

(1.2.24)
$$f(z) = \frac{1}{2\pi i}\int_{|\zeta_1 - a_1| = r_1} \frac{f(\zeta_1, z_2, \ldots, z_n)}{\zeta_1 - z_1} d\zeta_1.$$

The integrant of the right-hand side has the meaning for $z \in \Omega_2$, $|z_1 - a_1| < r_1$, and the function represented by the integration gives rise to a holomorphic function in

$$\{z = (z_j); |z_1 - a_1| < r_1, |z_j - a_j| < \gamma_j,\ 2 \le j \le n\}.$$

Letting $r_1 \nearrow \gamma_1$, we see that the function f consequently is analytically continued uniquely to a function $\tilde{f} \in \mathcal{O}(\mathrm{P}\Delta(a; \gamma))$, holomorphic in the whole polydisk $\mathrm{P}\Delta(a; \gamma) = \{z = (z_j); |z_j - a_j| < \gamma_j, 1 \le j \le n\}$.

This $\Omega_{\mathrm{H}}(a; \gamma)$ is called a *Hartogs domain* with center a. Every $f \in \mathcal{O}(\Omega_{\mathrm{H}}(a; \gamma))$ is simultaneously analytically continued to $\tilde{f} \in \mathcal{O}(\mathrm{P}\Delta(a; \gamma))$, where $\Omega_{\mathrm{H}}(a; \gamma) \subsetneq \mathrm{P}\Delta(a; \gamma)$ holds. We call this phenomenon *Hartogs' phenomenon*. Summarizing the above, we have:

Theorem 1.2.25 *All holomorphic functions in a Hartogs domain $\Omega_{\mathrm{H}}(a; \gamma)$ are simultaneously analytically continued uniquely to those in the polydisk $\mathrm{P}\Delta(a; \gamma)$.*

Theorem 1.2.26 (Hartogs' Extension) *Let $\mathrm{P}\Delta(0; \rho)(\subset \mathbf{C}^n)$ be a polydisk. Let $2 \le k \le n$, and set*

$$S = \{(z_j) \in \mathrm{P}\Delta(0; \rho); z_1 = \cdots = z_k = 0\}.$$

Then every $f \in \mathcal{O}(\mathrm{P}\Delta(0; \rho)\setminus S)$ is analytically continued uniquely to an element of $\mathcal{O}(\mathrm{P}\Delta(0; \rho))$.

Proof For any point $c \in S$ we can take a Hartogs domain $\Omega_{\mathrm{H}}(a; \gamma)$ inside $\mathrm{P}\Delta(0; \rho)$ such that

$$\mathrm{P}\Delta(a; \gamma) \subset \mathrm{P}\Delta(0; \rho), \quad c \in S \cap \mathrm{P}\Delta(a; \gamma) \subset \mathrm{P}\Delta(a; \gamma)\setminus\Omega_{\mathrm{H}}(a; \gamma).$$

Then Theorem 1.2.25 implies that f gives rise to a holomorphic function in a neighborhood $\mathrm{P}\Delta(a; \gamma)$ of c. Since $c \in S$ is arbitrary, the claim follows.

This is also seen directly by integral expression (1.2.24). For simplicity, but without loss of generality, we may put $a = 0$ and $\rho = (1, \ldots, 1)$. Taking $0 < r_1 < 1$, we set

$$\tilde{f}(z_1, z_2, \ldots, z_n) = \frac{1}{2\pi i}\int_{|\zeta_1| = r_1} \frac{f(\zeta_1, z_2, \ldots, z_n)}{\zeta_1 - z_1} d\zeta_1,$$
$$(z_j) \in \mathrm{P}\Delta(0; (r_1, 1, \ldots, 1)).$$

This is holomorphic in $\mathrm{P}\Delta(0; (r_1, 1, \ldots, 1))$. If $z_2 \neq 0$, the assumption implies that $\tilde{f}(z_1, \ldots, z_n) = f(z_1, \ldots, z_n)$. Letting $r_1 \nearrow 1$, we see that $\tilde{f} \in \mathcal{O}(\mathrm{P}\Delta(0; (1, \ldots, 1)))$. □

N.B. Hartogs' Extension Theorem 1.2.26 will be generalized to the case of an analytic subset S of codimension at least 2 (Theorem 6.5.15).

To see Hartogs' phenomenon from the viewpoint of the maximum principle, we define the *holomorphically convex hull* of a subset A of a general domain $\Omega \subset \mathbf{C}^n$ by

$$(1.2.27) \qquad \hat{A}_\Omega = \{z \in \Omega;\ |f(z)| \le \sup_A |f|,\quad {}^\forall f \in \mathscr{O}(\Omega)\}.$$

Now, let $s = (s_1, \ldots, s_n)$ with $\delta_j < s_j < \gamma_j$, $1 \le j \le n$, be arbitrarily fixed, and put

$$K = \{(z_j);\ |z_j - a_j| = s_j,\ 1 \le j \le n\} \Subset \Omega_{\mathrm{H}}(a;\gamma).$$

For this compact subset we consider two sets, $\hat{K}_{\Omega_{\mathrm{H}}(a;\gamma)}$ and $\hat{K}_{\mathrm{P}\Delta(a;\gamma)}$. Because of the above Hartogs' phenomenon, we see that

$$(1.2.28) \qquad \hat{K}_{\Omega_{\mathrm{H}}(a;\gamma)} = \hat{K}_{\mathrm{P}\Delta(a;\gamma)} \cap \Omega_{\mathrm{H}}(a;\gamma).$$

By the maximum principle Theorem 1.2.17, we see that

$$\hat{K}_{\mathrm{P}\Delta(a;\gamma)} = \overline{\mathrm{P}\Delta(a;s)}.$$

While $\hat{K}_{\mathrm{P}\Delta(a;\gamma)} \Subset \mathrm{P}\Delta(a;\gamma)$, (1.2.28) implies that $\hat{K}_{\Omega_{\mathrm{H}}(a;\gamma)}$ is *not* relatively compact in $\Omega_{\mathrm{H}}(a;\gamma)$.

Taking up this property, we give the following definition:

Definition 1.2.29 We say that Ω is *holomorphically convex* if for every compact subset $K \Subset \Omega$, $\hat{K}_\Omega \Subset \Omega$.

Note that in this definition it is the same to impose the condition $\hat{K}_\Omega \Subset \Omega$ for every $K \Subset \Omega$ (without assuming the closedness of K).

In the above examples, the polydisk $\mathrm{P}\Delta(a;\gamma)$ is holomorphically convex, but the Hartogs domain $\Omega_{\mathrm{H}}(a;\gamma)$ is *not* holomorphically convex.

Moreover, from the viewpoint of the analytic continuation of holomorphic functions we are necessarily led to consider a maximal domain as a natural existence domain of holomorphic functions:

Definition 1.2.30 (i) A domain $\Omega \subset \mathbf{C}^n$ is called a *domain of holomorphy* if for any domain $V(\subset \mathbf{C}^n)$ with $V \not\subset \Omega$ and a connected component $W \subset V \cap \Omega (\neq \emptyset)$ there is an element $f_0 \in \mathscr{O}(\Omega)$ for which there is no $g \in \mathscr{O}(V)$ with $g|_W = f|_W$; i.e., for any boundary point $a \in \partial\Omega$ of Ω, there is an element $f_0 \in \mathscr{O}(\Omega)$ which cannot be analytically continued over a.

(ii) If there exists an element $f \in \mathscr{O}(\Omega)$ such that f can be analytically continued over no boundary point of Ω, Ω is called the *domain of existence* of f.

Remark 1.2.31 In the case of dimension 1 ($n = 1$) any domain always satisfies all the above properties. Let $\Omega \subset \mathbf{C}$ be a domain.

(i) Ω is holomorphically convex.
(ii) Ω is a domain of holomorphy.
(iii) Ω is the domain of existence of some $f \in \mathscr{O}(\Omega)$.

∵) If $\partial\Omega = \emptyset$, then $\Omega = \mathbf{C}$, and (i), (ii) and (iii) hold (e.g., $f(z) = z$ will suffice).

Suppose that $\partial\Omega \neq \emptyset$. For each point $a \in \partial\Omega$, $f_a(z) = \frac{1}{z-a} \in \mathscr{O}(\Omega)$ cannot be analytically continued over a; thus, (ii) follows. (iii) is Theorem 1.1.16 itself.

We show (i). Take a compact $K \Subset \Omega$. Put $R = \max_K |z|$. Then, $\hat{K}_\Omega \subset \overline{\Delta(0; R)}$. We set

$$\delta = \min\left\{|z-a|; z \in K, a \in \partial\Omega \cap \overline{\Delta(0;R)}\right\} > 0.$$

For $a \in \partial\Omega \cap \overline{\Delta(0; R)}$, $f_a \in \mathscr{O}(\Omega)$, and hence

$$\hat{K}_\Omega \subset \left\{z \in \Omega;\ \inf_{a \in \partial\Omega} |z-a| \geq \delta\right\} \cap \overline{\Delta(0;R)} \Subset \Omega.$$

Therefore, (i) follows. △

The case of $n(\geq 2)$ variables gets even more involved. In this book we will deal with holomorphically convex domains in Chap. 3, and will see the concept of holomorphic convexity to be the central pillar of analytic function theory in several variables. In Chap. 4 we will deal with domains of holomorphy, and will show that they are equivalent.

1.2.5 Implicit Function Theorem

We recall the implicit function theorem of functions in real variables. We consider the following equations given by functions of class $C^r (r \geq 1)$ in real variables:

(1.2.32)
$$\psi_j(x_1, \ldots, x_n, y_1, \ldots, y_m) = 0, \quad 1 \leq j \leq m.$$

If $\psi_j(a, b) = 0$, $1 \leq j \leq m$, are satisfied at a point (a, b) of the defined domain, and the Jacobian satisfies

(1.2.33)
$$\det\left(\frac{\partial\psi_j}{\partial y_k}(a,b)\right)_{1\leq j,k\leq m} \neq 0,$$

then there are unique solutions of simultaneous equations (1.2.32) in some neighborhood of (a, b)

(1.2.34)
$$(y_j) = (\phi_j(x_1, \ldots, x_n)), \quad b = (\phi_j(a)),$$

which are of C^r-class.

Now, we consider simultaneous equations given by holomorphic functions

$$(1.2.35)\qquad f_j(z_1, \ldots, z_n, w_1, \ldots, w_m) = 0, \quad 1 \le j \le m.$$

Its *complex Jacobi matrix* and the *complex Jacobian* are defined as

$$(1.2.36)\qquad \left(\frac{\partial f_j}{\partial w_k}\right)_{1 \le j,k \le m}, \qquad \det\left(\frac{\partial f_j}{\partial w_k}\right)_{1 \le j,k \le m}.$$

Write f_j, w_k with real and imaginary parts as

$$f_j = f_{1j} + i f_{2j}, \quad w_k = w_{1k} + i w_{2k}.$$

Then, (1.2.35) is equivalent to

$$(1.2.37)\qquad \begin{aligned} &f_{1j}(z_1, \ldots, z_n, w_{11}, w_{21}, \ldots, w_{1m}, w_{2m}) = 0, \quad 1 \le j \le m, \\ &f_{2j}(z_1, \ldots, z_n, w_{11}, w_{21}, \ldots, w_{1m}, w_{2m}) = 0, \quad 1 \le j \le m. \end{aligned}$$

We denote the Jacobian of (1.2.37) by

$$(1.2.38)\qquad \frac{\partial(f_{1j}, f_{2j})}{\partial(w_{1k}, w_{2k})} = \begin{vmatrix} \frac{\partial f_{11}}{\partial w_{11}} & \frac{\partial f_{11}}{\partial w_{21}} & \frac{\partial f_{11}}{\partial w_{12}} & \frac{\partial f_{11}}{\partial w_{22}} & \cdots \\ \frac{\partial f_{21}}{\partial w_{11}} & \frac{\partial f_{21}}{\partial w_{21}} & \frac{\partial f_{21}}{\partial w_{12}} & \frac{\partial f_{21}}{\partial w_{22}} & \cdots \\ \vdots & \vdots & \vdots & \vdots & \end{vmatrix},$$

which will be called the *real Jacobian*.

Lemma 1.2.39 *Between the real Jacobian and the complex Jacobian of holomorphic functions $f_j(z, w)$, $1 \le j \le m$ in $z = (z_1, \ldots, z_n)$ and $w = (w_1, \ldots, w_m)$, the following relation holds:*

$$\frac{\partial(f_{1j}, f_{2j})}{\partial(w_{1k}, w_{2k})} = \left|\det\left(\frac{\partial f_j}{\partial w_k}\right)\right|^2.$$

Proof We first note that the next formulae follow from Definition 1.2.3 (i.e., Cauchy–Riemann equation):

$$(1.2.40)\qquad \frac{\partial f_j}{\partial w_k} = \frac{\partial(f_{1j} + i f_{2j})}{\partial w_{1k}}, \qquad i\frac{\partial f_j}{\partial w_k} = \frac{\partial(f_{1j} + i f_{2j})}{\partial w_{2k}}.$$

We continue to compute by making use of these formula. In the determinant (1.2.38) we multiply the even row by i, and add it to the row one line before:

$$
\begin{aligned}
&\frac{\partial(f_{1j}, f_{2j})}{\partial(w_{1k}, w_{2k})} \\
&= \begin{vmatrix}
\frac{\partial(f_{11}+if_{21})}{\partial w_{11}} & \frac{\partial(f_{11}+if_{21})}{\partial w_{21}} & \frac{\partial(f_{11}+if_{21})}{\partial w_{12}} & \frac{\partial(f_{11}+if_{21})}{\partial w_{22}} & \cdots \\
\frac{\partial f_{21}}{\partial w_{11}} & \frac{\partial f_{21}}{\partial w_{21}} & \frac{\partial f_{21}}{\partial w_{12}} & \frac{\partial f_{21}}{\partial w_{22}} & \cdots \\
\vdots & \vdots & \vdots & \vdots &
\end{vmatrix}
\end{aligned}
$$

$$\left[\text{multiply odd row by } \frac{i}{2} \text{ and add it to the next line}\right]$$

$$
= \begin{vmatrix}
\frac{\partial(f_{11}+if_{21})}{\partial w_{11}} & \frac{\partial(f_{11}+if_{21})}{\partial w_{21}} & \frac{\partial(f_{11}+if_{21})}{\partial w_{12}} & \frac{\partial(f_{11}+if_{21})}{\partial w_{22}} & \cdots \\
\frac{i}{2}\frac{\partial f_{11}}{\partial w_{11}} + \frac{1}{2}\frac{\partial f_{21}}{\partial w_{11}} & \frac{i}{2}\frac{\partial f_{11}}{\partial w_{21}} + \frac{1}{2}\frac{\partial f_{21}}{\partial w_{21}} & \frac{i}{2}\frac{\partial f_{11}}{\partial w_{12}} + \frac{1}{2}\frac{\partial f_{21}}{\partial w_{12}} & \frac{i}{2}\frac{\partial f_{11}}{\partial w_{22}} + \frac{1}{2}\frac{\partial f_{21}}{\partial w_{22}} & \cdots \\
\vdots & \vdots & \vdots & \vdots & \cdots
\end{vmatrix}
$$

[here use (1.2.40)]

$$
= \begin{vmatrix}
\frac{\partial f_1}{\partial w_1} & \frac{\partial f_1}{\partial w_1} & \frac{\partial f_1}{\partial w_2} & i\frac{\partial f_1}{\partial w_2} & \cdots \\
\frac{i}{2}\overline{\frac{\partial f_1}{\partial w_{11}}} & \frac{i}{2}\overline{\frac{\partial f_1}{\partial w_{21}}} & \frac{i}{2}\overline{\frac{\partial f_1}{\partial w_{12}}} & \frac{i}{2}\overline{\frac{\partial f_1}{\partial w_{22}}} & \cdots \\
\vdots & \vdots & \vdots & \vdots & \cdots
\end{vmatrix}
$$

$$
= \begin{vmatrix}
\frac{\partial f_1}{\partial w_1} & i\frac{\partial f_1}{\partial w_1} & \frac{\partial f_1}{\partial w_2} & i\frac{\partial f_1}{\partial w_2} & \cdots \\
\frac{i}{2}\overline{\frac{\partial f_1}{\partial w_1}} & \frac{1}{2}\overline{\frac{\partial f_1}{\partial w_1}} & \frac{i}{2}\overline{\frac{\partial f_1}{\partial w_2}} & \frac{1}{2}\overline{\frac{\partial f_1}{\partial w_2}} & \cdots \\
\vdots & \vdots & \vdots & \vdots & \cdots
\end{vmatrix}
$$

[multiply odd column by $-i$ and add it to the next]

$$
= \begin{vmatrix}
\frac{\partial f_1}{\partial w_1} & 0 & \frac{\partial f_1}{\partial w_2} & 0 & \cdots \\
\frac{i}{2}\overline{\frac{\partial f_1}{\partial w_1}} & \overline{\frac{\partial f_1}{\partial w_1}} & \frac{i}{2}\overline{\frac{\partial f_1}{\partial w_2}} & \overline{\frac{\partial f_1}{\partial w_2}} & \cdots \\
\vdots & \vdots & \vdots & \vdots & \cdots
\end{vmatrix}
$$

$$
= \begin{vmatrix}
\frac{\partial f_1}{\partial w_1} & 0 & \frac{\partial f_1}{\partial w_2} & 0 & \cdots \\
0 & \overline{\frac{\partial f_1}{\partial w_1}} & 0 & \overline{\frac{\partial f_1}{\partial w_2}} & \cdots \\
\vdots & \vdots & \vdots & \vdots & \cdots
\end{vmatrix}
= \begin{vmatrix}
\frac{\partial f_j}{\partial w_k} & \vdots & O \\
\cdots & & \cdots \\
O & \vdots & \overline{\frac{\partial f_j}{\partial w_k}}
\end{vmatrix}
= \left|\det\left(\frac{\partial f_j}{\partial w_k}\right)\right|^2 . \qquad \square
$$

Theorem 1.2.41 (Implicit function theorem) *Consider simultaneous equations* (1.2.35) *defined in a neighborhood of a point* $(a, b) \in \mathbf{C}^n \times \mathbf{C}^m$. *Suppose that* $f_j(a, b) = 0, 1 \le j \le m$. *If*

$$\det \left(\frac{\partial f_j}{\partial w_k}(a, b) \right)_{1 \le j,k \le m} \neq 0, \tag{1.2.42}$$

then there exists a unique holomorphic solution

$$(w_j) = (g_j(z_1, \ldots, z_n)), \quad b = (g_j(a)) \quad (1 \le j \le m).$$

of simultaneous equations (1.2.35) *in a neighborhood of* (a, b).

Proof We regard simultaneous equations (1.2.35) as that of (1.2.37) given by real functions. By condition (1.2.42) and Lemma 1.2.39 we can apply the implicit function theorem of real functions, so that there is a unique solution of class C^∞ given by (1.2.34). Express them in complex functions,

$$w_j = g_j(z_1, \ldots, z_n), \quad b_j = g_j(a), \quad 1 \le j \le m.$$

It is done if g_j are holomorphic. In a neighborhood of $z = a$,

$$f_j(z, g_1(z), \ldots, g_m(z)) = 0, \quad 1 \le j \le m,$$

are satisfied. Taking the partial differential by $\partial/\partial \bar{z}_k$, we get from (1.2.2)

$$\frac{\partial f_j}{\partial \bar{z}_k} + \sum_{l=1}^{m} \frac{\partial f_j}{\partial w_l} \cdot \frac{\partial g_l}{\partial \bar{z}_k} + \sum_{l=1}^{m} \frac{\partial f_j}{\partial \bar{w}_l} \cdot \frac{\partial \bar{g}_l}{\partial \bar{z}_k} = 0.$$

Since $f_j(z, w)$ are holomorphic in z and w,

$$\sum_{l=1}^{m} \frac{\partial f_j}{\partial w_l} \cdot \frac{\partial g_l}{\partial \bar{z}_k} = 0, \quad 1 \le j, k \le m.$$

The coefficient matrix $\left(\frac{\partial f_j}{\partial w_l} \right)_{1 \le j,l \le m}$ of these simultaneous equations is regular by condition (1.2.42) in a neighborhood of (a, b). Therefore we see that

$$\frac{\partial g_l}{\partial \bar{z}_k} = 0, \quad 1 \le k, l \le m,$$

and that $g_l(z)$ are holomorphic. □

Theorem 1.2.43 (Inverse function theorem) *Let U and V be neighborhoods of the origin* 0 *of* $\mathbf{C}^n$, *and let*

$$f : z = (z_k) \in U \to (f_j(z)) \in V$$

be a holomorphic map whose complex Jacobian $\det\left(\frac{\partial f_j}{\partial z_k}\right) \neq 0$. *Then, taking smaller U and V if necessary, we have a holomorphic inverse $f^{-1} : V \to U$ of f.*

Proof We consider the n simultaneous equations,

$$F_j = w_j - f_j(z) = 0, \quad 1 \le j \le n.$$

Since the complex Jacobian $\det\left(\frac{\partial F_j}{\partial z_k}(z)\right) \neq 0$ in a neighborhood of 0, Implicit Function Theorem 1.2.41 implies that there are holomorphic functions $z_k = g_k(w) \in \mathscr{O}(V)$, $1 \le k \le n$, satisfying

$$w_j - f_j(g_1(w), \ldots, g_n(w)) = 0, \quad 1 \le j \le n,$$

where V is taken smaller if necessary. Putting $z_h = g_h(w)$, we have that $z_h = g_h(f_1(z), \ldots, f_n(z))$, and then $f \circ g = \mathrm{id}_V$, $g \circ f = \mathrm{id}_U$. □

Definition 1.2.44 A closed subset M of an open set $U \subset \mathbf{C}^N$ is called a *complex submanifold* if for every point $a \in M$ there is a neighborhood $W \subset U$ and holomorphic functions $f_j \in \mathscr{O}(W)$, $1 \le j \le q$, satisfying the following two conditions:

(i) $M \cap W = \{f_j = 0, 1 \le j \le q\}$.

(ii) The matrix $\left(\frac{\partial f_j}{\partial z_k}(z)\right)_{1 \le j \le q, 1 \le k \le N}$ has rank q at every point $z \in M \cap W$.

In this case Theorem 1.2.41 implies that with W taken smaller if necessary, there are a neighborhood V of $0 \in \mathbf{C}^n$ $(n = N - q)$ and a holomorphic map

$$g : v \in V \to g(v) \in W \cap M (\subset \mathbf{C}^N)$$

such that g gives a homeomorphism between U and $W \cap M$. The coordinate system $v = (v_1, \ldots, v_n)$ of V is called a *holomorphic local coordinate system* of $W \cap M$. We write $n = \dim_a M$ and call it the *dimension* of M at a, and call $\dim M = \max_{a \in M} \dim_a M$ the *dimension* of M. If a function ϕ on M is holomorphic with respect to every holomorphic local coordinate system of M, ϕ is called a holomorphic function on M. In fact, in a neighborhood of a point $a \in M$ the property of ϕ being holomorphic is independent of the choice of a holomorphic local coordinate system of a, as is easily checked by formula (1.2.2). We denote by $\mathscr{O}(M)$ the set of all holomorphic functions on M.

Theorem 1.2.45 *Let $M \subset U$ $(\subset \mathbf{C}^N)$ be a complex submanifold, let $a \in M$ be a point, and set* $\dim_a M = n$. *Then there are a neighborhood $V \subset U$ of a, a polydisk* $\mathrm{P}\Delta$ *with center $0 \in \mathbf{C}^N$ and a biholomorphic map $\Psi : z \in V \to (\psi_j(z)) = (w_j) \in \mathrm{P}\Delta$ such that*

$$M \cap V = \Psi^{-1}\{w_{n+1} = \cdots = w_N = 0\},$$

so that with $v_j = \psi_j|_{V \cap M}$, $1 \le j \le n$, $(v_j)_{1 \le j \le n}$ gives rise to a holomorphic local coordinate system in $M \cap V$.

Proof We take a neighborhood W of a and holomorphic functions $f_j \in \mathscr{O}(W)$, $1 \le j \le N - n$, as in Definition 1.2.44. After changing the indices, we may assume that the rank of $\left(\frac{\partial f_j}{\partial z_k}(z)\right)_{1 \le j \le N-n, n+1 \le k \le N}$ is $N - n$. Consider the following holomorphic map:

$$\begin{aligned} \Psi : z = (z_1, \ldots, z_n, z_{n+1}, \ldots, z_N) \in W \\ \to (z_1, \ldots, z_n, f_1(z), \ldots, f_{N-n}(z)) \in \mathbf{C}^N. \end{aligned}$$

The construction implies that the complex Jacobian matric of Ψ has rank N. Since $\Psi(a) = 0$, inverse function Theorem 1.2.43 yields the existence of a polydisk neighborhood $\mathrm{P}\Delta$ about 0 and the holomorphic inverse map

$$\Psi^{-1} : \mathrm{P}\Delta \to V = \Psi^{-1}(\mathrm{P}\Delta) \subset W.$$

Thus, $\Psi(z) = (\psi_j(z))$ satisfies the required property. □

1.3 Sheaves

1.3.1 Definition of Sheaves

Let X be a general topological space. We begin with the definition of sheaves.

Definition 1.3.1 A *sheaf* $\mathscr{S}$ over X is defined by the following three conditions:

(i) $\mathscr{S}$ is a topological space.
(ii) There is a continuous surjection $\pi : \mathscr{S} \to X$.
(iii) π is locally homeomorphic.

Furthermore, $\mathscr{S}$ is called a *sheaf of abelian groups* if

(iv) for every $x \in X$, $\mathscr{S}_x := \pi^{-1}x$ has a structure of abelian group and the algebraic operation

(1.3.2) $(u, v) \in \mathscr{S} \times_X \mathscr{S} := \{(u, v) \in \mathscr{S} \times \mathscr{S}; \pi(u) = \pi(v)\} \to u \pm v \in \mathscr{S}$

is continuous.

We call π the projection, $\mathscr{S}_x$ the *stalk* of $\mathscr{S}$ at $x \in X$ and X the base space of the sheaf $\mathscr{S}$. Besides abelian groups, sheaves of rings and sheaves of fields are defined similarly with the condition that the algebraic operations are continuous.

Example 1.3.3 Endow $\mathbf{C}$ with discrete topology, and let $\pi : \mathbf{C}_X = \mathbf{C} \times X \to X$ be the natural projection. Then, $\mathbf{C}_X$ is a sheaf of fields over X. A sheaf like this is called a *constant sheaf*. Similarly, the constant sheaves $\mathbf{R}_X \to X$, $\mathbf{Z}_X \to X$ are defined. If X is given clearly, we simply write $\mathbf{C}$, $\mathbf{R}$, $\mathbf{Z}$ for them.

Let $\mathscr{S} \to X$ be a sheaf. A *section* f of $\mathscr{S}$ over an open subset $U \subset X$ is a continuous map $f : U \to \mathscr{S}$ such that $\pi \circ f = \mathrm{id}_U$. We shall consider discontinuous sections in some cases; in such cases, we shall state it explicitly. By $\Gamma(U, \mathscr{S})$ we denote the set of all sections of $\mathscr{S}$ over U.

Let $s \in \mathscr{S}$ be a point and put $\pi(s) = x$. Then there is a neighborhood $V \ni s$ such that

$$\pi|_V : V \to U = \pi(V)(\subset X)$$

is a homeomorphism. Therefore,

$$f := (\pi|_V)^{-1} : U \to V$$

is a section and $V = f(U)$ is a neighborhood of s. The family $\{V\}$ of such neighborhoods V forms a neighborhood base of s. If $\mathscr{S}$ is a sheaf of abelian groups, by Definition 1.3.1 (iv) and the above f, it is immediate to see that $f(x) - f(x) = 0_x$ (the zero-element of $\mathscr{S}_x$) $(x \in U)$ is continuous. Therefore we obtain:

Proposition 1.3.4 (i) *The induced topology in* $\mathscr{S}_x$ *is discrete.*
(ii) *For* $f, g \in \Gamma(U, \mathscr{S})$ *and* $x \in U$,

$$f(x) = g(x) \Longleftrightarrow {}^{\exists}\text{neighborhood } V \ni x \text{ such that } f|_V = g|_V.$$

(iii) *If* $\mathscr{S}$ *is a sheaf of abelian groups,* $f \pm g \in \Gamma(U, \mathscr{S})$ *is naturally defined for* $f, g \in \Gamma(U, \mathscr{S})$.
(iv) *If* $\mathscr{S}$ *is a sheaf of abelian groups, the zero-map* $0 : x \in X \to 0_x \in \mathscr{S}$ *is a section of* $\mathscr{S}$ *over* X.

Let $\pi : \mathscr{S} \to X$ and $\eta : \mathscr{T} \to X$ be sheaves over X. A continuous map $\phi : \mathscr{S} \to \mathscr{T}$ satisfying $\pi = \eta \circ \phi$ is called a *morphism of sheaves* (or a *sheaf morphism*):

$$\begin{array}{ccccc} \phi: & \mathscr{S} & \longrightarrow & \mathscr{T} & \\ & \pi \searrow & \circlearrowleft & \swarrow \eta & \\ & & X. & & \end{array}$$

If $\mathscr{S}$ and $\mathscr{T}$ are sheaves of abelian groups or other algebraic structures, and if ϕ is compatible with the algebraic structure, then ϕ is called a *homomorphism of sheaves* or a *sheaf homomorphism*. Furthermore, if ϕ is homeomorphic, ϕ is called an *isomorphism of sheaves* or a *sheaf isomorphism*, and $\mathscr{S}$ is said to be *isomorphic* to $\mathscr{T}$.

The image of ϕ, $\operatorname{Im}\phi := \phi(\mathscr{S})$ is a sheaf over X. If ϕ is injective, we call $\mathscr{S}$ a *subsheaf* of $\mathscr{T}$; in this case, one may regard $\mathscr{S}$ as a subspace or a subset of $\mathscr{T}$.

If $\mathscr{S}$ and $\mathscr{T}$ are sheaves of abelian groups, $\operatorname{Ker}\phi = \{s \in \mathscr{S}; \phi(s) = 0\}$ is called the *kernel* of ϕ, which is a sheaf over X (cf. Ex. 9 at the end of this chapter).

When the base spaces are distinct, sheaf homomorphisms are defined as follows: For sheaves $\pi : \mathscr{S} \to X$ and $\eta : \mathscr{T} \to Y$, a continuous map $\Psi : \mathscr{S} \to \mathscr{T}$ is called a sheaf morphism if there is a continuous map $\psi : X \to Y$ satisfying

$$\begin{array}{rccc} \Psi : & \mathscr{S} & \longrightarrow & \mathscr{T} \\ & \downarrow & \circlearrowleft & \downarrow \\ \psi : & X & \longrightarrow & Y. \end{array}$$

If $\mathscr{S}$ and $\mathscr{T}$ carry an algebraic structure and if $\Psi_x : \mathscr{S}_x \to \mathscr{T}_{\psi(x)}$ is compatible with the algebraic structure, then Ψ is called a *homomorphism of sheaves* or a *sheaf homomorphism*. If Ψ and ψ are homeomorphic, Ψ is called a *sheaf isomorphism* and $\mathscr{S}$ is said to be *isomorphic* to $\mathscr{T}$.

For a homomorphism $\Psi : \mathscr{S} \to \mathscr{T}$ of sheaves of, e.g., abelian groups, we call

$$\operatorname{Supp}\Psi := \overline{\{x \in X;\ \Psi_x \neq 0, \text{ i.e., } \Psi(\mathscr{S}_x) \neq \{0\} \subset \mathscr{T}_{\psi(x)}\}}$$

the *support* of Ψ.

1.3.2 Presheaves

We present a notion of presheaves which is one step before sheaves and is convenient for constructing sheaves.

Definition 1.3.5 A triple $(\{U_\alpha\}, \{\mathscr{S}_\alpha\}, \{\rho_{\alpha\beta}\})$ (α runs over an index set) is called a *presheaf* over X if:

(i) $\{U_\alpha\}$ is a base of open sets of X,
(ii) $\mathscr{S}_\alpha$ is a set,
(iii) for every pair of $U_\alpha \subset U_\beta$, there exists a so-called *restriction map* $\rho_{\alpha\beta} : \mathscr{S}_\beta \to \mathscr{S}_\alpha$ such that

$$\rho_{\alpha\gamma} = \rho_{\alpha\beta} \circ \rho_{\beta\gamma}$$

for every $U_\alpha \subset U_\beta \subset U_\gamma$.

When each $\mathscr{S}_\alpha$ carries an algebraic structure, e.g., an abelian group, and $\rho_{\alpha\beta}$ are homomorphisms of groups, then $(\{U_\alpha\}, \{\mathscr{S}_\alpha\}, \{\rho_{\alpha\beta}\})$ is called a *presheaf of abelian groups*.

A sheaf $\mathscr{S} \to X$ naturally yields a presheaf as follows:

(i) Let $\{U_\alpha\}$ be an arbitrary base of open sets of X.
(ii) Set $\mathscr{S}_\alpha := \Gamma(U_\alpha, \mathscr{S})$.
(iii) For $U_\alpha \subset U_\beta$, let $\rho_{\alpha\beta} : f \in \mathscr{S}_\beta \to f|_{U_\alpha} \in \mathscr{S}_\alpha$ be the restriction map.

Next, for a given presheaf $(\{U_\alpha\}, \{\mathscr{S}_\alpha\}, \{\rho_{\alpha\beta}\})$, we may construct a sheaf as follows. For $x \in X$ we consider a disjoint union

$$\Sigma(x) = \bigsqcup_{U_\alpha \ni x} \mathscr{S}_\alpha.$$

An equivalence relation is introduced so that for $f_\alpha, f_\beta \in \Sigma(x)$,

$$f_\alpha \sim f_\beta \tag{1.3.6}$$

holds if and only if there is a neighborhood $U_\gamma \subset U_\alpha \cap U_\beta$ of x satisfying

$$\rho_{\gamma\alpha}(f_\alpha) = \rho_{\gamma\beta}(f_\beta).$$

By Definition 1.3.5 (iii), this is in fact an equivalence relation (cf. Ex. 10 at the end of this chapter).

We take the quotient set,

$$\mathscr{S}_x = \Sigma(x)/\sim = \varinjlim_{U_\alpha \ni x} \mathscr{S}_\alpha. \tag{1.3.7}$$

The right-hand side is called an *inductive limit* or *direct limit*. We set

$$\mathscr{S} = \bigsqcup_{x \in X} \mathscr{S}_x$$

with the natural projection $\pi : \mathscr{S} \to X$. The equivalence class of $f_\alpha \in \mathscr{S}_\alpha$ is called a *germ* and is denoted by

$$\underline{f_\alpha}_x \in \mathscr{S}_x \subset \mathscr{S}.$$

We take a subset

$$\{\underline{f_\alpha}_y ; y \in U_\alpha\}$$

for an open neighborhood of $\underline{f_\alpha}_x$, and introduce a topology on $\mathscr{S}$ in this way (cf. Ex. 11 at the end of this chapter).

We call $\mathscr{S}$ the *induced sheaf* from the presheaf $(\{U_\alpha\}, \{\mathscr{S}_\alpha\}, \{\rho_{\alpha\beta}\})$.

Proposition 1.3.8 *A presheaf* $(\{U_\alpha\}, \{\mathscr{S}_\alpha\}, \{\rho_{\alpha\beta}\})$ *of abelian groups induces a sheaf of abelian groups.*

Proof It remains to show that the group operation is continuous. We observe how $s \pm t \in \mathscr{S}$ is determined by $(s, t) \in \mathscr{S} \times_X \mathscr{S}$ (cf. (1.3.2)). Using representatives, we write $s = \underline{f_\alpha}_x$ and $t = \underline{g_\beta}_x$ with $f_\alpha \in \mathscr{S}_\alpha$ and $g_\beta \in \mathscr{S}_\beta$, respectively. Take $(x \in)$ $U_\gamma \subset U_\alpha \cap U_\beta$. Then, in $\mathscr{S}_\gamma$ we have

$$\rho_{\gamma\alpha}(f_\alpha) \pm \rho_{\gamma\beta}(g_\beta),$$

and obtain

$$s \pm t = \underline{\rho_{\gamma\alpha}(f_\alpha) \pm \rho_{\gamma\beta}(g_\beta)}_x \in \mathscr{S}_x.$$

For every neighborhood

$$\mathscr{U}_\epsilon = \{\underline{h_\varepsilon}_y; y \in U_\varepsilon\}, \quad x \in U_\varepsilon, \quad h_\varepsilon \in \mathscr{S}_\varepsilon,$$
$$\underline{h_\varepsilon}_x = s \pm t$$

of $s \pm t$, there exists a neighborhood $U_\delta \subset U_\varepsilon \cap U_\gamma$ of x such that

$$\begin{aligned}\rho_{\delta\varepsilon}(h_\varepsilon) &= \rho_{\delta\gamma}(\rho_{\gamma\alpha}(f_\alpha) \pm \rho_{\gamma\beta}(g_\beta)) \\ &= \rho_{\delta\alpha}(f_\alpha) \pm \rho_{\delta\beta}(g_\beta).\end{aligned}$$

It follows that $\mathscr{V}_\delta = \{\underline{\rho_{\delta\alpha}(f_\alpha)}_y; y \in U_\delta\}$ (resp. $\mathscr{W}_\delta = \{\underline{\rho_{\delta\beta}(g_\beta)}_y; y \in U_\delta\}$) is a neighborhood of s (resp. t), and by the definition $\mathscr{V}_\delta \pm \mathscr{W}_\delta \subset \mathscr{U}_\epsilon$. Thus, the group operation of $\mathscr{S}$ is continuous. □

Now, by the procedure presented above we see the following in general.

- A sheaf $\mathscr{S} \Longrightarrow$ a presheaf $\Longrightarrow$ returning to the sheaf $\mathscr{S}$.
- A presheaf $(\{U_\alpha\}, \{\mathscr{S}_\alpha\}, \{\rho_{\alpha\beta}\}) \Rightarrow$ a sheaf $\Rightarrow$ not returning to the presheaf $(\{U_\alpha\}, \{\mathscr{S}_\alpha\}, \{\rho_{\alpha\beta}\})$.

Example 1.3.9 (1) Let $X = \mathbf{R}$ and let $\{U_\alpha\}$ be a base of open sets of X. We set all $\mathscr{S}_\alpha = \mathbf{Z}$ and $\rho_{\alpha\beta} = 0$. Then, $(\{U_\alpha\}, \{\mathscr{S}_\alpha\}, \{\rho_{\alpha\beta}\})$ satisfies the conditions of a presheaf given by Definition 1.3.5. The induced sheaf from this presheaf is a constant sheaf $0_X \to X$ with stalks consisting only of the zero element. Therefore, the presheaf obtained from this sheaf satisfies all $\Gamma(U_\alpha, 0_X) = 0 \neq \mathscr{S}_\alpha$, which is different to the original one.

(2) Let $\mathbf{R}$ be endowed with the standard metric-topology. For any open subsets $U \supset V$ of $\mathbf{R}$ we set

$$\Gamma(U) = \mathbf{Z},$$

$$\rho_{VU} : m \in \Gamma(U) \to \begin{cases} m \in \Gamma(V), & 0 \in V \subset U, \\ 0 \in \Gamma(V), & 0 \notin V \subset U. \end{cases}$$

The triple $(\{U\}, \{\Gamma(U)\}, \{\rho_{VU}\})$ forms a presheaf. It induces a sheaf $\mathscr{S}$ such that $\mathscr{S}_0 = \mathbf{Z}$ and $\mathscr{S}_x = 0$ ($^\forall x \in \mathbf{R}$). Thus,

$$\Gamma(U, \mathscr{S}) = \begin{cases} \mathbf{Z}, & 0 \in U, \\ 0, & 0 \notin U. \end{cases}$$

The presheaf obtained from $\mathscr{S}$ is clearly different to the original one.

Definition 1.3.10 A presheaf $(\{U_\alpha\}, \{\mathscr{S}_\alpha\}, \{\rho_{\alpha\beta}\})$ is said to be *complete* if for every

$$U_\alpha = \bigcup_{\beta \in \Phi'} U_\beta,$$

the following two conditions are fulfilled:

(i) If $f_\alpha, g_\alpha \in \mathscr{S}_\alpha$ satisfy $\rho_{\beta\alpha}(f_\alpha) = \rho_{\beta\alpha}(g_\alpha)$ for all $\beta \in \Phi'$, then $f_\alpha = g_\alpha$.
(ii) If there are $f_\beta \in \mathscr{S}_\beta$, $\beta \in \Phi'$ satisfying

$$\rho_{\gamma\beta_1}(f_{\beta_1}) = \rho_{\gamma\beta_2}(f_{\beta_2})$$

for all U_γ and $\beta_1, \beta_2 \in \Phi'$ with $U_\gamma \subset U_{\beta_1} \cap U_{\beta_2}$, then there exists an element $f_\alpha \in \mathscr{S}_\alpha$ such that

$$\rho_{\beta\alpha}(f_\alpha) = f_\beta, \quad ^\forall \beta \in \Phi'.$$

Proposition 1.3.11 *A presheaf is complete if and only if it is obtained from a sheaf.*

Proof It is immediate that a presheaf obtained from a sheaf is complete. We show the converse.

Let $(\{U_\alpha\}, \{\mathscr{S}_\alpha\}, \{\rho_{\alpha\beta}\})$ be a complete presheaf, and let $\mathscr{S} \to X$ be the induced sheaf. A natural morphism $\rho_\alpha : f_\alpha \in \mathscr{S}_\alpha \to \rho_\alpha(f_\alpha) \in \Gamma(U_\alpha, \mathscr{S})$ is defined by

$$\rho_\alpha(f_\alpha)(x) = \underline{f_\alpha}_x \in \mathscr{S}_x.$$

It suffices to show:
Claim $\rho_\alpha : \mathscr{S}_\alpha \to \Gamma(U_\alpha, \mathscr{S})$ *is an isomorphism.*

∵) Injectivity: For $f_\alpha, g_\alpha \in \mathscr{S}_\alpha$, we assume that $\rho_\alpha(f_\alpha) = \rho_\alpha(g_\alpha)$. By the definition of the direct limit, there is a neighborhood $U_\beta \subset U_\alpha$ of every point $x \in U_\alpha$ such that $\rho_{\beta\alpha}(f_\alpha) = \rho_{\beta\alpha}(g_\alpha)$. We cover U_α by such U_β's: $U_\alpha = \bigcup_{\beta \in \Phi'} U_\beta$. Definition 1.3.10 (i) implies $f_\alpha = g_\alpha$.

Surjectivity: Let $s \in \Gamma(U_\alpha, \mathscr{S})$ be any element. For every point $x \in U_\alpha$ there are a neighborhood $U_\beta \subset U_\alpha$ and $f_\beta \in \mathscr{S}_\beta$ with $s(x) = \underline{f_\beta}_x$. Since s is a section, we may take a smaller U_β if necessary, so that

$$s(y) = \underline{f_\beta}_y, \quad ^\forall y \in U_\beta.$$

Considering a family $\{U_\beta\}_{\beta\in\Phi'}$ of all such U_β's, we have

$$U_\alpha = \bigcup_{\beta\in\Phi'} U_\beta,$$
$$f_\beta \in \mathscr{S}_\beta,$$
$$s(x) = \underline{f_\beta}_x, \quad x \in U_\beta.$$

For a given $U_\gamma \subset U_{\beta_1} \cap U_{\beta_2}$ $(\gamma, \beta_1, \beta_2 \in \Phi')$ it follows that

$$\rho_\gamma(\rho_{\gamma\beta_1}(f_{\beta_1}))(x) = s(x) = \rho_\gamma(\rho_{\gamma\beta_2}(f_{\beta_2}))(x), \quad x \in U_\gamma.$$

By the injectivity shown already above, we get

$$\rho_{\gamma\beta_1}(f_{\beta_1}) = \rho_{\gamma\beta_2}(f_{\beta_2}).$$

Definition 1.3.10 (ii) implies the existence of an element $f_\alpha \in \mathscr{S}_\alpha$ such that

$$\rho_{\beta\alpha}(f_\alpha) = f_\beta, \quad {}^\forall\beta \in \Phi'.$$

Thus we see that $\rho_\alpha(f_\alpha) = s$. □

1.3.3 Examples of Sheaves

The method to construct a sheaf $\mathscr{S} \to X$ over a topological space X from a presheaf has an advantage in the sense that it is clearly seen how the neighborhoods of each point $s \in \mathscr{S}$ are defined.

(1) Let $\mathscr{S}_\alpha = \{f : U_\alpha \to \mathbf{Z};$ constant function$\}$ and let $\rho_{\alpha\beta}$ be the restriction map to $U_\beta \subset U_\alpha$. Then, they form a complete sheaf, which yields a constant sheaf $\mathbf{Z} = \mathbf{Z}_X \to X$, appeared already.

(2) Let $\mathscr{S}_\alpha = \{f : U_\alpha \to \mathbf{R};$ continuous (resp. nowhere vanishing) function$\}$, and set $\rho_{\alpha\beta}$ be the restriction map to the subset. Then, they form a complete sheaf yielding a sheaf $\mathscr{C}_X$(resp. $\mathscr{C}_X^*$) $\to X$, which is called the *sheaf of germs of (resp. nowhere vanishing) continuous functions* over X. This is a sheaf of (commutative) rings. The sheaf $\mathscr{C}_X$ is not Hausdorff in general: e.g., we take $X = \mathbf{R}$ with the standard metric topology. Let $s_0 = \underline{0}_0$ denote the germ of functions which are identically 0 in neighborhoods of $0 \in \mathbf{R}$. We take a continuous function defined by

$$f(x) = \begin{cases} 0, & x \le 0, \\ e^{-1/x}, & x > 0, \end{cases} \tag{1.3.12}$$

which defines a germ $s_1 \in \mathscr{C}_\mathbf{R}$ at 0. It is clear that $s_0 \ne s_1$, But, any neighborhoods $\mathscr{U}_i \ni s_i$, $i = 0, 1$, have the intersection $\mathscr{U}_0 \cap \mathscr{U}_1 \ne \emptyset$, since $f(x) = 0$ for $x < 0$.

(3) Let X be an open subset of $\mathbf{R}^n$ (more generally, X may be a differentiable manifold). Then, we set $\mathscr{S}_\alpha = \{f : U_\alpha \to \mathbf{R};\ C^\infty\text{-function}\}$,[2] and let $\rho_{\alpha\beta}$ be the restriction map to the subset. They form a complete sheaf, yielding a sheaf $\mathscr{E}_X \to X$. The sheaf $\mathscr{E}_X$ is called the *sheaf of germs of differentiable functions* and is a sheaf of rings. Since the function defined by (1.3.12) is of class C^∞, $\mathscr{E}_X$ is not Hausdorff.

It is similar to consider complex-valued C^∞ functions. We write $\mathscr{E}_X$ for it as above; no confusion will occur since it will be explicitly stated in which sense the notation will be used. We set $\mathscr{E}_X^* = \mathscr{E}_X \cap \mathscr{C}_X^*$, as $\mathscr{E}_X \subset \mathscr{C}_X$.

(4) Let X be an open subset of $\mathbf{C}^n$ (more generally, X may be a complex manifold[3]). Let $\mathscr{S}_\alpha = \{f : U_\alpha \to \mathbf{C};$ holomorphic function$\}$, and let $\rho_{\alpha\beta}$ be the restriction map to the subset. Then, they form a complete presheaf, yielding a sheaf $\mathscr{O}_X \to X$, which is called a *sheaf of germs of holomorphic functions* over X: $\mathscr{O}_X$ is a sheaf of rings and Hausdorff by the Identity Theorem 1.2.14. It also follows from Corollary 1.2.15 (ii) that each stalk $\mathscr{O}_{X,a}$ ($a \in X$) as a ring is an integral domain. We set $\mathscr{O}_X^* = \mathscr{O}_X \cap \mathscr{E}_X^*$, as $\mathscr{O}_X \subset \mathscr{E}_X$.

Let $0 \in X \subset \mathbf{C}^n$. Every $s \in \mathscr{O}_{X,0}$ is a germ $\underline{f}_0$ of a holomorphic function $f \in \mathscr{O}(\mathrm{P}\Delta(0; r))$ on a polydisk. Then $f(z)$ is expanded to a power series:

$$f(z) = \sum_\lambda c_\lambda z^\lambda.$$

We denote by $\mathbf{C}\{z_1, \ldots, z_n\} = \mathbf{C}\{(z_j)\}$ the set of all convergent power series about the origin 0. The element $\sum_\lambda c_\lambda z^\lambda \in \mathbf{C}\{(z_j)\}$ is uniquely determined by $s \in \mathscr{O}_{X,0}$. Therefore we have an isomorphism:

$$\mathscr{O}_{X,0} \cong \mathbf{C}\{(z_j)\},$$

which is a local ring[4] with the maximal ideal

$$\mathfrak{m}_{X,0} = \left\{ \sum\nolimits_{|\lambda| \geq 1} c_\lambda z^\lambda \in \mathbf{C}\{(z_j)\} \right\}.$$

For $k \in \mathbf{N}$ we consider the k-th power $\mathfrak{m}_{X,0}^k$ of $\mathfrak{m}_{X,0}$. Then,

$$\mathfrak{m}_{X,0}^k = \left\{ \sum\nolimits_{|\lambda| \geq k} c_\lambda z^\lambda \in \mathbf{C}\{(z_j)\} \right\}.$$

The quotient $\mathscr{O}_{X,0}/\mathfrak{m}_{X,0}^k$ is isomorphic to a vector space of all polynomials with degree at most $k-1$:

$$\mathscr{O}_{X,0}/\mathfrak{m}_{X,0}^k \cong \left\{ \sum\nolimits_{|\lambda| \leq k-1} c_\lambda z^\lambda \in \mathbf{C}[z_1, \ldots, z_n] \right\}.$$

---

[2]K. Oka denoted an element of $\mathscr{S}_\alpha$ by the pair (f, U_α) (Oka [62], VII).

[3]Those readers who do not know the notion of a complex manifold should just skip this comment.

[4]In general, a ring with a unique maximal ideal is called a local ring.

These facts will be used henceforth without specific comments.

(5) Let $X \subset \mathbf{C}^n$ be an open set. For a given subset $E \subset X$ we define a presheaf by setting

$$\mathscr{S}_\alpha = \{f \in \mathscr{O}(U_\alpha);\ f(x) = 0,\ {}^\forall x \in E \cap U_\alpha\},$$

which yields a sheaf denoted by $\mathscr{I}\langle E\rangle$. Each stalk $\mathscr{I}\langle E\rangle_x$ is an ideal of the ring $\mathscr{O}_{X,x}$. We call $\mathscr{I}\langle E\rangle$ the *ideal sheaf* of the subset E. In particular, if $E = \{a\}$ $(a \in X)$, then $\mathscr{I}\langle\{a\}\rangle_a = \mathfrak{m}_{X,a}$ and $\mathscr{I}\langle\{a\}\rangle_x = \mathscr{O}_{X,x}$ at $x \neq a$.

(6) Let $\mathscr{R} \to X$ be a sheaf of rings. A sheaf $\mathscr{S} \to X$ is called a *sheaf of modules* over $\mathscr{R}$ if $\mathscr{S}$ is a sheaf of abelian groups carrying a *continuous* algebraic operation:

$$(r, s) \in \mathscr{R} \times_X \mathscr{S} \longrightarrow rs \in \mathscr{S},$$

satisfying

$$\begin{aligned} r'(rs) &= (r'r)s, \\ (r + r')s &= rs + r's, \\ r(s + s') &= rs + rs'. \end{aligned}$$

If two sheaves $\mathscr{S}$, $\mathscr{T}$ of modules over $\mathscr{R}$ are given, we define a presheaf by

$$\mathscr{S}(U) \oplus \mathscr{T}(U)$$

for open subsets U, and then we obtain the sheaf of direct sum, $\mathscr{S} \oplus \mathscr{T}$, which is a sheaf of modules over $\mathscr{R}$. The sheaf of direct sum of p $(\in \mathbf{N})$ sheaves of $\mathscr{S}$ itself is denoted by $\mathscr{S}^p$.

Let $\mathscr{T} \subset \mathscr{S}$ be a sheaf of submodules. We define a presheaf of quotient modules by

$$\mathscr{S}(U)/\mathscr{T}(U)$$

for open subsets U, which yields the *quotient sheaf* denoted by $\mathscr{S}/\mathscr{T}$. There is a natural homomorphism $\mathscr{S} \to \mathscr{S}/\mathscr{T}$. At every $x \in X$

$$(\mathscr{S}/\mathscr{T})_x \cong \mathscr{S}_x/\mathscr{T}_x,$$

but in general,

$$(\mathscr{S}/\mathscr{T})(U) \not\cong \mathscr{S}(U)/\mathscr{T}(U).$$

Example 1.3.13 We consider $\mathbf{R}$ with the standard metric topology, and the constant sheaf $\mathbf{Z}_\mathbf{R} \to \mathbf{R}$, which is a sheaf of rings. For a given subset $E \subset \mathbf{R}$ we take the ideal sheaf $\mathscr{I}\langle E\rangle$ defined by

$$\mathscr{I}\langle E\rangle_x = \begin{cases} \mathbf{Z}, & x \notin E, \\ 0, & x \in E. \end{cases}$$

Now, we take the ideal sheaf $\mathscr{I}\langle\{0,1\}\rangle$ and $U = \mathbf{R}$. Then

$$(\mathbf{Z_R}/\mathscr{I}\langle\{0,1\}\rangle)(\mathbf{R}) \cong \mathbf{Z} \oplus \mathbf{Z},$$
$$\mathbf{Z_R}(\mathbf{R})/(\mathscr{I}\langle\{0,1\}\rangle(\mathbf{R})) \cong \mathbf{Z}/\{0\} = \mathbf{Z}.$$

Therefore,

$$(\mathbf{Z_R}/\mathscr{I}\langle\{0,1\}\rangle)(\mathbf{R}) \not\cong \mathbf{Z_R}(\mathbf{R})/(\mathscr{I}\langle\{0,1\}\rangle(\mathbf{R})).$$

Note that the presheaf $(\{\mathbf{Z_R}(U)/(\mathscr{I}\langle\{0,1\}\rangle(U))\}, \{\rho_{UV}\})$, where ρ_{VU} is the restriction map for $V \subset U$, is not complete; this is an example of a non-complete presheaf.

(7) We consider the sheaf $\mathscr{O}_X$ of germs of holomorphic functions over an open set $X \subset \mathbf{C}^n$. Note that by Corollary 1.2.15 $\mathscr{O}(U)$ is an integral domain for a subdomain $U \subset X$.

Definition 1.3.14 We define a presheaf by associating every subdomain $U \subset X$ with the quotient field $\mathscr{M}(U) = \mathscr{O}(U)/(\mathscr{O}(U)\setminus\{0\})$, which yields a sheaf $\mathscr{M}_X$ of fields over X; $\mathscr{M}_X$ is called the *sheaf of germs of meromorphic functions* over X. A section $f \in \Gamma(X, \mathscr{M}_X)$ is called a *meromorphic function* on X. If X is a domain, $\Gamma(X, \mathscr{M}_X)$ is a field.

(8) Let $f : X \to Y$ be a continuous map between topological spaces. Let $\mathscr{S} \to X$ be a sheaf over X. We define a presheaf $(\{V\}, \{\Gamma(f^{-1}V, \mathscr{S})\}, \{\rho_{VV'}\})$ by taking $\Gamma(f^{-1}V, \mathscr{S})$ for an open subset $V \subset Y$ and the restriction map $\rho_{VV'}$ for $V' \subset V$. This presheaf yields a sheaf denoted by $f_*\mathscr{S} \to Y$, which is called the *direct image sheaf* of $\mathscr{S}$ by f.

Exercises

1. Show Remark 1.1.5.
2. Show Remark 1.1.9.
3. (Finite interpolation). Let $a_\nu \in \mathbf{C}$, $1 \le \nu \le q$, be finitely many distinct points, and let $\alpha_\nu \in \mathbf{C}$, $1 \le \nu \le q$, be arbitrary numbers. Then, construct a polynomial $P(z)$ in z such that $P(a_\nu) = \alpha_\nu$ for all ν. (Hint: First consider the case of $q = 2$, and then generalize it to $q \ge 2$.)
4. Let $\alpha_\nu \in \mathbf{C}$, $\nu \in \mathbf{Z}$, be numbers such that for some $k \in \mathbf{N}$,

$$\sum_{\nu \in \mathbf{Z}} \frac{|\alpha_\nu|}{|\nu|^k} < \infty.$$

Show that the series

$$f(z) = (\sin \pi z)^k \sum_{\nu \in \mathbf{Z}} \frac{(-1)^{\nu k} \alpha_\nu}{\pi^k (z-\nu)^k}$$

converges normally to an entire function $f(z)$ satisfying $f(\nu) = \alpha_\nu$ for all $\nu \in \mathbf{Z}$.

5. Show (1.2.2).
6. Prove Liouville's theorem on $\mathbf{C}^n$: Any bounded holomorphic function on $\mathbf{C}^n$ is constant.
7. Let $\Omega \subset \mathbf{C}^n$ be a domain, and let $a \in \Omega$. Set $E = (a + \mathbf{R}^n) \cap \Omega$ (or, $(a + (i\mathbf{R})^n) \cap \Omega$). Show that if $f \in \mathscr{O}(\Omega)$ vanishes on E, then f vanishes on Ω.
8. Let f be a holomorphic function on $\mathbf{C}^n$ such that there are constants $C > 0$ and $k \in \mathbf{N}$ satisfying $|f(z)| \le C\|z\|^d$ for all $z \in \mathbf{C}^n$. Then, prove that f is a polynomial of degree at most k in z.
9. Let $\phi : \mathscr{S} \to \mathscr{T}$ be a morphism of sheaves of abelian groups over a topological space X. Show that $\operatorname{Im} \phi$ and $\operatorname{Ker} \phi$ are sheaves of abelian groups over X.
10. Prove that relation (1.3.6) is an equivalence relation.
11. Let $(\{U_\alpha\}, \{\mathscr{S}_\alpha\}, \{\rho_{\alpha\beta}\})$ be a presheaf over X, and let $f_\alpha \in \mathscr{S}_\alpha$. Show that $\{\underline{f_\alpha}_y ; y \in U_\alpha\}$ with all $U_\alpha \ni x$, satisfies the axioms of a base of neighborhoods.
12. Show that $\mathscr{O}_X$ is Hausdorff.
13. Show that the presheaf $(\{V\}, \{\Gamma(f^{-1}V, \mathscr{S})\}, \{\rho_{VV'}\})$ in Sect. 1.3.3 (8) is complete.

Chapter 2
Oka's First Coherence Theorem

We study the local properties of holomorphic functions. The main object is Oka's Coherence Theorem that plays the most fundamental and important role in analytic function theory in several variables. The notion of a coherent sheaf gives rise to a fundamental terminology in a broad area of modern Mathematics. Oka originally termed the notion as "idéal de domaines indéterminés". The wording "coherence" comes from H. Cartan's naming, "faisceau cohérent".

The word "cohérent" means "holding a logical compatibility", different from the Japanese counter-part "Rensetu", which is not a direct translation of "Coherent" and means "being contacted continuously". In view of the original notion of "de domaines indéterminés" or "Coherent", the Japanese wording "Rensetu" sounds appropriate for the meaning.

2.1 Weierstrass' Preparation Theorem

Let $\mathrm{P}\Delta(a; r) \subset \mathbf{C}^n$ be a polydisk and let $f \in \mathscr{O}(\mathrm{P}\Delta(a; r))$. Assume that $f \neq 0$ ($f(z) \not\equiv 0$). Then $f(z)$ is expanded to a power series as follows:

$$f(z) = \sum_{\lambda} c_\lambda (z-a)^\lambda = \sum_{\nu=\nu_0}^{\infty} P_\nu(z-a),$$
$$P_\nu(z-a) = \sum_{|\lambda|=\nu} c_\lambda (z-a)^\lambda$$
$$\text{(a homogeneous polynomial of degree } \nu),$$
$$P_{\nu_0}(z-a) \not\equiv 0.$$

The degree ν_0 of the first term $P_{\nu_0}(z-a)$ is called the *order of zero* of f at a, denoted by $\mathrm{ord}_a f$.

J. Noguchi, *Analytic Function Theory of Several Variables*,
DOI 10.1007/978-981-10-0291-5_2

For the sake of simplicity, we let $a = 0$ by translation. Assume that $f(0) = 0$ ($\nu_0 \geq 1$). Take a vector $v \in \mathbf{C}^n \setminus \{0\}$ such that $P_{\nu_0}(v) \neq 0$. For $\zeta \in \mathbf{C}$ we have

$$f(\zeta v) = \sum_{\nu=\nu_0}^{\infty} \zeta^{\nu} P_{\nu}(v) = \zeta^{\nu_0}(P_{\nu_0}(v) + \zeta P_{\nu_0+1}(v) + \cdots).$$

By a linear transformation of the coordinate system we choose a new coordinate system $z = (z_1, \ldots, z_n)$, so that $v = (0, \ldots, 0, 1)$. We write

$$\mathrm{P}\Delta(0; r) = \mathrm{P}\Delta_{n-1} \times \Delta(0; r_n) \subset \mathbf{C}^{n-1} \times \mathbf{C},$$

and for the coordinate system

$$\begin{aligned} z &= (z', z_n) \in \mathrm{P}\Delta_{n-1} \times \Delta(0; r_n), \\ 0 &= (0, 0). \end{aligned}$$

With respect to this coordinate system we assume the following conditions.

2.1.1 (i) f is holomorphic in a neighborhood of the closed polydisk $\overline{\mathrm{P}\Delta(0; r)}$, and the homogeneous polynomial expansion $f(z) = \sum_{\nu=\nu_0}^{\infty} P_{\nu}(z)$ satisfies that $P_{\nu_0}(0, 1) \neq 0$ and

$$f(0, z_n) = z_n^{\nu_0}(P_{\nu_0}(0, 1) + z_n P_{\nu_0+1}(0, 1) + \cdots).$$

(ii) Take $r_n > 0$ sufficiently small, so that $\{|z_n| \leq r_n;\ f(0, z_n) = 0\} = \{0\}$.

(iii) If we take small $r_1, \ldots, r_{n-1} > 0$, depending on r_n, the roots z_n of $f(z', z_n) = 0$ for every $z' \in \overline{\mathrm{P}\Delta_{n-1}}$ are contained in the disk $\Delta(0; r_n)$; in particular, $|f(z', z_n)| > 0$ for all $(z', z_n) \in \overline{\mathrm{P}\Delta_{n-1}} \times \{|z_n| = r_n\}$.

For $\underline{f}_0 \in \mathscr{O}_{\mathbf{C}^n,0}$, a polydisk $\mathrm{P}\Delta(0; r)$ satisfying 2.1.1 (i)–(iii) above is called the *standard polydisk* of $\underline{f}_0$ or f. The coordinate system $z = (z_1, \ldots, z_n)$ is called the *standard coordinate system* of $\underline{f}_0$.

Remark 2.1.2 (i) The standard polydisks of $\underline{f}_0$ form a basis of neighborhoods about 0, because $r_n > 0$ can be chosen arbitrarily small and then, depending on it, r_j, $1 \leq j \leq n-1$, are chosen arbitrarily small.

(ii) Since $\{v \in \mathbf{C}^n;\ P_{\nu_0}(v) = 0\}$ contains no interior point, the standard coordinate system and the standard polydisk can be chosen to be the same for finitely many $\underline{f_k}_0 \in \mathscr{O}_{\mathbf{C}^n,0} \setminus \{0\}$, $1 \leq k \leq l (< \infty)$, with $f_k(0) = 0$.

(iii) The direction vector $v \in \mathbf{C}^n \setminus \{0\}$ can be chosen to be the same for countably many $\underline{f_k}_0 \in \mathscr{O}_{\mathbf{C}^n,0} \setminus \{0\}$ with $f_k(0) = 0$, $k = 1, 2, \ldots$ For, with denoting $P_{k\nu_k}(z)$ the first non-zero term in the homogeneous polynomial expansion of $f_k(z)$, the set $A = \bigcup_{k=1}^{\infty} \{P_{k\nu_k}(v) = 0\}$ is a countable union of closed subsets containing no interior point. Baire's Category Theorem implies that A contains no interior point. Therefore, $\mathbf{C}^n \setminus A \neq \emptyset$, and one may take $v \in \mathbf{C}^n \setminus A$. It follows that for the countable family $\{\underline{f_{k}}_0\}$ one can take the same standard coordinate system of all $\underline{f_{k}}_0$. (It is not possible in general to take the common standard neighborhood of all $\underline{f_{k}}_0$.)

We write $\mathscr{O}(E)$ for the set of all holomorphic functions in neighborhoods of a closed subset $E \subset \mathbf{C}^n$. We define the sup-norm of a function g on a subset W of the domain of definition by

$$\|g\|_W = \sup_{z \in W} |g(z)|.$$

Theorem 2.1.3 (Weierstrass' Preparation Theorem) *Let* $\underline{f}_0 \in \mathscr{O}_{\mathbf{C}^n,0} \setminus \{0\}$, $f(0) = 0$, $p = \mathrm{ord}_0 f$, *and let* $\mathrm{P}\Delta = \mathrm{P}\Delta_{n-1} \times \Delta(0; r_n)$ $(\ni z = (z', z_n))$ *be the standard polydisk of* f.

(i) *There exist unique holomorphic functions,* $a_j \in \mathscr{O}(\overline{\mathrm{P}\Delta_{n-1}})$ *with* $a_j(0) = 0$, $1 \le j \le p$, *and zero-free* $u \in \mathscr{O}(\overline{\mathrm{P}\Delta})$ *such that*

$$f(z) = f(z', z_n) = u(z)\left(z_n^p + \sum_{j=1}^{p} a_j(z')z_n^{p-j}\right), \qquad (z', z_n) \in \overline{\mathrm{P}\Delta_{n-1}} \times \overline{\Delta(0; r_n)}. \tag{2.1.4}$$

(ii) *For every* $\varphi \in \mathscr{O}(\mathrm{P}\Delta)$ *there are unique holomorphic functions,* $a \in \mathscr{O}(\mathrm{P}\Delta)$ *and* $b_j \in \mathscr{O}(\mathrm{P}\Delta_{n-1})$, $1 \le j \le p$, *satisfying*

$$\varphi(z) = af + \sum_{j=1}^{p} b_j(z')z_n^{p-j}, \quad z = (z', z_n) \in \mathrm{P}\Delta_{n-1} \times \Delta(0; r_n). \tag{2.1.5}$$

(iii) *In* (ii) *there is a constant* $M > 0$ *depending only on* f, *independent of* φ, *such that*

$$\|a\|_{\mathrm{P}\Delta} \le M\|\varphi\|_{\mathrm{P}\Delta}, \qquad \|b_j\|_{\mathrm{P}\Delta_{n-1}} \le M\|\varphi\|_{\mathrm{P}\Delta}.$$

Proof (i) For $k \in \mathbf{Z}^+$ we set

$$\sigma_k(z') = \frac{1}{2\pi i}\int_{|z_n|=r_n} z_n^k \frac{\frac{\partial f}{\partial z_n}(z', z_n)}{f(z', z_n)} dz_n, \quad z' \in \overline{\mathrm{P}\Delta_{n-1}}. \tag{2.1.6}$$

It follows that $\sigma_k \in \mathscr{O}(\overline{\mathrm{P}\Delta_{n-1}})$. By the residue theorem $\sigma_0(z') \in \mathbf{Z}$ and hence the continuity implies

$$\sigma_0(z') \equiv \sigma_0(0) = \frac{1}{2\pi i}\int_{|z_n|=r_n} \frac{\frac{\partial f}{\partial z_n}(0, z_n)}{f(0, z_n)} dz_n = p.$$

Therefore, as $z' \in \mathrm{P}\Delta_{n-1}$ is fixed, the number of roots of $f(z', z_n) = 0$ with counting multiplicities is identically p. We write $\zeta_1(z'), \ldots, \zeta_p(z')$ for them with counting multiplicities. Again from the residue theorem we get

$$\sigma_k(z') = \sum_{j=1}^{p} (\zeta_j(z'))^k, \quad k = 0, 1, \ldots.$$

We set the elementary symmetric polynomial of degree ν in $\zeta_1(z'), \ldots, \zeta_p(z')$:

$$a_\nu(z') = (-1)^\nu \sum_{1 \le j_1 < \cdots < j_\nu \le p} \zeta_{j_1}(z') \cdots \zeta_{j_\nu}(z').$$

Then, $a_\nu(z') \in \mathbf{Q}[\sigma_1(z'), \ldots, \sigma_\nu(z')]$ (cf. Exercise 1 at the end of this chapter). For instance, when $p = 2$,

$$a_2(z') = \zeta_1(z')\zeta_2(z') = \frac{1}{2}\sigma_1(z')^2 - \frac{1}{2}\sigma_2(z').$$

Noting all $\zeta_j(0) = 0$, we see that $a_\nu \in \mathscr{O}(\overline{\mathrm{P}\Delta_{n-1}})$, $a_\nu(0) = 0$, $\nu \ge 1$. We put

$$W(z', z_n) = \prod_{j=1}^{p} \left(z_n - \zeta_j(z')\right) = z_n^p + \sum_{j=1}^{p} a_j(z') z_n^{p-j}. \tag{2.1.7}$$

Then, $W(z', z_n) \in \mathscr{O}(\overline{\mathrm{P}\Delta_{n-1}})[z_n] \subset \mathscr{O}(\overline{\mathrm{P}\Delta_{n-1}} \times \mathbf{C})$. With $s_n > r_n$ sufficiently close to r_n, and with every $z' \in \overline{\mathrm{P}\Delta_{n-1}}$ fixed, the roots of $W(z', z_n) = 0$ and $f(z', z_n) = 0$ in $|z_n| < s_n$ are the same with counting multiplicities,[1] so that the following integral formulae hold in $|z_n| < s_n$:

$$u(z', z_n) = \frac{f(z', z_n)}{W(z', z_n)} = \frac{1}{2\pi i} \int_{|\zeta_n| = s_n} \frac{f(z', \zeta_n)}{W(z', \zeta_n)} \cdot \frac{d\zeta_n}{\zeta_n - z_n},$$

$$v(z', z_n) = \frac{W(z', z_n)}{f(z', z_n)} = \frac{1}{2\pi i} \int_{|\zeta_n| = s_n} \frac{W(z', \zeta_n)}{f(z', \zeta_n)} \cdot \frac{d\zeta_n}{\zeta_n - z_n}.$$

It follows from these integral formulae that $u, v \in \mathscr{O}(\overline{\mathrm{P}\Delta_{n-1}} \times \{|z_n| \le r_n\})$. Furthermore, since in a neighborhood of $|z_n| = r_n$ the denominators of u, v are zero-free, $u \cdot v = 1$ holds. Therefore by the uniqueness of analytic continuation, $u \cdot v = 1$ on $\overline{\mathrm{P}\Delta}$. One sees that u is zero-free on $\overline{\mathrm{P}\Delta}$. There is a constant $C > 0$ such that

$$C^{-1} \le |u(z)| \le C, \qquad z \in \overline{\mathrm{P}\Delta}. \tag{2.1.8}$$

Thus it follows that

[1]By the arguments up to here one sees that the quotient $f(z', z_n)/W(z', z_n)$ with each fixed z' is a zero-free holomorphic function in $|z_n| < s_n$; however, as (z', z_n) runs freely, even its continuity is unclear.

$$f(z', z_n) = u(z)\Big(z_n^p + \sum_{j=1}^{p} a_j(z') z_n^{p-j}\Big) = u(z) W(z', z_n),$$

$u \in \mathscr{O}(\overline{\mathrm{P}\Delta})$, $a_\nu \in \mathscr{O}(\overline{\mathrm{P}\Delta_{n-1}})$, and $a_\nu(0) = 0$.

We confirm that $u(z)$ and $W(z', z_n)$ are uniquely determined as elements of $\mathscr{O}_{\mathbf{C}^n,0}$. Suppose that

$$\begin{aligned} \underline{f}_0 &= \underline{u}_0 \cdot \underline{\left(z_n^p + \textstyle\sum_{j=1}^p a_j(z') z_n^{p-j}\right)}_0 \\ &= \underline{\tilde{u}}_0 \cdot \underline{\left(z_n^p + \textstyle\sum_{j=1}^p \tilde{a}_j(z') z_n^{p-j}\right)}_0 . \end{aligned}$$

Let $\widetilde{\mathrm{P}\Delta}_{n-1} \times \Delta(0; \tilde{r}_n)$ be a standard polydisk for which the above expressions make sense. For each fixed $z' \in \widetilde{\mathrm{P}\Delta}_{n-1}$, the roots of two equations

$$z_n^p + \sum_{j=1}^{p} a_j(z') z_n^{p-j} = 0,$$

$$z_n^p + \sum_{j=1}^{p} \tilde{a}_j(z') z_n^{p-j} = 0$$

are identical with counting multiplicities, and then

$$a_j(z') = \tilde{a}_j(z'), \quad 1 \le j \le p.$$

Hence, $u(z) = \tilde{u}(z)$ follows.

(ii) We may assume that

$$\begin{aligned} f = W(z', z_n) &= z_n^p + \sum_{\nu=1}^{p} a_\nu(z') z_n^{p-\nu} \\ &= \sum_{\nu=0}^{p} a_\nu(z') z_n^{p-\nu} \in \mathscr{O}(\overline{\mathrm{P}\Delta_{n-1}})[z_n]. \end{aligned}$$

Here, we put $a_0(z') = 1$. For $\varphi \in \mathscr{O}(\mathrm{P}\Delta_{n-1} \times \Delta(0; r_n))$ we set

$$a(z', z_n) = \frac{1}{2\pi i} \int_{|\zeta_n| = t_n} \frac{\varphi(z', \zeta_n)}{W(z', \zeta_n)} \frac{d\zeta_n}{\zeta_n - z_n}, \tag{2.1.9}$$
$$(z', z_n) \in \mathrm{P}\Delta_{n-1} \times \Delta(0; r_n),$$

where $|z_n| < t_n < r_n$. Since $a(z', z_n)$ is independent of the choice of t_n close to r_n, $a(z', z_n) \in \mathscr{O}(\mathrm{P}\Delta_{n-1} \times \Delta(0; r_n))$ is determined. For $z' \in \mathrm{P}\Delta_{n-1}$, $|z_n| < t_n$ we write

$$
\begin{aligned}
(2.1.10)\quad & \varphi(z', z_n) - a(z', z_n)W(z', z_n) \\
&= \frac{1}{2\pi i}\int_{|\zeta_n|=t_n} \varphi(z', \zeta_n)\frac{d\zeta_n}{\zeta_n - z_n} - \frac{W(z', z_n)}{2\pi i}\int_{|\zeta_n|=t_n} \frac{\varphi(z', \zeta_n)}{W(z', \zeta_n)}\frac{d\zeta_n}{\zeta_n - z_n} \\
&= \frac{1}{2\pi i}\int_{|\zeta_n|=t_n} \varphi(z', \zeta_n)\left\{1 - \frac{W(z', z_n)}{W(z', \zeta_n)}\right\}\frac{d\zeta_n}{\zeta_n - z_n} \\
&= \frac{1}{2\pi i}\int_{|\zeta_n|=t_n} \varphi(z', \zeta_n)\frac{\sum_{\nu=0}^{p-1} a_\nu(z')\left(\zeta_n^{p-\nu} - z_n^{p-\nu}\right)}{W(z', \zeta_n)(\zeta_n - z_n)}d\zeta_n \\
&= \frac{1}{2\pi i}\int_{|\zeta_n|=t_n} \frac{\varphi(z', \zeta_n)}{W(z', \zeta_n)}\left\{\sum_{\nu=0}^{p-1} a_\nu(z')\left(\zeta_n^{p-\nu-1} + \zeta_n^{p-\nu-2}z_n + \cdots\right.\right. \\
&\qquad\qquad \left.\left. + z_n^{p-\nu-1}\right)\right\}d\zeta_n \\
&= b_1(z')z_n^{p-1} + b_2(z')z_n^{p-2} + \cdots + b_p(z'),
\end{aligned}
$$

where $b_\nu(z')$ are given by

$$
(2.1.11)\qquad b_\nu(z') = \frac{1}{2\pi i}\int_{|\zeta_n|=t_n} \frac{\varphi(z', \zeta_n)}{W(z', \zeta_n)}\left(\sum_{h=0}^{\nu-1} a_h(z')\zeta_n^{\nu-1-h}\right)d\zeta_n.
$$

It follows from this expression that $b_\nu(z') \in \mathscr{O}(\mathrm{P}\Delta_{n-1})$, $1 \leq \nu \leq p$ (independent of t_n). Therefore,

$$
(2.1.12)\qquad \varphi(z', z_n) = a(z', z_n)W(z', z_n) + \sum_{\nu=1}^{p} b_\nu(z')z_n^{p-\nu}.
$$

Next, we show the uniqueness. Suppose that

$$
(2.1.13)\qquad \varphi(z', z_n) = \tilde{a}(z', z_n)W(z', z_n) + \sum_{\nu=1}^{p} \tilde{b}_\nu(z')z_n^{p-\nu}.
$$

Subtracting the both sides of (2.1.12) and (2.1.13) and shifting terms, we assume that

$$
(a(z', z_n) - \tilde{a}(z', z_n))W(z', z_n) = \sum_{\nu=1}^{p}(\tilde{b}_\nu(z') - b_\nu(z'))z_n^{p-\nu} \not\equiv 0.
$$

Then for a fixed $z' \in \widetilde{\mathrm{P}\Delta}_{n-1}$ the left-hand side has at least p roots with counting multiplicities. The right-hand side has at most $p-1$ roots with counting multiplicities; this is absurd. Hence,

$$
\tilde{b}_\nu(z') = b_\nu(z'), \qquad \tilde{a}(z', z_n) = a(z', z_n).
$$

(iii) We show the estimates. In (2.1.9)–(2.1.11) there is a constant $\delta > 0$ such that

$$|W(z', \zeta_n)| \geq \delta > 0, \quad z' \in \mathrm{P}\Delta_{n-1}, \ |\zeta_n| = t_n \ (\nearrow r_n). \tag{2.1.14}$$

By (2.1.11) there is a constant $M > 0$, depending only on $\sup_{\mathrm{P}\Delta_{n-1}} |a_\nu|$, p and r_n such that

$$|b_\nu(z')z_n^{p-\nu}| \leq M\delta^{-1}\|\varphi\|_{\mathrm{P}\Delta}, \qquad \|b_\nu\|_{\mathrm{P}\Delta_{n-1}} \leq M\delta^{-1}\|\varphi\|_{\mathrm{P}\Delta}.$$

It follows from (2.1.10) that for each $z' \in \mathrm{P}\Delta_{n-1}$, $|z_n| = t_n (< r_n)$

$$|\varphi(z', z_n) - a(z', z_n)W(z', z_n)| \leq pM\delta^{-1}\|\varphi\|_{\mathrm{P}\Delta}.$$

From this we obtain

$$|a(z', z_n)W(z', z_n)| \leq (pM\delta^{-1} + 1)\|\varphi\|_{\mathrm{P}\Delta}.$$

By (2.1.14)

$$|a(z', z_n)| \leq (pM\delta^{-1} + 1)\delta^{-1}\|\varphi\|_{\mathrm{P}\Delta}.$$

Letting $t_n \nearrow r_n$, we see by the maximum principle (Theorem 1.2.17) that

$$\|a\|_{\mathrm{P}\Delta} \leq (pM\delta^{-1} + 1)\delta^{-1}\|\varphi\|_{\mathrm{P}\Delta}.$$

To obtain an estimate replacing W with the original f, we write

$$\begin{aligned} \varphi &= aW + \sum_{\nu=1}^{p} b_\nu(z')z_n^{p-\nu} \\ &= \left(\frac{a}{u}\right) \cdot f + \sum_{\nu=1}^{p} b_\nu(z')z_n^{p-\nu}. \end{aligned}$$

It follows from (2.1.8) that

$$C^{-1} \leq |u| \leq C.$$

Therefore, finally there is a positive constant $M' = M'(f)$ independent of φ such that

$$\begin{aligned} &\|a\|_{\mathrm{P}\Delta} \leq M'\|\varphi\|_{\mathrm{P}\Delta}, \\ &\|b_\nu\|_{\mathrm{P}\Delta_{n-1}} \leq M'\|\varphi\|_{\mathrm{P}\Delta}, \qquad 1 \leq \nu \leq p. \end{aligned}$$

□

Remark 2.1.15 Let the notation be as in Theorem 2.1.3. Moreover, assume that $p = 1$. Then, by the Implicit Function Theorem 1.2.41, equation $f(z', z_n) = 0$ has a unique solution $z_n = g(z')$ in a neighborhood of 0 with $g(0) = 0$. But this is just an existence theorem. Applying Theorem 2.1.3 for $\varphi = z_n$, we have an integral representation formula of the solution (cf. 2.1.6):

$$g(z') = \frac{1}{2\pi i} \int_{|z_n|=r_n} z_n \frac{\frac{\partial f}{\partial z_n}(z', z_n)}{f(z', z_n)} dz_n, \quad z' \in \overline{\mathrm{P}\Delta}_{n-1}.$$

Definition 2.1.16 (i) Letting $\mathrm{P}\Delta_{n-1} \subset \mathbf{C}^{n-1}$, we call the z_n-polynomial with coefficients in $\mathscr{O}(\overline{\mathrm{P}\Delta}_{n-1})$

$$W(z', z_n) = z_n^p + \sum_{\nu=1}^{p} a_\nu(z') \cdot z_n^{p-\nu},$$
$$a_\nu \in \mathscr{O}(\overline{\mathrm{P}\Delta}_{n-1}), \quad a_\nu(0) = 0$$

a *Weierstrass polynomial* (in z_n). Considering the induced germ

$$W = z_n^p + \sum_{\nu=1}^{p} \underline{a_\nu}_0 \cdot z_n^{p-\nu} \in \mathscr{O}_{\mathrm{P}\Delta_{n-1},0}[z_n],$$

we also call this a Weierstrass polynomial (in z_n).

(ii) Write (2.1.7) as $f(z', z_n) = uW(z', z_n)$ with unit u (i.e., $^{\exists}u^{-1}$) and Weierstrass polynomial $W(z', z_n)$. We call $f(z', z_n) = uW(z', z_n)$ the *Weierstrass decomposition* of f at 0, which is unique.

2.2 Local Rings

For the sake of simplicity we write $\mathscr{O}_{\mathbf{C}^n,a} = \mathscr{O}_{n,a}$ ($a \in \mathbf{C}^n$). This is an integral local ring (Sect. 1.3.3 (4)). In this section we investigate the algebraic properties in more detail.

2.2.1 Preparations from Algebra

Here we describe elementary properties on polynomial rings: Cf. Nagata [45], Morita [43], Lang [38] for general references.

Theorem 2.2.1 (Gauss) *The polynomial ring of a finite number of variables over a unique factorization domain is again a unique factorization domain.*

Theorem 2.2.2 (Hilbert) *The polynomial ring of a finite number of variables over a Noetherian ring is again Noetherian.*

A module M over a ring A is said to be Noetherian if every submodule of M is finitely generated over A.

Lemma 2.2.3 *If a ring A is Noetherian, then so is the p-th Cartesian product A^p for $p \in \mathbf{N}$.*

Proof We use induction on p. By the assumption, the case of $p = 1$ is trivial.

Suppose that $p > 1$ and that it folds for $p - 1$. Let $M \subset A^p$ be a submodule, and let

$$\pi : A^p \to A$$

be the projection for the first factor. The submodule $\pi(M) \subset A$ is generated by finitely many elements $u_i \in \pi(M)$, $1 \le i \le k$. Take $U_i \in \pi^{-1}u_i$ $(1 \le i \le k)$. Then it suffices to show that $M \cap \operatorname{Ker}\pi$ is finitely generated. Since $\operatorname{Ker}\pi \cong A^{p-1}$, the induction hypothesis implies the finite generation of $M \cap \operatorname{Ker}\pi$. □

We consider a polynomial ring $A[X]$ of a variable X over an integral domain A. Let $f(X)$ and $g(X)$ be two elements of $A[X]$ given by

$$\begin{aligned} &f(X) = a_0X^m + a_1X^{m-1} + \cdots + a_m, \quad m \ge 1, \\ &g(X) = b_0X^n + b_1X^{n-1} + \cdots + b_n, \quad n \ge 1, \\ &a_0b_0 \ne 0. \end{aligned}$$

We define the *resultant* $R(f, g) \in A$ of $f(X)$ and $g(X)$ to be the $(m+n)$-determinant

$$R(f, g) = \begin{vmatrix} a_0 & a_1 & \cdots\cdots & a_m & & & \\ & a_0 & a_1 & \cdots\cdots & a_m & & \\ & & \ddots & & & \ddots & \\ & & & a_0 & a_1 & \cdots\cdots & a_m \\ b_0 & b_1 & \cdots\cdots & b_n & & & \\ & b_0 & b_1 & \cdots\cdots & b_n & & \\ & & \ddots & & & \ddots & \\ & & & b_0 & b_1 & \cdots\cdots & b_n \end{vmatrix} \begin{matrix} \\ \\ n\text{-tuple} \\ \\ \\ \\ m\text{-tuple} \\ \\ \end{matrix} . \tag{2.2.4}$$

Here, the blank spaces are supposed to be filled with zeros.

Theorem 2.2.5 *For $f(X)$ and $g(X)$ above, there are $\varphi(X), \psi(X) \in A[X]$ such that $\deg\varphi < n$, $\deg\psi < m$, and*

$$\varphi(X)f(X) + \psi(X)g(X) = R(f, g) \ (\in A).$$

Proof In (2.2.4), multiply the jth column by X^{m+n-j} and then add it to the last column for $j = 1, 2, \ldots, m + n - 1$. Then we have

(2.2.6)
$$R(f,g) = \begin{vmatrix} a_0 & a_1 & \cdots\cdots & a_m & & & X^{n-1}f(X) \\ & a_0 & a_1 & \cdots\cdots & a_m & & X^{n-2}f(X) \\ & & \ddots & & & \ddots & \vdots \\ & & & a_0 & a_1 & \cdots\cdots & f(X) \\ b_0 & b_1 & \cdots\cdots & b_n & & & X^{m-1}g(X) \\ & b_0 & b_1 & \cdots\cdots & b_n & & X^{m-2}g(X) \\ & & \ddots & & & \ddots & \vdots \\ & & & b_0 & b_1 & \cdots\cdots & g(X) \end{vmatrix}.$$

Expanding this with respect to the last column, we get the required $\varphi(x)$ and $\psi(x)$. □

Let K be the algebraic closure of the quotient field of A.

Theorem 2.2.7 (i) *Two equations* $f(X) = 0$ *and* $g(X) = 0$ *share a common root in* K *if and only if* $R(f, g) = 0$.

(ii) *Let* A *be a unique factorization domain, Then* $f(X)$ *and* $g(X)$ *share a common prime factor if and only if* $R(f, g) = 0$.

Proof (i) Let $\alpha \in K$ be a common zero of f and g. Then Theorem 2.2.5 with the substitution $X = \alpha$ implies that

$$A \ni R(f, g) = \varphi(\alpha)f(\alpha) + \psi(\alpha)g(\alpha) = 0 \text{ in } K.$$

Since $A \hookrightarrow K$, $R(f, g) = 0$ in A.

On the other hand, if $R(f, g) = 0$, then $\varphi(X)f(X) = -\psi(X)g(X)$. Since $\deg \varphi < n$, some root of $g(X) = 0$ in K must be a root of $f(X) = 0$.

(ii) By the assumption and Theorem 2.2.1 $A[X]$ is a unique factorization domain. Suppose that $f(X)$ and $g(X)$ share a common prime factor $h(X)$. Then by Theorem 2.2.5, $h(X)$ must be a prime factor of $R(f, g) \in A$; this is absurd unless $R(f, g) = 0$. If $R(f, g) = 0$, then $\varphi(X)f(X) = -\psi(X)g(X)$. By the degree comparison as in (i), it is impossible that all prime factors of g are those of φ with counting multiplicities; i.e., there is a prime factor of g which divides $f(X)$. □

We take the roots $\alpha_1, \ldots, \alpha_m$ and $\beta_1, \ldots, \beta_n$ in K of $f(X) = 0$ and $g(X) = 0$, respectively with counting multiplicities. Then,

(2.2.8)
$$f(X) = a_0 \prod_{i=1}^{m} (X - \alpha_i),$$

(2.2.9)
$$g(X) = b_0 \prod_{j=1}^{n} (X - \beta_j).$$

Lemma 2.2.10 *We have*

$$R(f, g) = a_0^n b_0^m \prod_{i=1}^{m} \prod_{j=1}^{n} (\alpha_i - \beta_j) = a_0^n \prod_{i=1}^{m} g(\alpha_i) = b_0^m \prod_{j=1}^{n} f(\beta_j).$$

Proof Note that a_i/a_0 (resp. b_j/b_0) are expressed by elementary symmetric polynomials of $\alpha_1, \ldots, \alpha_m$ (resp. $\beta_1, \ldots, \beta_n$). It is also noted that $R(f, g)$ is a homogeneous polynomial of degree n (resp. m) in a_i (resp. b_j). Therefore, $R(f, g)$ is a polynomial in elementary symmetric polynomials in α_i and β_j multiplied with $a_0^n b_0^m$.

Consider α_i, β_j as undetermined elements. If $\alpha_i = \beta_j$, then f and g have a common factor $(x - \alpha_i)$, and so by Theorem 2.2.7, $R(f, g) = 0$. Therefore, as a polynomial in α_i and β_j, $R(f, g)$ can be divided by $(\alpha_i - \beta_j)$. It follows that $R(f, g)$ can be divided by $\prod_{i=1}^{m} \prod_{j=1}^{n} (\alpha_i - \beta_j)$. Now we check the coefficient of $(\beta_1 \cdots \beta_n)^n$ in each of $R(f, g)$ and $\prod_{i=1}^{m} \prod_{j=1}^{n} (\alpha_i - \beta_j)$. Note that

$$\frac{b_n}{b_0} = (-1)^n \beta_1 \cdots \beta_n,$$

$$R(f, g) = a_0^n b_0^m \left(\left(\frac{b_n}{b_0} \right)^m + \cdots \right) = a_0^n b_0^m \left((-1)^{mn} (\beta_1 \cdots \beta_n)^m + \cdots \right),$$

$$\prod_{i=1}^{m} \prod_{j=1}^{n} (\alpha_i - \beta_j) = (-1)^{mn} (\beta_1 \cdots \beta_n)^m + \cdots .$$

Thus we get the required coefficient $a_0^n b_0^m$ in the first equality. The rest follows from (2.2.8) and (2.2.9). □

The *discriminant* of $f(X)$ is defined by

$$\Delta(f) = a_0^{2m-2} \prod_{i<j} (\alpha_i - \alpha_j)^2.$$

The formal derivative of $f(X)$ is defined by

$$\begin{aligned} f'(X) &= m a_0 X^{m-1} + \cdots + a_{m-1} \\ &= a_0 \sum_{i=1}^{m} (X - \alpha_1) \cdots (X - \alpha_{i-1})(X - \alpha_{i+1}) \cdots (X - \alpha_m), \end{aligned}$$

where $(X - \alpha_0) = (X - \alpha_{m+1}) = 1$ as convention. Then we have:

Theorem 2.2.11 $R(f, f') = (-1)^{\frac{m(m-1)}{2}} a_0 \Delta(f)$.

Proof Since $f'(\alpha_i) = a_0 \prod_{j \neq i} (\alpha_i - \alpha_j)$, it follows from Lemma 2.2.10 that

$$R(f, f') = a_0^{m-1} \prod_{i=1}^{m} f'(\alpha_i) = a_0^{2m-1} \prod_{i=1}^{m} \prod_{j \neq i} (\alpha_i - \alpha_j)$$
$$= (-1)^{\frac{m(m-1)}{2}} a_0 \Delta(f). \qquad \square$$

2.2.2 Properties of $\mathscr{O}_{n,a}$

Theorem 2.2.12 *The ring $\mathscr{O}_{n,a}$ is a unique factorization domain.*

Proof We may assume $a = 0$. We proceed by induction on n.

(1) Assume that $n = 1$. Every $\underline{f}_0 \in \mathscr{O}_{1,0}$, $\underline{f}_0 \neq 0$ is uniquely represented in a neighborhood of 0 as

$$f(z) = z^p h(z), \qquad p \in \mathbf{Z}^+,\ h(0) \neq 0, \tag{2.2.13}$$

where $\underline{h}_0$ is a unit. The germ $\underline{z}_0$ is a prime element, and p is uniquely determined by $\underline{f}_0$. Therefore, $\mathscr{O}_{1,0}$ is a unique factorization domain.

(2) Suppose $n \geq 2$ and that the statement holds for $n - 1$. By Weierstrass' Preparation Theorem 2.1.3, every $\underline{f}_0 \in \mathscr{O}_{n,0}$ is reduced to a Weierstrass polynomial up to a unit:

$$\underline{f}_0 = z_n^p + \sum_{\nu=1}^{p} \underline{a_\nu}_0 \cdot z_n^{p-\nu} \in \mathscr{O}_{n-1,0}[z_n].$$

By the induction hypothesis, $\mathscr{O}_{n-1,0}$ is a unique factorization domain, and then by Theorem 2.2.1, the polynomial ring $\mathscr{O}_{n-1,0}[z_n]$ over it is a unique factorization domain. It remains to show the equivalence between the reducibility or irreducibility in $\mathscr{O}_{n-1,0}[z_n]$ and that in $\mathscr{O}_{n,0}$. We prove:

Lemma 2.2.14 *Let $\underline{f}_0 \in \mathscr{O}_{n-1,0}[z_n]$ be a Weierstrass polynomial. Assume that there are elements $\underline{g}_0, \underline{h}_0 \in \mathscr{O}_{n,0}$ satisfying $\underline{f}_0 = \underline{h}_0 \cdot \underline{g}_0$. Then there exist Weierstrass polynomials $W_1(z', z_n)$, $W_2(z', z_n)$ such that in a neighborhood of 0,*

$$f(z', z_n) = W_1(z', z_n) \cdot W_2(z', z_n), \qquad \frac{g}{W_1} \cdot \frac{h}{W_2} = 1.$$

$\because$) We may assume that g and h are non-units. We take a polydisk PΔ so that $f(z', z_n) = g(z', z_n) \cdot h(z', z_n) \in \mathscr{O}(\overline{\mathrm{P}\Delta})$. Note that $f(0, z_n) = z_n^p \not\equiv 0$. We may take P$\Delta$ to be the standard polydisk of f. Then,

$$g(0, z_n) \not\equiv 0, \qquad h(0, z_n) \not\equiv 0.$$

Therefore we may assume that $\mathrm{P}\Delta$ is the standard polydisk for g and h. Let

$$g(z', z_n) = u_1 W_1(z', z_n), \qquad h(z', z_n) = u_2 W_2(z', z_n).$$

be the Weierstrass decompositions of g and h at 0, respectively. We see that

$$f(z', z_n) = u_1 u_2 W_1(z', z_n) W_2(z', z_n),$$

and that $W_1 W_2$ is a Weierstrass polynomial. By the uniqueness of Weierstrass decomposition

$$u_1 u_2 = 1, \qquad f = W_1 W_2. \qquad \triangle$$

Continued Proof of Theorem 2.2.12: Lemma 2.2.14 implies that if a Weierstrass polynomial $\underline{f}_0$ is reducible in $\mathscr{O}_{n,0}$, then it is reducible in $\mathscr{O}_{n-1,0}[z_n]$, and the factors are again Weierstrass polynomials. □

The following lemma will be needed later.

Lemma 2.2.15 *Let $Q(z', z_n) \in \mathscr{O}_{n-1,0}[z_n]$ be a Weierstrass polynomial, and let $R \in \mathscr{O}_{n-1,0}[z_n]$. If $R = Q \cdot \underline{g}_0$ with $\underline{g}_0 \in \mathscr{O}_{n,0}$, then $\underline{g}_0 \in \mathscr{O}_{n-1,0}[z_n]$.*

Proof Assume that all the functions above are holomorphic in a neighborhood of a closed polydisk $\overline{\mathrm{P}\Delta}$. Since the leading coefficient of $Q(z', z_n)$ as a polynomial in z_n is 1, the Euclidean algorithm implies that

$$\begin{aligned} R = \varphi Q + \psi, &\qquad \varphi, \psi \in \mathscr{O}_{n-1,0}[z_n], \\ \deg_{z_n} \psi < p = \deg_{z_n} Q. & \end{aligned} \tag{2.2.16}$$

We may assume that $\mathrm{P}\Delta = \mathrm{P}\Delta_{n-1} \times \Delta_{(n)}$ ($\Delta_{(n)} = \{|z_n| < r_n\}$) is a standard polydisk for Q. With every $z' \in \mathrm{P}\Delta_{n-1}$ fixed, $Q(z', z_n) = 0$ has p zeros with counting multiplicities. Therefore, R has at least p zeros with counting multiplicities. By (2.2.16) ψ has at least p zeros with counting multiplicities, too. Since $\deg \psi < p$, $\psi \equiv 0$. Therefore we obtain

$$\begin{aligned} R &= \varphi Q = g Q, \\ (\varphi - g) Q &= 0, \qquad Q \neq 0. \end{aligned}$$

Since $\mathscr{O}_{n,0}$ is an integral domain, $\varphi - g = 0$. Thus we see that $g = \varphi \in \mathscr{O}_{n-1,0}[z_n]$. □

Lemma 2.2.17 *Let f, g be holomorphic functions in a neighborhood $0 \in \mathbf{C}^n$. If $\underline{f}_0$ and $\underline{g}_0$ are mutually prime (i.e., $(\underline{f}_0, \underline{g}_0) = 1$), then there is a neighborhood U of 0 such that $(\underline{f}_b, \underline{g}_b) = 1$ for $b \in U$.*

Proof Let $\mathrm{P}\Delta = \mathrm{P}\Delta_{n-1} \times \Delta_{(n)}$ ($\Delta_{(n)} = \Delta(0, r_n)$) be a standard polydisk for f and g. Then there are Weierstrass polynomials $P(z', z_n)$, $Q(z', z_n)$ and units $u, v \in \mathscr{O}(\overline{\mathrm{P}\Delta})$ such that

$$(2.2.18) \qquad f = uP, \qquad g = vQ.$$

It follows from the assumption that $\underline{P}_0$ and $\underline{Q}_0$ are mutually prime, By Lemma 2.2.14 this is equivalent to saying that they are mutually prime as elements of the ring $\mathscr{O}_{n-1,0}[z_n]$. By Theorem 2.2.7 this is again equivalent to the resultant

$$R(P(z', z_n), Q(z', z_n)) = R(z') \neq 0 \in \mathscr{O}_{n-1,0};$$

therefore $\underline{R}_{b'} \neq 0$ at every $b' \in \mathrm{P}\Delta_{n-1}$. This implies that P and Q are mutually prime in $\mathscr{O}_{n-1,b'}[z_n]$. Thus, $\underline{f}_b$ and $\underline{g}_b$ are mutually prime at every $b \in \mathrm{P}\Delta$. □

From this lemma the next follows immediately:

Proposition 2.2.19 *Let $U \subset \mathbf{C}^n$ be an open set and let $f, g \in \mathscr{O}(U)$. Then, $\{a \in U; (\underline{f}_a, \underline{g}_a) = 1\}$ is an open subset.*

N.B. Even if $\underline{f}_a$ is irreducible, $\underline{f}_b$ is not necessarily irreducible for b in a neighborhood of a. For example, we take $f(x, y, z) = x^2 - zy^2$. This is irreducible at the origin, but at any point $w = (0, 0, c)$ with $c \neq 0$ it is factorized as $\underline{f}_w = \underline{(x + \sqrt{z}y)}_w \cdot \underline{(x - \sqrt{z}y)}_w$.

Theorem 2.2.20 *The ring $\mathscr{O}_{n,a}$ is noetherian, and so is $\mathscr{O}_{n,a}^p$ ($p \geq 2$).*

Proof If $\mathscr{O}_{n,a}$ is noetherian, Lemma 2.2.3 implies the latter half.

We show that $\mathscr{O}_{n,a}$ is noetherian. We set $a = 0$ and use induction on n.

(1) $n = 1$. We take any ideal $\mathscr{I} \subset \mathscr{O}_{1,0}$. Assume that $\mathscr{I} \neq \{0\}, \mathscr{O}_{1,0}$. We decompose $\underline{f}_0 \in \mathscr{I} \setminus \{0\}$ as (2.2.13), and denote the least p by the same p. Then,

$$\mathscr{I} = z^p \cdot \mathscr{O}_{1,0}.$$

That is, $\mathscr{I}$ is a principal ideal. Let $\mathscr{I}_1 \subset \mathscr{I}_2 \subset \mathscr{I}_3 \subset \cdots$ be a sequence of increasing ideals. Then we have that

$$\mathscr{I}_\nu = z^{p_\nu} \mathscr{O}_{1,0}, \quad p_1 \geq p_2 \geq p_3 \geq \cdots.$$

Therefore there is a natural number ν_0 such that $p_{\nu_0} = p_{\nu_0+1} = \cdots$. Therefore the sequence stabilizes at $\mathscr{I}_{\nu_0} = \mathscr{I}_\nu$, $\nu \geq \nu_0$.

(2) $n \geq 2$. Assume that it holds for $n - 1$; i.e., $\mathscr{O}_{n-1,0}$ is noetherian. Theorem 2.2.2 implies that $\mathscr{O}_{n-1,0}[z_n]$ is noetherian. Let

$$\mathscr{I}_1 \subset \mathscr{I}_2 \subset \cdots \subset \mathscr{I}_\nu \subset \cdots$$

be an increasing sequence of ideals of $\mathscr{O}_{n,0}$. We show that there is a number $\nu_0 \in \mathbf{N}$ at which it stabilizes, $\mathscr{I}_{\nu_0} = \mathscr{I}_\nu$ $({}^\forall \nu \geq \nu_0)$.

Suppose that there exists no such ν_0. After taking a subsequence, we get

$$\mathscr{I}_1 \subsetneqq \mathscr{I}_2 \subsetneqq \cdots \subsetneqq \mathscr{I}_\nu \subsetneqq \cdots.$$

For every ν we take $\underline{f_\nu}_0 \in \mathscr{I}_\nu \setminus \mathscr{I}_{\nu-1}$ (we put $\mathscr{I}_0 = \{0\}$). By Remark 2.1.2 (iii) there is a standard coordinate system $(z', z_n) \in \mathbf{C}^{n-1} \times \mathbf{C}$ for all f_ν. Let

$$\underline{f_\nu}_0 = \underline{u_\nu}_0 \cdot W_\nu(z', z_n), \quad \nu = 1, 2, \ldots$$

be Weierstrass decompositions of $\underline{f_\nu}_0$ at 0. We may assume that $\underline{f_\nu}_0 = W_\nu(z', z_n)$, and then set

$$\mathscr{I}'_\nu = \sum_{\mu=1}^{\nu} W_\mu(z', z_n) \cdot \mathscr{O}_{n-1,0}[z_n],$$

which are ideals of $\mathscr{O}_{n-1,0}[z_n]$. By the definition, $\mathscr{I}'_\nu \subsetneqq \mathscr{I}'_{\nu+1}$, $\nu = 1, 2 \ldots$ Since $\mathscr{O}_{n-1,0}[z_n]$ is Noetherian, this is a contradiction. □

2.3 Analytic Subsets

Here we give the definition of an analytic subset, and study the elementary properties. Let $U \subset \mathbf{C}^n$ be an open set. We begin with a definition:

Definition 2.3.1 A subset $A \subset U$ is called an *analytic subset* or an *analytic set* if for every point $a \in U$ there are a neighborhood $V \subset U$ and finitely many holomorphic functions $f_j \in \mathscr{O}(V)$, $1 \leq j \leq l$, satisfying

$$A \cap V = \{z \in V;\ f_1(z) = \cdots = f_l(z) = 0\}.$$

By definition, analytic sets are closed.

Remark 2.3.2 Let $n = 1$ and let U be a domain. Then, a subset of U is analytic if and only if either it is U itself, or it is discrete without accumulation point in U.

Theorem 2.3.3 *Let U be a domain. If an analytic subset $A \subset U$ contains an interior point, then $U = A$.*

Proof Let A' be the subset of all interior points of A. Then, $A' \neq \emptyset$ and open in U. Take a point $a \in \overline{A'} \cap U$. Then there are a connected neighborhood V of a in U, and finitely many holomorphic functions $f_j \in \mathscr{O}(V)$, $1 \leq j \leq l$, such that

$$A \cap V = \{f_1 = \cdots = f_l = 0\}.$$

There exists $b \in V \cap A'$. The definition implies an existence of a neighborhood $W \subset A \cap V$ of b with $W \cap A = W$. That is, $f_j|_W(z) \equiv 0$, $1 \le j \le l$. By the Identity Theorem 1.2.14, $f_j(z) \equiv 0$, $1 \le j \le l$. Therefore $V \cap A = V$, and $a \in A'$. We saw that $A'(\subset U)$ is open and closed. Since U is connected, $A' = U$ follows. The inclusion relations, $A \supset A' = U \supset A$, imply $A = U$. □

Theorem 2.3.4 (Riemann's Extension Theorem) *Let U be a domain of $\mathbf{C}^n$ and let $A \subsetneqq U$ be a proper analytic subset. Let $f \in \mathscr{O}(U \setminus A)$ be a holomorphic function bounded around every point of A; i.e., there is a neighborhood V of a such that the restriction $f|_{V \setminus A}$ is bounded. Then, f is uniquely extended holomorphically over the whole U.*

Proof The case of $n = 1$ holds by Remark 2.3.2 and Theorem 1.1.11.

Let $n \ge 2$. Take any $a \in A$. By a translation we may assume $a = 0$. By the assumption there are a neighborhood V of 0 and $\phi \in \mathscr{O}(V) \setminus \{0\}$ such that

$$A \cap V \subset \{\phi = 0\}.$$

We may assume that V is a standard polydisk $\mathrm{P}\Delta = \mathrm{P}\Delta_{n-1} \times \Delta(0; r_n)$ for ϕ. For $z' \in \mathrm{P}\Delta_{n-1}$ with $|z_n| = r_n$, $\phi(z', z_n) \ne 0$; i.e., $(\mathrm{P}\Delta_{n-1} \times \{|z_n| = r_n\}) \cap A = \emptyset$. With fixed $z' \in \mathrm{P}\Delta_{n-1}$, $\phi(z', z_n) = 0$ has at most finitely many zeros in $\Delta(0; r_n)$, around which $f(z', z_n)$ is bounded. Theorem 1.1.11 implies that it is holomorphic over $\Delta(0; r_n)$. Therefore we may write

$$f(z', z_n) = \frac{1}{2\pi i} \int_{|\zeta_n| = r_n} \frac{f(z', \zeta_n)}{\zeta_n - z_n} d\zeta_n, \quad |z_n| < r_n.$$

Since $f(z', \zeta_n)$ with $|\zeta_n| = r_n$ is holomorphic in z', The above integral expression implies that $f(z', z_n)$ is, in fact, holomorphic in $\mathrm{P}\Delta$. □

Theorem 2.3.5 *Let U be a domain of $\mathbf{C}^n$ and let $A \subsetneqq U$ be a proper analytic subset. Then $U \setminus A$ is a domain.*

Proof Suppose that $U \setminus A$ is not connected. Then there are non-empty open subsets V_1, V_2 of U such that

$$U \setminus A = V_1 \cup V_2, \qquad V_1 \cap V_2 = \emptyset.$$

Define $f \in \mathscr{O}(U \setminus A)$ as follows:

$$f(z) = \begin{cases} 0, & z \in V_1, \\ 1, & z \in V_2. \end{cases}$$

By Theorem 2.3.4, f is uniquely extended to $\tilde{f} \in \mathscr{O}(U)$. Since $\tilde{f}|_{V_1} \equiv 0$, the Identity Theorem 1.2.14 implies $\tilde{f}(z) \equiv 0$, and this is a contradiction. □

Definition 2.3.6 (Geometric ideal sheaf) Let $U \subset \mathbf{C}^n$ be an open subset, and let $A \subset U$ be an analytic subset. Putting $E = A$ in Sect. 1.3.3 (5), we define the *ideal sheaf* $\mathscr{I}\langle A\rangle$ of the analytic subset A. We call $\mathscr{I}\langle A\rangle_z (\subset \mathscr{O}_{U,z})$ the *geometric ideal* of the germ $\underline{A}_z$ of the analytic subset A at $z \in U$. At $z \in U \setminus A$ the stalk $\mathscr{I}\langle A\rangle_z = \mathscr{O}_{U,z}$. In general, the ideal sheaf of an analytic subset is called a *geometric ideal sheaf* (*idéal géométrique de domaines indéterminés*[2]).

2.4 Coherent Sheaves

We begin with the definition of coherent sheaves in general.

Let X be a topological space, let $\mathscr{A} \to X$ be a sheaf of rings and let $\mathscr{S} \to X$ be a sheaf of $\mathscr{A}$-modules.

Definition 2.4.1 The sheaf $\mathscr{S}$ of $\mathscr{A}$-modules is said to be *locally finite* over $\mathscr{A}$ if for every $x \in X$ there exist a neighborhood $U \ni x$ and a finite number of sections $\sigma_j \in \Gamma(U, \mathscr{S})$, $1 \le j \le l$, satisfying

$$\mathscr{S}|_U = \sum_{j=1}^{l} \mathscr{A}|_U \cdot \sigma_j,$$

that is,

$$\mathscr{S}_y = \sum_{j=1}^{l} \mathscr{A}_y \cdot \sigma_j(y), \qquad {}^\forall y \in U.$$

In this case, $\{\sigma_j\}_{j=1}^l$ is called a *(locally) finite generator system* of the sheaf $\mathscr{S}$ of $\mathscr{A}$-modules on U, and we say that $\mathscr{S}$ is generated by $\{\sigma_j\}_{j=1}^l$ or by σ_j $(\in \Gamma(U, \mathscr{S}))$, $1 \le j \le l$, in U.

Definition 2.4.2 (*Relation sheaf*) A *relation sheaf* of $\mathscr{S}$ is a sheaf $\mathscr{R}((\tau_j)_{1\le j\le q})$ as follows:

(i) Let $U \subset X$ be an open subset.
(ii) Let $\tau_j \in \Gamma(U, \mathscr{S})$, $1 \le j \le q(<\infty)$, be finitely many sections.
(iii) Let $\mathscr{R}(\tau_1, \ldots, \tau_q) = \mathscr{R}((\tau_j)_{1\le j\le q}) \subset (\mathscr{A}|_U)^q$ be a sheaf of $\mathscr{A}|_U$-modules defined by

$$\mathscr{R}((\tau_j)_{1\le j\le q}) = \bigcup_{x\in U} \left\{ (a_1, \ldots, a_q) \in (\mathscr{A}_x)^q \,;\, \sum_{j=1}^{q} a_j \tau_j(x) = 0 \right\}. \tag{2.4.3}$$

---

[2] K. Oka referred to the notion in this way in Oka VII.

In the case when a finite number of linear relations (2.4.3) defined by $\tau_{(\lambda)} = (\tau_{(\lambda)j})_{1\le j\le q}$, $\lambda = 1, \ldots, l(<\infty)$, are imposed, we call

$$\mathscr{R}\Big((\tau_{(\lambda)j})_{1\le j\le q};\ 1 \le \lambda \le l\Big) = \bigcap_{\lambda=1}^{l} \mathscr{R}\Big((\tau_{(\lambda)j})_{1\le j\le q}\Big)$$

a *simultaneous relation sheaf*; this is a relation sheaf of the product sheaf $\mathscr{S}^l$.

Definition 2.4.4 (*Coherence*) We say that a sheaf $\mathscr{S}$ of $\mathscr{A}$-modules is a *coherent sheaf* of ($\mathscr{A}$-) modules over $\mathscr{A}$, or *coherent over* $\mathscr{A}$ if the following two conditions are satisfied:

(i) $\mathscr{S}$ is locally finite over $\mathscr{A}$.
(ii) Every relation sheaf of $\mathscr{S}$ is locally finite over $\mathscr{A}$.

N.B. In the definition of coherence, the coherence of the base sheaf of rings, $\mathscr{A}$ itself is not required.

Definition 2.4.5 (*Coherent sheaf*) Let $\Omega \subset \mathbf{C}^n$ be an open subset. We call a coherent sheaf of modules over $\mathscr{O}_\Omega$ simply a *coherent sheaf* over Ω.

N.B. In some references a coherent sheaf defined as above is called a "coherent analytic sheaf". In this book, a "coherent sheaf" over a complex domain means a coherent sheaf of modules over a sheaf of germs of holomorphic functions, unless otherwise mentioned.

We first study the general properties.

Proposition 2.4.6 (Point–local generation) *Let $\mathscr{S}$ be a locally finite sheaf of $\mathscr{A}$-modules over X. If finitely many elements $\underline{\gamma_j}_a \in \mathscr{S}_a$, $1 \le j \le l$, generate $\mathscr{S}_a$ over $\mathscr{A}_a$ at a point $a \in X$, then there is a neighborhood $U \ni a$ such that each $\underline{\gamma_j}_a$ has a representative $\gamma_j \in \Gamma(U, \mathscr{S})$, and the equality*

$$\mathscr{S}_x = \sum_{j=1}^{l} \mathscr{A}_x \cdot \gamma_j(x), \qquad {}^\forall x \in U$$

holds; i.e., γ_j, $1 \le j \le l$, generate $\mathscr{S}$ over U. In particular, this holds for a coherent sheaf $\mathscr{S}$ of modules over $\mathscr{A}$.

Proof Because of the local finiteness of $\mathscr{S}$ there is a locally finite generator system $\{\sigma_k\}_{k=1}^m \subset \Gamma(V, \mathscr{S})$ over a neighborhood $V(\ni a)$. By assumption we may write

$$\sigma_k(a) = \sum_{j=1}^{l} \underline{f_{kj}}_a \underline{\gamma_j}_a, \quad \underline{f_{kj}}_a \in \mathscr{A}_a,\ 1 \le k \le m.$$

Take a neighborhood $U \subset V$ of a such that all $\underline{\gamma_j}_a$ and $\underline{f_{kj}}_a$ have the representatives γ_j, f_{kj} over V. After shrinking U if necessary, it follows from Proposition 1.3.4 (ii) that

$$\sigma_k(x) = \sum_{j=1}^{l} \underline{f_{kj}}_x \gamma_j(x), \quad {}^{\forall}x \in U, \ 1 \le k \le m.$$

Since $\{\sigma_k(x)\}$ generates $\mathscr{S}_x$ over $\mathscr{A}_x$ ($x \in U$), $\{\gamma_j(x)\}$ generates $\mathscr{S}_x$ over $\mathscr{A}_x$. □

Proposition 2.4.7 *Let $\mathscr{S}$ be a coherent sheaf of modules over $\mathscr{A}$.*

(i) *A subsheaf of $\mathscr{S}$ (i.e., subsheaf of $\mathscr{A}$-submodules of $\mathscr{S}$) is coherent over $\mathscr{A}$ if and only if it is locally finite over $\mathscr{A}$.*
(ii) *$\mathscr{S}^N$ ($N = 2, 3, \ldots$) are coherent over $\mathscr{A}$. That is, every simultaneous relation sheaf of $\mathscr{S}$ is coherent over $\mathscr{A}$.*

Proof (i) It remains to show the local finiteness of relation sheaves, but they are so since $\mathscr{S}$ is coherent.

(ii) We use induction on N. The case of $N = 1$ is the assumption.

Let $N \ge 2$, and suppose that it holds for $N - 1$. The local finiteness of $\mathscr{S}^N$ follows from that of $\mathscr{S}$. Let $U \subset X$ be an open subset, and let $F_i \in \Gamma(U, \mathscr{S}^N)$, $1 \le i \le q$, be a finite number of sections. It suffices to show the local finiteness of the relation sheaf

$$\mathscr{R} = \left\{ (a_i) \in \mathscr{A}_x^q \subset \mathscr{A}^q; \ \sum_{i=1}^{q} a_i \underline{F_i}_x = 0, \ x \in U \right\}.$$

With the expressions $F_i = (F_{i1}, \ldots, F_{iN})$, $\mathscr{R}$ ($\subset (\mathscr{A}|_U)^q$) is determined by

$$(a_i) \in \mathscr{A}_x^q, \quad \sum_{i=1}^{q} a_i \underline{F_{ij}}_x = 0, \quad 1 \le j \le N, \ x \in U.$$

We first consider the case for $j = 1$. Denote by $\mathscr{R}_1 \subset (\mathscr{A}|_U)^q$ the relation sheaf defined by

$$(a_i) \in \mathscr{A}_x^q, \quad \sum_{i=1}^{q} a_i \underline{F_{i1}}_x = 0, \quad x \in U.$$

Then $\mathscr{R} \subset \mathscr{R}_1$, and since $\mathscr{S}$ is coherent over $\mathscr{A}$, $\mathscr{R}_1$ is locally finite over $\mathscr{A}$. For every point $a \in U$ there are a neighborhood $V \subset U$ of a and a locally finite generator system $\{\phi^{(\lambda)}\}_{\lambda=1}^{L}$ of $\mathscr{R}_1|_V$ with $\phi^{(\lambda)} \in \Gamma(V, \mathscr{R}_1)$. Set $\phi^{(\lambda)} = (\phi_i^{(\lambda)})_{1 \le i \le q}$. At every point $x \in V$ an element of $\mathscr{R}_{1x}$,

$$(a_i) = \left(\sum_{\lambda} \underline{c_\lambda}_x \cdot \phi_i^{(\lambda)}(x) \right), \qquad \underline{c_\lambda}_x \in \mathscr{A}_x$$

belongs to $\mathscr{R}_x$ if and only if

(2.4.8) $$\sum_i \sum_\lambda \underline{c_{\lambda}}_x \cdot \phi_i^{(\lambda)}(x) \cdot \underline{F_{ij}}_x = 0, \quad 1 \le j \le N.$$

We consider this as a linear relation on $\left(\underline{c_\lambda}_x\right)$. For $j = 1$ it is already satisfied because of the choice of $\phi_i^{(\lambda)}$. Therefore, simultaneous relation (2.4.8), in fact, consists of $N - 1$ relations. The induction hypothesis implies that in a neighborhood $W(\subset V)$ of a such $\left(\underline{c_\lambda}_x\right)$ is written by a linear sum of finitely many sections $\gamma^{(\nu)} = \left(\gamma_\lambda^{(\nu)}\right)$ with $\gamma_\lambda^{(\nu)} \in \Gamma(W, \mathscr{A})$. Therefore, the sections $(a_i^{(\nu)}) = \sum_\lambda \gamma_\lambda^{(\nu)} \cdot \phi_i^{(\lambda)}$ generate $\mathscr{R}_x$ at every $x \in W$ over $\mathscr{A}_x$. □

Proposition 2.4.9 *Let $\mathscr{S}$ be a coherent sheaf of modules over $\mathscr{A}$. If $\mathscr{F}_i \subset \mathscr{S}$, $i = 1, 2$, are coherent subsheaves of modules over $\mathscr{A}$, so is the intersection $\mathscr{F}_1 \cap \mathscr{F}_2$.*

Proof For every point $a \in X$ there are a neighborhood $U(\subset X)$ and locally finite generator systems of $\mathscr{F}_i$, $i = 1, 2$,

$$\alpha_j \in \Gamma(U, \mathscr{F}_1), \quad 1 \le j \le l,$$
$$\beta_k \in \Gamma(U, \mathscr{F}_2), \quad 1 \le k \le m.$$

At any point $b \in U$, $\gamma \in \mathscr{S}_b$ belongs to $\mathscr{F}_{1b} \cap \mathscr{F}_{2b}$ if and only if it is written as

$$\gamma = \sum_j a_j \alpha_j(b) = \sum_k b_k \beta_k(b),$$
$$a_j, b_k \in \mathscr{A}_b.$$

This is equivalent to

(2.4.10) $$\sum_j a_j \alpha_j(b) + \sum_k b_k(-\beta_k(b)) = 0,$$
$$\gamma = \sum_j a_j \alpha_j(b).$$

The above expression defines a relation sheaf of $\mathscr{S}$ with (a_j, b_k) being unknowns, which is denoted by $\mathscr{R} := \mathscr{R}(\ldots, \alpha_j, \ldots, -\beta_k, \ldots)$. Since $\mathscr{S}$ is coherent over $\mathscr{A}$, $\mathscr{R}$ is locally finite over $\mathscr{A}$. Taking a smaller $U \ni a$ if necessary, we may assume that $\mathscr{R}|_U$ is generated by a finite number of $\eta^{(h)} \in \Gamma(U, \mathscr{R})$, $1 \le h \le L$. With $\eta^{(h)} := (a_j^{(h)}, b_k^{(h)})$, $(\mathscr{F}_1 \cap \mathscr{F}_2)|_U$ is generated by

$$\xi^{(h)} := \sum_j a_j^{(h)} \alpha_j = \sum_k b_k^{(h)} \beta_k \in \Gamma(U, \mathscr{F}_1 \cap \mathscr{F}_2), \quad 1 \le h \le L.$$ □

Example 2.4.11 We introduce a non-coherent example given in Oka VII. Consider a hypersurface $X = \{z = w\}$ in $\mathbf{C}^2$ with variables z, w. Taking two concentric balls

$B_i = \{|z|^2 + |w|^2 < r_i^2\}$ $(r_1 < r_2)$, we set $X_0 = X \cap B_2 \setminus B_1$. Let $\Gamma_1 = \partial B_1$ be the boundary hypersphere. Let $\mathscr{B}(U)$ be the set of all holomorphic functions $f(z, w)$ on an open subset $U \subset B_2$ such that $f(z, w)/(z - w)$ is holomorphic at every point of $U \cap X_0$. With the natural restrictions $\{\mathscr{B}(U)\}$ forms a presheaf, which defines a sheaf $\mathscr{B}$ of $\mathscr{O}_{B_2}$-modules; in fact, the sheaf $\mathscr{B}$ is an ideal sheaf of $\mathscr{O}_{B_2}$. By the construction we have

$$
\mathscr{B}_a = \begin{cases} \mathscr{O}_{n,a} \cdot \underline{(z - w)}_a, & a \in X_0, \\ \mathscr{O}_{n,a}, & a \in B_2 \setminus X_0. \end{cases} \tag{2.4.12}
$$

Then $\mathscr{B}$ is not locally finite in any neighborhood of a point $a \in X_0 \cap \Gamma_1$. Suppose that it is locally finite in a neighborhood of a. Then there are a neighborhood U of a and finitely many $f_j \in \mathscr{B}(U)$, $1 \leq j \leq N$, such that

$$
\mathscr{B}_b = \sum_{j=1}^{N} \mathscr{O}_{n,b} \cdot \underline{f_j}_b, \quad {}^\forall b \in U.
$$

But, because of $f_j(z, z) \equiv 0$, $\mathscr{B}_b \neq \mathscr{O}_{n,b}$ for $b \in U \cap X \setminus X_0$, which contradicts (2.4.12).

Example 2.4.13 We may construct a non-coherent relation sheaf as follows. We take an open ball B of $\mathbf{C}^n$ with center at the origin. Denote by Γ its boundary. Let $\chi(a)$ denote the characteristic function of B; that is, on B, $\chi = 1$, and $\chi = 0$ on $\mathbf{C}^n \setminus B$. Set

$$
\mathscr{R} = \left\{ \underline{f}_a \in \mathscr{O}_{n,a};\ \underline{f}_a \cdot \underline{\chi}_a = 0, \quad a \in \mathbf{C}^n \right\}.
$$

Then,

$$
\mathscr{R}_a = \begin{cases} 0, & a \in B \cup \Gamma, \\ \mathscr{O}_{n,a}, & a \notin B \cup \Gamma. \end{cases}
$$

Therefore $\mathscr{R}$ is not coherent about $a \in \Gamma$.

In the above example, the length of the relation is one. To make the length two, we set $\phi(a) = 1 - \chi(a)$ and

$$
\mathscr{S} = \left\{ \underline{f}_a \oplus \underline{g}_a \in \mathscr{O}_{n,a} \oplus \mathscr{O}_{n,a};\ \underline{f}_a \cdot \underline{\chi}_a + \underline{g}_a \cdot \underline{\phi}_a = 0, \quad a \in \mathbf{C}^n \right\} \subset \mathscr{O}_n^2.
$$

Then,

$$
\mathscr{S}_a = \begin{cases} 0 \oplus \mathscr{O}_{n,a}, & a \in B, \\ 0 \oplus 0, & a \in \Gamma, \\ \mathscr{O}_{n,a} \oplus 0, & a \notin B \cup \Gamma. \end{cases}
$$

Therefore, $\mathscr{S}$ is not locally finite about any point of Γ.

If one requires the differentiability for χ and ϕ, it suffices to take a C^∞ function such that

$$\chi(a) > 0, \quad a \in B; \ \chi(a) = 0, \quad a \notin B.$$

Then, the same conclusion is obtained.

Example 2.4.14 For examples of coherent sheaves, we have $\mathscr{O}_{\mathbf{C}^n}$ which we are going to show, and a geometric ideal sheaf $\mathrm{i}\langle A\rangle$ (cf. Definition 2.3.6 and Sect. 6.5), but the proofs are not easy.

2.5 Oka's First Coherence Theorem

The following theorem is called *Oka's Coherence Theorem*, but in the present book we call this *Oka's First Coherence Theorem.*[3] It is impossible to explain the meaning of this theorem in a few lines.

Reinhold Remmert, who is a well-known German complex analysist, describes it in *Encyclopedia of Mathematics* [69] published in 1994 as follows:

> It is no exaggeration to claim that Oka's theorem became a landmark in the development of function theory of several complex variables.

Theorem 2.5.1 (Oka's First Coherence Theorem 1948, Oka [62] VII) *The sheaf $\mathscr{O}^N_{\mathbf{C}^n}$ $(N \geq 1)$ is coherent.*

Proof We proceed by induction on $n \geq 0$ with general $N \geq 1$. We write $\mathscr{O}_{\mathbf{C}^n} = \mathscr{O}_n$.

(a) $n = 0$: In this case, it is a matter of a finite-dimensional vector space over $\mathbf{C}$.
(b) $n \geq 1$: Suppose that $\mathscr{O}^N_{n-1}$ is coherent for every $N \geq 1$.

By Proposition 2.4.7 (ii) it suffices to show the case of $N = 1$.

The problem is local and it is sufficient to prove Definition 2.4.4 (ii). Taking an open subset $\Omega \subset \mathbf{C}^n$ and $\tau_j \in \mathscr{O}(\Omega) \cong \Gamma(\Omega, \mathscr{O}_n)$, $1 \leq j \leq q$, we consider the relation sheaf $\mathscr{R}(\tau_1, \ldots, \tau_q)$ defined by

$$\text{(2.5.2)} \qquad \underline{f_1}_z \underline{\tau_1}_z + \cdots + \underline{f_q}_z \underline{\tau_q}_z = 0, \quad \underline{f_j}_z \in \mathscr{O}_{n,z}, \ z \in \Omega.$$

What we want to show is:

Claim 2.5.3 *For every point $a \in \Omega$ there are a neighborhood $V \subset \Omega$ of a and finitely many sections $s_k \in \Gamma(V, \mathscr{R}(\tau_1, \ldots, \tau_q))$, $1 \leq k \leq l$, such that*

$$\mathscr{R}(\tau_1, \ldots, \tau_q)_b = \sum_{k=1}^{l} \mathscr{O}_{n,b} \cdot s_k(b), \quad {}^\forall b \in V.$$

[3] For the reason, cf. "Historical supplements" at the end of this chapter and Chap. 9.

For the proof we may assume that $a = 0$. If an element $\underline{\tau_j}_0 = 0$, then the j-th component of $\mathscr{R}(\tau_1, \ldots, \tau_q)|_V \subset (\mathscr{O}_V)^q$ is just $\mathscr{O}_V$ in a neighborhood V of 0, and so we may assume that $\underline{\tau_j}_0 \neq 0$, $1 \leq j \leq q$.

By Theorem 2.2.12, $\mathscr{O}_{n,0}$ is a unique factorization domain. We may divide $\underline{\tau_j}_0$, $1 \leq j \leq q$, by the common factors, and may assume that there is no common factor among them (this procedure is not necessary for the proof itself, but may possibly decrease the computational complexity of the algorithm to obtain the finite generator system).

Let p_j be the order of zero of τ_j at 0, and set

$$p = \max_{1 \leq j \leq q} p_j,$$
$$p' = \min_{1 \leq j \leq q} p_j \geq 0.$$

After reordering the indices, we may assume that

$$p' = p_1.$$

The following sections are clearly solutions of (2.5.2) and are sections of $\mathscr{R}$:

(2.5.4) $$t_i = (\tau_i, 0, \ldots, 0, \overset{i\text{-th}}{-\tau_1}, 0, \ldots, 0) \in \Gamma(\mathrm{P}\Delta, \mathscr{R}), \quad 2 \leq i \leq q,$$

which we call the *trivial solutions*.

Take a common standard polydisk $\mathrm{P}\Delta = \mathrm{P}\Delta_{n-1} \times \Delta_{(n)}$ with $\Delta_{(n)} = \{|z_n| < r_n\}$ for all τ_j. By Weierstrass's Preparation Theorem 2.1.3 at 0 one can transfer a unit factor of τ_j to f_j in (2.5.2) so that all τ_j may be assumed to be Weierstrass polynomials:

(2.5.5) $$\tau_j = P_j(z', z_n) = \sum_{\nu=0}^{p_j} a_{j\nu}(z') z_n^{\nu} = \sum_{\nu=0}^{p} a_{j\nu}(z') z_n^{\nu} \in \mathscr{O}(\mathrm{P}\Delta_{n-1})[z_n],$$
$$a_{j\nu}(0) = 0 \ (\nu < p_j), \quad a_{jp_j} = 1, \quad a_{j\nu} = 0 \ (p_j < \nu \leq p).$$

If $p_j = 0$ ($\underline{\tau_j}_0$ is a unit), $P_j = 1$. We set

(2.5.6) $$\mathscr{R} = \mathscr{R}(P_1, \ldots, P_q).$$

Here, the trivial solutions are

$$T_i = (P_i, 0, \ldots, 0, \overset{i\text{-th}}{-P_1}, 0, \ldots, 0) \in \Gamma(\mathrm{P}\Delta, \mathscr{R}), \quad 2 \leq i \leq q.$$

It suffices to show the local finiteness of $\mathscr{R}$. We perform a division algorithm for an unknown vector $\alpha = (\alpha_j) \in \mathscr{R}$ with respect to the trivial solutions T_i ($2 \leq i \leq q$);

more precisely, we perform a division algorithm for α_j with respect to P_1 (cf. (2.5.11) and Remark 2.5.23 below).

Take an arbitrary point $b = (b', b_n) \in \mathrm{P}\Delta_{n-1} \times \Delta_{(n)}$. We call an element of $\mathscr{O}_{n-1,b'}[z_n]$ a *z_n-polynomial-like germ*. In the same way, we call $\alpha = (\alpha_1, \ldots, \alpha_q) \in \mathscr{O}^q_{n,b}$ consisting of z_n-polynomial-like germs α_j a *polynomial-like element*, and $f = (f_j)$ with $(f_j)_{1 \le j \le q} \in (\mathscr{O}(\mathrm{P}\Delta_{n-1})[z_n])^q$ a *z_n-polynomial-like section*. We set

$$\deg \alpha = \deg_{z_n} \alpha = \max_j \deg_{z_n} \alpha_j,$$

$$\deg f = \deg_{z_n} f = \max_j \deg_{z_n} f_j.$$

Then we have:

2.5.7 The trivial solutions T_i are z_n-polynomial-like sections of $\deg T_i \le p$.

We now show the following:

Lemma 2.5.8 (Degree structure) *Let the notation be as above. Then an element of $\mathscr{R}_b$ is written as a finite linear sum of the trivial solutions, T_i, $2 \le i \le q$, and z_n-polynomial-like elements $\alpha = (\alpha_1, \alpha_2, \ldots, \alpha_q)$ of $\mathscr{R}_b$ with coefficients in $\mathscr{O}_{n,b}$ such that*

$$\deg \alpha_1 < p,$$

$$\deg \alpha_j < p', \quad 2 \le j \le q.$$

N.B. If $p' = 0$, then there is no term of α.

∵) By making use of Weierstrass' Preparation Theorem at b we decompose P_1 into a unit u and a Weierstrass polynomial Q:

$$P_1(z', z_n) = u \cdot Q(z', z_n - b_n), \qquad \deg Q = d \le p_1.$$

Lemma 2.2.15 implies $u \in \mathscr{O}_{n-1,b'}[z_n]$. Therefore,

$$\deg_{z_n} u = p_1 - d. \tag{2.5.9}$$

Take an arbitrary $f = \left(f_1, \ldots, f_q\right) \in \mathscr{R}_b$. By Weierstrass' Preparation Theorem 2.1.3 (ii) we have

$$\begin{aligned} f_i &= c_i Q + \beta_i, \quad 1 \le i \le q, \\ & c_i \in \mathscr{O}_{n,b}, \quad \beta_i \in \mathscr{O}_{n-1,b'}[z_n], \\ & \deg_{z_n} \beta_i \le d - 1. \end{aligned} \tag{2.5.10}$$

Since $u \in \mathscr{O}_{n,b}$ is a unit, with $\tilde{c}_i := c_i u^{-1}$ we get

$$f_i = \tilde{c}_i P_1 + \beta_i, \qquad 1 \le i \le q. \tag{2.5.11}$$

By making use of this we perform the following calculation:

$$(2.5.12)\qquad \begin{aligned}(f_1, \ldots, f_q) &+ \tilde{c}_2 T_2 + \cdots + \tilde{c}_q T_q \\ &= (\tilde{c}_1 P_1 + \beta_1, \tilde{c}_2 P_1 + \beta_2, \ldots, \tilde{c}_q P_1 + \beta_q) \\ &\quad + (\tilde{c}_2 P_2, -\tilde{c}_2 P_1, 0, \ldots, 0) \\ &\quad + \cdots \\ &\quad + (\tilde{c}_q P_q, 0, \ldots, 0, -\tilde{c}_q P_1) \\ &= \left(\sum_{i=1}^{q} \tilde{c}_i P_i + \beta_1, \beta_2, \ldots, \beta_q \right) \\ &= \left(g_1, \beta_2, \ldots, \beta_q \right).\end{aligned}$$

Here we put $g_1 = \sum_{i=1}^{q} \tilde{c}_i P_i + \beta_1 \in \mathscr{O}_{n,b}$. Note that $\beta_i \in \mathscr{O}_{n-1,b'}[z_n]$, $2 \leq i \leq q$. Since $\left(g_1, \beta_2, \ldots, \beta_q \right) \in \mathscr{R}_b$,

$$(2.5.13)\qquad g_1 P_1 = -\beta_2 P_2 - \cdots - \beta_q P_q \in \mathscr{O}_{n-1,b'}[z_n].$$

Remark 2.5.14 It should be noticed that if $p_1 = 0$, then $P_1 = 1$, $\beta_i = 0$, $1 \leq i \leq q$, and $g_1 = 0$; the proof is finished in this case.

In general, it follows from the expression of the right-hand side of (2.5.13) that

$$\deg_{z_n} g_1 P_1 \leq \max_{2 \leq i \leq q} \deg_{z_n} \beta_i + \max_{2 \leq i \leq q} \deg_{z_n} P_i \leq d + p - 1.$$

On the other hand, $g_1 P_1 = g_1 u Q$ and Q is a Weierstrass polynomial at b. Again by Lemma 2.2.15 we see that

$$(2.5.15)\qquad \begin{aligned} \alpha_1 &:= g_1 u \in \mathscr{O}_{n-1,b'}[z_n], \\ \deg_{z_n} \alpha_1 &= \deg_{z_n} g_1 P_1 - \deg_{z_n} Q \\ &\leq d + p - 1 - d = p - 1. \end{aligned}$$

Setting $\alpha_i = u \beta_i$ for $2 \leq i \leq q$. we have by (2.5.9) and (2.5.10) that

$$(2.5.16)\qquad \deg_{z_n} \alpha_i \leq p_1 - d + d - 1 = p_1 - 1 = p' - 1, \quad 2 \leq i \leq q,$$

and then by (2.5.12) that

$$(2.5.17)\qquad f = -\sum_{i=2}^{q} \tilde{c}_i T_i + u^{-1}(\alpha_1, \alpha_2, \ldots, \alpha_q). \qquad \triangle$$

Until now we have not used the induction hypothesis. Now we are going to use it to prove the existence of a locally finite generator system of those $(\alpha_1, \ldots, \alpha_q)$ appearing in (2.5.17). We write

$$\text{(2.5.18)} \qquad \alpha_1 = \sum_{\nu=0}^{p-1} \underline{c_{1\nu}(z')}_{b'} z_n^\nu, \quad \underline{c_{1\nu}(z')}_{b'} \in \mathscr{O}_{n-1,b'},$$

$$\alpha_i = \sum_{\nu=0}^{p'} \underline{c_{i\nu}(z')}_{b'} z_n^\nu, \quad \underline{c_{i\nu}(z')}_{b'} \in \mathscr{O}_{n-1,b'}, \ 2 \le i \le q.$$

By $\mathscr{S}$ we denote the sheaf of all these $(\alpha_1, \ldots, \alpha_q)$ over $\mathrm{P}\Delta = \mathrm{P}\Delta_{n-1} \times \Delta_{(n)}$ satisfying

$$\text{(2.5.19)} \qquad \alpha_1 P_1 + \alpha_2 P_2 + \cdots + \alpha_q P_q = 0.$$

The left-hand side above is a z_n-polynomial-like element of degree at most $p+p'-1$, and relation (2.5.19) is equivalent to the nullity of all $p+p'$ coefficients. With the expression in (2.5.5) we have

$$\text{(2.5.20)} \qquad \sum_{i=1}^{q} \sum_{k+h=\nu}{}' \underline{a_{ik}(z')}_{b'} \cdot \underline{c_{ih}(z')}_{b'} = 0 \in \mathscr{O}_{n-1,b'}, \quad 0 \le \nu \le p+p'-1,$$

where $\sum'$ stands for the sum over those indices h, k to which some elements $\underline{a_{ik}(z')}_{b'}$, $\underline{c_{ih}(z')}_{b'}$ correspond. Then (2.5.20) defines a $(p+p')$-simultaneous linear relation sheaf $\widetilde{\mathscr{S}}$ in $\mathscr{O}_{\mathrm{P}\Delta_{n-1}}^{p+p'}$ with $p+p'(q-1)$ unknowns, c_{ih}'s. The induction hypothesis implies that $\widetilde{\mathscr{S}}$ is coherent, and hence there is a locally finite generator system of $\widetilde{\mathscr{S}}$ over a polydisk neighborhood $\widetilde{\mathrm{P}\Delta}_{n-1} \subset \mathrm{P}\Delta_{n-1}$ of 0. Therefore we infer from (2.5.18) that $\mathscr{S}$ has a locally finite generator system $\{\pi_\mu\}_{\mu=1}^M$ over $\widetilde{\mathrm{P}\Delta} := \widetilde{\mathrm{P}\Delta}_{n-1} \times \Delta_{(n)} (\subset \mathrm{P}\Delta_{n-1} \times \Delta_{(n)} = \mathrm{P}\Delta)$.

Thus, the finite system $\{T_i\}_{i=2}^q \cup \{\pi_\mu\}_{\mu=1}^M$ generates $\mathscr{R}$ over $\widetilde{\mathrm{P}\Delta}$. □

Definition 2.5.21 A sheaf $\mathscr{S}$ of $\mathscr{A}$-modules over X is said to be *locally free with finite rank* if for every point $x \in X$ there are a neighborhood U of x and $p \in \mathbf{N}$ such that $\mathscr{S}|_U \cong \mathscr{A}_U^p$.

The following statement is immediate from Oka's First Coherence Theorem 2.5.1.

Corollary 2.5.22 *A sheaf of $\mathscr{O}_\Omega$-modules which is locally free with finite rank is coherent.*

Remark 2.5.23 In (2.5.17) we obtained the degree estimates such that $\deg_{z_n} T_i \le p = \max_j \deg_{z_n} P_j$, and $\deg_{z_n}(\alpha_1, \alpha_2, \ldots, \alpha_q) < p$, where, furthermore, only the first element α_1 is of $\deg_{z_n} < p$ and for the others,

$$\deg_{z_n} \alpha_i < p' = \min_j \deg_{z_n} P_j, \quad 2 \le i \le q$$

(see (2.5.15) and (2.5.16)). Therefore, as noticed in the proof, if $p_1 = 0$, the trivial solutions T_i, $2 \le i \le q$, form a local finite generator system of $\mathscr{R}$; if $p_1 = 1$, these α_i are constants. The argument presented here is due to [54]. This seems not to have been widely observed before and gives a slight improvement of the proof of Oka VII (1948). In the most-known references such as K. Oka [62] VII, H. Cartan [10], R. Narasimhan [48], L. Hörmander [33], T. Nishino [49], etc., division algorithm (2.5.11) is performed with respect to an element P_{j_0} with the maximum degree $p(= p_{j_0})$, so that the degree estimate for α_i is less than p; it is, however, more natural to use P_1 in (2.5.11) than P_{j_0}. For the proof with P_{j_0} does not reduce to the easiest one when $p' = 0$, where the trivial solutions already form a local generator system of the relation sheaf (cf. Remark 2.5.14). As seen in Exercise 4 at the end of this chapter, when $n = 1$, it is reduced to the case of $p' = 0$. The proof presented above reflects some merit as p' is small.

Example 2.5.24 Let $(z, w) \in \mathbf{C}^2$ and set

$$F_1(z, w) = w + z,$$
$$F_2(z, w) = w^2 + z^2 w + z^3 e^z,$$
$$F_3(z, w) = w^3 + zw^2 + z^2 \tan z.$$

These are Weierstrass polynomials in w about the origin 0 without common factor. Let $\mathscr{R}(F_1, F_2, F_3)$ be the relation sheaf defined by

$$f_1 F_1 + f_2 F_2 + f_3 F_3 = 0. \tag{2.5.25}$$

We shall obtain a locally finite generator system of $\mathscr{R}(F_1, F_2, F_3)$ about 0. If one uses the division algorithm of the maximum degree, that is, by F_3, the computation is rather involved. But, by the division algorithm of the minimum degree, $\deg_w F_1 = \min\{\deg_w F_i\} = 1$, it is carried out easily as follows.

By Weierstrass' Preparation Theorem 2.1.3 we set

$$f_i = c_i F_1 + \beta_i, \quad \beta_i \in \mathscr{O}_z, \ 1 \le i \le 3,$$

where $\mathscr{O}_z$ stands for the set of holomorphic functions only in variable z. Let

$$T_2 = \begin{pmatrix} -F_2 \\ F_1 \\ 0 \end{pmatrix}, \quad T_3 = \begin{pmatrix} -F_3 \\ 0 \\ F_1 \end{pmatrix}$$

be the trivial solutions. Then it follows that

$$\begin{pmatrix} f_1 \\ f_2 \\ f_3 \end{pmatrix} - c_2 T_2 - c_3 T_3 = \begin{pmatrix} \sum_{i=1}^{3} c_i F_i + \beta_1 \\ \beta_2 \\ \beta_3 \end{pmatrix} = \begin{pmatrix} g_1 \\ \beta_2 \\ \beta_3 \end{pmatrix},$$

where $g_1 = \sum_{i=1}^{3} c_i F_i + \beta_1$. We have

$$g_1 F_1 + \beta_2 F_2 + \beta_3 F_3 = 0. \tag{2.5.26}$$

We infer from the degree comparison that $\deg g_1 \leq 2$, and then set

$$g_1(z, w) = g_{12}(z)w^2 + g_{11}(z)w + g_{10}(z).$$

Substituting this in (2.5.26), we get

$$\begin{aligned}(g_{12}w^2 + g_{11}w + g_{10})(w + z) + \beta_2(w^2 + z^2 w + z^3 e^z) \\ + \beta_3(w^3 + zw^2 + z^2 \tan z) = 0.\end{aligned}$$

As a polynomial of degree 3 in w, we get

$$\begin{aligned}&(g_{12} + \beta_3)w^3 + (g_{12}z + g_{11} + \beta_2 + \beta_3 z)w^2 \\ &\quad + (g_{11}z + g_{10} + \beta_2 z^2)w + g_{10}z + \beta_2 z^3 e^z + \beta_3 z^2 \tan z = 0.\end{aligned}$$

All the coefficients are 0:

$$\begin{aligned} g_{12} + \beta_3 &= 0, \\ g_{12}z + g_{11} + \beta_2 + \beta_3 z &= 0, \\ g_{11}z + g_{10} + \beta_2 z^2 &= 0, \\ g_{10} + \beta_2 z^2 e^z + \beta_3 z \tan z &= 0. \end{aligned}$$

Here the last one is already divided by z. With a matrix we obtain that

$$\begin{pmatrix} 1 & 0 & 0 & 0 & 1 \\ z & 1 & 0 & 1 & z \\ 0 & z & 1 & z^2 & 0 \\ 0 & 0 & 1 & z^2 e^z & z \tan z \end{pmatrix} \begin{pmatrix} g_{12} \\ g_{11} \\ g_{10} \\ \beta_2 \\ \beta_3 \end{pmatrix} = \begin{pmatrix} 0 \\ 0 \\ 0 \\ 0 \\ 0 \end{pmatrix}. \tag{2.5.27}$$

Elementary transforms of matrices yield that

$$\begin{pmatrix} 1 & 0 & 0 & 0 & 1 \\ 0 & 1 & 0 & 1 & 0 \\ 0 & 0 & 1 & z^2 - z & -z^2 \\ 0 & 0 & 0 & 1 & \frac{z + \tan z}{1 - z + ze^z} \end{pmatrix} \begin{pmatrix} g_{12} \\ g_{11} \\ g_{10} \\ \beta_2 \\ \beta_3 \end{pmatrix} = \begin{pmatrix} 0 \\ 0 \\ 0 \\ 0 \\ 0 \end{pmatrix}.$$

Here it is noted that $\frac{z+\tan z}{1-z+ze^z}$ is holomorphic in a neighborhood of 0. Therefore we have

$$\begin{pmatrix} g_1 \\ \beta_2 \\ \beta_3 \end{pmatrix} = \begin{pmatrix} -w^2 + \frac{z+\tan z}{1-z+ze^z} w + z^2 + (z^2 - z) \frac{z+\tan z}{1-z+ze^z} \\ -\frac{z+\tan z}{1-z+ze^z} \\ 1 \end{pmatrix} \beta_3.$$

Thus, $\mathscr{R}(F_1, F_2, F_3)$ is generated in a neighborhood of 0 by the finite system

$$\left\{ T_2, T_3, \begin{pmatrix} -w^2 + \frac{z+\tan z}{1-z+ze^z} w + z^2 + (z^2 - z) \frac{z+\tan z}{1-z+ze^z} \\ -\frac{z+\tan z}{1-z+ze^z} \\ 1 \end{pmatrix} \right\}.$$

Historical Supplements

L. Bers, well-known in Teichmüller moduli theory, closes the preface of his lecture notes on the theory of several complex variables at Courant Institute, New York University [6] (1964) with the following sentence:

> Every account of the theory of several complex variables is largely a report on the ideas of Oka. This one is no exception.

The pillar of Oka's ideas is "Oka's First Coherence Theorem 2.5.1". There is no way to describe the proof of Theorem 2.5.1 other than "really marvelous".

In the preface of their basic book [28], H. Grauert and R. Remmert, the German luminaries in complex analysis of 20th century, write:

> Of greatest importance in Complex Analysis is the concept of a coherent analytic sheaf.

And they list the following as the *four fundamental coherence theorems*:

(i) The structure sheaf $\mathscr{O}_X$ of a complex space X is coherent (cf. Sect. 6.9 (of the present book)).
(ii) The geometric ideal sheaf $\mathscr{I}\langle A\rangle$ of an analytic subset A is coherent (cf. Sect. 6.5).
(iii) The normalization sheaf $\hat{\mathscr{O}}_X$ of a complex space is coherent (cf. Sect. 6.10).
(iv) The direct image sheaf of a coherent sheaf through a proper holomorphic map is coherent.

In Oka VII (1950) and VIII (1951) K. Oka proved the first three Coherence Theorems (the first (i) is a direct consequence of the coherence of $\mathscr{O}_{\mathbf{C}^n}$ and (ii)). As for the second one, it is often attributed to H. Cartan in many literatures (cf., e.g., [29]), but as discussed in Chap. 9, this result had been clearly announced in Oka VII (received 1948), so that it would be proved "*without any additional assumption*" in the next paper (i.e., Oka VIII, where in fact Oka gave the proof). As one reads the papers, one naturally finds that Oka VII and VIII form one set of papers. As already mentioned in the Preface, a key part of the proof of (ii) was already discussed and proved in Oka VII: In Sect. 3 it was formulated and discussed as Problème (K) and proved in Sect. 6. When he completed the paper, Oka VII, he had the proof of the Second Coherence Theorem in hand.

Because of these historical developments of mathematical comprehension, we call here the first three in order:

- *Oka's First Coherence Theorem* (for $\mathscr{O}_{\mathbf{C}^n}$ with $X = \mathbf{C}^n$),
- *Oka's Second Coherence Theorem* (for $\mathscr{I}\langle A\rangle$),
- *Oka's Third Coherence Theorem* (for $\hat{\mathscr{O}}_X$).

It would be the closest to the actual history that between Oka VII and VIII, H. Cartan gave his own proof to Oka's Second Coherence Theorem. For more details, see Chap. 9.

Exercises

1. For $t_j \in \mathbf{C}$, $1 \le j \le n$, set

$$s_\nu = (-1)^\nu \sum_{1 \le j_1 < \cdots < j_\nu \le n} t_{j_1} \cdots t_{j_\nu}, \quad 1 \le \nu \le n,$$

$$\sigma_\mu = \sum_{j=1}^{n} t_i^\mu, \quad \mu \ge 1.$$

Show that

$$\sigma_m + \sigma_{m-1}s_1 + \cdots + \sigma_1 s_{m-1} + m s_m = 0, \quad m \le n,$$
$$\sigma_m + \sigma_{m-1}s_1 + \cdots + \sigma_{m-n}s_n = 0, \quad m > n.$$

(Hint: For the first, set $f(X) = \prod_{j=1}^{n}(1 - a_j X)$. Then, $\frac{f'(X)}{f(X)} = -\sum_{j=1}^{n} \frac{t_j}{1-t_j X} = -\sum_{\nu=0}^{\infty} \sigma_{\nu+1} X^\nu$. Then, use $f'(X) = -f(X)\sum_{\nu=0}^{\infty} \sigma_{\nu+1}X^\nu$. For the second, use $X^m + s_1 X^{m-1} + \cdots + s_m X^{m-n} = X^{m-n}\prod_{j=1}^{n}(X - a_j)$. Then, substitute $X = t_j$, $1 \le j \le n$.)

2. (1) Let $f(z, w) = \sin w + z^2$ with $(z, w) \in \mathbf{C}^2$.
   a. Show that $\mathrm{ord}_0 f = 1$.
   b. Obtain the Weierstrass decomposition of $f(z, w)$ at 0.

   (2) Consider the same for $g(z, w) = w + \sin w + z^2$.

3. Consider the implicit function $w = w(z)$ defined by $g(z, w) = 0$ given in (2) just above. By making use of (2.1.6), obtain an integral formula of $w(w(z))$ in a neighborhood of 0 in terms of $g(z, w)$.
4. Show directly Claim 2.5.3, when $n = 1$.
5. Write down the simultaneous relation (2.5.20) in the form of a matrix equation as in (2.5.27).

6. Set

$$P_1(z, w) = w + z,$$
$$P_2(z, w) = w^2 + (z^2 + z)w + z^3,$$
$$P_3(z, w) = w^3 + zw^2 + z^3.$$

Obtain a locally finite generator system of the relation sheaf $\mathscr{R}(P_1, P_2, P_3)$ about the origin.

7. Set

$$F_1(z, w) = w + ze^z,$$
$$F_2(z, w) = w^2 + z^2 w,$$
$$F_3(z, w) = w^3 + w^2 \sin z,$$
$$F_4(z, w) = w^3 + z^3 e^z.$$

Obtain a locally finite generator system of the relation sheaf $\mathscr{R}(F_1, F_2, F_3, F_4)$ about the origin.

Chapter 3
Sheaf Cohomology

In Sect. 1.3 of Chap. 1 we introduced the notion of sheaves. In this chapter we present the basic theory of sheaf cohomologies, which we owe to Cartan and Serre. Here, sheaves are assumed at least to have a structure of abelian groups. The sheaf cohomology makes it easier to understand the theory of Oka and to use it. As we shall see in the next chapter, "Oka's Jôku-Ikô" due to Oka I is clarified by it, and we will be led to understand the depth of the implication of Oka's First Coherence Theorem.

3.1 Exact Sequences

Let X be a topological space, and suppose that there is a sequence of homomorphisms (or simply called, morphism) of sheaves (of abelian groups),

$$\mathscr{R} \xrightarrow{\phi} \mathscr{S} \xrightarrow{\psi} \mathscr{T}.$$

We say that the sequence is *exact* (at $\mathscr{S}$) if

$$\operatorname{Im} \phi = \operatorname{Ker} \psi.$$

It follows from the definition that:

(i) ϕ is injective $\iff$ $0 \to \mathscr{R} \xrightarrow{\phi} \mathscr{S}$ is exact.

(ii) ψ is surjective $\iff$ $\mathscr{S} \xrightarrow{\psi} \mathscr{T} \to 0$ is exact.

A *short exact sequence of sheaves* is a sequence,

$$0 \to \mathscr{R} \xrightarrow{\phi} \mathscr{S} \xrightarrow{\psi} \mathscr{T} \to 0, \tag{3.1.1}$$

J. Noguchi, *Analytic Function Theory of Several Variables*,
DOI 10.1007/978-981-10-0291-5_3

such that ϕ is injective, ψ is surjective, and $\operatorname{Im}\phi = \operatorname{Ker}\psi$.

In general, for an exact sequence, $0 \to \mathscr{R} \xrightarrow{\phi} \mathscr{S}$,

$$0 \to \mathscr{R} \xrightarrow{\phi} \mathscr{S} \longrightarrow \mathscr{S}/\mathscr{R} \to 0$$

is exact. Therefore, (3.1.1) is exact if and only if ϕ is injective and $\mathscr{T} \cong \mathscr{S}/\mathscr{R}$.

By the compositions of maps, (3.1.1) induces the following exact sequence:

$$\text{(3.1.2)} \qquad \begin{array}{ccccccc} 0 \longrightarrow & \Gamma(X, \mathscr{R}) & \xrightarrow{\phi_*} & \Gamma(X, \mathscr{S}) & \xrightarrow{\psi_*} & \Gamma(X, \mathscr{T}). \\ & \Psi & & \Psi & & \Psi \\ & f & \longmapsto & \phi \circ f; \; g & \longmapsto & \psi \circ g \end{array}$$

For, with regarding $\mathscr{R} \subset \mathscr{S}$ through ϕ, every $g \in \Gamma(X, \mathscr{S})$ such that $\psi_*(f) = 0$ satisfies $\psi(g(x)) = 0$ $(x \in X)$, so that $g(x) \in \mathscr{R}_x$, and hence $g \in \Gamma(X, \mathscr{R})$.

In (3.1.2), however, ψ_* is not necessarily surjective. It is the sheaf cohomology theory dealt with in the present chapter to discuss how to make the sequence exact by adding sequences on the right-hand side of (3.1.2).

Example 3.1.3 As in Example 1.3.13, we take $X = \mathbf{R}$, $\mathscr{I}\langle \{0, 1\}\rangle \subset \mathbf{Z}_X$ and set $\mathscr{J} = \mathbf{Z}_X/\mathscr{I}\langle \{0, 1\}\rangle$. The sequence with the natural morphisms,

$$0 \to \mathscr{I}\langle \{0, 1\}\rangle \to \mathbf{Z}_X \xrightarrow{\psi} \mathscr{J} \to 0$$

is exact, and we have that

$$0 \to \Gamma(X, \mathscr{I}\langle \{0, 1\}\rangle) \to \Gamma(X, \mathbf{Z}_X) \xrightarrow{\psi_*} \Gamma(X, \mathscr{J}) \cong \mathbf{Z} \oplus \mathbf{Z}.$$

On the other hand, $\Gamma(X, \mathscr{I}\langle \{0, 1\}\rangle) = 0$, and $\Gamma(X, \mathbf{Z}_X) = \mathbf{Z}$ so that ψ_* cannot be surjective.

Example 3.1.4 Let $X = \mathbf{C}^* = \mathbf{C} \setminus \{0\}$. Then the following is exact:

$$0 \to \mathbf{Z}_X \to \mathscr{O}_X \xrightarrow{\mathbf{e}} \mathscr{O}_X^* \to 0.$$

Here $\mathbf{e}(f) = e^{2\pi i f}$ for $f \in \mathscr{O}_X$. Then,

$$0 \to \Gamma(X, \mathbf{Z}_X) = \mathbf{Z} \to \Gamma(X, \mathscr{O}_X) \xrightarrow{\mathbf{e}_*} \Gamma(X, \mathscr{O}_X^*)$$

is exact. For the holomorphic function $z \in \Gamma(X, \mathscr{O}_X^*)$, $\frac{1}{2\pi i}\log z$ is locally a section of $\mathscr{O}_X$, but $\frac{1}{2\pi i}\log z$ cannot be one-valued on X, and hence is not an element of $\Gamma(X, \mathscr{O}_X)$. Therefore $z \notin \operatorname{Im}\mathbf{e}_*$; thus, $\mathbf{e}_*$ is *not* surjective.

3.2 Tensor Product

3.2.1 Tensor Product

Let R be a ring (always assumed to be commutative) with $1 \neq 0$. Let A, B be (left) R-modules. We denote by $[A \times B]$ the free group (**Z**-module) generated by $(a, b) \in A \times B$. We denote by $[A \times B]_0$ the subgroup of $[A \times B]$ generated by the following elements:

(3.2.1)
$$\begin{aligned}&(a + a', b) - (a, b) - (a', b),\\ &(a, b + b') - (a, b) - (a, b'),\\ &(ra, b) - (a, rb),\\ &a, a' \in A, \quad b, b' \in B, \quad r \in R.\end{aligned}$$

We write $A \otimes_R B = [A \times B]/[A \times B]_0$, which is called a tensor product of A and B (over R). The equivalence class represented by $(a, b) \in A \times B$ is denoted by $a \otimes b \in A \otimes B$. Note that $R \otimes_R A \cong A$.

Theorem 3.2.2 *Let A_i $(i = 1, 2, 3)$ and B be R-modules. Let*

$$A_1 \xrightarrow{\phi} A_2 \xrightarrow{\psi} A_3 \to 0$$

be an exact sequence of R-modules. Then the sequence naturally induced with $\tilde{\phi} = \phi \otimes 1$ and $\tilde{\psi} = \psi \otimes 1$,

$$A_1 \otimes_R B \xrightarrow{\tilde{\phi}} A_2 \otimes_R B \xrightarrow{\tilde{\psi}} A_3 \otimes_R B \to 0$$

is exact.

Proof For simplicity we write $A_i \otimes B$ without the subscript R. The assumption and the definition of tensor product imply that

$$A_2 \otimes B \longrightarrow A_3 \otimes B \longrightarrow 0$$

is exact. We then show the exactness at $A_2 \otimes B$. The inclusion $\operatorname{Ker} \tilde{\psi} \supset \operatorname{Im} \tilde{\phi}$ follows from the assumption. For the converse, we set

$$\pi : A_2 \otimes B \longrightarrow (A_2 \otimes B)/\operatorname{Im} \tilde{\phi} = A_4.$$

For $a' \otimes b \in A_3 \otimes B$ we take $a \in A_2$ with $\psi(a) = a'$, and set

$$f(a' \otimes b) = [a \otimes b + \operatorname{Im} \tilde{\phi}] \in A_4.$$

We check that a homomorphism $f : A_3 \otimes B \to A_4$ is well-defined by this. In place of a, we take $a_1 \in A_2$ with $\psi(a_1) = a'$. There is $c \in \operatorname{Ker}\psi = \operatorname{Im}\phi$ such that $a_1 = a + c$. Therefore,

$$a_1 \otimes b + \operatorname{Im}\tilde{\phi} = a \otimes b + c \otimes b + \operatorname{Im}\tilde{\phi} = a \otimes b + \operatorname{Im}\tilde{\phi}.$$

Here $c \otimes b \in \operatorname{Im}\tilde{\phi}$ was used. Thus f is well-defined, and the following is exact:

$$\begin{array}{ccc} A_2 \otimes B & \xrightarrow{\tilde{\psi}} & A_3 \otimes B \\ \pi \searrow & \circlearrowleft & \swarrow f \\ & A_4 & \end{array}$$

It is deduced from this that $\operatorname{Ker}\tilde{\psi} \subset \operatorname{Ker} f \circ \tilde{\psi} = \operatorname{Ker}\pi = \operatorname{Im}\tilde{\phi}$. We finally see that $\operatorname{Im}\tilde{\phi} = \operatorname{Ker}\tilde{\psi}$. □

3.2.2 *Tensor Product of Sheaves*

Caution is required in the tensor product of sheaves of modules. In what follows, a sheaf of rings $\mathscr{R} \to X$ is assumed to satisfy

$$\Gamma(X, \mathscr{R}) \ni 1 \neq 0.$$

Let $\mathscr{S}$ and $\mathscr{T}$ be sheaves of $\mathscr{R}$-modules over X. Then the tensor product $\mathscr{S} \otimes_{\mathscr{R}} \mathscr{T}$ is defined to be a sheaf induced from the presheaf,

$$\{\mathscr{S}(U) \otimes_{\mathscr{R}(U)} \mathscr{T}(U);\ U \subset X \text{ is an open subset}\}.$$

Then the stalks at every point $x \in X$ satisfy

(3.2.3) $$(\mathscr{S} \otimes_{\mathscr{R}} \mathscr{T})_x = \mathscr{S}_x \otimes_{\mathscr{R}_x} \mathscr{T}_x.$$

In general, however, for an open subset $U \subset X$

$$\mathscr{S}(U) \otimes_{\mathscr{R}(U)} \mathscr{T}(U) \to (\mathscr{S} \otimes_{\mathscr{R}} \mathscr{T})(U)$$

is not an isomorphism.

Example 3.2.4 (1) We consider $X = \mathbf{R}$ endowed with the standard metric topology. Let $\mathbf{Z}_X \to X$ be the constant sheaf over X. With the same notation as in Example 1.3.13, we consider the ideal sheaves, $\mathscr{I}\langle \{0\}\rangle$ and $\mathscr{I}\langle \{0, 1\}\rangle$.

Take the quotient sheaf, $\mathscr{S} = \mathbf{Z}_X / \mathscr{I}\langle \{0, 1\}\rangle$. Then we have

$$\mathscr{S}_x = \begin{cases} \mathbf{Z}, & x = 0, 1, \\ 0, & x \neq 0, 1. \end{cases}$$

$$\mathscr{S}(X) \cong \mathbf{Z} \oplus \mathbf{Z}.$$

Consider the tensor product, $\mathscr{T} = \mathscr{I}\langle \{0\}\rangle \otimes_{\mathbf{Z}_X} \mathscr{S}$, so that

$$\mathscr{T}_x = \mathscr{I}\langle \{0\}\rangle_x \otimes_{\mathbf{Z}} \mathscr{S}_x = \begin{cases} \mathbf{Z}, & x = 1, \\ 0, & x \neq 1. \end{cases}$$

Therefore, $\mathscr{T}(X) \cong \mathbf{Z}$. On the other hand, because $\mathscr{I}\langle \{0\}\rangle(X) = 0$,

$$\mathscr{I}\langle \{0\}\rangle(X) \otimes_{\mathbf{Z}} \mathscr{S}(X) = 0.$$

(2) One may construct a similar example for coherent sheaves. For example, let X be a Riemann surface.[1] Let $\mathscr{O}_X$ denote the sheaf of germs of holomorphic functions over X. Let $a, b \in X$ be distinct points. Consider a geometric ideal sheaf $\mathscr{I}\langle \{a, b\}\rangle$ (Example 2.4.14). Take the quotient sheaf $\mathscr{S} = \mathscr{O}_X / \mathscr{I}\langle \{a, b\}\rangle$ and the tensor product $\mathscr{T} = \mathscr{I}\langle \{a\}\rangle \otimes_{\mathscr{O}_X} \mathscr{S}$. Then it follows that

$$\mathscr{T}(X) \cong \mathbf{C} \neq \mathscr{I}\langle \{a\}\rangle(X) \otimes_{\mathscr{O}_X(X)} \mathscr{S}(X) = \{0\} \otimes_{\mathscr{O}_X(X)} \mathscr{S}(X) = 0.$$

Theorem 3.2.5 *Let*

$$\mathscr{A}_1 \longrightarrow \mathscr{A}_2 \longrightarrow \mathscr{A}_3 \longrightarrow 0$$

be an exact sequence of sheaves of $\mathscr{R}$-modules over X, and let $\mathscr{B} \to X$ be a sheaf of $\mathscr{R}$-modules. Then the sequence

$$\mathscr{A}_1 \otimes_{\mathscr{R}} \mathscr{B} \longrightarrow \mathscr{A}_2 \otimes_{\mathscr{R}} \mathscr{B} \longrightarrow \mathscr{A}_3 \otimes_{\mathscr{R}} \mathscr{B} \longrightarrow 0$$

is again exact.

Proof It suffices to show that for every stalk,

$$\mathscr{A}_{1x} \otimes_{\mathscr{R}_x} \mathscr{B}_x \longrightarrow \mathscr{A}_{2x} \otimes_{\mathscr{R}_x} \mathscr{B}_x \longrightarrow \mathscr{A}_{3x} \otimes_{\mathscr{R}_x} \mathscr{B}_x \longrightarrow 0, \quad x \in X$$

is exact. This follows from (3.2.3) and Theorem 3.2.2. □

[1] If readers know the concept of "complex manifold", X may be a complex manifold of general dimension $\dim X \geq 1$.

3.3 Exact Sequences of Coherent Sheaves

Keeping the notation as before, we let X be a topological space, and let $\mathscr{A}$ be a sheaf of rings over X. The following is due to J.-P. Serre.

Theorem 3.3.1 (Serre's Theorem) *Assume that there is an exact sequence of sheaves of $\mathscr{A}$-modules,*

$$0 \to \mathscr{R} \xrightarrow{\phi} \mathscr{S} \xrightarrow{\psi} \mathscr{T} \to 0.$$

If any two of the sheaves above are coherent over $\mathscr{A}$, then the remaining one is coherent over $\mathscr{A}$, too.

Proof Below, through the injection ϕ we regard $\mathscr{R}$ as a subsheaf contained in $\mathscr{S}$. Since $\mathscr{A}$ is fixed, we mean by the "coherence" simply the coherence over $\mathscr{A}$.

(a) If $\mathscr{S}$ and $\mathscr{T}$ are coherent, by Proposition 2.4.7 (i) it is sufficient to show that $\mathscr{R}$ is locally finite. Take any point $x \in X$. By the coherence of $\mathscr{S}$ there are a neighborhood $U \ni x$ and a locally finite generator system of $\mathscr{S}$ over U, $\{\sigma_j \in \Gamma(U, \mathscr{S});\ 1 \leq j \leq p\}$, so that

$$\mathscr{S}|_U = \sum_{j=1}^{p} \mathscr{A}|_U \cdot \sigma_j.$$

An element $\sum_{j=1}^{p} a_{jx}\sigma_j(x) \in \mathscr{S}_x$ $(a_{jx} \in \mathscr{A}_x)$ belongs to $\mathscr{R}_x$ if and only if

$$\psi\left(\sum_{j=1}^{p} a_{jx}\sigma_j(x)\right) = \sum_{j=1}^{p} a_{jx}(\psi_*\sigma_j)(x) = 0.$$

This is equivalent to say that $(a_{jx}) \in (\mathscr{A}|_U)^p$ is an element of the relation sheaf $\mathscr{R}\left((\psi_*\sigma_j)_j\right)$ of $\mathscr{T}$. Since $\mathscr{T}$ is coherent, $\mathscr{R}\left((\psi_*\sigma_j)_j\right)$ is finitely generated over a neighborhood $V \subset U$ of x. Therefore, $\mathscr{R}$ is finitely generated over V.

(b) Suppose that $\mathscr{R}$ and $\mathscr{S}$ are coherent. Since $\mathscr{S}$ is locally finite and ψ is surjective, $\mathscr{T}$ is locally finite.

Take an open subset $U \subset X$ and sections $\tau_j \in \Gamma(U, \mathscr{T})$, $1 \leq j \leq l (< \infty)$. We show the coherence of the relation sheaf $\mathscr{R}((\tau_j)_j)$. At any point $x \in U$ an element $(a_{jx}) \in (\mathscr{A}_x)^l$ belongs to $\mathscr{R}((\tau_j)_j)_x$ if and only if

$$\sum_{j=1}^{l} a_{jx}\tau_j(x) = 0.$$

Since ψ is surjective, there are a neighborhood $V \subset U$ of x and $\sigma_j \in \Gamma(V, \mathscr{S})$ with $\psi_*\sigma_j = \tau_j|_V$, such that

$$(3.3.2) \qquad (a_{jy}) \in \mathscr{A}_y^l \text{ is an element of } \mathscr{R}((\tau_j)_j)_y$$
$$\Longleftrightarrow \psi\left(\sum_{j=1}^{l} a_{jy}\sigma_j(y)\right) = 0, \quad y \in V.$$

Since $\operatorname{Ker}\psi = \mathscr{R}$ and $\mathscr{R}$ is locally finite, by taking V smaller if necessary, there are finitely many $\lambda_i \in \Gamma(V, \mathscr{R})$, $1 \le i \le m$, which generate $\mathscr{R}|_V$. Then, (3.3.2) is equivalent to

$$\sum_{j=1}^{l} a_{jy}\sigma_j(y) = \sum_{k=1}^{m} b_{ky}\lambda_k(y), \qquad {}^{\exists}(b_{ky}) \in (\mathscr{A}_y)^m.$$

This means $((a_{jy}), (b_{ky})) \in \mathscr{R}((\sigma_j)_j, (-\lambda_k)_k)$. The coherence of $\mathscr{S}$ implies that the relation sheaf $\mathscr{R}((\sigma_j)_j, (-\phi_*\lambda_k)_k)$ is locally coherent, and hence such (a_{jy})'s form a locally finite sheaf of modules. Therefore follows the coherence of $\mathscr{R}((\tau_j)_j)$.

(c) Suppose that $\mathscr{R}$ and $\mathscr{T}$ are coherent. We first show the local finiteness of $\mathscr{S}$. Let $x \in X$ be any point. Then there are a neighborhood U and a locally finite generator system of $\mathscr{T}$ with $\tau_j \in \Gamma(U, \mathscr{T})$, $1 \le j \le t$:

$$\mathscr{T}|_U = \sum_j \mathscr{A}|_U \cdot \tau_j.$$

Since ψ is surjective, there are some $\sigma_{jx} \in \mathscr{S}_x$ such that $\psi(\sigma_{jx}) = \tau_j(x)$. By the definition of germs there are a neighborhood $V \subset U$ of z and $\sigma_j \in \Gamma(V, \mathscr{S})$ such that $\psi_*\sigma_j = \tau_j|_V$. Since $\mathscr{R}$ is locally finite, with smaller V chosen if necessary, there are finitely many $\lambda_i \in \Gamma(V, \mathscr{R})$, $1 \le i \le s$, such that

$$\mathscr{R}|_U = \sum_i \mathscr{A}|_U \cdot \lambda_i.$$

Take an arbitrary element $f \in \mathscr{S}_y$, $y \in V$. We may write

$$\psi(f) = \textstyle\sum_j a_{jy}\tau_j(y) = \psi\left(\sum_j a_{jy}\sigma_j(y)\right), \quad a_{jy} \in \mathscr{A}_y.$$

Because $f - \sum_j a_{jy}\sigma_j(y) \in \operatorname{Ker}\psi = \mathscr{R}$, there exist $b_{iy} \in \mathscr{A}_y$ such that

$$f - \sum_j a_{jy}\sigma_j(y) = \sum_i b_{iy}\lambda_i(y).$$

Thus, $f = \sum_j a_{jy}\sigma_j(y) + \sum_i b_{iy}\lambda_i(y)$. We see that $\mathscr{S}$ is generated by $\{\sigma_j, \lambda_i\}_{j,i}$ over V.

We next check a relation sheaf of $\mathscr{S}$. Let $U \subset X$ be an open subset and let $\mathscr{R}((\sigma_i)_i) \to U$ be a relation sheaf of $\mathscr{S}$ defined by finitely many sections $\sigma_i \in \Gamma(U, \mathscr{S})$, $1 \le i \le l$. If $(a_{iy}) \in \mathscr{R}((\sigma_i)_i)$, then $\sum_i a_{iy} \cdot (\psi_*\sigma_i)(y) = 0$. Since the relation sheaf $\mathscr{R}((\psi_*\sigma_i)_i)$ of $\mathscr{T}$ is locally finite, for every $x \in U$ there are a neighborhood $V \subset U$ and $\alpha_h \in \Gamma(V, \mathscr{O}_V^l)$, $1 \le h \le k$, such that $\mathscr{R}((\psi_*\sigma_i)_i)$ is generated by $\{\alpha_h\}$ over V. We may write

$$(a_{iy}) = \sum_{h=1}^{k} b_{hy}\alpha_h(y). \tag{3.3.3}$$

We look for a condition of (b_{hy}) for (a_{iy}) belonging to $\mathscr{R}((\sigma_i)_i)_y$. Put $\alpha_h = (\alpha_{hi})$. Then,

$$\psi\left(\sum_i \alpha_{hi}(y)\sigma_i(y)\right) = 0,$$

and hence $\lambda_h := \sum_i \alpha_{hi} \cdot \sigma_i \in \Gamma(V, \mathscr{R})$. Therefore it is necessary and sufficient for $(a_{iy}) \in \mathscr{R}((\sigma_i)_i)_y$ that (3.3.3) holds with (b_{hy}) satisfying

$$\sum b_{hy}\lambda_h(y) = 0. \tag{3.3.4}$$

This is the same as $(b_{hy}) \in \mathscr{R}((\lambda_h)_h)$, and the relation sheaf $\mathscr{R}((\lambda_h)_h)$ of $\mathscr{R}$ is locally finite. Thus, there are a neighborhood $W \subset V$ of x, and a locally finite generator system $\{(b_{h\nu})\}_\nu$ of $\mathscr{R}((\lambda_h)_h)$ over V, so that a finite system $\{(a_{i\nu})\}_\nu$ given by (3.3.3) generates $\mathscr{R}((\sigma_i)_i)|_W$. □

The contents up to now do not require the coherence of the base sheaf $\mathscr{A}$ of rings. Here, we let $\Omega \subset \mathbf{C}^n$ be an open set. We describe the notion of coherence of sheaves over Ω in terms of exact sequences. In the sequel, it is essential that $\mathscr{O}_\Omega$ is coherent by Oka's First Coherence Theorem 2.5.1.

Proposition 3.3.5 *For a sheaf $\mathscr{S}$ of $\mathscr{O}_\Omega$-modules over Ω the following are equivalent:*

(i) *$\mathscr{S}$ is coherent.*
(ii) *For every $x \in \Omega$ there are a neighborhood $U \subset \Omega$ and an exact sequence over U,*

$$\mathscr{O}_U^q \xrightarrow{\phi} \mathscr{O}_U^p \xrightarrow{\psi} \mathscr{S}|_U \to 0.$$

Proof (i)⇒(ii): By the local finiteness there are a neighborhood U of every $x \in \Omega$ with a locally finite generator system $\{\sigma_j \in \Gamma(U, \mathscr{S});\ 1 \le j \le p\}$ of $\mathscr{S}$ over U, so that

$$\psi : (\underline{f_j}_y) \in \mathscr{O}_{U,y}^p \to \sum_{j=1}^{p} \underline{f_j}_y \cdot \sigma_j(y) \in \mathscr{S}_y, \quad y \in U$$

is surjective. The kernel Ker ψ consists of those elements $(\underline{f_j}_y) \in (\mathscr{O}_U)^p$ such that

$$\sum_{j=1}^{p} \underline{f_j}_y \cdot \sigma_j(y) = 0,$$

which is nothing but the relation sheaf $\mathscr{R}((\sigma_j)_j)$. Since $\mathscr{S}$ is assumed to be coherent, $\operatorname{Ker}\psi = \mathscr{R}((\sigma_j)_j)$ is locally finite. Shrinking U if necessary, we may construct a surjection, $\phi : \mathscr{O}_U^q \to \operatorname{Ker}\psi$, as in the case of ψ.

(ii)⇒(i): We set $\mathscr{R} = \operatorname{Im}\phi$, which is finitely generated subsheaf of modules of $\mathscr{O}_U^p$. By Proposition 2.4.7 (i), $\mathscr{R}$ is coherent. Therefore we get the following short exact sequence:

$$0 \longrightarrow \mathscr{R} \longrightarrow \mathscr{O}_U^p \longrightarrow \mathscr{S}|_U \longrightarrow 0.$$

Since $\mathscr{R}$ and $\mathscr{O}_U^p$ are coherent, $\mathscr{S}$ is coherent by Theorem 3.3.1. □

This proposition implies the following basic property of coherence.

Theorem 3.3.6 *Let $\mathscr{S}$ and $\mathscr{T}$ be coherent sheaves over Ω. Then the tensor product $\mathscr{S} \otimes_{\mathscr{O}_\Omega} \mathscr{T}$ is coherent, too.*

Proof It follows from Proposition 3.3.5 that for every $x \in \Omega$ there is an exact sequence over a neighborhood U of x,

$$\mathscr{O}_U^q \xrightarrow{\phi} \mathscr{O}_U^p \xrightarrow{\psi} \mathscr{S}|_U \to 0.$$

Setting $\tilde{\phi} = \phi \otimes 1$, $\tilde{\psi} = \psi \otimes 1$, we see by Theorem 3.2.5 that

$$\mathscr{O}_U^q \otimes \mathscr{T}|_U \xrightarrow{\tilde{\phi}} \mathscr{O}_U^p \otimes \mathscr{T}|_U \xrightarrow{\tilde{\psi}} \mathscr{S}|_U \otimes \mathscr{T}|_U \to 0$$

is exact. Since

$$\mathscr{O}_U^q \otimes \mathscr{T}|_U \cong (\mathscr{T}|_U)^q, \qquad \mathscr{O}_U^p \otimes \mathscr{T}|_U \cong (\mathscr{T}|_U)^p,$$

the sequence

$$(\mathscr{T}|_U)^q \xrightarrow{\tilde{\phi}} (\mathscr{T}|_U)^p \xrightarrow{\tilde{\psi}} \mathscr{S}|_U \otimes \mathscr{T}|_U \to 0$$

is exact. Proposition 2.4.7 (ii) implies the coherence of $(\mathscr{T}|_U)^p$ and $(\mathscr{T}|_U)^q$. Since $\operatorname{Im}\tilde{\phi}$ is a finitely generated subsheaf of $(\mathscr{T}|_U)^p$, it is coherent by Proposition 2.4.7 (i). Therefore the sequence

$$0 \longrightarrow \operatorname{Im}\tilde{\phi} \longrightarrow (\mathscr{T}|_U)^p \longrightarrow \mathscr{S}|_U \otimes \mathscr{T}|_U \to 0$$

is exact. From Theorem 3.3.1 the coherence of $\mathscr{S}|_U \otimes \mathscr{T}|_U$ is deduced. □

3.4 Sheaf Cohomology

3.4.1 Čech cohomology

Let X be a topological space and let $\mathscr{U} = \{U_\alpha\}_{\alpha\in\Phi}$ be be an open covering of X. We set

$$N_q(\mathscr{U}) = \{\sigma = (U_0, \ldots, U_q); U_i \in \mathscr{U}\}, \qquad q \geq 0,$$
$$N(\mathscr{U}) = \bigcup_{q=0}^{\infty} N_q(\mathscr{U}),$$

and call $N(\mathscr{U})$ (resp. $N_q(\mathscr{U})$) the (resp. q-)*nerve* of $\mathscr{U}$. An element $\sigma = (U_0, \ldots, U_q) \in N_q(\mathscr{U})$ is called a *simplex* or *q-simplex*, and $|\sigma| = \bigcap_{i=0}^q U_i$ is called the support of σ.

Let $\mathscr{S} \to X$ be a sheaf. A map

$$f : \sigma \in N_q(\mathscr{U}) \to f(\sigma) \in \Gamma(|\sigma|, \mathscr{S})$$

satisfying the alternating property

$$\text{(3.4.1)} \quad f(U_0, \ldots, U_i, \ldots, U_j, \ldots, U_q) = -f(U_0, \ldots, U_j, \ldots, U_i, \ldots, U_q),$$
$$i \neq j, \quad (U_0, \ldots, U_i, \ldots, U_j, \ldots, U_q) \in N_q(\mathscr{U})$$

is called an $\mathscr{S}$-valued *(q-)cochain* of $\mathscr{U}$. Here, we assume that $\Gamma(\emptyset, \mathscr{S}) = 0$. We write $C^q(\mathscr{U}, \mathscr{S})$ for all such maps f. Note that $C^q(\mathscr{U}, \mathscr{S})$ carries a structure of abelian group, naturally induced from $\mathscr{S}$.

The *coboundary* $\delta f \in C^{q+1}(\mathscr{U}, \mathscr{S})$ of $f \in C^q(\mathscr{U}, \mathscr{S})$ is defined by

$$\text{(3.4.2)} \qquad (\delta f)(\sigma) = \sum_{i=0}^{q+1} (-1)^i f(U_0, \ldots, \check{U}_i, \ldots, U_{q+1})|_{|\sigma|}$$

for $\sigma = (U_0, \ldots, U_{q+1}) \in N_{q+1}(\mathscr{U})$, where $\check{U}_i$ means the deletion. Then the map $\delta : C^q(\mathscr{U}, \mathscr{S}) \to C^{q+1}(\mathscr{U}, \mathscr{S})$ is a homomorphism of groups.

Lemma 3.4.3 *Let the notation be as above. No matter whether the alternation condition (3.4.1) is imposed for $C^q(\mathscr{U}, \mathscr{S})$ or not, the following holds:*

$$\text{(3.4.4)} \qquad C^q(\mathscr{U}, \mathscr{S}) \xrightarrow{\delta} C^{q+1}(\mathscr{U}, \mathscr{S}) \xrightarrow{\delta} C^{q+2}(\mathscr{U}, \mathscr{S}),$$
$$\delta \circ \delta = 0.$$

Proof Following the definition, we have

$$
\begin{aligned}
(\delta^2 f)(U_0, \ldots, U_{q+2}) &= \delta(\delta f)(U_0, \ldots, U_{q+2}) \\
&= \sum_{i=0}^{q+2} (-1)^i \delta f(U_0, \ldots, \check{U}_i, \ldots, U_{q+2}) \\
&= \sum_{i=0}^{q+2} (-1)^i \Big\{ \sum_{j=0}^{i-1} (-1)^j f(U_0, \ldots, \check{U}_j, \ldots, \check{U}_i, \ldots, U_{q+2}) \\
&\quad + \sum_{j=i+1}^{q+2} (-1)^{j-1} f(U_0, \ldots, \check{U}_i, \ldots, \check{U}_j, \ldots, U_{q+2}) \Big\} \\
&= \sum_{i=0}^{q+2} \sum_{j=0}^{i-1} (-1)^{i+j} f(U_0, \ldots, \check{U}_j, \ldots, \check{U}_i, \ldots, U_{q+2}) \\
&\quad - \sum_{i=0}^{q+2} \sum_{j=i+1}^{q+2} (-1)^{i+j} f(U_0, \ldots, \check{U}_i, \ldots, \check{U}_j, \ldots, U_{q+2}) \\
&= \sum_{0 \le j < i \le q+2} (-1)^{i+j} f(U_0, \ldots, \check{U}_j, \ldots, \check{U}_i, \ldots, U_{q+2}) \\
&\quad - \sum_{0 \le i < j \le q+2} (-1)^{i+j} f(U_0, \ldots, \check{U}_i, \ldots, \check{U}_j, \ldots, U_{q+2}) \\
&= 0.
\end{aligned}
$$

□

We define

$$
\begin{aligned}
Z^q(\mathscr{U}, \mathscr{S}) &= \{f \in C^q(\mathscr{U}, \mathscr{S}); \delta f = 0\}, \quad q \ge 0, \\
B^q(\mathscr{U}, \mathscr{S}) &= \delta C^{q-1}(\mathscr{U}, \mathscr{S}), \quad C^{-1}(\mathscr{U}, \mathscr{S}) = 0.
\end{aligned}
$$

An element of $Z^q(\mathscr{U}, \mathscr{S})$ is called a *q-cocycle*, and that of $B^q(\mathscr{U}, \mathscr{S})$ is called a *q-coboundary*. It follows from (3.4.4) that

$$B^q(\mathscr{U}, \mathscr{S}) \subset Z^q(\mathscr{U}, \mathscr{S}).$$

The *q-th cohomology group* $H^q(\mathscr{U}, \mathscr{S})$ of $\mathscr{S}$ with respect to $\mathscr{U}$ is defined by

(3.4.5) $$H^q(\mathscr{U}, \mathscr{S}) = Z^q(\mathscr{U}, \mathscr{S}) / B^q(\mathscr{U}, \mathscr{S}), \quad q \ge 0.$$

Lemma 3.4.6 *With $\mathscr{S}$ a sheaf, $H^0(\mathscr{U}, \mathscr{S}) \cong \Gamma(X, \mathscr{S})$.*

Proof Let $f \in C^0(\mathscr{U}, \mathscr{S})$. For each $U_\alpha \in \mathscr{U}$, $f(U_\alpha) = \sigma_\alpha \in \Gamma(U_\alpha, \mathscr{S})$ is assigned. Since $\delta f = 0$, for $U_\alpha \cap U_\beta \neq \emptyset$ we have

$$\delta f(U_\alpha, U_\beta) = \sigma_\beta|_{U_\alpha \cap U_\beta} - \sigma_\alpha|_{U_\alpha \cap U_\beta} = 0.$$

That is, $\sigma_\beta|_{U_\alpha \cap U_\beta} = \sigma_\alpha|_{U_\alpha \cap U_\beta}$, and hence there exists an element $\sigma \in \Gamma(X, \mathscr{S})$ such that $\sigma|_{U_\alpha} = \sigma_\alpha$.

Conversely, let $\sigma \in \Gamma(X, \mathscr{S})$. Setting $f(U_\alpha) = \sigma|_{U_\alpha}$, we get $f \in C^0(\mathscr{U}, \mathscr{S})$ with $\delta f = 0$. Therefore, $f \in Z^0(\mathscr{U}, \mathscr{S}) = H^0(\mathscr{U}, \mathscr{S})$. □

An open covering $\mathscr{V} = \{V_\lambda\}_{\lambda \in \Lambda}$ of X is called a *refinement* of $\mathscr{U} = \{U_\alpha\}_{\alpha \in \Phi}$ ($\mathscr{U} \prec \mathscr{V}$) if there is a map $\mu : \mathscr{V} \to \mathscr{U}$ (or a map $\mu : \Lambda \to \Phi$ between the index sets) satisfying $V_\lambda \subset \mu(V_\lambda)(= U_{\mu(\lambda)})$. This refinement naturally induces a homomorphism

$$\mu^* : f \in C^q(\mathscr{U}, \mathscr{S}) \to C^q(\mathscr{V}, \mathscr{S}),$$
$$\mu^*(f)(V_0, \ldots, V_q) = f(\mu(V_0), \ldots, \mu(V_q))|_{V_0 \cap \cdots \cap V_q}.$$

Henceforth, to avoid complication, we omit the restriction symbol, $|_{V_0 \cap \cdots \cap V_q}$. Since μ^* and δ commute ($\delta \circ \mu^* = \mu^* \circ \delta$), by definition (3.4.5) μ naturally induces a homomorphism between cohomologies:

$$\mu^* : H^q(\mathscr{U}, \mathscr{S}) \to H^q(\mathscr{V}, \mathscr{S}).$$

Lemma 3.4.7 *Assume that $\mathscr{V}$ is a refinement of $\mathscr{U}$ by two maps $\mu : \mathscr{V} \to \mathscr{U}$ and $\nu : \mathscr{V} \to \mathscr{U}$. Then,*

$$\mu^* = \nu^* : H^q(\mathscr{U}, \mathscr{S}) \to H^q(\mathscr{V}, \mathscr{S}).$$

holds.

Proof The case of $q = 0$: By Lemma 3.4.6 both are isomorphic to $\Gamma(X, \mathscr{S})$, so that the same $f \in \Gamma(X, \mathscr{S})$ induces the restrictions, $(f_\alpha) = (f|_{U_\alpha}) \in H^0(\mathscr{U}, \mathscr{S})$ and $(g_\lambda) = (f|_{V_\lambda}) \in H^0(\mathscr{V}, \mathscr{S})$. Therefore, $\mu^* = \nu^*$.

The case of $q > 0$: We define a homomorphism

$$\theta : C^q(\mathscr{U}, \mathscr{S}) \to C^{q-1}(\mathscr{V}, \mathscr{S})$$

as follows. For $f \in C^q(\mathscr{U}, \mathscr{S})$ and $\tau = (V_0, \ldots, V_{q-1}) \in N_{q-1}(\mathscr{V})$, we set

$$\theta(f)(\tau) = \theta(f)(V_0, \ldots, V_{q-1})$$
$$= \sum_{j=0}^{q-1} (-1)^j f(\mu(V_0), \ldots, \mu(V_j), \nu(V_j), \ldots, \nu(V_{q-1})).$$

For $\tau' = (V_0, \ldots, V_q) \in N_q(\mathscr{V})$ we have

$$\begin{aligned}
(\delta \circ \theta(f))(V_0, \ldots, V_q) &= \sum_{i=0}^{q} (-1)^i \theta(f)(V_0, \ldots, \check{V}_i, \ldots, V_q) \\
&= \sum_{i=0}^{q} \Big\{ (-1)^i \sum_{j=0}^{i-1} (-1)^j f(\mu(V_0), \ldots, \mu(V_j), \nu(V_j), \ldots, \underset{(i+1)\text{th}}{\nu(\check{V}_i)}, \ldots, \nu(V_q)) \\
&\quad + (-1)^i \sum_{j=i+1}^{q} (-1)^{j+1} f(\mu(V_0), \ldots, \mu(\check{V}_i), \ldots, \mu(V_j), \nu(V_j), \ldots, \nu(V_q)) \Big\} \\
&= \sum_{j=0}^{q} (-1)^{j+1} (\delta f)(\mu(V_0), \ldots, \mu(V_j), \nu(V_j), \ldots, \nu(V_q)) \\
&\quad + f(\nu(V_0), \ldots, \nu(V_q)) - f(\mu(V_0), \ldots, \mu(V_q)).
\end{aligned}$$

Therefore, if $\delta f = 0$,

$$\delta(\theta(f)) = \nu(f) - \mu(f) \in B^q(\mathscr{V}, \mathscr{S}).$$

Thus, $\mu^* = \nu^* : H^q(\mathscr{U}, \mathscr{S}) \to H^q(\mathscr{V}, \mathscr{S})$. □

The set of all open coverings of X forms an ordered (or directed) set by the relation $\mathscr{U} \prec \mathscr{V}$ $(\mu : \mathscr{V} \to \mathscr{U})$. For $\mathscr{U} \prec \mathscr{V} \prec \mathscr{W}$ the following commutative diagram holds:

$$\begin{array}{ccc}
H^q(\mathscr{U}, \mathscr{S}) & \longrightarrow & H^q(\mathscr{V}, \mathscr{S}) \\
& \circlearrowleft & \Big\downarrow \\
& \searrow & \\
& & H^q(\mathscr{W}, \mathscr{S}).
\end{array}$$

Therefore as in (1.3.7) we can take a inductive limit. That is, we first take a disjoint union

$$\Xi^q(\mathscr{S}) = \bigsqcup_{\mathscr{U}} H^q(\mathscr{U}, \mathscr{S}).$$

We define an equivalence relation, $f \sim g$ for two elements $f \in H^q(\mathscr{U}, \mathscr{S}) \subset \Xi^q(\mathscr{S})$ and $g \in H^q(\mathscr{V}, \mathscr{S}) \subset \Xi^q(\mathscr{S})$ if there is a common refinement $\mathscr{W}$ of $\mathscr{U}$ and $\mathscr{V}$,

$$\mu : \mathscr{W} \to \mathscr{U}, \qquad \nu : \mathscr{W} \to \mathscr{V}$$

satisfying $\mu^*(f) = \nu^*(g) \in H^q(\mathscr{W}, \mathscr{S})$. The inductive limit is defined as a quotient of $\Xi^q(\mathscr{S})$ by this equivalence relation:

(3.4.8) $$H^q(X, \mathscr{S}) = \varinjlim_{\mathscr{U}} H^q(\mathscr{U}, \mathscr{S}) = \Xi^q(\mathscr{S}) / \sim .$$

We call this the *q-th (Čech) cohomology (group)*.

Remark 3.4.9 It is possible to define $H^q(X, \mathscr{S})$ without assuming the alternation (3.4.1) for $C^q(\mathscr{U}, \mathscr{S})$. As seen later (Theorem 3.4.38), the cohomology group $H^q(X, \mathscr{S})$ is isomorphic to the cohomology defined by a resolution in either case, and hence there is no difference.

There is a natural homomorphism

$$H^q(\mathscr{U}, \mathscr{S}) \longrightarrow H^q(X, \mathscr{S}), \qquad q \geq 0.$$

By Lemma 3.4.6 in the case of $q = 0$,

$$H^0(X, \mathscr{S}) \cong H^0(\mathscr{U}, \mathscr{S}) \cong \Gamma(X, \mathscr{S}). \tag{3.4.10}$$

Furthermore, the case of $q = 1$ is special:

Proposition 3.4.11 *With the notation above, $H^1(\mathscr{U}, \mathscr{S}) \to H^1(X, \mathscr{S})$ is injective.*

Proof Let $f = (f_{\alpha\beta}) \in Z^1(\mathscr{U}, \mathscr{S})$ be any 1-cocycle. Since $\delta f = 0$, the following holds over $U_\alpha \cap U_\beta \cap U_\gamma$:

$$f_{\beta\gamma} - f_{\alpha\gamma} + f_{\alpha\beta} = 0. \tag{3.4.12}$$

Suppose that the image $[(f_{\alpha\beta})]$ of this $(f_{\alpha\beta})$ in $H^1(X, \mathscr{S})$ satisfies $[(f_{\alpha\beta})] = 0$. Then there are a refinement $\mathscr{V} = \{V_\lambda\}_{\lambda \in \Lambda} \succ \mathscr{U} = \{U_\alpha\}_{\alpha \in \Phi}$ of $\mathscr{U}$ ($\varphi : \Lambda \to \Phi$) and $(g_\lambda) \in C^0(\mathscr{V}, \mathscr{S})$ such that $\delta(g_\lambda) = (f_{\varphi(\lambda)\varphi(\mu)})$. That is,

$$f_{\varphi(\lambda)\varphi(\mu)} = g_\mu - g_\lambda \quad \text{on } V_\lambda \cap V_\mu. \tag{3.4.13}$$

Noting that $U_\alpha = \bigcup_\lambda (V_\lambda \cap U_\alpha)$, we set on each $V_\lambda \cap U_\alpha$

$$h_{\alpha\lambda} = g_\lambda + f_{\varphi(\lambda)\alpha}.$$

It follows from (3.4.12) and (3.4.13) that on $V_\lambda \cap V_\mu \cap U_\alpha$,

$$\begin{aligned} h_{\alpha\lambda} - h_{\alpha\mu} &= g_\lambda - g_\mu + f_{\varphi(\lambda)\alpha} - f_{\varphi(\mu)\alpha} \\ &= f_{\varphi(\mu)\varphi(\lambda)} + f_{\varphi(\lambda)\alpha} - f_{\varphi(\mu)\alpha} \\ &= \delta f(U_{\varphi(\mu)}, U_{\varphi(\lambda)}, U_\alpha) = 0. \end{aligned}$$

Therefore, $(h_{\alpha\lambda})$ determines $h_\alpha \in \Gamma(U_\alpha, \mathscr{S})$ so that $h_\alpha|_{U_\alpha \cap V_\lambda} = h_{\alpha\lambda}$. Taking any $x \in U_\alpha \cap U_\beta$, we take $V_\lambda \ni x$, and then have that

$$\begin{aligned} h_\beta(x) - h_\alpha(x) &= h_{\beta\lambda}(x) - h_{\alpha\lambda}(x) \\ &= g_\lambda(x) + f_{\varphi(\lambda)\beta}(x) - g_\lambda(x) - f_{\varphi(\lambda)\alpha}(x) \\ &= f_{\varphi(\lambda)\beta}(x) - f_{\varphi(\lambda)\alpha}(x) \\ &= f_{\alpha\beta}(x). \end{aligned}$$

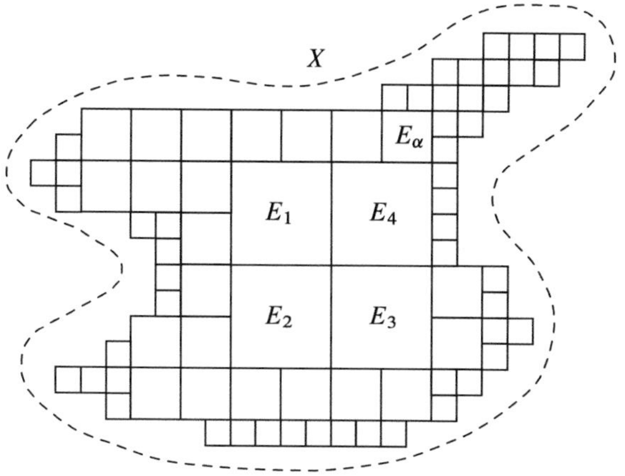

Fig. 3.1 Covering X with closed cubes

This means that $(h_\alpha) \in C^0(\mathscr{U}, \mathscr{S})$ and $\delta(h_\alpha) = (f_{\alpha\beta})$. Thus for the cohomology class, $[(f_{\alpha\beta})] = 0 \in H^1(\mathscr{U}, \mathscr{S})$. □

Theorem 3.4.14 *Let $X \subset \mathbf{R}^n$ be any open set. For every sheaf $\mathscr{S} \to X$*

$$H^q(X, \mathscr{S}) = 0, \quad q \geq 2^n.$$

Proof We consider covering X with open cubes in $\mathbf{R}^n$. We divide $\mathbf{R}^n$ into closed cubes with edges parallel to the coordinate axes (real and imaginary), and cover X with countably many such closed cubes E_α with no common interior point, where the lengths of the edges of E_α are not fixed and converge to zero as E_α approaches to the boundary of X (cf. Fig. 3.1). We then take open cubes U_α slightly larger than E_α such that $E_\alpha \Subset U_\alpha \Subset X$, so that the number of U_α with non-empty intersection is at most 2^n. Let $\mathscr{U} = \{U_\alpha\}$ be such an open covering of X. We take an element $f \in C^q(\mathscr{U}, \mathscr{S})$ with $q \geq 2^n$. Then for $\sigma = (U_0, \ldots, U_q) \in N_q(\mathscr{U})$ we consider $f(\sigma)$; by the alternation (3.4.1), $U_0, \ldots, U_q$ may be assumed to be all distinct. Then, $|\sigma| = \emptyset$, and by definition, $f = 0$. □

We define here a property of a topological space characterized by open coverings, which will be used in the next section.

Definition 3.4.15 (i) An open covering $\{U_\alpha\}$ of X is said to be *locally finite* if for every point $x \in X$ there is a neighborhood V such that the number of U_α with $U_\alpha \cap V \neq \emptyset$ is finite.

(ii) We say that X is *paracompact* if for every open covering $\mathscr{U}$ of X there is a locally finite refinement of $\mathscr{U}$.

3.4.2 Long Exact Sequences

Let

$$0 \to \mathscr{R} \xrightarrow{\phi} \mathscr{S} \xrightarrow{\psi} \mathscr{T} \to 0 \tag{3.4.16}$$

be a short exact sequence of sheaves over X. For any open $U \subset X$ the following sequence is naturally induced:

$$0 \to \Gamma(U, \mathscr{R}) \xrightarrow{\phi_*} \Gamma(U, \mathscr{S}) \xrightarrow{\psi_*} \Gamma(U, \mathscr{T}).$$

This is exact; i.e., ϕ_* is injective, and $\operatorname{Im} \phi_* = \operatorname{Ker} \psi_*$. But, ψ_* is not surjective in general (cf. Examples 3.1.3 and 3.1.4).

Theorem 3.4.17 *Let X be a paracompact Hausdorff topological space. For a given short exact sequence (3.4.16) there exists a "long exact sequence" of the following type:*

$$\begin{aligned} 0 \to H^0(X, \mathscr{R}) &\xrightarrow{\phi_*} H^0(X, \mathscr{S}) \xrightarrow{\psi_*} H^0(X, \mathscr{T}) \\ &\xrightarrow{\delta_*} H^1(X, \mathscr{R}) \longrightarrow H^1(X, \mathscr{S}) \longrightarrow H^1(X, \mathscr{T}) \\ &\xrightarrow{\delta_*} H^2(X, \mathscr{R}) \longrightarrow \cdots \\ &\vdots \end{aligned}$$

Here δ_* will be defined in the proof below.

Proof Let $\mathscr{U} = \{U_\alpha\}_{\alpha \in \Phi}$ be a locally finite open covering of X. For every simplex $\sigma \in N_q(\mathscr{U})$ there are exact sequences:

$$\begin{aligned} 0 &\to \Gamma(|\sigma|, \mathscr{R}) \xrightarrow{\phi_*} \Gamma(|\sigma|, \mathscr{S}) \xrightarrow{\psi_*} \Gamma(|\sigma|, \mathscr{T}), \\ 0 &\to C^q(\mathscr{U}, \mathscr{R}) \xrightarrow{\phi_*} C^q(\mathscr{U}, \mathscr{S}) \xrightarrow{\psi_*} C^q(\mathscr{U}, \mathscr{T}). \end{aligned}$$

Set $\bar{C}^q(\mathscr{U}, \mathscr{S}) = \operatorname{Im} \psi_* \subset C^q(\mathscr{U}, \mathscr{S})$. Then we obtain the following commutative diagram, in which all row sequences are exact:

$$
\begin{array}{ccccccccc}
 & & \vdots & & \vdots & & \vdots & & \\
 & & \downarrow & & \downarrow & & \downarrow & & \\
0 \to & & C^{q-1}(\mathscr{U}, \mathscr{R}) & \xrightarrow{\phi_*} & C^{q-1}(\mathscr{U}, \mathscr{S}) & \xrightarrow{\psi_*} & \bar{C}^{q-1}(\mathscr{U}, \mathscr{T}) & \to 0 & \\
 & & \downarrow \delta & \circlearrowleft & \downarrow \delta & \circlearrowleft & \downarrow \delta & & \\
0 \to & & C^{q}(\mathscr{U}, \mathscr{R}) & \xrightarrow{\phi_*} & C^{q}(\mathscr{U}, \mathscr{S}) & \xrightarrow{\psi_*} & \bar{C}^{q}(\mathscr{U}, \mathscr{T}) & \to 0 & \\
 & & \downarrow \delta & \circlearrowleft & \downarrow \delta & \circlearrowleft & \downarrow \delta & & \\
0 \to & & C^{q+1}(\mathscr{U}, \mathscr{R}) & \xrightarrow{\phi_*} & C^{q+1}(\mathscr{U}, \mathscr{S}) & \xrightarrow{\psi_*} & \bar{C}^{q+1}(\mathscr{U}, \mathscr{T}) & \to 0 & \\
 & & \downarrow & & \downarrow & & \downarrow & & \\
 & & \vdots & & \vdots & & \vdots & &
\end{array}
\tag{3.4.18}
$$

Here, δ is the one defined in (3.4.2). Now, we put

$$
\begin{aligned}
\bar{Z}^q(\mathscr{U}, \mathscr{T}) &= \{f \in \bar{C}^q(\mathscr{U}, \mathscr{T}); \delta f = 0\}, \\
\bar{H}^q(\mathscr{U}, \mathscr{T}) &= \bar{Z}^q(\mathscr{U}, \mathscr{T}) / \delta \bar{C}^{q-1}(\mathscr{U}, \mathscr{T}).
\end{aligned}
$$

We write $[f] \in \bar{H}^q(\mathscr{U}, \mathscr{T})$ for the equivalence class of $f \in \bar{C}^q(\mathscr{U}, \mathscr{T})$. For each q,

$$
H^q(\mathscr{U}, \mathscr{R}) \xrightarrow{\phi_*} H^q(\mathscr{U}, \mathscr{S}) \xrightarrow{\psi_*} \bar{H}^q(\mathscr{U}, \mathscr{T}) \tag{3.4.19}
$$

is exact.

By definition, for $f \in \bar{Z}^q(\mathscr{U}, \mathscr{T})$ there is a $g \in C^q(\mathscr{U}, \mathscr{S})$ with $\psi_*(g) = f$. Since $\psi_* \circ \delta(g) = \delta \circ \psi_*(g) = \delta f = 0$, there is an element $h \in C^{q+1}(\mathscr{U}, \mathscr{R})$ such that

$$
\phi_*(h) = \delta(g). \tag{3.4.20}
$$

(It may help comprehension to follow how elements correspond each other in (3.4.18).) Since $\phi_*(\delta(h)) = \delta(g) = 0$ and ϕ_* is injective, $\delta(h) = 0$, and hence $[h] \in H^{q+1}(\mathscr{U}, \mathscr{R})$ is determined. We define

$$
\delta_*[f] = [h] \in H^{q+1}(\mathscr{U}, \mathscr{R}). \tag{3.4.21}
$$

We have to confirm that this is defined independently from the choice of the representative of $[f] \in \bar{H}^q(\mathscr{U}, \mathscr{T})$. Assume $[f] = [f']$ with another representative f'. Then, $f - f' = \delta(f'')$, $f'' \in \bar{C}^{q-1}(\mathscr{U}, \mathscr{T})$. By definition there is $g'' \in C^{q-1}(\mathscr{U}, \mathscr{S})$ such that $f'' = \psi_*(g'')$. Take an element $g' \in C^q(\mathscr{U}, \mathscr{S})$ with $\psi_*(g') = f'$. By choice, $\psi_*(g) - \psi_*(g') = \delta\psi_*(g'')$, and hence

$$
g - g' - \delta(g'') \in \operatorname{Ker} \psi_*.
$$

Therefore, there is $h'' \in C^q(\mathscr{U}, \mathscr{R})$ such that $g - g' - \delta(g'') = \phi_*(h'')$. It follows that

$$\delta(g) - \delta(g') = \delta\phi_*(h'') = \phi_*(\delta h'').$$

Note that $h, h' \in C^{q+1}(\mathscr{U}, \mathscr{R})$ satisfy $\phi_*(h) = \delta(g)$ and $\phi_*(h') = \delta(g')$. The above equation implies

$$\phi_*(h) - \phi_*(h') = \phi_*(\delta h'').$$

Since ϕ_* is injective,

$$h - h' = \delta h''.$$

Therefore, $[h] = [h'] \in H^{q+1}(\mathscr{U}, \mathscr{R})$ is deduced.

Thus, we have the following sequence:

(3.4.22)
$$\begin{aligned} \cdots \longrightarrow H^q(\mathscr{U}, \mathscr{R}) &\xrightarrow{\phi_*} H^q(\mathscr{U}, \mathscr{S}) \xrightarrow{\psi_*} \bar{H}^q(\mathscr{U}, \mathscr{T}) \\ &\xrightarrow{\delta_*} H^{q+1}(\mathscr{U}, \mathscr{R}) \xrightarrow{\phi_*} H^{q+1}(\mathscr{U}, \mathscr{S}) \xrightarrow{\psi_*} \bar{H}^{q+1}(\mathscr{U}, \mathscr{T}) \\ &\xrightarrow{\delta_*} \cdots. \end{aligned}$$

Claim (3.4.22) *is exact.*

∵) By (3.4.19) it is exact at $H^q(\mathscr{U}, \mathscr{S})$.

We shall show the exactness at $\bar{H}^q(\mathscr{U}, \mathscr{T})$. Take $[f] \in \bar{H}^q(\mathscr{U}, \mathscr{T})$ with $\delta_*[f] = 0$. Let g, h be those chosen in (3.4.20) for f:

$$\psi_*(g) = f, \qquad \phi_*(h) = \delta(g).$$

Since $[h] = \delta_*[f] = 0$, there is $h' \in C^q(\mathscr{U}, \mathscr{R})$ such that $h = \delta h'$. It follows that $\delta\phi_*(h') = \phi_*(\delta h') = \phi_*(h) = \delta(g)$. Put

$$g' = g - \phi_*(h').$$

Since

(3.4.23)
$$\psi_*(g') = \psi_*(g) - \psi_* \circ \phi_*(h') = \psi_*(g) = f,$$

g' fulfills the role of g. Since

$$\delta g' = \delta g - \delta\phi_*(h') = \delta(g) - \delta(g) = 0,$$

$[g'] \in H^q(\mathscr{U}, \mathscr{S})$ is defined, and from (3.4.23) follows $\psi_*[g'] = [f]$.

Next, we see the exactness at $H^{q+1}(\mathscr{U}, \mathscr{R}) \ni [h]$. Suppose $\phi_*[h] = 0$. Then $h \in Z^{q+1}(\mathscr{U}, \mathscr{R})$, and the condition implies that

$$\phi_*(h) = \delta(g), \qquad {}^{\exists}g \in C^q(\mathscr{U}, \mathscr{S}),$$
$$\psi_*(g) = f \in \bar{C}^q(\mathscr{U}, \mathscr{T}).$$

It follows from the correspondences that $\delta_*[f] = [h]$. Thus, we see that (3.4.22) is exact. △

Let $\mathscr{U} \prec \mathscr{V}$ ($\mu : \mathscr{V} \to \mathscr{U}$) be a refinement of the open covering $\mathscr{U}$ of X. We obtain the naturally induced homomorphism

$$\bar{\mu}^* : \bar{H}^q(\mathscr{U}, \mathscr{T}) \to \bar{H}^q(\mathscr{V}, \mathscr{T}).$$

Similarly for $\mathscr{R}$, $\mathscr{S}$, we obtain homomorphisms:

$$\mu^* : H^q(\mathscr{U}, \mathscr{R}) \to H^q(\mathscr{V}, \mathscr{R}), \quad \mu^* : H^q(\mathscr{U}, \mathscr{S}) \to H^q(\mathscr{V}, \mathscr{S}).$$

These are compatible with commutative diagram (3.4.18). Therefore, taking direct images $\varinjlim_{\mathscr{U}}$, we have the following exact sequence by (3.4.22):

(3.4.24)
$$\begin{aligned} \cdots \longrightarrow H^q(X, \mathscr{R}) &\xrightarrow{\phi_*} H^q(X, \mathscr{S}) \xrightarrow{\psi_*} \bar{H}^q(X, \mathscr{T}) \\ &\xrightarrow{\delta_*} H^{q+1}(X, \mathscr{R}) \xrightarrow{\phi_*} H^{q+1}(X, \mathscr{S}) \xrightarrow{\psi_*} \bar{H}^{q+1}(X, \mathscr{T}) \\ &\xrightarrow{\delta_*} \cdots . \end{aligned}$$

It remains to show that $\bar{H}^q(X, \mathscr{T}) = H^q(X, \mathscr{T})$. It suffices to see the following:

Claim 3.4.25 *For every $f \in C^q(\mathscr{U}, \mathscr{T})$ there are a refinement $\mathscr{V} \succ \mathscr{U}$ ($\mu : \mathscr{V} \to \mathscr{U}$) and $g \in C^q(\mathscr{V}, \mathscr{S})$ such that $\mu^*(f) = \psi_*(g)$.*

∵) Here we use the paracompactness. We take and fix f. We may assume that $\mathscr{U} = \{U_\alpha\}_{\alpha \in \Phi}$ is locally finite. Since X is normal because of being paracompact and Hausdorff, there is an open covering $\mathscr{W} = \{W_\alpha\}_{\alpha \in \Phi}$ such that $\bar{W}_\alpha \subset U_\alpha$ (${}^{\forall}\alpha \in \Phi$). For each point $x \in X$ we may take a small neighborhood V_x of x satisfying the following properties:

3.4.26 (i) There is a W_α such that $V_x \subset W_\alpha$.
(ii) If $V_x \cap W_\beta \neq \emptyset$, then $V_x \subset U_\beta$.
(iii) For every $\sigma = (U_0, \ldots, U_q) \in N_q(\mathscr{U})$ with $x \in \bigcap_{j=0}^q W_j$ ((ii) implies that $V_x \subset |\sigma|$), $f(\sigma)|_{V_x}$ always belongs to the image of $\psi_* : \Gamma(V_x, \mathscr{S}) \to \Gamma(V_x, \mathscr{T})$.

∵) (i) Since $\{W_\alpha\}$ is an open covering, there exists such W_α if V_x is taken sufficiently small.

(ii) Since $\mathscr{U}$ is locally finite, with V_x being made smaller if necessary, there are only finitely many α with $U_\alpha \cap V_x \neq \emptyset$. Let them be $\{1, 2, \ldots, l\}$. After reordering if necessary, we may assume that

$$x \in \bar{W}_j, \quad 1 \le j \le k,$$
$$x \notin \bar{W}_j, \quad k+1 \le j \le l.$$

Take V_x smaller so that $V_x \cap (\bigcup_{j=k+1}^{l} \bar{W}_j) = \emptyset$. Then, those W_α with $V_x \cap W_\alpha \neq \emptyset$ are only W_j, $1 \le j \le k$. We may take V_x so that $V_x \subset \bigcap_{j=1}^{k} U_j$. Since $\bar{W}_j \subset U_j$, the required property holds.

(iii) For a point $x \in X$ we choose a neighborhood $V_x \ni x$ taken in (i) and (ii) above. Because of the choice, there are only a finite number of β such that $V_x \cap U_\beta \neq \emptyset$. There are only finitely many simplexes $\sigma' \in N(\mathscr{U})$ with those β's. Since

$$\mathscr{S}_x \longrightarrow \mathscr{T}_x \longrightarrow 0$$

is exact, for sufficiently small V_x, and for the above finitely many σ', $f(\sigma')|_{V_x}$ belong to $\phi_*(\Gamma(V_x, \mathscr{S}))$. Therefore, (iii) is satisfied.

Continued proof of Claim 3.4.25: For each V_x, we take $W_\alpha \supset V_x$ by 3.4.26 (i), which is denoted by W_x. We have

$$V_x \subset W_x \subset U_\alpha.$$

In this way, the open covering $\mathscr{V} = \{V_x\}$ gives rise to a refinement of $\mathscr{U} = \{U_\alpha\}$. Write $\mu : \mathscr{V} \to \mathscr{U}$ for the correspondence. For every $\tau = (V_0, \ldots, V_q) \in N_q(\mathscr{V})$, $|\tau| = V_0 \cap \cdots \cap V_q \subset W_0 \cap \cdots \cap W_q$, and if $|\tau| \neq \emptyset$, $V_0 \cap W_i \neq \emptyset$, $0 \le i \le q$. By 3.4.26 (ii), $V_0 \subset U_i$, $0 \le i \le q$. Therefore,

$$|\tau| \subset V_0 \subset U_0 \cap \cdots \cap U_q = |\mu(\tau)|,$$

and so

$$\mu(f)(\tau) = f(\mu(\tau))|_{|\tau|} = f(U_0, \ldots, U_q)|_{|\tau|}.$$

Here it follows from 3.4.26 (iii) that

$$f(U_0, \ldots, U_q)|_{V_0} \in \psi_*(\Gamma(V_0, \mathscr{S})).$$

Hence there exists $g(\tau) \in \Gamma(|\tau|, \mathscr{S})$ such that

$$\mu(f)(\tau) = \big(f(U_0, \ldots, U_q)|_{V_0}\big)\Big|_{|\tau|} = \psi_*(g(\tau)).$$

Setting $g = (g(\tau)) \in C^q(\mathscr{V}, \mathscr{S})$, we have $\mu(f) = \psi_*(g)$. △

Thus, the proof of the theorem is completed. □

Example 3.4.27 We consider the short exact sequence of Example 3.1.4 over $X \subset \mathbf{C}^n$:

$$0 \to \mathbf{Z} \to \mathscr{O}_X \xrightarrow{\mathbf{e}} \mathscr{O}_X^* \to 0.$$

We obtain the following long exact sequence:

$$
\begin{aligned}
0 &\longrightarrow \mathbf{Z} \longrightarrow H^0(X, \mathscr{O}_X) \xrightarrow{\mathbf{e}_*} H^0(X, \mathscr{O}_X^*) \\
&\xrightarrow{\delta_*} H^1(X, \mathbf{Z}) \longrightarrow H^1(X, \mathscr{O}_X) \longrightarrow H^1(X, \mathscr{O}_X^*) \\
&\xrightarrow{\delta_*} H^2(X, \mathbf{Z}) \longrightarrow H^2(X, \mathscr{O}_X) \longrightarrow \cdots .
\end{aligned}
$$

Here we regard $\Gamma(X, \mathbf{Z}) \cong \mathbf{Z}$.

3.4.3 Resolutions of Sheaves and Cohomology

We consider a sequence of sheaves (or abelian groups) with homomorphisms d_p ($p \in \mathbf{Z}^+$):

$$
\mathscr{S}_0 \xrightarrow{d_0} \mathscr{S}_1 \xrightarrow{d_1} \cdots \longrightarrow \mathscr{S}_p \xrightarrow{d_p} \cdots . \tag{3.4.28}
$$

If $d_{p+1} \circ d_p = 0$ for all $p \geq 0$, then sequence (3.4.28) or the family $\{\mathscr{S}_p\}_{p\geq 0}$ is called a *complex* of sheaves (or abelian groups). In this case, we define the q-th *cohomology group* of (3.4.28) or of $\{\mathscr{S}_p\}_{p\geq 0}$ by

$$
\mathscr{H}^q\left(X, \{\mathscr{S}_p\}_{p\geq 0}\right) = \begin{cases} \Gamma(X, \operatorname{Ker} d_0) = \Gamma(X, \mathscr{S}), & q = 0, \\ \Gamma(X, \operatorname{Ker} d_q)/d_{q-1}\Gamma(X, \mathscr{S}_{q-1}), & q \geq 1. \end{cases}
$$

Let $\mathscr{S} \to X$ be a sheaf over X. Assume that there is a complex of sheaves

$$
0 \longrightarrow \mathscr{S} \longrightarrow \mathscr{S}_0 \xrightarrow{d_0} \mathscr{S}_1 \xrightarrow{d_1} \cdots \longrightarrow \mathscr{S}_p \xrightarrow{d_p} \cdots . \tag{3.4.29}
$$

If this is exact, then complex (3.4.29) or the family $\{\mathscr{S}_p\}_{p\in\mathbf{Z}^+}$ is called a *resolution* of $\mathscr{S}$. It is sometimes called a right resolution of $\mathscr{S}$.

Theorem 3.4.30 *Let X be paracompact and Hausdorff. Assume that (3.4.29) is a resolution of a sheaf $\mathscr{S}$ over X. If $H^q(X, \mathscr{S}_p) = 0$ for all $q \geq 1$ and $p \geq 0$, then*

$$
H^q(X, \mathscr{S}) \cong \mathscr{H}^q\left(X, \{\mathscr{S}_p\}_{p\geq 0}\right), \quad q \geq 0.
$$

Proof Set $\mathscr{K}_p = \operatorname{Ker} d_p$, $p \geq 0$. The both of the following are exact by the assumption:

$$
\begin{aligned}
&0 \longrightarrow \mathscr{S} \longrightarrow \mathscr{S}_0 \longrightarrow \mathscr{K}_1 \longrightarrow 0, \\
&0 \longrightarrow \mathscr{K}_p \longrightarrow \mathscr{S}_p \longrightarrow \mathscr{K}_{p+1} \longrightarrow 0, \quad p \geq 1.
\end{aligned}
$$

The case of $q = 0$ is clear. Assume $q \geq 1$. By Theorem 3.4.17 the following sequence is exact:

$$\begin{aligned}(3.4.31)\qquad \cdots \longrightarrow H^{q-1}(X, \mathscr{S}_0) &\xrightarrow{d_{0*}} H^{q-1}(X, \mathscr{K}_1) \xrightarrow{\delta_*} H^q(X, \mathscr{S}) \\ &\longrightarrow H^q(X, \mathscr{S}_0) \xrightarrow{d_{0*}} H^q(X, \mathscr{K}_1) \longrightarrow \cdots .\end{aligned}$$

Since $H^q(X, \mathscr{S}_0) = 0$ by the assumption, we see that for $q = 1$,

$$H^0(X, \mathscr{S}_0) \xrightarrow{d_{0*}} H^0(X, \mathscr{K}_1) \xrightarrow{\delta_*} H^1(X, \mathscr{S}) \longrightarrow 0$$

is exact. Therefore,

$$H^1(X, \mathscr{S}) \cong \Gamma(X, \mathscr{K}_1)/d_0\Gamma(X, \mathscr{S}_0) = \mathscr{H}^1\left(X, \{\mathscr{S}_p\}_{p\geq 0}\right).$$

This proves the case of $q = 1$.

In the sequel, we let $q \geq 2$ and use induction on q. Suppose that the theorem holds for $q - 1$. It follows from (3.4.31) and the assumption that

$$0 \longrightarrow H^{q-1}(X, \mathscr{K}_1) \xrightarrow{d_{0*}} H^q(X, \mathscr{S}) \longrightarrow 0$$

is exact. Therefore,

$$(3.4.32)\qquad H^q(X, \mathscr{S}) \cong H^{q-1}(X, \mathscr{K}_1).$$

There is a resolution of $\mathscr{K}_1$ as follows:

$$0 \longrightarrow \mathscr{K}_1 \longrightarrow \mathscr{S}_1 \xrightarrow{d_1} \mathscr{S}_2 \xrightarrow{d_2} \cdots \longrightarrow \mathscr{S}_p \xrightarrow{d_p} \cdots .$$

The induction hypothesis applied for $\mathscr{K}_1$ implies that

$$H^{q-1}(X, \mathscr{K}_1) \cong \Gamma(X, \mathscr{K}_q)/d_{q-1}\Gamma(X, \mathscr{S}_{q-1}).$$

Combining this with (3.4.32), we obtain

$$H^q(X, \mathscr{S}) \cong \Gamma(X, \mathscr{K}_q)/d_{q-1}\Gamma(X, \mathscr{S}_{q-1}) = \mathscr{H}^q\left(X, \{\mathscr{S}_p\}_{p\geq 0}\right). \qquad \square$$

Here it arises as a problem if there exists a resolution (3.4.29) satisfying

$$H^q(X, \mathscr{S}_p) = 0, \qquad q \geq 1,\ p \geq 0.$$

Definition 3.4.33 In general, a sheaf $\mathscr{S} \to X$ is called a *fine sheaf* if for every locally finite open covering $X = \bigcup_\alpha U_\alpha$ there is a so-called *partition of unity* $\{\rho_\alpha\}$ as follows:

(i) $\rho_\alpha : \mathscr{S} \to \mathscr{S}$ is a homomorphism of sheaves.
(ii) $\mathrm{Supp}\, \rho_\alpha \subset U_\alpha$.
(iii) $\sum_\alpha \rho_\alpha(s) = s, \ {}^\forall s \in \mathscr{S}$.

In resolution (3.4.29), if all $\mathscr{S}_p$ $(p \geq 0)$ are fine sheaves, it is called a *fine resolution* of $\mathscr{S}$.

Example 3.4.34 (Important) Let X be an open subset of $\mathbf{R}^m$ (in general, it may be assumed to be a paracompact differentiable manifold). For a locally finite open covering $X = \bigcup_\alpha U_\alpha$, there exists a partition of unity $\varphi_\alpha \in C_0^\infty(U_\alpha)$ (the set of all C^∞-functions with compact supports in U_α) subordinated to $\{U_\alpha\}$. For the sheaf $\mathscr{E}_X$ of germs of differentiable functions over X we define

$$\rho_\alpha : \underline{f}_x \in \mathscr{E} \to \underline{\varphi_\alpha \cdot f}_x \in \mathscr{E}.$$

Then $\{\rho_\alpha\}$ is a partition of unity of $\mathscr{E}_X$. Therefore, every sheaf of $\mathscr{E}_X$-modules over X is a fine sheaf, and of course, so is the sheaf $\mathscr{C}_X$ of germs of continuous functions over X.

Theorem 3.4.35 *If $\mathscr{S} \to X$ is a fine sheaf, then for every locally finite open covering $\mathscr{U} = \{U_\alpha\}$ of X,*

$$H^q(\mathscr{U}, \mathscr{S}) = 0, \qquad q \geq 1.$$

In particular, if X is paracompact, $H^q(X, \mathscr{S}) = 0$, $q \geq 1$.

Proof Let $\{\rho_\alpha\}$ be a partition of unity subordinated to $\mathscr{U}$. For every section $f \in \Gamma(U, \mathscr{S})$ over an open subset U we define $\rho_\alpha f \in \Gamma(U, \mathscr{S})$ as follows:

$$\rho_\alpha f(x) = \begin{cases} \rho(f(x)), & x \in U \cap U_\alpha, \\ 0, & x \in U \setminus U_\alpha. \end{cases}$$

Take any $f \in Z^q(\mathscr{U}, \mathscr{S})$. We define $g \in C^{q-1}(\mathscr{U}, \mathscr{S})$ by

$$g(\tau) = g(U_0, \ldots, U_{q-1}) = \sum_\alpha \rho_\alpha f(U_\alpha, U_0, \ldots, U_{q-1})$$

for $\tau = (U_0, \ldots, U_{q-1}) \in N_{q-1}(\mathscr{U})$. Then we see $\delta g = f$ as follows:

$$\begin{aligned}\delta g(U_0,\ldots,U_q) &= \sum_{i=0}^{q}(-1)^i g(U_0,\ldots,\check{U}_i,\ldots,U_q)\\ &= \sum_{i=0}^{q}(-1)^i \sum_{\alpha} \rho_\alpha f(U_\alpha,U_0,\ldots,\check{U}_i,\ldots,U_q)\\ &= \sum_{\alpha}\sum_{i=0}^{q}(-1)^i \rho_\alpha f(U_\alpha,U_0,\ldots,\check{U}_i,\ldots,U_q)\\ &= \sum_{\alpha} \rho_\alpha f(U_0,\ldots,\ldots,U_q) = f(U_0,\ldots,\ldots,U_q).\end{aligned}$$

At the last step we used that by $\delta f = 0$,

$$f(\check{U}_\alpha,U_0,\ldots,\ldots,U_q) - \sum_{i=0}^{q}(-1)^i \rho_\alpha f(U_\alpha,U_0,\ldots,\check{U}_i,\ldots,U_q) = 0. \quad \square$$

Theorem 3.4.36 *For every sheaf $\mathscr{S} \xrightarrow{\pi} X$ there exists a fine resolution $\{\mathscr{S}_p\}_{p\geq 0}$ of $\mathscr{S}$. Moreover, it may be taken so that for any open subset U the restriction $\{\mathscr{S}_p|_U\}_{p\geq 0}$ is a fine resolution of $\mathscr{S}|_U$.*

Proof On every open subset V we consider an abelian group of all sections, $\mathscr{S}^*(V) = \{f : V \to \mathscr{S}; \pi \circ f(x) = x, x \in V\}$, without assuming the continuity. Then, $\{(V, \mathscr{S}^*(V))\}$ forms a complete presheaf. We denote by $\mathscr{S}^* \to X$ the sheaf induced from the presheaf.

Here we prove:

Lemma 3.4.37 *For any open subset $U \subset X$, the restriction $\mathscr{S}^*|_U$ is a fine sheaf over U.*

$\because$) Let $\mathscr{W} = \{W_\lambda\}_{\lambda\in\Lambda}$ be a locally finite open covering of U. We introduce a total order in Λ and write $\Lambda = \{1, 2, \ldots\}$. Put $F_1 = W_1$, and inductively $F_\mu = W_\mu \setminus \bigcup_{\lambda<\mu} F_\lambda$, $\mu > 1$. Then,

$$F_\lambda \subset W_\lambda; \quad F_\lambda \cap F_\mu = \emptyset,\ \lambda \neq \mu; \quad U = \bigcup_\lambda F_\lambda.$$

By making use of these, we obtain

$$\begin{aligned}&\rho_\lambda : \mathscr{S}^*|_U \longrightarrow \mathscr{S}^*|_U,\\ &\rho_\lambda(s) = \begin{cases} s, & s \in \mathscr{S}^*_x, \quad x \in F_\lambda,\\ 0, & s \in \mathscr{S}^*_x, \quad x \notin F_\lambda,\end{cases}\\ &\sum_\lambda \rho_\lambda = 1.\end{aligned}$$

Thus, $\{\rho_\lambda\}$ gives rise to a partition of unity of $\mathscr{S}^*|_U$, so that $\mathscr{S}^*|_U$ is a fine sheaf. △

Continued Proof of Theorem: Set $\mathscr{S}_0 = \mathscr{S}^*$. Then, $\mathscr{S}_0$ is a fine sheaf, and

$$0 \longrightarrow \mathscr{S} \longrightarrow \mathscr{S}_0$$

is exact. Put $\mathscr{S}'_1 = \mathscr{S}_0/\mathscr{S}$. Similarly to the above, making use of discontinuous sections $\mathscr{S}'_1$, we get $\mathscr{S}'^*_1$, which is a fine sheaf, and

$$0 \longrightarrow \mathscr{S} \longrightarrow \mathscr{S}_0 \longrightarrow \mathscr{S}'^*_1$$

is exact. We set $\mathscr{S}_1 = \mathscr{S}'^*_1$. Repeating this process, we obtain a fine resolution:

$$0 \longrightarrow \mathscr{S} \longrightarrow \mathscr{S}_0 \longrightarrow \mathscr{S}_1 \longrightarrow \mathscr{S}_2 \longrightarrow \cdots.$$

By Lemma 3.4.37, $\mathscr{S}_p|_U$ ($p \geq 0$) are fine for any open subset U of X. □

The following is deduced from Theorems 3.4.30, 3.4.35 and 3.4.36:

Theorem 3.4.38 *Let X be paracompact and Hausdorff.*

(i) *The cohomology $\mathscr{H}^q(X, \{\mathscr{S}_p\})$ of a fine resolution $\{\mathscr{S}_p\}_{p\geq 0}$ of a sheaf $\mathscr{S}$ is independent of the choice of $\{\mathscr{S}_p\}_{p\geq 0}$ up to isomorphism.*
(ii) *In the definition of Čech cohomology $H^q(X, \mathscr{S})$ of a sheaf $\mathscr{S}$ we may exclude the alternation (3.4.1) to obtain a cohomology, which is isomorphic to $H^q(X, \mathscr{S})$.*

Proof (i) This is because all cohomologies $\mathscr{H}^q(X, \{\mathscr{S}_p\})$ are isomorphic to Čech cohomologies $H^q(X, \mathscr{S})$.

(ii) Both Čech cohomologies $H^q(X, \mathscr{S})$ with alternation (3.4.1) and without it are isomorphic to a cohomology $\mathscr{H}^q(X, \{\mathscr{S}_p\})$ of a fine resolution of $\mathscr{S}$. □

Definition 3.4.39 An open covering $\mathscr{U}$ of X is called a *Leray covering* with respect to a sheaf $\mathscr{S}$ if

$$H^q(|\sigma|, \mathscr{S}) = 0, \quad {}^\forall\sigma \in N(\mathscr{U}),\ q \geq 1.$$

Theorem 3.4.40 *Let $\mathscr{U} = \{U_\alpha\}$ be a Leray covering with respect to a sheaf $\mathscr{S}$ over X. Then,*

$$H^q(\mathscr{U}, \mathscr{S}) \cong H^q(X, \mathscr{S}), \quad q \geq 0.$$

Proof In the case of $q = 0$, this is immediate from Lemma 3.4.6.

Let $q \geq 1$. By Theorem 3.4.36 we take a fine resolution of $\mathscr{S}$:

$$0 \longrightarrow \mathscr{S} \longrightarrow \mathscr{S}_0 \xrightarrow{d_0} \mathscr{S}_1 \xrightarrow{d_1} \cdots \longrightarrow \mathscr{S}_p \xrightarrow{d_p} \cdots. \tag{3.4.41}$$

For a simplex $\sigma \in N(\mathscr{U})$, the sequence (3.4.41) restricted to the support $|\sigma|$ gives a fine resolution:

(3.4.42) $$0 \longrightarrow \mathscr{S}|_{|\sigma|} \longrightarrow \mathscr{S}_0|_{|\sigma|} \xrightarrow{d_0} \mathscr{S}_1|_{|\sigma|} \xrightarrow{d_1} \cdots \longrightarrow \mathscr{S}_p|_{|\sigma|} \xrightarrow{d_p} \cdots.$$

It follows from Theorem 3.4.30 that

$$\begin{aligned}&\{f \in \Gamma(|\sigma|, \mathscr{S}_p);\, d_p f = 0\}/d_{p-1}\Gamma(|\sigma|, \mathscr{S}_{p-1})\\ &\cong H^p(|\sigma|, \mathscr{S}) = 0, \quad p \geq 1.\end{aligned}$$

Hence, the induced sequence by (3.4.42)

$$\begin{aligned}0 \longrightarrow \Gamma(|\sigma|, \mathscr{S}) \longrightarrow \Gamma(|\sigma|, \mathscr{S}_0) \xrightarrow{d_{0*}} \Gamma(|\sigma|, \mathscr{S}_1) \xrightarrow{d_{1*}} \cdots\\ \longrightarrow \Gamma(|\sigma|, \mathscr{S}_p) \xrightarrow{d_{p*}} \cdots\end{aligned}$$

is exact. Thus the following sequence is exact:

(3.4.43) $$\begin{aligned}0 \longrightarrow C^q(\mathscr{U}, \mathscr{S}) \longrightarrow C^q(\mathscr{U}, \mathscr{S}_0) \xrightarrow{(-1)^q d_{0*}} C^q(\mathscr{U}, \mathscr{S}_1) \xrightarrow{(-1)^q d_{1*}} \cdots\\ \longrightarrow C^q(\mathscr{U}, \mathscr{S}_p) \xrightarrow{(-1)^q d_{p*}} \cdots.\end{aligned}$$

Then we consider the so-called double complex (3.4.44) below. By (3.4.43) and the assumption for $\mathscr{U}$ being Leray we see that all column and row sequences in (3.4.44) except for the sequences of the first column and the first row are exact:

(3.4.44)

$$\begin{array}{ccccccccc}
 & 0 & & 0 & & 0 & & 0 & \\
 & \downarrow & & \downarrow & & \downarrow & & \downarrow & \\
0 \to & \Gamma(X, \mathscr{S}) & \to & \Gamma(X, \mathscr{S}_0) & \xrightarrow{d_{0*}} & \Gamma(X, \mathscr{S}_1) & \xrightarrow{d_{1*}} \cdots \to & \Gamma(X, \mathscr{S}_q) & \xrightarrow{d_{q*}} \cdots \\
 & \downarrow & & \downarrow & & \downarrow & & \downarrow & \\
0 \to & C^0(\mathscr{U}, \mathscr{S}) & \to & C^0(\mathscr{U}, \mathscr{S}_0) & \xrightarrow{d_{0*}} & C^0(\mathscr{U}, \mathscr{S}_1) & \xrightarrow{d_{1*}} \cdots \to & C^0(\mathscr{U}, \mathscr{S}_q) & \xrightarrow{d_{q*}} \cdots \\
 & \downarrow \delta & & \downarrow \delta & & \downarrow \delta & & \downarrow \delta & \\
0 \to & C^1(\mathscr{U}, \mathscr{S}) & \to & C^1(\mathscr{U}, \mathscr{S}_0) & \xrightarrow{-d_{0*}} & C^1(\mathscr{U}, \mathscr{S}_1) & \xrightarrow{-d_{1*}} \cdots \to & C^1(\mathscr{U}, \mathscr{S}_q) & \xrightarrow{-d_{q*}} \cdots \\
 & \downarrow \delta & & \downarrow \delta & & \downarrow \delta & & \downarrow \delta & \\
0 \to & C^2(\mathscr{U}, \mathscr{S}) & \to & C^2(\mathscr{U}, \mathscr{S}_0) & \xrightarrow{d_{0*}} & C^2(\mathscr{U}, \mathscr{S}_1) & \xrightarrow{d_{1*}} \cdots \to & C^2(\mathscr{U}, \mathscr{S}_q) & \xrightarrow{d_{q*}} \cdots \\
 & \downarrow \delta & & \downarrow \delta & & \downarrow \delta & & \downarrow \delta & \\
 & \vdots & & \vdots & & \vdots & & \vdots & \\
0 \to & C^q(\mathscr{U}, \mathscr{S}) & \to & C^q(\mathscr{U}, \mathscr{S}_0) & \xrightarrow{(-1)^q d_{0*}} & C^q(\mathscr{U}, \mathscr{S}_1) & \xrightarrow{(-1)^q d_{1*}} \cdots \to & C^q(\mathscr{U}, \mathscr{S}_q) & \xrightarrow{(-1)^q d_{q*}} \cdots \\
 & \downarrow \delta & & \downarrow \delta & & \downarrow \delta & & \downarrow \delta & \\
 & \vdots & & \vdots & & \vdots & & \vdots &
\end{array}$$

The following combined with Theorem 3.4.30 finishes the proof.

Claim 3.4.45 $\mathscr{H}^q(X, \{\mathscr{S}_p\}) \cong H^q(\mathscr{U}, \mathscr{S})$.

∵) For every $f \in \{f \in \Gamma(X, \mathscr{S}_q);\, d_{q*}f = 0\}$ we take $d_{q*}f = 0$. In commutative diagram (3.4.44), we denote the image of f by $\Gamma(X, \mathscr{S}_q) \to C^0(\mathscr{U}, \mathscr{S}_q)$ by $f_q^0 =$

$(f|_{U_\alpha})$. Chasing (3.4.44) vertically and horizontally, we inductively take f^i_{q-i} and f^i_{q-i-1}, so that

$$f^i_{q-i-1} \in C^i(\mathscr{U}, \mathscr{S}_{q-i-1}), \qquad (-1)^i d_{q-i-1*} f^i_{q-i-1} = f^i_{q-i},$$
$$f^{i+1}_{q-i-1} = \delta f^i_{q-i-1} \in C^{i+1}(\mathscr{U}, \mathscr{S}_{q-i-1}).$$

For $i = q$ we have that $f^q_0 \in C^q(\mathscr{U}, \mathscr{S}_0)$, $(-1)^q d_{0*} f^q_0 = 0$. Hence, $f^q_0 \in C^q(\mathscr{U}, \mathscr{S})$. Moreover, since $\delta f^q_0 = 0$, $f^q_0 \in Z^q(\mathscr{U}, \mathscr{S})$. Thus we obtain a correspondence

$$f \in \{f \in \Gamma(X, \mathscr{S}_q);\ d_{q*} f = 0\} \to f^q_0 \in Z^q(\mathscr{U}, \mathscr{S}). \tag{3.4.46}$$

In this correspondence, if $f \in d_{q-1*}\Gamma(X, \mathscr{S}_{q-1})$, there is a $g \in \Gamma(X, \mathscr{S}_{q-1})$ such that $f = d_{q-1*} g$. Let g^0_{q-1} denote the image of g through $\Gamma(X, \mathscr{S}_{q-1}) \to C^0(\mathscr{U}, \mathscr{S}_{q-1})$. We have $d_{q-1*} g^0_{q-1} = f^0_q$. In the same way as above, chasing (3.4.44) vertically and horizontally, we inductively define $g^i_{q-1-i} \in C^i(\mathscr{U}, \mathscr{S}_{q-1-i})$ so that

$$f^i_{q-i} = \delta g^{i-1}_{q-i}, \qquad i = 1, 2, \ldots,$$
$$(-1)^i d_{q-i-1*} g^i_{q-i-1} = g^i_{q-i}.$$

At $i = q$, $f^q_0 = \delta g^{q-1}_0$ with $g^{q-1}_0 \in C^{q-1}(\mathscr{U}, \mathscr{S})$, and (3.4.46) defines a homomorphism

$$\Phi_q : \{f \in \Gamma(X, \mathscr{S}_q);\ d_{q*} f = 0\}/d_{q-1*}\Gamma(X, \mathscr{S}_{q-1}) \to H^q(\mathscr{U}, \mathscr{S}).$$

Chasing commutative diagram (3.4.44) conversely to the above, we obtain a homomorphism

$$\begin{aligned}\Psi_q : H^q(\mathscr{U}, \mathscr{S}) &\longrightarrow \{f \in \Gamma(X, \mathscr{S}_q);\ d_{q*} f = 0\}/d_{q-1*}\Gamma(X, \mathscr{S}_{q-1}) \\ &= \mathscr{H}^q(X, \{\mathscr{S}_p\}).\end{aligned}$$

By the construction, $\Phi_q \circ \Psi_q = \mathrm{id}$ and $\Psi_q \circ \Phi_q = \mathrm{id}$. Therefore, Φ_q and Ψ_q are isomorphisms. This proves

$$\mathscr{H}^q(X, \{\mathscr{S}_p\}) \cong H^q(\mathscr{U}, \mathscr{S}). \qquad \square$$

3.5 De Rham Cohomology

In the present section we describe de Rham cohomology of a domain of $\mathbf{R}^n$, and show the de Rham Theorem. Let $\mathscr{E}_X \to X$ be the sheaf of germs of C^∞ (real-valued or complex-valued) functions over an open set $X \subset \mathbf{R}^n$. We sometimes abbreviate $\mathscr{E}_X$ to $\mathscr{E}$, when X is fixed.

3.5.1 Differential Forms and Exterior Products

Let $(x^1, \ldots, x^n)$ be the standard coordinate system of $\mathbf{R}^n$ and let $X \subset \mathbf{R}^n$ be an open set. As in Sect. 1.2.1 a vector field $\xi = \sum_{i=1}^n a^i \frac{\partial}{\partial x^i}$ and a differential form $\eta = \sum_{j=1}^n b_j dx^j$ are defined over X, where the coefficients a^i and b_j are functions in X. The duality between ξ and η is given by

$$(\eta, \xi) = \sum_{i=1}^n b_i a^i;$$

in particular, $\left(dx^j, \frac{\partial}{\partial x^i}\right) = \delta_{ji}$ (Kronecker's symbol). Here we consider the case where a^i and b_j are functions of $\mathscr{E}(X)$ $(:= \Gamma(X, \mathscr{E}_X))$.

Let q be a natural number. For q numbers, $1 \le i_1, \ldots, i_q \le n$, we write symbols

$$dx^{i_1} \wedge \cdots \wedge dx^{i_q},$$

which are related by the alternative rule

$$dx^{i_{\sigma(1)}} \wedge \cdots \wedge dx^{i_{\sigma(q)}} = (\operatorname{sgn} \sigma)\, dx^{i_1} \wedge \cdots \wedge dx^{i_q}$$

for permutations σ of q indices, where $\operatorname{sgn} \sigma$ denotes the signature of σ. Thus, for instance, $dx^i \wedge dx^i = 0$ $(1 \le i \le n)$. A sum of type

$$f = \sum_{1 \le i_1, \ldots, i_q \le n} b_{i_1 \cdots i_q} dx^{i_1} \wedge \cdots \wedge dx^{i_q}, \quad b_{i_1 \cdots i_q} \in \mathscr{E}(X) \tag{3.5.1}$$

is called a *differential q-form* on X. We denote by $\mathscr{E}^{(q)}(X)$ the set of all differential q-forms on X, which is a module over the ring $\mathscr{E}(X)$.

We write I for a multi-index $1 \le i_1, \ldots, i_q \le n$ and call $|I| := q$ the *length of the multi-index I*. We put

$$dx^I = dx^{i_1} \wedge \cdots \wedge dx^{i_q}.$$

We say that I is strictly increasing if $1 \le i_1 < \cdots < i_q \le n$. The elements dx^I with strictly increasing indices I with $|I| = q$ form a free basis of $\mathscr{E}^{(q)}(X)$ over $\mathscr{E}(X)$. Thus, every $f \in \mathscr{E}^{(q)}(X)$ is uniquely written as

$$f = \sum_{|I|=q}{}' f_I dx^I, \tag{3.5.2}$$

where $\sum'_{|I|=q}$ stands for the sum with strictly increasing multi-indices I of length q. It follows from the definition that

$$\mathscr{E}^{(q)}(X) = 0, \qquad q > n.$$

For two multi-indices $I = (i_1, \ldots, i_q)$ and $J = (j_1, \ldots, j_r)$ $(1 \leq r \leq n)$ we define a product, which is a differential $(q + r)$-form, by

$$\left(dx^{i_1} \wedge \cdots \wedge dx^{i_q}\right) \wedge \left(dx^{j_1} \wedge \cdots \wedge dx^{j_r}\right) = dx^{i_1} \wedge \cdots \wedge dx^{i_q} \wedge dx^{j_1} \wedge \cdots \wedge dx^{j_r},$$

and extend this by linearity to

$$(f, g) \in \mathscr{E}^{(q)}(X) \times \mathscr{E}^{(r)}(X) \to f \wedge g \in \mathscr{E}^{(q+r)}(X), \tag{3.5.3}$$

which is called the *exterior* or *wedge product* of differential forms. For convenience we write $\mathscr{E}^{(0)}(X) = \mathscr{E}(X)$.

Let $Y \subset \mathbf{R}^m$ be another open set and let $\varphi : X \to Y$ be a C^∞ map. Let $y = (y^1, \ldots, y^m)$ be the standard coordinate system of $\mathbf{R}^m$ and write $\varphi(x) = (\varphi^1(x), \ldots, \varphi^m(x))$ with $\varphi^j \in \mathscr{E}(X)$. We consider a C^∞ function $y^j = \varphi^j(x)$ on X and take its exterior derivative, which defines the pull-back of differential form dy^j by φ:

$$\varphi^* dy^j = d(y^j \circ \varphi) = d\varphi^j = \sum_{i=1}^n \frac{\partial \varphi^j}{\partial x^i} dx^i.$$

This extends linearly to the *pull-back map*

$$\varphi^* : \eta \in \mathscr{E}^{(q)}(Y) \to \varphi^* \eta \in \mathscr{E}^{(q)}(X). \tag{3.5.4}$$

3.5.2 Real Domains

We keep the notation as before. Let $f = \sum'_{|I|=q} f_I dx^I$ be a locally defined differential q-form. Then the *exterior differential* df of f is defined by

$$df = \sum_{|I|=q}' \sum_{j=1}^n \frac{\partial f_I}{\partial x^j} dx^j \wedge dx^I.$$

Let $\mathscr{E}_X^{(q)} \to X$ denote the sheaf of germs of differential q-forms (of C^∞-class) over X. By Example 3.4.34, $\mathscr{E}_X^{(q)}$ is a fine sheaf.

The exterior differential induces a homomorphism of sheaves, which is $\mathbf{R}$ (or $\mathbf{C}$)-linear (but not $\mathscr{E}_X$-linear),

$$d : \underline{f}_x \in \mathscr{E}_X^{(q)} \to \underline{df}_x \in \mathscr{E}_X^{(q+1)}, \qquad q = 0, 1, \ldots.$$

It is immediate from the alternative property that

$$d \circ d = 0.$$

Let $\varphi^* : \mathscr{E}^{(q)}(Y) \to \mathscr{E}^{(q)}(X)$ be as in (3.5.4). Then we have by computation

$$d \circ \varphi^* = \varphi^* \circ d.$$

For simplicity we write **R** (resp. **C**) for the constant sheaf $\mathbf{R}_X \to X$ (resp. $\mathbf{C}_X \to X$). Depending on $\mathscr{E}_X$ taken as **R**-valued or **C**-valued, we have the following two complexes of sheaves:

$$\text{(3.5.5)} \qquad \begin{aligned} &0 \to \mathbf{R} \to \mathscr{E}_X^{(0)} \xrightarrow{d} \mathscr{E}_X^{(1)} \xrightarrow{d} \cdots \xrightarrow{d} \mathscr{E}_X^{(n)} \xrightarrow{d} 0, \\ &0 \to \mathbf{C} \to \mathscr{E}_X^{(0)} \xrightarrow{d} \mathscr{E}_X^{(1)} \xrightarrow{d} \cdots \xrightarrow{d} \mathscr{E}_X^{(n)} \xrightarrow{d} 0. \end{aligned}$$

Since the argument is the same for both cases, we consider the **C**-valued case.

We say that X is *starlike* with respect to a point $x_0 \in X$ if the line segment connecting x_0 and every $x \in X$ is contained in X. If X is convex, then X is starlike with respect to any $x \in X$.

Lemma 3.5.6 (Poincaré) *Assume that X is starlike with respect to some point $x_0 \in X$. Then, for every $f \in \Gamma(X, \mathscr{E}_X^{(q)})$ with $df = 0$ there exists an element $g \in \Gamma(X, \mathscr{E}_X^{(q-1)})$ such that $f = dg$.*

In particular, sequence (3.5.5) is exact over a general open set X, and a fine resolution of the constant sheaf **C**.

Proof We may set $x_0 = 0$ by a translation.

The case of $q = 0$: If $df = 0$, then $f(x) \equiv f(0) \in \mathbf{C}$.

The case of $q > 0$: Take a monotone increasing C^∞ function $\phi : \mathbf{R} \to \mathbf{R}$ such that

$$\phi(t) = 0 \quad (t \le 0), \qquad \phi(t) = 1 \quad (t \ge 1).$$

We set a C^∞ map by

$$\Phi : (t, x) \in \mathbf{R} \times X \to \phi(t)x \in X.$$

Let $f \in \Gamma(X, \mathscr{E}_X^{(q)})$ and write

$$f = \sum_{|I|=q}{}' f_I dx^I.$$

Then we have

$$\begin{aligned}\Phi^* f &= \sum_{|I|=q}{}' f_I(\phi(t)x) d(\phi(t)x^{i_1}) \wedge \cdots \wedge d(\phi(t)x^{i_q}) \\ &= \sum_{|I|=q}{}' f_I(\phi(t)x)(\phi(t))^q dx^I + \sum_{|J|=q-1}{}' \beta_J(t,x) dt \wedge dx^J \\ &= \sum_{|I|=q}{}' \alpha_I(t,x) dx^I + \sum_{|J|=q-1}{}' \beta_J(t,x) dt \wedge dx^J .\end{aligned}$$

It follows that

$$\alpha_I(0,x) = 0, \qquad \alpha_I(1,x) = f_I(x). \tag{3.5.7}$$

For this f we set a differential $(q-1)$-form by

$$\theta f = \sum_{|J|=q-1}{}' \left(\int_0^1 \beta_J(t,x) dt \right) dx^J \in \Gamma(X, \mathscr{E}_X^{(q-1)}) \quad (q-1 \geq 0).$$

It follows from the definition that

$$d\theta f = \sum_{|J|=q-1}{}' \sum_{h=1}^{n} \left(\int_0^1 \frac{\partial \beta_J(t,x)}{\partial x^h} dt \right) dx^h \wedge dx^J . \tag{3.5.8}$$

On the other hand,

$$\begin{aligned}\Phi^* df = d\Phi^* f &= \sum_{|I|=q}{}' \sum_{h=1}^{n} \frac{\partial \alpha_I}{\partial x^h} dx^h \wedge dx^I \\ &+ \sum_{|I|=q}{}' \frac{\partial \alpha_I}{\partial t} dt \wedge dx^I - \sum_{|J|=q-1}{}' \sum_{h=1}^{n} \frac{\partial \beta_J}{\partial x^h} dt \wedge dx^h \wedge dx^J .\end{aligned}$$

Therefore, applying θ for this, we obtain

$$\begin{aligned}\theta df &= \sum_{|I|=q}{}' \left(\int_0^1 \frac{\partial \alpha_I}{\partial t} dt \right) dx^I - \sum_{|J|=q-1}{}' \sum_{h=1}^{n} \left(\int_0^1 \frac{\partial \beta_J}{\partial x^h} dt \right) dx^h \wedge dx^J \\ &= \sum_{|I|=q}{}' (\alpha_I(1,x) - \alpha_I(0,x)) dx^I \\ &\quad - \sum_{|J|=q-1}{}' \sum_{h=1}^{n} \left(\int_0^1 \frac{\partial \beta_J}{\partial x^h} dt \right) dx^h \wedge dx^J .\end{aligned}$$

Now, making use of (3.5.7) and (3.5.8), we deduce that $\theta df = f - d\theta f$, i.e.,

$$f = \theta df + d\theta f.$$

If $df = 0$, then $f = dg$ with $g = \theta f$. □

By Lemma 3.5.6, the long sequence (3.5.5) is a fine resolution of $\mathbf{C}$, which is called the *de Rham resolution*. From (3.5.5) we obtain the so-called *de Rham complex*

$$(3.5.9)\quad 0 \to \mathbf{C} \to \Gamma(X, \mathscr{E}_X^{(0)}) \xrightarrow{d} \Gamma(X, \mathscr{E}_X^{(1)}) \xrightarrow{d} \cdots \xrightarrow{d} \Gamma(X, \mathscr{E}_X^{(n)}) \xrightarrow{d} 0.$$

For $q \geq 0$, the (q-th) *de Rham cohomology* is defined by

$$(3.5.10)\qquad H^q_{\mathrm{DR}}(X, \mathbf{C}) = \{f \in \Gamma(X, \mathscr{E}_X^{(q)}); df = 0\}/d\Gamma(X, \mathscr{E}_X^{(q-1)}).$$

We obtain the following from Theorem 3.4.30.

Theorem 3.5.11 (de Rham) *For a domain X of $\mathbf{R}^n$ we have an isomorphism*

$$H^q(X, \mathbf{C}) \cong H^q_{\mathrm{DR}}(X, \mathbf{C}) \quad (q \geq 0).$$

In particular, $H^q(X, \mathbf{C}) = 0$ for $q > n$.

Corollary 3.5.12 *Let the notation be as above.*

(i) *If X is simply connected, $H^1(X, \mathbf{C}) = 0$.*
(ii) *If X is convex, $H^0(X, \mathbf{C}) \cong \mathbf{C}$ and $H^q(X, \mathbf{C}) = 0$, $q \geq 1$.*

Proof (i) It follows from the de Rham Theorem 3.5.11 that

$$H^1(X, \mathbf{C}) \cong \{f \in \Gamma(X, \mathscr{E}_X^{(1)}); df = 0\}/d\Gamma(X, \mathscr{E}_X^{(0)}).$$

Take any $f \in \Gamma(X, \mathscr{E}_X^{(1)})$ with $df = 0$. Fix an arbitrary point $x_0 \in X$. For $x \in X$ we take a piecewise C^1-curve $C(x)$ from x_0 and to x, and set

$$g(x) = \int_{C(x)} f.$$

This is invariant under the homotopy of $C(x)$. Since $\pi_1(x_0, X) = 0$, $g(x)$ $(x \in X)$ gives rise to a C^∞ function on X. Because $dg = f$, $[f] = 0 \in H^1(X, \mathbf{C})$.

(ii) Note that a convex domain X is starlike with respect to every point of X. Thus, the assertion follows from Poincaré's Lemma 3.5.6 and the de Rham Theorem 3.5.11.

□

Remark 3.5.13 (i) Taking $\mathbf{R}$ instead of $\mathbf{C}$ and real-valued functions for all coefficients of differential forms, we similarly define the *q-th de Rham cohomology* $H^q_{\mathrm{DR}}(X, \mathbf{R})$ as in (3.5.10), and obtain an isomorphism (*de Rham Theorem*)

$$H^q(X, \mathbf{R}) \cong H^q_{\mathrm{DR}}(X, \mathbf{R}).$$

Because of the extension relation of coefficients, $H^q(X, \mathbf{C}) = H^q(X, \mathbf{R}) \otimes_{\mathbf{R}} \mathbf{C}$.

(ii) Theorem 3.5.11 holds for a paracompact differentiable manifold because of the existence of a partition of unity.

3.5.3 Complex Domains

Let $z = (z^1, \ldots, z^n)$ be the standard complex coordinate system of $\mathbf{C}^n$. As in the real case, for a multi-index $I = (i_1, \ldots, i_p)$ we set

$$\begin{aligned} dz^I &= dz^{i_1} \wedge \cdots \wedge dz^{i_p}, \\ d\bar{z}^I &= d\bar{z}^{i_1} \wedge \cdots \wedge d\bar{z}^{i_p}. \end{aligned}$$

Let $U \subset \mathbf{C}^n$ be an open subset. For integers $p, q \geq 0$, we denote by $\mathscr{E}^{(p,q)}(U)$ the set of all differential $(p+q)$-forms f of type

$$f = \sum_{|I|=p, |J|=q} f_{I\bar{J}} \, dz^I \wedge d\bar{z}^J, \quad f_{I\bar{J}} \in \mathscr{E}(U), \tag{3.5.14}$$

which is called a *differential form of type* (p, q) or *differential* (p, q)*-form* Then $\mathscr{E}^{(p,q)}(U)$ is an $\mathscr{E}(U)$-module. It follows from an elementary computation that

$$\mathscr{E}^{(r)}(U) = \bigoplus_{p+q=r} \mathscr{E}^{(p,q)}(U), \quad r \geq 0. \tag{3.5.15}$$

Let $\Omega \subset \mathbf{C}^n$ be a domain. It is clear that $\{(U, \mathscr{E}^{(p,q)}(U))\}$ with open subsets $U \subset \Omega$ forms a presheaf inducing $\mathscr{E}^{(p,q)}_\Omega$, a *sheaf of germs of differential forms of type* (p, q) over Ω. It follows from (3.5.15) that

$$\mathscr{E}^{(r)}_\Omega = \bigoplus_{p+q=r} \mathscr{E}^{(p,q)}_\Omega, \quad r \geq 0. \tag{3.5.16}$$

We sometimes simply write $\mathscr{E}^{(p,q)}$ for $\mathscr{E}^{(p,q)}_\Omega$.

Remark 3.5.17 The sheaf $\mathscr{E}^{(p,q)}$ is a fine sheaf, since it is a sheaf of $\mathscr{E}$-modules.

For f in (3.5.14) we set

$$\begin{aligned}\partial f &= \sum_{|I|=p,|J|=q}\sum_{i=1}^{n}\frac{\partial f_{I\bar{J}}}{\partial z^i}dz^i\wedge dz^I\wedge d\bar{z}^J,\\ \bar{\partial} f &= \sum_{|I|=p,|J|=q}\sum_{i=1}^{n}\frac{\partial f_{I\bar{J}}}{\partial \bar{z}^i}d\bar{z}^i\wedge dz^I\wedge d\bar{z}^J\\ &= \sum_{|I|=p,|J|=q}\sum_{i=1}^{n}(-1)^p\frac{\partial f_{I\bar{J}}}{\partial \bar{z}^i}dz^I\wedge d\bar{z}^i\wedge d\bar{z}^J.\end{aligned}$$

Then,

$$df = \partial f + \bar{\partial} f.$$

Assume $f \in \Gamma\left(U, \mathscr{E}_{\Omega}^{(p,0)}\right)$ and write $f = \sum_{|I|=p} f_I dz^I$. Then the following equivalence holds:

$$\begin{aligned}\bar{\partial} f = \sum_{|I|=p}\sum_{i=1}^{n}\frac{\partial f_I}{\partial \bar{z}^i}d\bar{z}^i\wedge dz^I = 0\\ \Longleftrightarrow \frac{\partial f_I}{\partial \bar{z}^i} = 0, \quad 1 \le i \le n\\ \Longleftrightarrow f_I \in \mathscr{O}(U).\end{aligned}$$

Differential forms of type $(p, 0)$ with coefficients of holomorphic functions are called *holomorphic p-forms* and the sheaf of germs of holomorphic p-forms over Ω is denoted by

(3.5.18) $$\mathscr{O}_{\Omega}^{(p)} = \sum_{|I|=p}\mathscr{O}_{\Omega}dz^I.$$

The following lemma will be necessary later for the analytic de Rham theorem.

Lemma 3.5.19 (Analytic Poincaré's Lemma) *If Ω is starlike with respect to a point $x_0 \in \Omega$, then for every $f \in \Gamma(\Omega, \mathscr{O}_{\Omega}^{(q)})$ $(q \ge 1)$ with $df = 0$ there is an element $g \in \Gamma(\Omega, \mathscr{O}_{\Omega}^{(q-1)})$ such that $f = dg$.*

Proof The proof is the same as that of Poincaré's Lemma 3.5.6, but here dx^i are replaced by dz^i. Although a real parameter t is involved, all coefficients are holomorphic functions in z. Therefore, $\theta f \in \Gamma(\Omega, \mathscr{O}_{\Omega}^{(q-1)})$, and

$$f = \theta df + d\theta f.$$

Thus, if $df = 0$, then $f = dg$ with $g = \theta f$. □

The following is obtained from this lemma.

Corollary 3.5.20 *The sequence below is a resolution of the constant sheaf* **C***:*

$$0 \to \mathbf{C} \to \mathscr{O}_\Omega = \mathscr{O}_\Omega^{(0)} \xrightarrow{d} \mathscr{O}_\Omega^{(1)} \xrightarrow{d} \cdots \xrightarrow{d} \mathscr{O}_\Omega^{(n)} \xrightarrow{d} 0.$$

N.B. In this case, $d = \partial$. The above resolution is called the *analytic de Rham resolution*. We have

$$\text{(3.5.21)}\quad 0 \to \mathbf{C} \to \Gamma(\Omega, \mathscr{O}_\Omega) \xrightarrow{d} \Gamma(\Omega, \mathscr{O}_\Omega^{(1)}) \xrightarrow{d} \cdots \xrightarrow{d} \Gamma(\Omega, \mathscr{O}_\Omega^{(n)}) \xrightarrow{d} 0,$$

which is called the *analytic de Rham complex*. Without having $H^p(\Omega, \mathscr{O}_\Omega^{(q)}) = 0$ for $p \geq 1$, the cohomology $\mathscr{H}^p(\Omega, \{\mathscr{O}_\Omega^{(q)}\})$ defined by this sequence is not known to be isomorphic to $H^p(\Omega, \mathbf{C})$.

Example 3.5.22 Let $\Omega = \mathbf{C}^2 \setminus \{0\}$. By (1.2.28) Ω is neither holomorphically convex nor a domain of holomorphy by Theorem 1.2.26. Because of dimension 2 we have that $\mathscr{O}_\Omega^{(3)} = 0$, so that

$$\mathscr{H}^3(\Omega, \{\mathscr{O}_\Omega^{(q)}\}) = 0.$$

On the other hand, we set (cf. 1.2.1)

$$\begin{aligned} f &= dd^c \log \|z\|^2 \wedge d^c \log \|z\|^2, \\ z &= (z_1, z_2) \in \Omega, \quad \|z\| = \sqrt{|z_1|^2 + |z_2|^2}. \end{aligned}$$

Put $w = z_2/z_1$ with $z_1 \neq 0$. Then we have

$$\begin{aligned} & dd^c \log(1 + |w|^2) \wedge d^c \log(1 + |w|^2) \\ & = \frac{i}{2\pi} \frac{dw \wedge d\bar{w}}{(1 + |w|^2)^2} \wedge \frac{i}{4\pi} \left(\frac{w d\bar{w}}{1 + |w|^2} - \frac{\bar{w} dw}{1 + |w|^2} \right) = 0, \\ & dd^c \log(1 + |w|^2) \wedge dd^c \log(1 + |w|^2) = 0, \\ f & = dd^c \log(1 + |w|^2) \wedge \left(d^c \log(1 + |w|^2) + d^c \log |z_1|^2\right) \\ & = dd^c \log(1 + |w|^2) \wedge d^c \log |z_1|^2. \end{aligned}$$

Therefore, $df = 0$. With $z_1 = r_1 e^{i\theta_1}$ we have

$$\begin{aligned} d^c \log |z_1|^2 &= \frac{1}{4\pi} \left\{ \frac{\partial \log r_1^2}{\partial(\log r_1)} d\theta_1 - \frac{\partial \log r_1^2}{\partial \theta_1} d(\log r_1) \right\} = \frac{1}{2\pi} d\theta_1, \\ f &= \frac{i}{2\pi} \frac{dw \wedge d\bar{w}}{(1 + |w|^2)^2} \wedge \frac{1}{2\pi} d\theta_1. \end{aligned}$$

Let $S = \{\|z\| = 1\} \subset \Omega$, and put $S' := S \setminus \{z_1 = 0\}$. Introducing a coordinate system $(w, \theta_1) \in \mathbf{C} \times [0, 2\pi)$ in S', we have

$$\int_S f = \int_{S'} f = \int_{w \in \mathbf{C}} \int_{0 \leq \theta_1 2\pi} \frac{i}{2\pi} \frac{dw \wedge d\bar{w}}{(1 + |w|^2)^2} \wedge \frac{1}{2\pi} d\theta_1 = 1.$$

If the cohomology class $[f] = 0$ in $H^3_{\mathrm{DR}}(\Omega, \mathbf{C}) \cong H^3(\Omega, \mathbf{C})$ (de Rham Theorem 3.5.11), then there is a 2-form $g \in \Gamma(\Omega, \mathscr{E}^2)$ with $f = dg$, so that $\int_S f = \int_S dg = 0$. Thus, $[f] \neq 0$, and

$$H^3(\Omega, \mathbf{C}) \neq 0.$$

3.6 Dolbeault Cohomology

We next describe the fundamental result of Dolbeault cohomology. We keep the notation of the previous section.

We consider the following complex:

$$(3.6.1) \qquad 0 \to \mathscr{O}^{(p)}_\Omega \to \mathscr{E}^{(p,0)}_\Omega \xrightarrow{\bar\partial} \mathscr{E}^{(p,1)}_\Omega \xrightarrow{\bar\partial} \cdots \xrightarrow{\bar\partial} \mathscr{E}^{(p,n)}_\Omega \to 0.$$

We are going to show that this is exact. Sequence (3.6.1) is called the *Dolbeault resolution.*

An *integral transform* $\mathrm{T}\psi(z)$ of a function ψ over $\mathbf{C}$ is defined by

$$(3.6.2) \qquad \mathrm{T}\psi(z) = \frac{1}{2\pi i}\int_{\mathbf{C}} \frac{\psi(\zeta)}{\zeta - z} d\zeta \wedge d\bar\zeta,$$

where the integrability condition is assumed.

Lemma 3.6.3 *Let ψ be a C^∞ function on $\mathbf{C}$ with compact support. Then, $\mathrm{T}\psi$ is of C^∞-class and satisfies*

$$\frac{\partial \mathrm{T}\psi}{\partial \bar z} = \psi.$$

Proof Rewriting ζ for $\zeta - z$ in (3.6.2), we have

$$\mathrm{T}\psi(z) = \frac{1}{2\pi i}\int_{\mathbf{C}} \frac{\psi(\zeta + z)}{\zeta} d\zeta \wedge d\bar\zeta.$$

With the polar coordinate $\zeta = re^{i\theta}$ it follows that

$$\begin{aligned}
\mathrm{T}\psi(z) &= \frac{1}{2\pi i}\int_0^\infty \int_0^{2\pi} \frac{\psi(re^{i\theta} + z)}{re^{i\theta}}(-2i) r dr d\theta \\
&= \frac{-1}{\pi}\int_0^\infty dr \int_0^{2\pi} d\theta\, \psi(re^{i\theta} + z)e^{-i\theta}.
\end{aligned}$$

By this we see that $T\psi$ is of C^∞-class. We calculate the following by making use of Stokes' Theorem:

$$\begin{aligned}
\frac{\partial T\psi}{\partial \bar z} &= \frac{1}{2\pi i}\int_{\mathbf{C}} \frac{1}{\zeta}\frac{\partial \psi(\zeta+z)}{\partial \bar z} d\zeta \wedge d\bar\zeta = \frac{1}{2\pi i}\int_{\mathbf{C}} \frac{\frac{\partial}{\partial \bar\zeta}\psi(\zeta+z)}{\zeta} d\zeta \wedge d\bar\zeta \\
&= \frac{-1}{2\pi i}\int_{\mathbf{C}} \frac{\partial}{\partial \bar\zeta}\left(\frac{\psi(\zeta+z)}{\zeta}\right) d\bar\zeta \wedge d\zeta = \frac{i}{2\pi}\int_{\mathbf{C}} \bar\partial_\zeta\left(\frac{\psi(\zeta+z)}{\zeta}\right) \wedge d\zeta \\
&= \frac{i}{2\pi}\int_{\mathbf{C}} d_\zeta\left(\frac{\psi(\zeta+z)}{\zeta} d\zeta\right) \quad \left[\text{since } \operatorname{Supp}\psi \Subset \mathbf{C}, \text{ for } R \gg 0\right] \\
&= \frac{i}{2\pi}\int_{|\zeta|=R} \frac{\psi(\zeta+z)}{\zeta} d\zeta - \lim_{\varepsilon\to+0}\frac{i}{2\pi}\int_{|\zeta|=\varepsilon} \frac{\psi(\zeta+z)}{\zeta} d\zeta \\
&= \lim_{\varepsilon\to+0}\frac{1}{2\pi i}\int_{|\zeta|=\varepsilon} \frac{\psi(\zeta+z)}{\zeta} d\zeta = \lim_{\varepsilon\to+0}\frac{1}{2\pi}\int_0^{2\pi} \psi(z+\varepsilon e^{i\theta}) d\theta \\
&= \psi(z).
\end{aligned}$$

□

Lemma 3.6.4 (Dolbeault) *Let $\Omega = \prod_{j=1}^n \Omega_j \Subset \mathbf{C}^n$ be a bounded convex cylinder domain. Let $U = \prod_{j=1}^n U_j$ be a neighborhood of the closure $\bar\Omega$ and let $f \in \Gamma(U, \mathscr{E}_U^{(p,q)})$ such that $\bar\partial f = 0$. Then there is an element $g \in \Gamma(\mathbf{C}^n, \mathscr{E}_{\mathbf{C}^n}^{(p,q-1)})$ such that*

$$\bar\partial g|_\Omega = f|_\Omega .$$

In particular, sequence (3.6.1) with a general domain $\Omega(\subset \mathbf{C}^n)$ is exact, and gives a fine resolution of $\mathscr{O}_\Omega^{(p)}$.

Proof Take a C^∞ function $\chi(z) \geq 0$ of product type $\chi(z) = \prod_{j=1}^n \chi_j(z_j)$ such that $\operatorname{Supp}\chi_j \Subset U_j$ and $\chi_j|_{\Omega_j} = 1$, and put $\hat f = \chi f \in \Gamma(\mathbf{C}^n, \mathscr{E}^{(p,q)})$. Write

$$\hat f = \sum_{|I|=p, |J|=q} \hat f_{I\bar J}\, dz^I \wedge d\bar z^J .$$

Let $1 \leq k \leq n$, and let $J = \{j_1, \ldots, j_q\}$ be a multi-index such that

$$k \leq j_1 < \cdots < j_q \leq n.$$

We proceed by induction on k.

The case of $k = n$: In this case, $q = 1$ and we get

$$\hat f = \sum_{|I|=p} \hat f_{I\bar n}\, dz^I \wedge d\bar z^n .$$

Taking the integral transform in variable z^n, we have

$$\mathrm{T}_n \hat{f}_{I\bar{n}}(z^1, \ldots, z^n) = \frac{1}{2\pi i} \int_{\zeta^n \in \mathbf{C}} \frac{\hat{f}_{I\bar{n}}(z^1, \ldots, z^{n-1}, \zeta^n)}{\zeta^n - z^n} d\zeta^n \wedge d\bar{\zeta}^n.$$

Since $\chi|_\Omega = 1$,

$$\hat{f}(z) = f, \quad \bar{\partial}\hat{f}(z) = 0, \qquad z \in \Omega.$$

On Ω we see that

$$\begin{aligned}
\bar{\partial}\hat{f} &= \sum_{|I|=p} \sum_{i=1}^{n} \frac{\partial \hat{f}_{I\bar{n}}}{\partial \bar{z}^i} d\bar{z}^i \wedge dz^I \wedge d\bar{z}^n \\
&= \sum_{|I|=p} \sum_{i=1}^{n-1} \frac{\partial \hat{f}_{I\bar{n}}}{\partial \bar{z}^i} (-1)^p dz^I \wedge d\bar{z}^i \wedge d\bar{z}^n \\
&= 0 \Longleftrightarrow \frac{\partial \hat{f}_{I\bar{n}}}{\partial \bar{z}^i} = 0, \quad 1 \leq i \leq n-1, \\
&\Longleftrightarrow \hat{f}_{I\bar{n}}(z^1, \ldots, z^n) \text{ is holomorphic in } z^1, \ldots, z^{n-1}.
\end{aligned}$$

Therefore, $\mathrm{T}_n \hat{f}_{I\bar{n}}(z^1, \ldots, z^n)$ is holomorphic in $(z^1, \ldots, z^{n-1})$ on Ω. Set

$$g = \sum_{|I|=p} (-1)^p (\mathrm{T}_n \hat{f}_{I\bar{n}}) dz^I \in \Gamma(\mathbf{C}^n, \mathscr{E}^{(p,0)}).$$

By Lemma 3.6.3 we have

$$\begin{aligned}
\bar{\partial} g &= \sum_{|I|=p} (-1)^p \frac{\partial \mathrm{T}_n \hat{f}_{I\bar{n}}}{\partial \bar{z}^n} d\bar{z}^n \wedge dz^I \\
&= \sum_{|I|=p} f_{I\bar{n}} dz^I \wedge \bar{z}^n = f \quad (\text{on } \Omega).
\end{aligned}$$

The case of $k \leq n-1$: Assume that the statement holds for $k+1$, and let $J = \{j_1, \ldots, j_q\}$ with $k \leq j_1 < \cdots < j_q \leq n$. We take a convex cylinder domain Ω' so that $\Omega \Subset \Omega' \Subset U$, and we may assume that $\chi|_{\Omega'} = 1$. Decompose $\hat{f}$ as follows:

$$(3.6.5) \qquad \begin{aligned}
\hat{f} &= \sum_{\substack{|I|=p \\ |J'|=q-1, k \notin J'}} \hat{f}_{I\bar{J}'} dz^I \wedge d\bar{z}^k \wedge d\bar{z}^{J'} + \sum_{\substack{|I|=p \\ |J|=q, k \notin J}} \hat{f}_{I\bar{J}} \, dz^I \wedge d\bar{z}^J \\
&= f_{(1)} + f_{(2)}.
\end{aligned}$$

Since $\bar{\partial}\hat{f} = \bar{\partial}f = 0$ on Ω', we have

$$\bar{\partial}\hat{f} = \sum_{\substack{|I|=p \\ |J'|=q-1, k\notin J'}} \sum_{i=1}^{n} \frac{\partial \hat{f}_{I\bar{J}'}}{\partial \bar{z}^i} d\bar{z}^i \wedge dz^I \wedge d\bar{z}^k \wedge d\bar{z}^{J'}$$
$$+ \sum_{\substack{|I|=p \\ |J|=q, k\notin J}} \sum_{i=1}^{n} \frac{\partial \hat{f}_{I\bar{J}}}{\partial \bar{z}^i} d\bar{z}^i \wedge dz^I \wedge d\bar{z}^J = 0.$$

Hence, $\frac{\partial \hat{f}_{I\bar{J}'}}{\partial \bar{z}^i} = 0$, $1 \leq i \leq k-1$. We see that $\hat{f}_{I\bar{J}'}$ is holomorphic in $(z^1, \ldots, z^{k-1})$ on Ω'. By making use of the integral transform we set

$$G = \sum_{\substack{|I|=p \\ |J'|=q-1, k\notin J'}} \mathrm{T}_k \hat{f}_{I\bar{J}'} (-1)^p dz^I \wedge d\bar{z}^{J'}.$$

Note that $\mathrm{T}_k \hat{f}_{I\bar{J}'}$ is holomorphic in $(z^1, \ldots, z^{k-1})$. Using Lemma 3.6.3, we compute as follows:

$$\begin{aligned}
\bar{\partial}G &= \sum_{\substack{|I|=p \\ |J'|=q-1, k\notin J'}} \sum_{i=1}^{n} \frac{\mathrm{T}_k \hat{f}_{I\bar{J}'}}{\partial \bar{z}^i} (-1)^p d\bar{z}^i \wedge dz^I \wedge d\bar{z}^{J'} \\
&= \sum_{\substack{|I|=p \\ |J'|=q-1, k\notin J'}} \frac{\mathrm{T}_k \hat{f}_{I\bar{J}'}}{\partial \bar{z}^k} dz^I \wedge d\bar{z}^k \wedge d\bar{z}^{J'} \\
&\quad + \sum_{\substack{|I|=p \\ |J'|=q-1, k\notin J'}} \sum_{i=k+1}^{n} \frac{\mathrm{T}_k \hat{f}_{I\bar{J}'}}{\partial \bar{z}^i} dz^I \wedge d\bar{z}^i \wedge d\bar{z}^{J'} \\
&= f_{(1)} + h_{(1)}.
\end{aligned}$$

Here, $h_{(1)}$ represents the second term in the last equation, and $f_{(1)}$ is the one set in (3.6.5). Put

$$h_{(2)} := \hat{f} - \bar{\partial}G = f_{(2)} - h_{(1)}. \tag{3.6.6}$$

Then, $h_{(2)}$ does not contain $d\bar{z}^1, \ldots, d\bar{z}^k$, and $\bar{\partial}h_{(2)} = \bar{\partial}\hat{f} - \bar{\partial}\bar{\partial}G = \bar{\partial}f = 0$ on Ω'. Therefore, by the induction hypothesis, there is an element $g_{(2)} \in \Gamma(\mathbf{C}^n, \mathscr{E}^{(p,q-1)})$ such that $\bar{\partial}g_{(2)} = h_{(2)}$ on Ω. Setting $g = G + g_{(2)} \in \Gamma(\mathbf{C}^n, \mathscr{E}^{(p,q-1)})$, we have by (3.6.6)

$$\bar{\partial}g = \bar{\partial}G + \bar{\partial}g_2 = f - h_{(2)} + h_{(2)} = f$$

on Ω. This finishes the proof. □

For $p, q \geq 0$ we have by (3.6.1)

(3.6.7)
$$0 \to \Gamma(\Omega, \mathscr{O}_\Omega^{(p)}) \to \Gamma(\Omega, \mathscr{E}_\Omega^{(p,0)}) \xrightarrow{\bar\partial} \Gamma(\Omega, \mathscr{E}_\Omega^{(p,1)}) \xrightarrow{\bar\partial} \cdots \xrightarrow{\bar\partial} \Gamma(\Omega, \mathscr{E}_\Omega^{(p,n)}) \to 0,$$

which is called the *Dolbeault complex*. We define the *Dolbeault cohomology* (of type (p, q)) by

(3.6.8)
$$H^q_{\bar\partial}(\Omega, \mathscr{O}_\Omega^{(p)}) = \{f \in \Gamma(\Omega, \mathscr{E}_\Omega^{(p,q)}); \bar\partial f = 0\}/\bar\partial\Gamma(\Omega, \mathscr{E}_\Omega^{(p,q-1)}).$$

Lemma 3.6.4 and Theorem 3.4.30 imply the following:

Theorem 3.6.9 (Dolbeault's Theorem) *For a domain $\Omega \subset \mathbf{C}^n$ we have*

$$H^q(\Omega, \mathscr{O}_\Omega^{(p)}) \cong H^q_{\bar\partial}(X, \mathscr{O}_\Omega^{(p)}).$$

In particular, if $q > n$ or $p > n$, then $H^q(\Omega, \mathscr{O}_\Omega^{(p)}) = 0$.

By making use of this theorem we show a prototype of the fundamental vanishing theorem of cohomology.

Theorem 3.6.10 *For a convex cylinder domain $\Omega \subset \mathbf{C}^n$ we have*

$$H^q(\Omega, \mathscr{O}_\Omega) = 0, \qquad q \geq 1.$$

Proof Here we abbreviate Ω in the notation $\mathscr{E}_\Omega^{(p,q)}$ for simplicity. Theorem 3.6.9 says that

$$H^q(\Omega, \mathscr{O}_\Omega) \cong \{f \in \Gamma(\Omega, \mathscr{E}^{(0,q)}); \bar\partial f = 0\}/\bar\partial\Gamma(\Omega, \mathscr{E}^{(0,q-1)}).$$

Let $\{\Omega_\nu\}$ be an increasing open covering by convex cylinder subdomains of Ω such that $\Omega_\nu \Subset \Omega_{\nu+1}$ and $\Omega = \bigcup_{\nu=1}^\infty \Omega_\nu$.

Take any $f \in \Gamma(\Omega, \mathscr{E}^{(0,q)})$ with $\bar\partial f = 0$. It suffices to find $g \in \Gamma(\Omega, \mathscr{E}^{(0,q-1)})$ such that $\bar\partial g = f$. Applying Lemma 3.6.4 on each Ω_ν, we get

(3.6.11)
$$h_\nu \in \Gamma(\mathbf{C}^n, \mathscr{E}^{(0,q-1)}), \quad \bar\partial h_\nu|_{\Omega_\nu} = f|_{\Omega_\nu}, \quad \nu = 2, 3, \ldots.$$

The argument henceforth is divided into two cases, depending on whether $q \geq 2$ or $q = 1$. The case of $q = 1$ is essentially more difficult and needs Runge's approximation theorem. (This situation does not change in extending Ω and $\mathscr{O}_\Omega$ to Stein spaces and coherent sheaves over them.)

The case of $q \geq 2$: Set $g_2 = h_2$. Assume that $g_2, \ldots, g_\nu$ have been defined so that

(3.6.12)
$$\bar\partial g_\mu|_{\Omega_\mu} = f|_{\Omega_\mu}, \quad \mu = 2, 3, \ldots, \nu,$$
$$g_\mu|_{\Omega_{\mu-2}} = g_{\mu-1}|_{\Omega_{\mu-2}}, \quad \mu = 3, 4, \ldots, \nu.$$

Since $\bar{\partial}(h_{\nu+1} - g_\nu)|_{\Omega_\nu} = 0$, by Lemma 3.6.4 we can take $\alpha_{\nu+1} \in \Gamma(\Omega, \mathscr{E}^{(0,q-2)})$ such that $\bar{\partial}\alpha_{\nu+1}|_{\Omega_{\nu-1}} = (h_{\nu+1} - g_\nu)|_{\Omega_{\nu-1}}$. Set

$$g_{\nu+1} = h_{\nu+1} - \bar{\partial}\alpha_{\nu+1} \in \Gamma(\Omega, \mathscr{E}^{(0,q-1)}).$$

It follows from the construction that

$$\begin{aligned} \bar{\partial}g_{\nu+1}|_{\Omega_{\nu+1}} &= (\bar{\partial}h_{\nu+1} - \bar{\partial}\bar{\partial}\alpha_{\nu+1})|_{\Omega_{\nu+1}} = f|_{\Omega_{\nu+1}}, \\ g_{\nu+1}|_{\Omega_{\nu-1}} &= h_{\nu+1}|_{\Omega_{\nu-1}} - \bar{\partial}\alpha_{\nu+1}|_{\Omega_{\nu-1}} \\ &= h_{\nu+1}|_{\Omega_{\nu-1}} - (h_{\nu+1} - g_\nu)|_{\Omega_{\nu-1}} = g_\nu|_{\Omega_{\nu-1}}. \end{aligned}$$

Therefore, we obtain $g_{\nu+1}$ satisfying (3.6.12) with $\mu = \nu + 1$. Setting

$$g = \lim_{\nu \to \infty} g_\nu \in \Gamma(\Omega, \mathscr{E}^{(0,q-1)}),$$

we get $\bar{\partial}g = f$.

The case of $q = 1$: Let $h_\nu \in \Gamma(\Omega, \mathscr{E})$ ($\mathscr{E} = \mathscr{E}^{(0,0)}$) be those taken in (3.6.11) such that $\bar{\partial}h_\nu|_{\Omega_\nu} = f|_{\Omega_\nu}$, $\nu = 2, 3, \ldots$.

Set $g_2 = h_2$. Assume that $g_\mu \in \Gamma(\Omega_\mu, \mathscr{E})$, $\mu = 2, \ldots, \nu$, are taken as follows:

(3.6.13)
$$\begin{aligned} &\bar{\partial}g_\mu|_{\Omega_\mu} = f|_{\Omega_\mu}, \quad \mu = 2, 3, \ldots, \nu, \\ &\|g_\mu - g_{\mu-1}\|_{\Omega_{\mu-2}} := \sup\{|g_\mu(z) - g_{\mu-1}(z)|; z \in \Omega_{\mu-2}\} < \frac{1}{2^{\mu-2}}, \\ &\qquad \mu = 3, 4, \ldots, \nu. \end{aligned}$$

Since $\bar{\partial}(h_{\nu+1} - g_\nu)|_{\Omega_\nu} = (f - f)|_{\Omega_\nu} = 0$, $h_{\nu+1} - g_\nu$ is holomorphic in Ω_ν. By Corollary 1.2.22 (Runge's Approximation) there are polynomials $P_{\nu+1}(z^1, \ldots, z^n)$ such that

$$\|h_{\nu+1} + P_{\nu+1} - g_\nu\|_{\Omega_{\nu-1}} < \frac{1}{2^{\nu-1}}.$$

Set $g_{\nu+1} = h_{\nu+1} + P_{\nu+1}$. Then (3.6.13) is satisfied with $\mu = \nu + 1$. The following series,

$$\begin{aligned} g &= g_2 + (g_3 - g_2) + \cdots + (g_{\mu+1} - g_\mu) + \cdots \\ &= g_\nu + \sum_{\mu=\nu}^{\infty} (g_{\mu+1} - g_\mu) \end{aligned}$$

converges normally in Ω and defines $g \in \Gamma(\Omega, \mathscr{E})$, which satisfies $\bar{\partial}g = f$. □

Remark 3.6.14 In fact, Theorem 3.6.10 can be proved in the same way as above for an arbitrary cylinder domain; however, it is not necessary here. It is just sufficient to show this easy case of convex cylinder domains, and the rest is dealt with by making use of Oka's First Coherence Theorem to prove the cohomology vanishing of coherent sheaves over holomorphically convex domains and domains of holomorphy as well. This is the path that we take in this book.

3.7 Cousin Problems

With the preparations as above we now come to the situation of the eve of the day when Kiyoshi Oka wrote the first paper Oka I (1938). It is interesting to know what is the original motivation of the study, and it may be a good reference to know it; it will also be helpful for those who wish to begin with something new. For more details, cf. Sect. 5.5.

3.7.1 Cousin I Problem

Let $\Omega \subset \mathbf{C}^n$ be a domain, and let $\Omega = \bigcup_\alpha U_\alpha$ be an open covering. On each U_α a meromorphic function $f_\alpha \in \Gamma(U_\alpha, \mathscr{M}_\Omega)$ is given so that

$$f_\alpha|_{U_\alpha \cap U_\beta} - f_\beta|_{U_\alpha \cap U_\beta} \in \Gamma(U_\alpha \cap U_\beta, \mathscr{O}_\Omega)$$

is satisfied. Under this assumption it was the *Cousin I Problem* to ask for the existence of $F \in \Gamma(\Omega, \mathscr{M}_\Omega)$ such that for every U_α

$$F|_{U_\alpha} - f_\alpha \in \Gamma(U_\alpha, \mathscr{O}_\Omega).$$

In the case of one variable this was nothing but Mittag-Leffler's Theorem (cf. Theorem 5.5.1).

Setting $f_{\alpha\beta} = f_\alpha|_{U_\alpha \cap U_\beta} - f_\beta|_{U_\alpha \cap U_\beta}$, we have $(f_{\alpha\beta}) \in Z^1(\{U_\alpha\}, \mathscr{O}_\Omega)$. If Ω is a convex cylinder domain Theorem 3.6.10 implies that $[(f_{\alpha\beta})] \in H^1(\Omega, \mathscr{O}_\Omega) = 0$, and hence by Proposition 3.4.11, there is an element $(g_\alpha) \in C^0(\{U_\alpha\}, \mathscr{O}_\Omega)$ such that

$$f_{\alpha\beta} = g_\beta - g_\alpha.$$

Therefore, $f_\alpha - f_\beta = g_\beta - g_\alpha$, and on $U_\alpha \cap U_\beta$, $f_\alpha + g_\alpha = f_\beta + g_\beta$ holds. Defining $F = f_\alpha + g_\alpha$ on U_α, we obtain a meromorphic function F in Ω such that

$$F|_{U_\alpha} - f_\alpha = g_\alpha \in \Gamma(U_\alpha, \mathscr{O}_\Omega).$$

Thus, this gives a solution of the Cousin I Problem.

In fact, P. Cousin had solved this problem for any cylinder domain.

3.7.2 Cousin II Problem

We consider the sheaf $\mathscr{M}^*_\Omega$ of germs of non-zero meromorphic functions over $\Omega \subset \mathbf{C}^n$, which is a sheaf of (multiplicative) abelian groups. There is a natural inclusion, $\mathscr{O}^*_\Omega \hookrightarrow \mathscr{M}^*_\Omega$. The quotient sheaf $\mathscr{D}_\Omega := \mathscr{M}^*_\Omega / \mathscr{O}^*_\Omega$ is called the *sheaf of germs of divisors*, and a section of $\mathscr{D}$ over Ω is called a *divisor* or a *Cousin II distribution* on Ω. There is a short exact sequence:

$$
(3.7.1) \qquad \begin{array}{ccccccccc} 0 & \longrightarrow & \mathscr{O}^*_\Omega & \longrightarrow & \mathscr{M}^*_\Omega & \longrightarrow & \mathscr{D}_\Omega & \longrightarrow & 0. \\ & & & & \Psi & & \Psi & & \\ & & & & \phi & \longmapsto & [\phi] & & \end{array}
$$

Then the *Cousin II Problem* was to ask if the sequence

$$
\Gamma(\Omega, \mathscr{M}^*_\Omega) \longrightarrow \Gamma(\Omega, \mathscr{D}_\Omega) \longrightarrow 0
$$

is exact. In the case of one variable, this was Weierstrass' Theorem (cf. Theorem 5.5.4). That is, for any given divisor $\gamma \in \Gamma(\Omega, \mathscr{D}_\Omega)$ there are an open covering $\mathscr{U} = \{U_\alpha\}$ of Ω and $\phi_\alpha \in \Gamma(U_\alpha, \mathscr{M}^*_\Omega)$ such that

$$
(3.7.2) \qquad \phi_{\alpha\beta} := \phi_\beta / \phi_\alpha \in \Gamma(U_\alpha \cap U_\beta, \mathscr{O}^*_\Omega), \quad [\phi_\alpha] = \gamma|_{U_\alpha}.
$$

The Cousin II Problem asks for the existence of a meromorphic function $G \in \Gamma(\Omega, \mathscr{M}^*_\Omega)$ over Ω such that for every U_α

$$
\frac{G|_{U_\alpha}}{\phi_\alpha} \in \Gamma(U_\alpha, \mathscr{O}^*_\Omega).
$$

The short exact sequence (3.7.1) induces the following long exact sequence:

$$
(3.7.3) \qquad \begin{aligned} 0 \to H^0(\Omega, \mathscr{O}^*_\Omega) &\longrightarrow H^0(\Omega, \mathscr{M}^*_\Omega) \longrightarrow H^0(\Omega, \mathscr{D}_\Omega) \\ &\xrightarrow{\delta_*} H^1(\Omega, \mathscr{O}^*_\Omega) \longrightarrow H^1(\Omega, \mathscr{M}^*_\Omega) \longrightarrow \cdots . \end{aligned}
$$

By (3.7.2), $\phi_{\alpha\beta}$'s define $(\phi_{\alpha\beta}) \in Z^1(\mathscr{U}, \mathscr{O}^*_\Omega)$. The cohomology class $[(\phi_{\alpha\beta})] \in H^1(\mathscr{U}, \mathscr{O}^*_\Omega)$ satisfies $\delta_*(\gamma) = [(\phi_{\alpha\beta})] \in H^1(\mathscr{U}, \mathscr{O}^*_\Omega)$ in (3.7.3). Since $H^1(\mathscr{U}, \mathscr{O}^*_\Omega) \hookrightarrow H^1(\Omega, \mathscr{O}^*_\Omega)$ is injective by Proposition 3.4.11, the condition

$$
(3.7.4) \qquad \beta := \delta_*(\gamma) = 0 \in H^1(\Omega, \mathscr{O}^*_\Omega)
$$

gives rise to a necessary and sufficient condition for the Cousin II Problem to be solvable. In fact, the necessity follows from the definition. For the converse, if $\beta = 0$, then there is an element $\eta_\alpha \in \Gamma(U_\alpha, \mathscr{O}^*)$ on each U_α such that

$$\text{(3.7.5)} \qquad \frac{\eta_\beta|_{U_\alpha\cap U_\beta}}{\eta_\alpha|_{U_\alpha\cap U_\beta}} = \phi_{\alpha\beta} = \frac{\phi_\beta}{\phi_\alpha} \in \Gamma(U_\alpha \cap U_\beta, \mathscr{O}^*).$$

Therefore, setting $G = \phi_\alpha \eta_\alpha^{-1}$, we obtain a solution $G \in \Gamma(\Omega, \mathscr{M}^*_\Omega)$.

Here we consider the long exact sequence of Example 3.4.27:

$$\text{(3.7.6)} \qquad \begin{array}{ccccccc} H^1(\Omega, \mathscr{O}_\Omega) \to & H^1(\Omega, \mathscr{O}^*_\Omega) & \xrightarrow{\delta_*} & H^2(\Omega, \mathbf{Z}) & \to H^2(\Omega, \mathscr{O}_\Omega) \to \cdots. \\ & \Psi & & \Psi & \\ & \beta & \longmapsto & \delta_*(\beta) = c_1(\beta) & \end{array}$$

The second cohomology $c_1(\gamma) := -\delta_*(\beta) \in H^2(\Omega, \mathbf{Z})$ (minus sign "−" is attached) is called the *first Chern class* of γ. The first Chern class $c_1(\gamma)$ is a topological invariant. If there is a solution G, then $\beta = 0$, and so $c_1(\gamma) = 0$.

Now, as in the case of convex cylinder domains, we assume that $H^q(\Omega, \mathscr{O}_\Omega) = 0$, $q \geq 1$ (Theorem 3.6.10). Then by (3.7.6) the following is exact:

$$0 \longrightarrow H^1(\Omega, \mathscr{O}^*_\Omega) \longrightarrow H^2(\Omega, \mathbf{Z}) \longrightarrow 0.$$

Therefore, $H^1(\Omega, \mathscr{O}^*_\Omega) \cong H^2(\Omega, \mathbf{Z})$. In this case the necessity condition $c_1(\gamma) = 0$ implies $\beta = 0$, and hence the Cousin II Problem is solvable. The *Oka Principle* states that the existence of analytic solutions is determined by a topological condition just as in the above case.

Cousin had solved this problem on a cylinder domain $\Omega = \prod \Omega_j$ such that except for one Ω_j all the other Ω_j are simply connected; in fact, $H^2(\Omega, \mathbf{Z}) = 0$ in this case (cf. Remark 3.6.14).

The Cousin II Problem originally came from the problem of the fractional presentation of meromorphic functions due to J.H. Poincaré: That is, for a given non-zero meromorphic function f on a domain Ω, it is asked if there exist holomorphic functions $g, h \in \mathscr{O}(\Omega)$ such that the germs $\underline{g}_a, \underline{h}_a$ are coprime at every point $a \in \Omega$ (cf. Theorem 2.2.12) and

$$\text{(3.7.7)} \qquad f = \frac{g}{h}.$$

In Chap. 4 we will prove Theorem 3.6.10 for every coherent sheaf over holomorphically convex domains, and then in Chap. 5 it will be proved over domains of holomorphy. The proof does not follow the way that Oka wrote (Oka I, Oka II, ...) but rather the opposite, starting with Oka VII and then going backwards; such a way was already mentioned in the introduction of Oka VII.

Remark 3.7.8 In the cohomology theory, of importance is the first cohomology $H^1(\Omega, \mathscr{S})$. This determines whether the problem is solvable globally. Readers may wonder why one deals with higher cohomologies $H^q(\Omega, \mathscr{S})$, $q = 2, 3, \ldots$, which look complicated at a glance. Here is the good point of the cohomology theory; after preparing the cohomology theory for general degrees, the problem is reduced to that of a higher cohomology of another sheaf that is easier to handle, and then it leads to a solution.

Historical Supplements

H. Cartan observed that Oka's concept of "ideals of undetermined domains" ("idéal de domaines indéterminés") was the same as the concept of "sheaf" in J. Leray's papers published around the same time (C.R. Paris **222** (1946), J. Math. pure appl. **29** (1950)), and developed the theory of sheaf cohomology together with J.-P. Serre. The Cousin Problems are those of $H^1(X, \mathscr{S})$ ($\mathscr{S} = \mathscr{O}_X, \mathscr{O}_X^*$). In the paper of the proceedings of a meeting at Bruxelles 1953 ([11] Sect. 7) Cartan was writing "it is due to Serre that one should consider $H^q(X, \mathscr{S})$ for general $q \geq 1$".

The theory of sheaf cohomology will be applied to "Oka's Syzygies" (cf. (4.3.4)) due to Oka VII (1948, [62]) in the proof of the Oka–Cartan Fundamental Theorem 4.4.2. K. Oka did not use the theory of sheaf cohomology, but rather directly worked on his "Syzygies" to obtain solutions.

Exercises

1. Show that $g \wedge f = (-1)^{qr} f \wedge g$ in (3.5.3).
2. For differential q-form dx^I with multi-index $I = (i_1, \ldots, i_q)$ and q vector fields, $\frac{\partial}{\partial x^{j_1}}, \ldots, \frac{\partial}{\partial x^{j_q}}$ with $1 \leq j_1, \ldots, j_q \leq n$, we put

$$(3.7.9) \qquad dx^I \left(\frac{\partial}{\partial x^{j_1}}, \ldots, \frac{\partial}{\partial x^{j_q}} \right) = \det \left(\delta_{i_\nu j_\mu} \right)_{\nu, \mu}$$

$$= \begin{cases} \operatorname{sgn} \begin{pmatrix} i_1 & \cdots & i_q \\ j_1 & \cdots & j_q \end{pmatrix}, & \text{if } I = J \text{ as sets;} \\ 0, & \text{otherwise.} \end{cases}$$

Let $\mathscr{X}(X)$ denote the $\mathscr{E}(X)$-module of all C^∞ vector fields over X.

Show that $\mathscr{E}^{(q)}(X)$ is isomorphic as $\mathscr{E}(X)$-module to the module of all alternative multi-linear maps extended linearly from (3.7.9),

$$\eta(\xi_1, \ldots, \xi_q) \in \mathscr{E}^{(q)}(X) \times (\mathscr{X}(X))^q \to \mathscr{E}(X).$$

3. Let X be a starlike domain of $\mathbf{R}^n$ with respect to a point $a \in X$. Then, show that

$$H^q(X, \mathbf{R}) = 0, \quad q \geq 1.$$

4. Give a complete proof of Lemma 3.5.19.
5. Show that the existence of expression (3.7.7) follows from the Cousin II Problem (cf. Proposition 2.2.19).

6. Show that

$$\dim_{\mathbf{C}} H^1(\mathbf{C}^2 \setminus \{0\}) = \infty,$$

as follows:

a. With the natural coordinate system $(z_1, z_2) \in \mathbf{C}^2$ we set $X = \mathbf{C}^2 \setminus \{0\}$ and an open covering $\{U_1, U_2\}$ of X by

$$U_j = \{(z_1, z_2) \in \mathbf{C}^2 \setminus \{0\};\ z_j \neq 0\}, \quad j = 1, 2.$$

Show that

$$Z^1(\mathscr{U}, \mathscr{O}_X) = O(U_1 \cap U_2),$$
$$C^0(\mathscr{U}, \mathscr{O}_X) = \mathscr{O}(U_1) \oplus \mathscr{O}(U_2).$$

b. Write down the coboundary operator

$$\delta : C^0(\mathscr{U}, X) \to B^1(\mathscr{U}, \mathscr{O}_X) \subset C^1(\mathscr{U}, \mathscr{O}_X).$$

c. Show the following expressions by convergent Laurent series:

$$\mathscr{O}(U_1) \ni f_1(z_1, z_2) = \sum_{\mu,\nu \in \mathbf{Z}} a_{\mu\nu} z_1^{\mu} z_2^{\nu}, \quad a_{\mu\nu} = 0 \text{ for all } \nu < 0,$$
$$\mathscr{O}(U_2) \ni f_2(z_1, z_2) = \sum_{\mu,\nu \in \mathbf{Z}} b_{\mu\nu} z_1^{\mu} z_2^{\nu}, \quad b_{\mu\nu} = 0 \text{ for all } \mu < 0,$$
$$\mathscr{O}(U_1 \cap U_2) \ni f(z_1, z_2) = \sum_{\mu,\nu \in \mathbf{Z}} c_{\mu\nu} z_1^{\mu} z_2^{\nu}.$$

d. Show that

$$B^1(\mathscr{U}, \mathscr{O}_X) = \Big\{ f(z_1, z_2) = \sum_{\mu,\nu \in \mathbf{Z}} c_{\mu\nu} z_1^{\mu} z_2^{\nu} \in \mathscr{O}(U_1 \cap U_2);$$
$$c_{\mu\nu} = 0 \text{ for all } \mu, \nu \text{ with } \mu < 0 \text{ and } \nu < 0 \Big\}.$$

e. Show that

$$H^1(\mathscr{U}, \mathscr{O}_X) \cong \Big\{ g(z_1, z_2) = \sum_{\mu,\nu \in \mathbf{Z}} d_{\mu\nu} z_1^{\mu} z_2^{\nu} \in \mathscr{O}(U_1 \cap U_2);$$
$$d_{\mu\nu} = 0 \text{ for all } \mu, \nu \text{ with } \mu \geq 0 \text{ or } \nu \geq 0 \Big\}.$$

It is noticed that $H^1(\mathscr{U}, \mathscr{O}_X) \hookrightarrow H^1(X, \mathscr{O}_X)$.

Chapter 4
Holomorphically Convex Domains and the Oka–Cartan Fundamental Theorem

In this chapter we prove the Oka–Cartan Fundamental Theorem on holomorphically convex domain Ω of $\mathbf{C}^n$; that, is, it is proved that $H^q(\Omega, \mathscr{F}) = 0$ $(q \geq 1)$ for every coherent sheaf $\mathscr{F}$ over holomorphically convex domains Ω. The method of the proof is a rewriting of the very ingenious method to solve the Cousin I Problem in Oka I and II by a double induction in terms of cohomologies of coherent sheaves. As a way to use coherent sheaves, we follow what was written in the introduction of Oka VII by means of the cohomology theory in a comprehensive way. In the course an essential role is played by "*Oka's Jôku-Ikô*" (pronounced "dzóuku ikou") through "*Oka maps*" that were invented in Oka I. In that sense, the essential part of the Fundamental Theorem is finished by Oka VII and I. The formulation of the Fundamental Theorem in terms of the cohomology theory of general sheaves is due to Cartan and Serre.

4.1 Holomorphically Convex Domains

The notion of a holomorphically convex domain was introduced and discussed a little at the end of Sect. 1.2. Here we think of it, going back to the origin of "convexity". Let $A \subset \mathbf{C}^n$ be a subset. Then A is convex if for every two points $z, w \in A$ and for every $t \in [0, 1]$,

$$tz + (1 - t)w \in A.$$

In general, the closed convex hull $\mathrm{co}(A)$ of A is defined as the smallest closed convex set containing A, but it is also defined as

$$\mathrm{co}(A) = \{z \in \mathbf{C}^n \cong \mathbf{R}^{2n};\ L(z) \leq \sup_A L \text{ for every linear functional } L : \mathbf{R}^{2n} \to \mathbf{R}\}. \tag{4.1.1}$$

J. Noguchi, *Analytic Function Theory of Several Variables*,
DOI 10.1007/978-981-10-0291-5_4

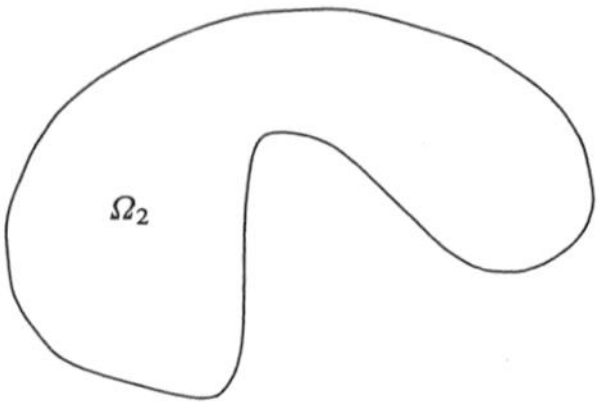

Fig. 4.1 Non-convex domain

A closed set A is convex if and only if $A = \mathrm{co}(A)$. If A is compact, so is $\mathrm{co}(A)$. It is equivalent for a domain $\Omega \subset \mathbf{C}^n$ to be convex that

$$\mathrm{co}(A) \Subset \Omega$$

for every subset $A \Subset \Omega$.

This convexity is invariant under affine transformations of $\mathbf{C}^n \cong \mathbf{R}^{2n}$, but not under biholomorphic maps. That is, for two domains $\Omega_i \subset \mathbf{C}^n$, $i = 1, 2$, with a biholomorphic map $\varphi : \Omega_1 \to \Omega_2$, Ω_2 is not necessarily convex even though Ω_1 is convex. For instance, with $n = 1$ we let $\Omega_1 = \Delta(0, 1)$ be the unit disk, and let Ω_2 be a non-convex simply connected domain (cf., e.g., Fig. 4.1). By Riemann's Mapping Theorem there is a biholomorphic map $\varphi : \Delta(0; 1) \to \Omega_2$.

We are interested in a convexity which is naturally invariant under biholomorphisms. We write (4.1.1), using holomorphic functions. Let $z_j = x_j + iy_j$, $1 \le j \le n$, and set

$$L(z) = L(x_1, y_1, \ldots, x_n, y_n) = \sum_{j=1}^{n} (a_j x_j + b_j y_j), \quad a_j, b_j \in \mathbf{R}.$$

Put $L_0(z) = \sum_{j=1}^{n} (a_j - ib_j) z_j$. Then this is holomorphic and

$$L(z) = \Re L_0(z).$$

We have $|e^{L_0(z)}| = e^{L(z)}$. Therefore, it follows that

$$\text{(4.1.2)} \qquad \mathrm{co}(A) = \Big\{ z \in \mathbf{C}^n;\ |e^{L_0(z)}| \le \sup_{w \in A} |e^{L_0(w)}| \text{ for every linear functional } L_0 : \mathbf{C}^n \to \mathbf{C} \Big\}.$$

Functions $e^{L_0(z)}$ are holomorphic in $\mathbf{C}^n$. It is noticed that in definition (1.2.27) of the holomorphic hull $\hat{A}_{\Omega}$, functions $e^{L_0(z)}$ are replaced by all holomorphic functions in the domain, so that the hull is biholomorphically invariant.

Proposition 4.1.3 *Let $\Omega \subset \mathbf{C}^n$ be a domain and let $A \subset \Omega$ be a subset. Then the following holds:*

(i) $\widehat{(\hat{A}_\Omega)}_\Omega = \hat{A}_\Omega$.
(ii) $\hat{A}_\Omega \subset \mathrm{co}(A)$; *in particular, if A is bounded, so is $\hat{A}_\Omega$.*

The proof is immediate.

Theorem 4.1.4 (i) *The holomorphic convexity is biholomorphically invariant. That is, let $\Omega_i \subset \mathbf{C}^n$, $i = 1, 2$, be domains, and let $\varphi : \Omega_1 \to \Omega_2$ be a biholomorphic map. Then, Ω_1 is holomorphically convex if and only if so is Ω_2.*
(ii) *Let Ω_μ, $1 \leq \mu \leq l$, be finitely many holomorphically convex domains. Then the intersection $\Omega = \bigcap_{\mu=1}^{l} \Omega_\mu$ is holomorphically convex.*

Proof (i) For a subset $K \subset \Omega_1$, K is compact if and only if so is $\varphi(K)$, and

$$\varphi\left(\hat{K}_{\Omega_1}\right) = \widehat{\varphi(K)}_{\Omega_2}$$

holds. Thus the claim holds.

(ii) For a compact subset $K \Subset \Omega$,

$$\hat{K}_\Omega \subset \hat{K}_{\Omega_\mu} \Subset \Omega_\mu, \quad 1 \leq \mu \leq l.$$

Therefore, $\hat{K}_\Omega \Subset \Omega$. □

Theorem 4.1.5 *In the case of $n = 1$, every domain $\Omega \subset \mathbf{C}$ is holomorphically convex.*

Proof Let $K \Subset \Omega$ be any compact subset. By definition,

$$\hat{K}_\Omega \subset \left\{ z \in \Omega;\, |z| \leq \max_K |z| \right\}.$$

Thus, $\hat{K}_\Omega$ is bounded. If $\hat{K}_\Omega \not\Subset \Omega$, then there would be a sequence $\zeta_\nu \in \hat{K}_\Omega$, $\nu = 1, 2, \ldots$, which converges to a boundary point $a \in \partial\Omega$. Because of a holomorphic function $f(z) = \frac{1}{z-a} \in \mathscr{O}(\Omega)$, we have

$$\infty > \max_K |f| \geq \lim_{\nu\to\infty} |f(\zeta_\nu)| = \infty.$$

This is a contradiction. □

Theorem 4.1.6 *A convex domain $\Omega \subset \mathbf{C}^n$ is holomorphically convex.*

Proof If Ω is convex, $\mathrm{co}(K) \Subset \Omega$ for a compact subset $K \Subset \Omega$. From Proposition 4.1.3, $\hat{K}_\Omega \Subset \Omega$ follows. □

Example 4.1.7 (i) The ball $B(a; \rho)$ ($a \in \mathbf{C}^n$, $\rho > 0$) of $\mathbf{C}^n$ is convex, and hence holomorphically convex.
(ii) For the same reason as above, $\mathrm{P}\Delta(a; r)$ is holomorphically convex.
(iii) A convex cylinder domain (Definition 1.2.5) is holomorphically convex, and an intersection of a finite number of convex cylinder domains is holomorphically convex.

For a convex cylinder domain Ω, the vanishing $H^q(\Omega, \mathscr{O}_\Omega) = 0$ $(q \geq 1)$ was shown by the fundamental Theorem 3.6.10. The purpose of this chapter is firstly to show

$$H^q(\Omega, \mathscr{F}) = 0, \quad q \geq 1 \tag{4.1.8}$$

for every coherent sheaf $\mathscr{F}$ on convex cylinder domain Ω, and then extend it for a general holomorphically convex domain by making use of "Oka's Jôku-Ikô", or "Lifting Principle" (Nishino [51]). Here we remark the following four points.

Remark 4.1.9 (i) A convex cylinder domain is biholomorphic to a product space of any simply connected domains by Riemann's Mapping Theorem; in particular, it is biholomorphic to a polydisk. It is the same to take any of them to describe a result which is biholomorphically invariant. When we use it, however, we have to consider finite intersections of such domains; it is because we need a Leray covering (cf. Definition 3.4.39) to deal with Čech cohomology. The "convexity" may be the simplest shape which is preserved under taking arbitrary finite intersections. This is the reason why we work on convex cylinder domains.
(ii) There is one more reason why we take convex cylinder domains. If readers read and understand well the proofs of the results here, they will find that all arguments work without using the elaborate Riemann's Mapping Theorem once they are dealt with over convex cylinder domains from the beginning. But the treatments will get rather involved; it will be a good exercise for readers to check this point after reading through to the end of the chapter.
(iii) To obtain (4.1.8) for $\mathscr{F} = \mathscr{O}_\Omega$ on a holomorphically convex domain, we prove that (4.1.8) holds for every coherent sheaf over a convex cylinder domain.
(iv) Lastly, it may be a question why "holomorphically convex domains" matter, because they are the most natural and necessary domains from the viewpoint of the existence domain of holomorphic functions. In Chap. 5 we will see the equivalence between holomorphically convex domains and domains of holomorphy, and furthermore, they are equivalent to pseudoconvex domains as shown in Chap. 7; but this equivalence fails for general complex manifolds. On the other hand, holomorphically convex domains are easily generalized to Stein manifolds and Stein spaces, where the Oka–Cartan Fundamental Theorem holds. In this sense, a "holomorphically convex domain" is the most basic pillar case.

Some preparations will continue for a moment in the sequel.

4.2 Cartan's Merging Lemma

We consider a coherent sheaf $\mathscr{F} \to \Omega$ over a domain $\Omega \subset \mathbf{C}^n$. As finite local generator systems of $\mathscr{F}$ over adjoining closed domains E' and E'' are given, it is necessary to form a local finite generator system of $\mathscr{F}$ over $E' \cup E''$ by merging them. We here assume the following:

4.2.1 (Closed cubes) A *closed cube* or a *closed rectangle* is a closed subset of $\mathbf{C}^n$ bounded with all edges parallel to real or imaginary axes of the complex coordinates; here we include the case when the widths of some edges degenerate to zero.

Assume that two closed cubes $E', E'' \Subset \Omega$ are represented as follows. There are a closed cube $F \Subset \mathbf{C}^{n-1}$, and two closed rectangles $E'_n, E''_n \Subset \mathbf{C}$ sharing an edge ℓ, such that (cf. Fig. 4.2)

$$E' = F \times E'_n, \quad E'' = F \times E''_n, \quad \ell = E'_n \cap E''_n.$$

Let $GL(p; \mathbf{C})$ denote the general linear group of degree p, and let $\mathbf{1}_p$ denote the unit matrix of degree p. The following is due to H. Cartan [8].

Lemma 4.2.2 (Cartan's matrix decomposition) *Let the notation be as above. Then there is a neighborhood $V_0 \subset GL(p; \mathbf{C})$ of $\mathbf{1}_p$ such that for a matrix-valued holomorphic function $A : U \to V_0$ on a neighborhood U of $F \times \ell$, there is a matrix-valued holomorphic function $A' : U' \to GL(p; \mathbf{C})$ (resp. $A'' : U'' \to GL(p; \mathbf{C})$) on a neighborhood U' (resp. U'') of E' (resp. E'') satisfying $A = A' \cdot A''^{-1}$ on a neighborhood of $F \times \ell$.*

Proof (cf. p. 395) We will determine a neighborhood V_0 of $\mathbf{1}_p$ in the arguments below. Firstly in general, the exponential function of a square p-matrix P is defined by

$$\exp P = \sum_{\mu=0}^{\infty} \frac{1}{\mu!} P^{\mu}. \tag{4.2.3}$$

Here, the convergence is normal (cf. 1.1.3). Conversely, if a square p-matrix S belongs to a sufficiently small neighborhood V_0 of $\mathbf{1}_p$, the logarithm of S is given by

$$\begin{aligned} \log S &= \log(\mathbf{1}_p + (S - \mathbf{1}_p)) \\ &= \sum_{\mu=1}^{\infty} \frac{(-1)^{\mu-1}}{\mu} (S - \mathbf{1}_p)^{\mu}, \end{aligned}$$

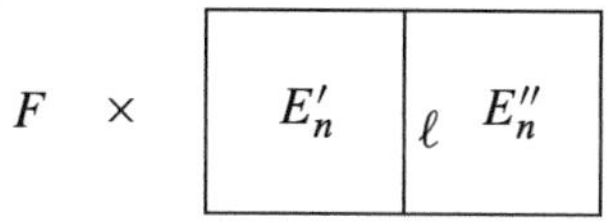

Fig. 4.2 Adjoining closed cubes

which converges normally. We choose V_0 so that this holds. Then, the restricted map, $\log : V_0 \to \log V_0$, gives a biholomorphism from V_0, a neighborhood of $\mathbf{1}_p$ onto $\log V_0$, a neighborhood of the zero p-matrix $\mathbf{O}_p$. If $P = \log S$, then $\exp P = S$, and the equivalence

$$S \to \mathbf{1}_p \iff P \to \mathbf{O}_p$$

holds. If p-matrices P, Q are sufficiently close to $\mathbf{O}_p$, then

$$M(P, Q) = \exp(-P) \exp(P + Q) \exp(-Q)$$

is close to $\mathbf{1}_p$. If $PQ = QP$, then $M(P, Q) = \mathbf{1}_p$, so that $M(P, Q)$ measures the error of the commutativity of P and Q.

Let $|P|$ stand for the maximum of absolute values of all entries of P. Then there is an $\varepsilon_1 > 0$ such that for $|P| < \varepsilon_1$ and $|Q| < \varepsilon_1$,

$$\exp(-P),\ \exp(-Q),\ \exp(P + Q),\ M(P, Q) \in V_0.$$

Expanding $M(P, Q)$ as in definition (4.2.3), we have

$$\begin{aligned} M(P, Q) &= \left(\mathbf{1}_p + \sum_{\mu=1}^{\infty} \frac{1}{\mu!}(-P)^{\mu}\right) \cdot \left(\mathbf{1}_p + \sum_{\mu=1}^{\infty} \frac{1}{\mu!}(P + Q)^{\mu}\right) \\ &\quad \times \left(\mathbf{1}_p + \sum_{\mu=1}^{\infty} \frac{1}{\mu!}(-Q)^{\mu}\right) \\ &= \mathbf{1}_p + (-P) + (P + Q) + (-Q) + (\text{terms of order } \geq 2 \text{ in } P, Q) \\ &= \mathbf{1}_p + (\text{terms of order } \geq 2 \text{ in } P, Q). \end{aligned}$$

The last series above is a power series in P and Q, which is absolutely convergent. Since $M(P, Q) \in V_0$,

$$\log M(P, Q) = \sum_{\mu=1}^{\infty} \frac{(-1)^{\mu-1}}{\mu} (\text{terms of degree } \geq 2 \text{ in } P, Q)^{\mu}.$$

With $\varepsilon_1 > 0$ taken smaller if necessary, there is $K_1 > 0$ such that

$$|\log M(P, Q)| \leq K_1 \max\{|P|^2, |Q|^2\}. \tag{4.2.4}$$

It is noted that $B(z) = \log A(z)$ is holomorphic in $z \in U$, if $A(z) \in V_0$ for $z \in U$.

Let $0 < \varepsilon_2 < \varepsilon_1$ be a positive constant determined in the following arguments, and assume

$$|B(z)| < \varepsilon_2. \tag{4.2.5}$$

Note that this implies $A(z) \in V_0$ because of the choice of ε_1.

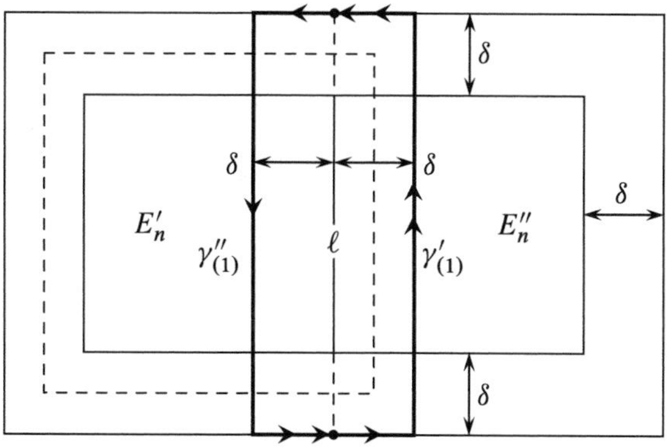

Fig. 4.3 δ-closed neighborhoods of the adjoining closed cubes

We widen each edge of F, E'_n, E''_n by the same length, $\delta > 0$ outward and denote the resulting closed cube and closed rectangles by $\tilde{F}$, $\tilde{E}'_{n(1)}$, $\tilde{E}''_{n(1)}$, respectively. Taking $\delta > 0$ sufficiently small, we have

$$F \times \ell \subset \tilde{F} \times \left(\tilde{E}'_{n(1)} \cap \tilde{E}''_{n(1)}\right) \Subset U.$$

Set the boundaries as in Fig. 4.3:

(4.2.6)
$$\begin{aligned} \partial\left(\tilde{E}'_{n(1)} \cap \tilde{E}''_{n(1)}\right) &= \gamma_{(1)} = \gamma'_{(1)} + \gamma''_{(1)}, \\ \gamma'_{(1)} &= \left(\partial \tilde{E}'_{n(1)}\right) \cap \tilde{E}''_{n(1)}, \\ \gamma''_{(1)} &= \tilde{E}'_{n(1)} \cap \partial \tilde{E}''_{n(1)}. \end{aligned}$$

Similarly, keeping the inner $\frac{\delta}{2}$ of the width δ as E'_n is widened to $\tilde{E}'_{n(1)}$ we successively shrink inward by dividing in half the outer $\frac{\delta}{2}$. That is, $\tilde{E}'_{n(2)}$ denotes the closed cube shrunk inward by $\frac{\delta}{4}$ from $\tilde{E}'_{n(1)}$. Assuming $\tilde{E}'_{n(k)}$ determined, we denote by $\tilde{E}'_{n(k+1)}$ the closed cube shrunk inward by $\frac{\delta}{2^{k+1}}$ from $\tilde{E}'_{n(k)}$ (cf. Fig. 4.4). Since

$$\frac{\delta}{4} + \frac{\delta}{8} + \cdots = \frac{\delta}{2},$$

$$\bigcap_{k=1}^{\infty} \tilde{E}'_{n(k)} = \text{ the closed cube widened from } E'_n \text{ by } \frac{\delta}{2}.$$

We set $\tilde{E}''_{n(k)}$, similarly. As in (4.2.6) we write

$$\partial\left(\tilde{E}'_{n(k)} \cap \tilde{E}''_{n(k)}\right) = \gamma_{(k)} = \gamma'_{(k)} + \gamma''_{(k)}.$$

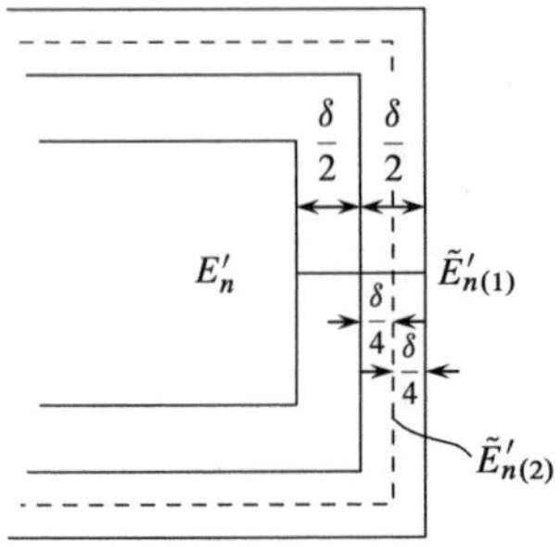

Fig. 4.4 Closed $\frac{\delta}{2^k}$-neighborhoods of closed cubes

Let

$$\tilde{E}'_{(k)} = \tilde{F} \times \tilde{E}'_{n(k)}, \quad \tilde{E}''_{(k)} = \tilde{F} \times \tilde{E}''_{n(k)}$$

be the closed cube neighborhoods of E' and E'', respectively.

For $(z', z_n) \in \tilde{E}'_{(2)} \cap \tilde{E}''_{(2)}$ we set

$$B_1(z', z_n) = B(z', z_n) = \log A(z', z_n).$$

Using Cauchy's integral expression, we write

$$(4.2.7) \qquad \begin{aligned} B(z', z_n) &= \frac{1}{2\pi i} \int_{\gamma_{(1)}} \frac{B_1(z', \zeta)}{\zeta - z_n} d\zeta \\ &= \frac{1}{2\pi i} \int_{\gamma'_{(1)}} \frac{B_1(z', \zeta)}{\zeta - z_n} d\zeta + \frac{1}{2\pi i} \int_{\gamma''_{(1)}} \frac{B_1(z', \zeta)}{\zeta - z_n} d\zeta \\ &= B'_1(z', z_n) + B''_1(z', z_n). \end{aligned}$$

Here, $B'_1(z', z_n)$ (resp. $B''_1(z', z_n)$) is holomorphic in $(z', z_n) \in \tilde{E}'_{(2)}$ (resp. $(z', z_n) \in \tilde{E}''_{(2)}$), and

$$(4.2.8) \qquad |z_n - \zeta| \geq \frac{\delta}{4}, \quad {}^\forall(z', z_n) \in \tilde{E}'_{(2)}, {}^\forall \zeta \in \gamma'_{(1)}.$$

Letting L be the length of the curve $\gamma'_{(1)}$, we get for $k \geq 1$

$$L = \text{the length of } \gamma''_{(1)} \geq \text{the length of } \gamma'_{(k)} \text{ (also, the length of } \gamma''_{(k)}).$$

We choose ε_2 so that

$$(4.2.9) \qquad 0 < \varepsilon_2 < \frac{\pi \delta}{2L} \varepsilon_1.$$

Then, by (4.2.5), (4.2.7) and (4.2.8) we have that for $(z', z_n) \in \tilde{E}'_{(2)}$

$$|B'_1(z', z_n)| \le \frac{1}{2\pi} \cdot \frac{4}{\delta} L \cdot \max_{\gamma'_{(1)}} |B_1(z', \zeta)| \le \frac{2L}{\pi\delta}\varepsilon_2 < \varepsilon_1. \tag{4.2.10}$$

For $(z', z_n) \in \tilde{E}''_{(2)}$ we have similarly

$$|B''_1(z', z_n)| \le \frac{2L}{\pi\delta}\varepsilon_2 < \varepsilon_1. \tag{4.2.11}$$

We take $\varepsilon_3 > 0$ so that

$$\frac{2L\varepsilon_2}{\pi\delta} < \varepsilon_3 < \varepsilon_1. \tag{4.2.12}$$

As seen in the sequel, it suffices to take

$$0 < \varepsilon_3 < \min\left\{\frac{\pi\delta}{8LK_1}, \varepsilon_1\right\}, \tag{4.2.13}$$

where K_1 is the constant in (4.2.4); we then choose ε_2 satisfying (4.2.9) and (4.2.12).

In what follows, we assume that B'_k and B''_k are inductively defined to be holomorphic in the interiors of $\tilde{E}'_{(k)}$ and $\tilde{E}''_{(k)}$, respectively, and to satisfy

$$\begin{aligned} &\max_{\tilde{E}'_{(k+1)}} |B'_k| \le \frac{\varepsilon_3}{2^{k-1}}, \quad \max_{\tilde{E}''_{(k+1)}} |B''_k| \le \frac{\varepsilon_3}{2^{k-1}}, \\ &B_k = B'_k + B''_k \quad \text{(in the commonly defined domain).} \end{aligned} \tag{4.2.14}$$

Set $B_{k+1} = \log M(B'_k, B''_k)$. This is holomorphic in the interior of $\tilde{E}'_{(k+1)} \cap \tilde{E}''_{(k+1)}$, and satisfies by (4.2.4)

$$|B_{k+1}| \le K_1 \left(\frac{\varepsilon_3}{2^{k-1}}\right)^2. \tag{4.2.15}$$

Decompose B_{k+1} into the sum of B'_{k+1} and B''_{k+1} as follows:

$$\begin{aligned} B_{k+1}(z', z_n) &= \frac{1}{2\pi i}\int_{\gamma_{(k+1)}} \frac{B_{k+1}(z', \zeta)}{\zeta - z_n} d\zeta \\ &= \frac{1}{2\pi i}\int_{\gamma'_{(k+1)}} \frac{B_{k+1}(z', \zeta)}{\zeta - z_n} d\zeta + \frac{1}{2\pi i}\int_{\gamma''_{(k+1)}} \frac{B_{k+1}(z', \zeta)}{\zeta - z_n} d\zeta \\ &= B'_{k+1}(z', z_n) + B''_{k+1}(z', z_n). \end{aligned}$$

Then, $B'_{k+1}(z', z_n)$ and $B''_{k+1}(z', z_n)$ are holomorphic in the interiors of $\tilde{E}'_{(k+1)}$ and $\tilde{E}''_{(k+1)}$, respectively. It follows from (4.2.15) and (4.2.12) that for $(z', z_n) \in \tilde{E}'_{(k+2)}$

$$(4.2.16) \qquad \begin{aligned} |B'_{k+1}(z', z_n)| &\le \frac{L}{2\pi} \frac{2^{k+2}}{\delta} K_1 \frac{\varepsilon_3^2}{2^{2k-2}} \\ &\le \frac{8LK_1}{\pi\delta} \varepsilon_3 \cdot \frac{\varepsilon_3}{2^k} \le \frac{\varepsilon_3}{2^k}. \end{aligned}$$

Similarly, we get

$$(4.2.17) \qquad \max_{\tilde{E}''_{(k+2)}} |B''_{k+1}(z', z_n)| \le \frac{\varepsilon_3}{2^k}.$$

The convergences of the following limits are deduced from (4.2.16) and (4.2.17) and the majorant test, and the convergences are absolute and uniform:

$$\begin{aligned} A'(z', z_n) &= \lim_{k\to\infty} \exp B'_1(z', z_n) \exp B'_2(z', z_n) \cdots \exp B'_k(z', z_n), \\ &\qquad (z', z_n) \in \bigcap_{k=1}^{\infty} \tilde{E}'_{(k)}; \\ A''(z', z_n) &= \lim_{k\to\infty} \exp(-B''_1(z', z_n)) \exp(-B''_2(z', z_n)) \cdots \exp(-B''_k(z', z_n)), \\ &\qquad (z', z_n) \in \bigcap_{k=1}^{\infty} \tilde{E}''_{(k)}. \end{aligned}$$

We see that $A'(z', z_n)$ (resp. $A''(z', z_n)$) is holomorphic in the interior of $\bigcap_{k=1}^{\infty} \tilde{E}'_{(k)}$ (resp. $\bigcap_{k=1}^{\infty} \tilde{E}''_{(k)}$), which contains E' (resp. E''). It follows from the construction that for points of $(\bigcap_{k=1}^{\infty} \tilde{E}'_{(k)}) \cap (\bigcap_{k=1}^{\infty} \tilde{E}''_{(k)})$

$$\begin{aligned} \exp B_{k+1} &= M(B'_k, B''_k) = \exp(-B'_k) \exp(B'_k + B''_k) \exp(-B''_k) \\ &= \exp(-B'_k) \exp(B_k) \exp(-B''_k) \\ &= \exp(-B'_k) \exp(-B'_{k-1}) \exp(B_{k-1}) \exp(-B''_{k-1}) \exp(-B''_k) \\ &= \cdots = \exp(-B'_k) \exp(-B'_{k-1}) \cdots \exp(-B'_1) \\ &\qquad \cdot \exp(B_1) \exp(-B''_1) \cdots \exp(-B''_{k-1}) \exp(-B''_k). \end{aligned}$$

Since $A = \exp B = \exp B_1$,

$$(4.2.18) \qquad \begin{aligned} A &= \exp(B'_1) \exp(B'_2) \cdots \exp(B'_k) \exp(B_{k+1}) \\ &\qquad \cdot \left(\exp(-B''_1) \exp(-B''_2) \cdots \exp(-B''_k)\right)^{-1}. \end{aligned}$$

Using $\exp B_{k+1} \to \mathbf{1}_p$ $(k \to \infty)$, we let $k \to \infty$ in (4.2.18), and then obtain

$$A(z) = A'(z) \cdot A''(z)^{-1}, \quad z \in \left(\bigcap_{k=1}^{\infty} \tilde{E}'_{(k)} \right) \cap \left(\bigcap_{k=1}^{\infty} \tilde{E}''_{(k)} \right). \qquad \square$$

The following is Cartan's Merging Lemma in [81] (1940). In a footnote of the introduction of Oka VII, K. Oka describes a comment such that we owe a lot also to the theorems in [8].[1]

Lemma 4.2.19 (Cartan's Merging Lemma) *Let $E' \subset U'$ and $E'' \subset U''$ be those in Lemma 4.2.2. Let $\mathscr{F} \to \Omega$ be a coherent sheaf.*

Assume that finitely many sections $\sigma'_j \in \Gamma(U', \mathscr{F})$, $1 \le j \le p'$, generate $\mathscr{F}$ over U', and similarly $\sigma''_k \in \Gamma(U'', \mathscr{F})$, $1 \le k \le p''$, generate $\mathscr{F}$ over U''. Furthermore, assume the existence of $a_{jk}, b_{kj} \in \mathscr{O}(U' \cap U'')$, $1 \le j \le p'$, $1 \le k \le p''$, such that

$$\sigma'_j = \sum_{k=1}^{p''} a_{jk}\sigma''_k, \quad \sigma''_k = \sum_{j=1}^{p'} b_{kj}\sigma'_j.$$

Then there are a neighborhood $W \supset E' \cup E''$ with $W \subset U' \cup U''$ and finitely many sections σ_l on W, $1 \le l \le p = p' + p''$, which generate $\mathscr{F}$ over W.

Proof We set column vectors and matrices as follows: $\sigma' = {}^t(\sigma'_1, \ldots, \sigma'_{p'})$, $\sigma'' = {}^t(\sigma''_1, \ldots, \sigma''_{p''})$, $A = (a_{jk})$, $B = (b_{kj})$. Then we have

$$\text{(4.2.20)} \qquad \sigma' = A\sigma'', \quad \sigma'' = B\sigma'.$$

Adding 0 to σ' and σ'' to form vectors of the same degree p, we put

$$\tilde{\sigma}' = \begin{pmatrix} \sigma'_1 \\ \vdots \\ \sigma'_{p'} \\ \hline 0 \\ \vdots \\ 0 \end{pmatrix}, \quad \tilde{\sigma}'' = \begin{pmatrix} 0 \\ \vdots \\ 0 \\ \hline \sigma''_1 \\ \vdots \\ \sigma''_{p''} \end{pmatrix}.$$

We also put

$$\tilde{A} = \left(\begin{array}{c|c} \mathbf{1}_{p'} & A \\ \hline -B & \mathbf{1}_{p''} - BA \end{array} \right).$$

Since $BA\sigma'' = \sigma''$ by (4.2.20),

[1]In the original version of Oka VII (Iwanami) K. Oka wrote after the citation of [8], "dont nous devons beaucoup aussi aux théorèmes". In the version of Bull. Soc. Math. France, it is "Nous devons beaucoup aux théorèmes de ce Mémoire".

$$\tilde{\sigma}' = \tilde{A}\,\tilde{\sigma}''. \tag{4.2.21}$$

We take the following matrices consisting of the repetition of elementary transformations:

$$\begin{aligned} P &= \left(\begin{array}{c|c} \mathbf{1}_{p'} & A \\ \hline 0 & \mathbf{1}_{p''} \end{array}\right), \quad P^{-1} = \left(\begin{array}{c|c} \mathbf{1}_{p'} & -A \\ \hline 0 & \mathbf{1}_{p''} \end{array}\right), \\ Q &= \left(\begin{array}{c|c} \mathbf{1}_{p'} & 0 \\ \hline B & \mathbf{1}_{p''} \end{array}\right), \quad Q^{-1} = \left(\begin{array}{c|c} \mathbf{1}_{p'} & 0 \\ \hline -B & \mathbf{1}_{p''} \end{array}\right). \end{aligned} \tag{4.2.22}$$

Transforming $\tilde{A}$ from right and left, we get

$$Q\,\tilde{A}\,P^{-1} = \left(\begin{array}{c|c} \mathbf{1}_{p'} & 0 \\ \hline 0 & \mathbf{1}_{p''} \end{array}\right) = \mathbf{1}_p. \tag{4.2.23}$$

Since the entries a_{jk} and b_{kj} of A and B are holomorphic in a neighborhood of $E' \cap E'' = F \times \ell$, by Corollary 1.2.22 they are approximated uniformly on a suitable neighborhood $W_0 (\Subset U' \cap U'')$ of $E' \cap E''$ by polynomials $\tilde{a}_{jk}$ and $\tilde{b}_{kj}$, respectively. Let $\tilde{P}$ and $\tilde{Q}$ be matrices with entries $\tilde{a}_{jk}$ and $\tilde{b}_{kj}$, respectively. By a sufficiently close approximation it is deduced from (4.2.23) that for neighborhood V_0 of $\mathbf{1}_p$ in Lemma 4.2.2,

$$\hat{A}(z) = \tilde{Q}(z)\,\tilde{A}(z)\,\tilde{P}(z)^{-1} \in V_0, \quad z \in W_0. \tag{4.2.24}$$

Then, by Lemma 4.2.2 there are a neighborhood W' (resp. W'') of E' (resp. E'') and a regular p-matrix-valued holomorphic function $\hat{A}'$ (resp. $\hat{A}''$) defined there such that on $W' \cap W'' (\subset W_0)$,

$$\hat{A} = \left(\hat{A}'\right)^{-1} \hat{A}'' \tag{4.2.25}$$

holds. By this and (4.2.24), $\tilde{A} = \tilde{Q}^{-1} (\hat{A}')^{-1} \hat{A}''\,\tilde{P}$, and hence by (4.2.21)

$$\hat{A}'\,\tilde{Q}\,\tilde{\sigma}' = \hat{A}''\,\tilde{P}\,\tilde{\sigma}'' \tag{4.2.26}$$

on $W' \cap W''$. Therefore, $\tau_j \in \Gamma(W' \cup W'', \mathscr{F})$, $1 \le j \le p$, are well-defined by setting

$$\begin{pmatrix} \tau_1 \\ \vdots \\ \tau_p \end{pmatrix} = \begin{cases} \hat{A}'\,\tilde{Q}\,\tilde{\sigma}', & \text{on } W', \\ \hat{A}''\,\tilde{P}\,\tilde{\sigma}'', & \text{on } W''. \end{cases}$$

Since $\hat{A}'\tilde{Q}$ and $\hat{A}''\tilde{P}$ are regular matrices, τ_j, $1 \le j \le p$, generate $\mathscr{F}$ over $W' \cup W''$. □

We call the above-obtained (τ_j) a locally finite generator system of $\mathscr{F}$ by *merging* (σ'_j) and (σ''_k).

4.3 Oka's Fundamental Lemma

The aim of this section is to prove the vanishing of higher cohomologies of coherent sheaves $\mathscr{F}$ over convex cylinder domains $\Omega = \prod \Omega_j \subset \mathbf{C}^n$. For that purpose we construct a finite generator system of $\mathscr{F}$ over a compact subset of Ω, and then extend it over a larger subset, successively. To make the arguments comprehensive, we let each Ω_j be a rectangle by Riemann's Mapping Theorem.

Remark 4.3.1 If the descriptions of arguments are allowed to be complicated, it is not necessary to use Riemann's Mapping Theorem and it is possible to deal with them over convex domains directly. But, in that case it is necessary, e.g., to modify properly Cartan's Merging Lemma in the previous section.

4.3.1 Steps of Proof

Our purpose is to prove the following theorem in the next section:

Oka–Cartan Fundamental Theorem. For every coherent sheaf $\mathscr{F} \to \Omega$ over a holomorphically convex domain $\Omega \subset \mathbf{C}^n$ we have

$$H^q(\Omega, \mathscr{F}) = 0, \quad q \ge 1.$$

Here we describe the steps of that proof; there appear some undefined terms, but readers should continue to read. It is not necessary for readers to understand the contents completely on the first reading, but they may find the path where the theory heads. After reading to the end of this chapter, it is recommended for readers to return here. It will be a good exercise for readers to read these steps and give a complete proof of each step by themselves. They are sure to obtain a better understanding, and they will find how well the concept and the theorem of Oka's Coherence.

(i) For an arbitrary sheaf $\mathscr{S} \to \Omega$ over any domain Ω, $H^q(\Omega, \mathscr{S}) = 0$, $q \ge 2^{2n}$ (Theorem 3.4.14).

(ii) Let $\mathscr{F}$ be a coherent sheaf over any domain Ω. Then the local finiteness implies that for every point $a \in \Omega$ there are a polydisk neighborhood $\mathrm{P}\Delta = \mathrm{P}\Delta(a; r)$ and the following exact sequence over $\mathrm{P}\Delta$:

$$\mathscr{O}_{\mathrm{P}\Delta}^{N_1} \xrightarrow{\varphi_1} \mathscr{F}|_{\mathrm{P}\Delta} \to 0.$$

(iii) Let Ω be a convex cylinder domain. Moreover, one may assume Ω to be an open cube for the simplicity of arguments (cf. Remark 4.3.1). Take an increasing covering of Ω by relatively compact open cubes $\Omega_\nu \Subset \Omega_{\nu+1}$, $\nu = 1, 2, \ldots$. For a moment, the aim is to show over relatively compact Ω_ν,

$$H^q(\Omega_\nu, \mathscr{F}) = 0, \quad q \geq 1.$$

Here, $\mathscr{F}$ is the one restricted to Ω_ν.

(iv) We divide each closed cube $\bar{\Omega}_\nu$ by real hyperplanes parallel to coordinate axes, and write $\bar{\Omega}_\nu = \bigcup_\mu E_{\nu\mu}$ with a finite number of such small closed cubes $E_{\nu\mu}$. If the division is taken sufficiently fine, by (ii) above there is an exact sequence over a neighborhood $U_{\nu\mu}$ of $E_{\nu\mu}$:

$$\mathscr{O}_{U_{\nu\mu}}^{N_{\nu\mu}} \xrightarrow{\varphi_{\nu\mu}} \mathscr{F}|_{U_{\nu\mu}} \to 0. \tag{4.3.2}$$

(v) For adjoining $E_{\nu\mu}$ and $E_{\nu\mu'}$, we merge the exact sequences (4.3.2) by using Cartan's Merging Lemma 4.2.19 to make an exact sequence on a neighborhood of $E_{\nu\mu} \cup E_{\nu\mu'}$. Repeating this procedure, we construct an exact sequence on a neighborhood U_ν of $\bar{\Omega}_\nu$:

$$\mathscr{O}_{U_\nu}^{N_1} \xrightarrow{\varphi_1} \mathscr{F}|_{U_\nu} \to 0.$$

Thus, we obtain the following short exact sequence,

$$0 \to \operatorname{Ker} \varphi_1 \to \mathscr{O}_{U_\nu}^{N_1} \xrightarrow{\varphi_1} \mathscr{F}|_{U_\nu} \to 0. \tag{4.3.3}$$

Letting $q \geq 1$ and restricting (4.3.3) to Ω_ν, we get a long exact sequence,

$$\begin{aligned}\cdots \to H^q(\Omega_\nu, \mathscr{O}_{\Omega_\nu}^{N_1}) &\to H^q(\Omega_\nu, \mathscr{F}) \to H^{q+1}(\Omega_\nu, \operatorname{Ker} \varphi_1) \\ &\to H^{q+1}(\Omega_\nu, \mathscr{O}_{\Omega_\nu}^{N_1}) \to \cdots .\end{aligned}$$

Since Ω_ν is a convex cylinder domain, Dolbeault's Theorem 3.6.10 implies that $H^q(\Omega_\nu, \mathscr{O}_{\Omega_\nu}^{N_1}) = 0$, $q \geq 1$. Therefore,

$$H^q(\Omega_\nu, \mathscr{F}) \cong H^{q+1}(\Omega_\nu, \operatorname{Ker} \varphi_1).$$

By this we have increased the degree of the cohomology by one. The coherence of $\mathscr{F}$ implies that of $\operatorname{Ker} \varphi_1$.

(vi) To reduce the problem to (i), we make the same arguments for the coherent sheaf $\operatorname{Ker} \varphi_1$, and repeat it further to obtain an exact sequence of coherent sheaves with length $p\,(\in \mathbf{Z}^+)$:

$$0 \to \operatorname{Ker} \varphi_p \to \mathscr{O}_{\Omega_\nu}^{N_{p-1}} \xrightarrow{\varphi_{p-1}} \cdots \xrightarrow{\varphi_2} \mathscr{O}_{\Omega_\nu}^{N_1} \xrightarrow{\varphi_1} \mathscr{F}|_{\Omega_\nu} \to 0. \tag{4.3.4}$$

We call this *Oka's Syzygies* of the coherent sheaf $\mathscr{F}$ with length $p \geq 1$. By making use of this process we increase the degree of the cohomology by one each time, and taking $p = 2^{2n} - 1$ and using (i), we finally obtain

$$
\begin{aligned}
(4.3.5) \qquad H^q(\Omega_\nu, \mathscr{F}) &\cong H^{q+1}(\Omega_\nu, \operatorname{Ker} \varphi_1) \cong \cdots \\
&\cong H^{2^{2n}}(\Omega_\nu, \operatorname{Ker} \varphi_{2^{2n}-q}) = 0, \quad q \geq 1.
\end{aligned}
$$

(vii) **[Oka's Fundamental Lemma]** Letting $\nu \to \infty$, we deduce that $H^q(\Omega, \mathscr{F}) = 0$, $q \geq 1$.

(viii) **[Oka's Jôku-Ikô]** This method was invented in Oka I to solve the Cousin I Problem. A difficult problem by itself can be solved by increasing the dimension of the domain, which enables us to reduce the problem over a simple domain, such as a polydisk. K. Oka called this "*Principle of Jôku-Ikô*". Oka's Jôku-Ikô is used in a number of places, notably in the proofs of Lemmas 4.4.9, 4.4.11, 4.4.15, 4.4.17, and Theorem 5.4.3.

Let Ω be a holomorphically convex domain. Take an increasing open covering $\Omega_\mu \Subset \Omega_{\mu+1} \nearrow \Omega$ by analytic polyhedra Ω_μ. Each analytic polyhedron Ω_μ can be embedded into a polydisk $\mathrm{P}\Delta_{(\mu)}$ of some (supposedly very high) dimension as a closed complex submanifold. Hence we may regard Ω_μ as an analytic subset of $\mathrm{P}\Delta_{(\mu)}$, and denote by $\mathscr{I}\langle \Omega_\mu \rangle$ its ideal sheaf. Thus, the following short exact sequence is deduced:

$$0 \to \mathscr{I}\langle \Omega_\mu \rangle \to \mathscr{O}_{\mathrm{P}\Delta_{(\mu)}} \to \mathscr{O}_{\Omega_\mu} := \mathscr{O}_{\mathrm{P}\Delta_{(\mu)}} / \mathscr{I}\langle \Omega_\mu \rangle \to 0.$$

Let $\hat{\mathscr{F}}_\mu$ denote the simple extension of $\mathscr{F}|_{\Omega_\mu}$, defined as 0 outside Ω_μ. Taking the tensor product with the above exact sequence, we have an exact sequence by Theorem 3.2.5,

$$\mathscr{I}\langle \Omega_\mu \rangle \otimes \hat{\mathscr{F}}_\mu \xrightarrow{\psi} \hat{\mathscr{F}}_\mu \xrightarrow{\phi} \mathscr{O}_{\mathrm{P}\Delta_{(\mu)}} / \mathscr{I}\langle \Omega_\mu \rangle \cong \mathscr{O}_{\Omega_\mu} \otimes \hat{\mathscr{F}}_\mu \cong \mathscr{F}|_{\Omega_\mu} \to 0.$$

Since the image of ψ is locally finite, it is coherent. Therefore, we obtain the following short exact sequence of coherent sheaves:

$$0 \to \operatorname{Im} \psi \to \hat{\mathscr{F}}_\mu \xrightarrow{\phi} \mathscr{F}|_{\Omega_\mu} \to 0.$$

Since the Fundamental Theorem holds over $\mathrm{P}\Delta_{(\mu)}$ by (iv),

$$H^q(\Omega_\mu, \mathscr{F}) \cong H^q(\mathrm{P}\Delta_{(\mu)}, \hat{\mathscr{F}}_\mu) = 0, \quad q \geq 1.$$

(ix) Letting $\mu \to \infty$, we have $H^q(\Omega, \mathscr{F}) = 0$, $q \geq 1$.

In the present section, (i)–(vii) are shown, and (viii) and (ix) will be dealt with in the next section.

Remark 4.3.6 The content given in Remark 3.7.8 appears, e.g., at steps (v) and (vi). As seen in (4.3.5), the problem is reduced to some higher cohomology, and then it is trivially solved.

4.3.2 Oka's Syzygies

Let $E \Subset \mathbf{C}^n$ be a closed cube (cf. 4.2.1). The dimension of E is defined as the number of edges with positive length, and denoted by $\dim E$; $0 \le \dim E \le 2n$.

For a coherent sheaf $\mathscr{F}$ over a neighborhood of E, we show the existence of Oka's Syzygies (4.3.4). The following lemma claims the existence of a quasi-global finite generator system for a coherent sheaf.

Lemma 4.3.7 (Oka's Syzygy, Oka VII) *Let $E \Subset \mathbf{C}^n$ be any closed cube.*

(i) *For a coherent sheaf $\mathscr{F}$ over a neighborhood of E, there is a finite generator system of $\mathscr{F}$ over a neighborhood of E. That is, there are a neighborhood $U \supset E$, where $\mathscr{F}$ is defined, a number $N \in \mathbf{N}$ and an exact sequence,*

$$\mathscr{O}_U^N \xrightarrow{\varphi} \mathscr{F}|_U \to 0.$$

(ii) *Let $\mathscr{F}$ be a coherent sheaf over a neighborhood U of E, let $\{\sigma_j\}_{1 \le j \le N}$ be a finite generator system of $\mathscr{F}$ over U, and let $\sigma \in \Gamma(U, \mathscr{F})$ be any given section. Then there are holomorphic functions $a_j \in \mathscr{O}(U')$, $1 \le j \le N$, on a neighborhood $U' (\subset U)$ of E such that*

$$\sigma = \sum_{j=1}^{N} a_j \sigma_j \quad (\text{on } U').$$

Proof We prove (i) and (ii), simultaneously by induction on $\nu := \dim E$.

(a) Case of $\nu = 0$: These follow from the definition of coherence.
(b) Case of $\nu \ge 1$: Suppose that (i) and (ii) hold for every closed cube of dimension $\nu - 1$.

(i) Let E be a closed cube of dimension ν, and let $\mathscr{F}$ be a coherent sheaf over a neighborhood of E. We write $(z_j) = (x_j + iy_j) \in \mathbf{C}^n$ for the standard coordinate system, and order it as $(x_1, y_1, \ldots, x_n, y_n)$. By a translation and a change of coordinate order we may assume without loss of generality that

$$E = F \times [0, T],$$

where F is a closed cube of dimension $\nu - 1$ and $T > 0$. Take an arbitrary point $t \in [0, T]$ and take the slice $E_t := F \times \{t\}$. Since E_t is a closed cube of dimension $\nu - 1$, the induction hypothesis for (i) implies the existence of a finite generator system of $\mathscr{F}$ on a neighborhood of E_t. By the Heine–Borel theorem there is a partition of the interval

$$0 = t_0 < t_1 < \cdots < t_L = T, \tag{4.3.8}$$

such that there is a finite generator system $\{\sigma_{\alpha j}\}_j$ of $\mathscr{F}$ on a neighborhood of each $E_\alpha = F \times [t_{\alpha-1}, t_\alpha]$. Since $E_\alpha \cap E_{\alpha+1} = E_{t_\alpha}$ is a closed cube of dimension $\nu - 1$, by the induction hypothesis for (ii) there are holomorphic functions a_{jk}, b_{kj} in a neighborhood of $E_\alpha \cap E_{\alpha+1}$ such that

$$\sigma_{\alpha j} = \sum_k a_{jk}\sigma_{\alpha+1k}, \quad \sigma_{\alpha+1k} = \sum_j b_{kj}\sigma_{\alpha j}.$$

Applying Cartan's Merging Lemma 4.2.19, we obtain a finite generator system of $\mathscr{F}$ on a neighborhood of $E_\alpha \cup E_{\alpha+1}$. Merging firstly the finite generator system on neighborhood of E_1 with that of E_2, we obtain a finite generator system of $\mathscr{F}$ on a neighborhood of $E_1 \cup E_2$ (Fig. 4.5).

Similarly, merging it with the finite generator system $\{\sigma_{3j}\}_j$ on a neighborhood of E_3, we obtain a finite generator system of $\mathscr{F}$ on a neighborhood $\bigcup_{\alpha=1}^{3} E_\alpha$. Repeating this, we obtain a finite generator system of $\mathscr{F}$ on a neighborhood of $\bigcup_{\alpha=1}^{L} E_\alpha = E$.

(ii) Let $\mathscr{F}$, $\{\sigma_j\}$, and σ be as given. Take the closed cube E as above in (i), and use the same notation.

For every $t \in [0, T]$, E_t is a closed cube of dimension $\nu - 1$. By the induction hypothesis for (ii) there are holomorphic functions a_{tj} in a neighborhood of E_t such that

$$\sigma = \sum_j a_{tj}\sigma_j.$$

Similarly to the arguments in (i) we get a finite partition (4.3.8) such that in a neighborhood U_α of E_α there are holomorphic functions $a_{\alpha j} \in \mathscr{O}(U_\alpha)$ satisfying

$$\sigma = \sum_j a_{\alpha j}\sigma_j \quad (\text{in } U_\alpha).$$

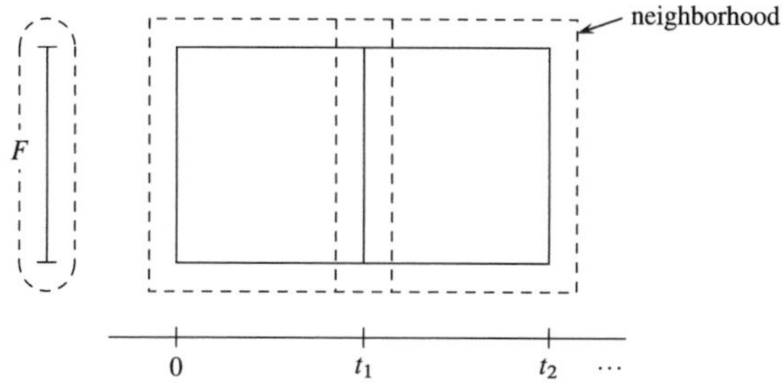

Fig. 4.5 $F \times [0, t_\alpha]$

Let $\mathscr{R} := \mathscr{R}((\sigma_j)_j)$ be the relation sheaf defined by $(\sigma_j)_j$. By Oka's First Coherence Theorem 2.5.1 $\mathscr{R}$ is coherent in a neighborhood of E. For another index β we have

$$\sum_j (a_{\alpha j} - a_{\beta j})\sigma_j = 0$$

on $U_\alpha \cap U_\beta$. Therefore,

$$(b_{\alpha\beta j})_j := (a_{\alpha j} - a_{\beta j})_j \in \Gamma(U_\alpha \cap U_\beta, \mathscr{R}).$$

Since $\mathscr{R}$ is coherent, by (i) above in the case of $\dim E = \nu$, $\mathscr{R}$ has a finite generator system $\{\tau_h\}$ on a neighborhood of E. Again, since $E_\alpha \cap E_\beta$ (assumed $\neq \emptyset$) is a closed cube of dimension $\nu - 1$, there are holomorphic functions $c_{\alpha\beta h}$ in a neighborhood of it such that

$$(b_{\alpha\beta j})_j = \sum_h c_{\alpha\beta h}\tau_h.$$

After taking U_α smaller, we may assume that this holds in $U_\alpha \cap U_\beta$. Choosing a sufficiently small convex cylinder neighborhood Ω of E, and considering U_α intersected with Ω, we may assume that $\{U_\alpha\}$ is an open covering of Ω. We take $c_{\beta\alpha h}$ so that $c_{\beta\alpha h} = -c_{\alpha\beta h}$ holds. For three distinct indices α, β, γ, it may be assumed that $U_\alpha \cap U_\beta \cap U_\gamma = \emptyset$. Thus we get a 1-cocycle $(c_{\alpha\beta h})_{\alpha,\beta} \in Z^1(\{U_\alpha\}, \mathscr{O}_\Omega)$. It follows from Theorem 3.6.10 and Proposition 3.4.11 that $H^1(\{U_\alpha\}, \mathscr{O}_\Omega) = 0$. Therefore, there is an element $(d_{\alpha h}) \in \mathscr{O}(U_\alpha)$ such that

$$c_{\alpha\beta h} = d_{\beta h} - d_{\alpha h} \quad (\text{on } U_\alpha \cap U_\beta)$$

for every α, β.

Thus, we see that

$$(a_{\alpha j} - a_{\beta j})_j = \sum_h (d_{\beta h} - d_{\alpha h})\tau_h.$$

Writing $\tau_h = (\tau_{hj})$ with entries, we deduce that

$$a_{\alpha j} + \sum_h d_{\alpha h}\tau_{hj} = a_{\beta j} + \sum_h d_{\beta h}\tau_{hj}.$$

Hence this defines a holomorphic function $a_j \in \mathscr{O}(\Omega)$ on Ω. Since τ_h are sections of $\mathscr{R}$, it has been shown that

$$\sigma = \sum_j a_j\sigma_j \quad (\text{on } \Omega).$$

□

Lemma 4.3.9 (Oka's Syzygies, Oka VII) *Let $\mathscr{F}$ be a coherent sheaf over a neighborhood of a closed cube $E \Subset \mathbf{C}^n$. Then there exist Oka's Syzygies of $\mathscr{F}$ of an arbitrary length p on a neighborhood U of E; i.e., an exact sequence*

(4.3.10) $$0 \to \operatorname{Ker}\varphi_p \to \mathscr{O}_U^{N_p} \xrightarrow{\varphi_p} \cdots \xrightarrow{\varphi_2} \mathscr{O}_U^{N_1} \xrightarrow{\varphi_1} \mathscr{F}|_U \to 0$$

exists.

Proof Lemma 4.3.7 (i) guarantees the existence of a short exact sequence on a neighborhood U_1 of E,

(4.3.11) $$0 \to \operatorname{Ker}\varphi_1 \to \mathscr{O}_{U_1}^{N_1} \xrightarrow{\varphi_1} \mathscr{F}|_{U_1} \to 0.$$

By the coherence assumption of $\mathscr{F}$, $\operatorname{Ker}\varphi_1$ is coherent. By applying again Lemma 4.3.7 (i) for $\operatorname{Ker}\varphi_1$, there are a neighborhood U_2 with $E \subset U_2 \subset U_1$ and a short exact sequence of coherent sheaves over U_2

(4.3.12) $$0 \to \operatorname{Ker}\varphi_2 \to \mathscr{O}_{U_2}^{N_2} \xrightarrow{\varphi_2} \operatorname{Ker}\varphi_1|_{U_2} \to 0.$$

Connecting (4.3.11) and (4.3.12) over U_2, one gets the following exact sequence on U_2:

(4.3.13) $$0 \to \operatorname{Ker}\varphi_2 \to \mathscr{O}_{U_2}^{N_2} \xrightarrow{\varphi_2} \mathscr{O}_{U_2}^{N_1} \xrightarrow{\varphi_1} \mathscr{F}|_{U_2} \to 0.$$

Repeating this p times, we obtain Oka's Syzygies of $\mathscr{F}$ of an arbitrary length. □

4.3.3 Oka's Fundamental Lemma

It is the idea of *Oka's Jôku-Ikô* to transfer a problem over a domain of n variables to the one over a simple *polydisk*, but with increasing the number of variables more (Oka I). We here prepare a key lemma for coherent sheaves over polydisks or equivalently convex cylinder domains (cf. Lemma 4.3.15), which is a generalization of Theorem 3.6.10 for $\mathscr{O}_{\mathrm{P}\Delta}$ to the one for any coherent sheaf over $\mathrm{P}\Delta$: This seems to be simple in a glance, but in fact it is the most innovative part of Oka's idea to establish the present goal, the Oka–Cartan Fundamental Theorem 4.4.2 for arbitrary holomorphically convex domains Ω (or Stein manifolds M as in Theorem 4.5.8) and for coherent sheaves, even just for the sheaf $\mathscr{O}_\Omega$ (or $\mathscr{O}_M$) itself.

This generalization enables us to use *Oka's Jôku-Ikô* freely, so that the logical distance from the case of polydisks (equivalently, convex cylinder domains) to the case of holomorphically convex domains is the same as that to Stein manifolds (provided that the notion of Stein manifolds is known). In this sense, the essential part of the Oka–Cartan Fundamental Theorem 4.4.2 is contained in Oka's Fundamental Lemma 4.3.15: It was was first proved in Oka VII.[2]

[2] In an essay, K. Oka writes that the *Jôku-Ikô* in Oka I, and the notion and the proof of *Coherence* in Oka VII each took seven years to invent. This is the reason why we call Lemma 4.3.15 *Oka's Fundamental Lemma.*

We keep the same notation as before. In Lemma 4.3.7 of Oka's Syzygy we saw that every short exact sequence of coherent sheaves

$$0 \to \operatorname{Ker}\varphi \to \mathscr{O}_U^N \xrightarrow{\varphi} \mathscr{F}|_U \to 0$$

leads to a surjection φ_* and a long exact sequence with $\mathscr{O}_n = \mathscr{O}_{\mathbf{C}^n}$ as follows:

$$\Gamma(E, \mathscr{O}_n^N) \xrightarrow{\varphi_*} \Gamma(E, \mathscr{F}) \to 0 \to H^1(E, \operatorname{Ker}\varphi) \to H^1(E, \mathscr{O}_n^N) \to \cdots.$$

Since $H^1(E, \mathscr{O}_n^N) = 0$ (cf. Theorem 3.6.10), this implies that $H^1(E, \operatorname{Ker}\varphi) = 0$. On the other hand, if $H^1(E, \operatorname{Ker}\varphi) = 0$, we obtain a surjection

$$\Gamma(E, \mathscr{O}_n^N) \xrightarrow{\varphi_*} \Gamma(E, \mathscr{F}) \to 0.$$

It is noted that this surjection already suffices to obtain a solution of problems of Cousin type (from local to global) by applying Oka's Jôku-Ikô (e.g., Cousin I Problem with respect to $\mathscr{F}$, cf. Sect. 3.7). In what follows we formalize these facts more consistently in terms of cohomologies, the theory of which was developed by H. Cartan with J.-P. Serre, and by which the theory in all gets more comprehensive.

Lemma 4.3.14 *Let $\mathscr{F}$ be a coherent sheaf over a neighborhood of a closed cube E. Then for the interior E° of E we have*

$$H^q(E^\circ, \mathscr{F}) = 0, \quad q \geq 1.$$

Proof We use Lemma 4.3.9 with $p = 2^{2n} - 1$, and restrict the obtained Oka's Syzygies to E°. It follows from the arguments of Sect. 4.3.1 (v)–(vii) that

$$H^q(E^\circ, \mathscr{F}) \cong H^{2^{2n}}(E^\circ, \operatorname{Ker}\varphi_{2^{2n}-q}) = 0, \quad q \geq 1. \qquad \square$$

Thus we have seen that the Oka–Cartan Fundamental Theorem 4.4.2 holds in the interior of E, when the coherent sheaf is defined in a neighborhood of E. It requires one more step of arguments to deal with the case where the coherent sheaf is defined only in the interior of E.

Lemma 4.3.15 (Oka's Fundamental Lemma) *Let $\Omega \subset \mathbf{C}^n$ be a convex cylinder domain, and let $\mathscr{F} \to \Omega$ be a coherent sheaf. Then,*

$$H^q(\Omega, \mathscr{F}) = 0, \quad q \geq 1.$$

Proof (a) By Riemann's Mapping Theorem Ω may be assumed to be an open cube. We take an increasing open covering of Ω by relatively compact open cubes:

$$\Omega_\nu \Subset \Omega_{\nu+1}, \quad \nu = 1, 2, \ldots, \quad \bigcup_{\nu=1}^{\infty} \Omega_\nu = \Omega.$$

Let $\mathscr{U} = \{U_\alpha\}$ be a locally finite open covering of Ω by relatively compact open cubes U_α. Since the support $|\sigma|$ of every q-simplex $\sigma \in N_q(\mathscr{U})$ is an open cube ($\Subset \Omega$), it follows from Lemma 4.3.14 that

$$H^q(|\sigma|, \mathscr{F}) = 0, \quad q \geq 1.$$

Therefore, $\mathscr{U}$ is a Leray covering with respect to $\mathscr{F}$; in the same way, the restriction $\mathscr{U}_\nu = \{U_\alpha \cap \Omega_\nu\}_\alpha$ of $\mathscr{U}$ to every Ω_ν is a Leray covering with respect to $\mathscr{F}|_{\Omega_\nu}$. We thus see by Theorem 3.4.40 and Lemma 4.3.14 that

$$\text{(4.3.16)} \qquad \begin{aligned} &\text{(i) } H^q(\Omega, \mathscr{F}) \cong H^q(\mathscr{U}, \mathscr{F}), \quad q \geq 0; \\ &\text{(ii) } H^q(\Omega_\nu, \mathscr{F}) \cong H^q(\mathscr{U}_\nu, \mathscr{F}) = 0, \quad q \geq 1. \end{aligned}$$

Now we take any $[f] \in H^q(\mathscr{U}, \mathscr{F})$ with $f \in Z^q(\mathscr{U}, \mathscr{F})$. Consider the restriction $f|_{\Omega_\nu} \in Z^q(\mathscr{U}_\nu, \mathscr{F})$. By (4.3.16) (ii) there is an element $g_\nu \in C^{q-1}(\mathscr{U}_\nu, \mathscr{F})$ such that

$$\text{(4.3.17)} \qquad f|_{\Omega_\nu} = \delta g_\nu, \quad \nu = 1, 2, \ldots.$$

(b) Case of $q \geq 2$: Set $\tilde{g}_1 = g_1$. Assume that elements up to $\tilde{g}_\nu \in C^{q-1}(\mathscr{U}_\nu, \mathscr{F})$ are set so that

$$\text{(4.3.18)} \qquad \begin{aligned} &f|_{\Omega_\nu} = \delta \tilde{g}_\nu, \\ &\tilde{g}_\nu(\sigma) = \tilde{g}_{\nu-1}(\sigma), \quad \sigma \in N_{q-1}(\mathscr{U}) \text{ with } |\sigma| \subset \Omega_{\nu-1}. \end{aligned}$$

By (4.3.17), $\delta(\tilde{g}_\nu - g_{\nu+1}|_{\Omega_\nu}) = 0$, and then there is an $h_{\nu+1} \in C^{q-2}(\mathscr{U}_\nu, \mathscr{F})$ satisfying

$$\text{(4.3.19)} \qquad \tilde{g}_\nu - g_{\nu+1}|_{\Omega_\nu} = \delta h_{\nu+1}.$$

We extend $h_{\nu+1}$ over the whole Ω as follows:

$$\text{(4.3.20)} \qquad \tilde{h}_{\nu+1}(\sigma) = \begin{cases} h_{\nu+1}(\sigma), & \sigma \in N_{q-2}(\mathscr{U}),\ |\sigma| \subset \Omega_\nu, \\ 0, & \text{otherwise.} \end{cases}$$

Thus, $\tilde{h}_{\nu+1} \in C^{q-2}(\mathscr{U}, \mathscr{F})$. Set

$$\tilde{g}_{\nu+1} = g_{\nu+1} + \delta \tilde{h}_{\nu+1}|_{\Omega_{\nu+1}}.$$

It follows from (4.3.19) and (4.3.20) that

$$\begin{aligned} &\delta \tilde{g}_{\nu+1} = \delta g_{\nu+1} = f|_{\Omega_{\nu+1}}, \\ &\tilde{g}_{\nu+1}(\sigma) = \tilde{g}_\nu(\sigma), \quad \sigma \in N_{q-1}(\mathscr{U}),\ |\sigma| \subset \Omega_\nu. \end{aligned}$$

This implies that the limit $\tilde{g} = \lim_{\nu\to\infty} \tilde{g}_\nu \in C^{q-1}(\mathscr{U}, \mathscr{F})$ can be defined so that $\delta\tilde{g} = f$. Hence, $[f] = 0$ is shown.

(c) Case of $q = 1$: In the case of $q \geq 2$, it was possible to adjust $\tilde{g}_{\nu+1}$ so that it is equal to $\tilde{g}_\nu$ on $\Omega_{\nu-1}$ by choosing a suitable element of $C^{q-2}(\mathscr{U}, \mathscr{F})$. In the case of $q = 1$, there is no "$C^{-1}(\mathscr{U}, \mathscr{F})$", and this method cannot be applied. In place of it we will use a more substantial approximation theorem.

In the sequel, we set the following convention: "on a closed set $\bar{\Omega}_\nu$" means "on an open neighborhood (say, some open cube) of $\bar{\Omega}_\nu$", and "in $\bar{\Omega}_\nu$" means "just for points of $\bar{\Omega}_\nu$".

For $f \in Z^1(\mathscr{U}, \mathscr{F})$ given above we consider the restriction $f|_{\bar{\Omega}_\nu}$. By making use of (4.3.16) (ii) we re-choose $g_\nu \in C^0(\mathscr{U}|_{\bar{\Omega}_\nu}, \mathscr{F})$ so that on $\bar{\Omega}_\nu$,

(4.3.21) $$\delta g_\nu = f|_{\bar{\Omega}_\nu}.$$

By Lemma 4.3.7 there is a finite generator system $\{\sigma_{(\nu)j}\}_{j=1}^{M_\nu}$ of $\mathscr{F}$ on each $\bar{\Omega}_\nu$, which is fixed. By Lemma 4.3.7 (ii) there are holomorphic functions

$$\alpha_{(\nu,\nu-1)jk}, \quad 1 \leq j \leq M_\nu, 1 \leq k \leq M_{\nu-1}$$

on $\bar{\Omega}_{\nu-1}$ ($\nu \geq 2$) such that

(4.3.22) $$\sigma_{(\nu)j} = \sum_{k=1}^{M_{\nu-1}} \alpha_{(\nu,\nu-1)jk}\sigma_{(\nu-1)k} \quad \text{on } \bar{\Omega}_{\nu-1}.$$

By repeating use of (4.3.22) we have that for $\nu > \mu \geq 1$

(4.3.23) $$\sigma_{(\nu)j} = \sum_{k_h, \mu\leq h\leq\nu-1} \alpha_{(\nu,\nu-1)jk_{\nu-1}}\alpha_{(\nu-1,\nu-2)k_{\nu-1}k_{\nu-2}} \cdots \alpha_{(\mu+1,\mu)k_{\mu+1}k_\mu}\sigma_{(\mu)k_\mu}$$

on $\bar{\Omega}_\mu$. Here, with setting

(4.3.24) $$\alpha_{(\nu,\mu)jk} = \sum_{k_h, \mu+1\leq h\leq\nu-1} \alpha_{(\nu,\nu-1)jk_{\nu-1}}\alpha_{(\nu-1,\nu-2)k_{\nu-1}k_{\nu-2}} \cdots \alpha_{(\mu+1,\mu)k_{\mu+1}k},$$

$\alpha_{(\nu,\mu)jk}$ are holomorphic functions on $\bar{\Omega}_\mu$, and the following holds on $\bar{\Omega}_\mu$:

(4.3.25) $$\sigma_{(\nu)j} = \sum_{k=1}^{M_\mu} \alpha_{(\nu,\mu)jk}\sigma_{(\mu)k}.$$

We set $\tilde{g}_1 = g_1$ on $\bar{\Omega}_1$. We are going to define inductively $\tilde{g}_\nu$ on $\bar{\Omega}_\nu$ so that $\delta\tilde{g}_\nu = f|_{\bar{\Omega}_\nu}$ and some convergence condition is satisfied.

Assume that elements up to $\tilde{g}_\nu$ ($\nu \geq 1$) are defined. Since $\delta(g_{\nu+1}|_{\bar{\Omega}_\nu} - \tilde{g}_\nu) = f|_{\bar{\Omega}_\nu} - f|_{\bar{\Omega}_\nu} = 0$, $g_{\nu+1}|_{\bar{\Omega}_\nu} - \tilde{g}_\nu \in \Gamma(\bar{\Omega}_\nu, \mathscr{F})$. By Lemma 4.3.7 (ii) there are holomorphic functions $a_{\nu+1j} \in \mathscr{O}(\bar{\Omega}_\nu)$, $1 \leq j \leq M_{\nu+1}$, such that

$$g_{\nu+1}|_{\bar{\Omega}_\nu} - \tilde{g}_\nu = \sum_{j=1}^{M_{\nu+1}} a_{\nu+1j}\sigma_{(\nu+1)j} \quad (\text{on } \bar{\Omega}_\nu). \tag{4.3.26}$$

By Corollary 1.2.22 (Runge's Approximation) the holomorphic functions $a_{\nu+1j}$ on $\bar{\Omega}_\nu$ are uniformly approximated in $\bar{\Omega}_\nu$ by holomorphic functions $\tilde{a}_{\nu+1j}$ on $\bar{\Omega}_{\nu+1}$ (e.g., in $\Omega_{\nu+2}$), so that

$$\|a_{\nu+1j} - \tilde{a}_{\nu+1j}\|_{\bar{\Omega}_\nu} = \max_{\bar{\Omega}_\nu}\{|a_{\nu+1j} - \tilde{a}_{\nu+1j}|\} < \varepsilon_\nu \tag{4.3.27}$$

for all $j = 1, \ldots, M_{\nu+1}$, where $\varepsilon_\nu > 0$ will be determined later. Set

$$\tilde{g}_{\nu+1} = g_{\nu+1} - \alpha_{\nu+1}, \quad \alpha_{\nu+1} = \sum_{j=1}^{M_{\nu+1}} \tilde{a}_{\nu+1j}\sigma_{(\nu+1)j} \in \Gamma(\bar{\Omega}_{\nu+1}, \mathscr{F}). \tag{4.3.28}$$

Then $\delta\tilde{g}_{\nu+1} = \delta g_{\nu+1} = f|_{\bar{\Omega}_{\nu+1}}$. It is deduced from (4.3.28), (4.3.26) and (4.3.23) that

$$\begin{aligned}\tilde{g}_{\nu+1} - \tilde{g}_\nu &= \sum_{j=1}^{M_{\nu+1}} (a_{\nu+1j} - \tilde{a}_{\nu+1j})\,\sigma_{(\nu+1)j} \quad (\text{on } \bar{\Omega}_\nu)\\ &= \sum_{j,k} (a_{\nu+1j} - \tilde{a}_{\nu+1j})\alpha_{(\nu+1,\mu)jk}\,\sigma_{(\mu)k} \quad (\text{on } \bar{\Omega}_\mu),\\ &\qquad 1 \leq \mu \leq \nu.\end{aligned}$$

Taking ε_ν sufficiently small, we have that

$$\sum_{j,k} \|(a_{\nu+1j} - \tilde{a}_{\nu+1j})\alpha_{(\nu+1,\mu)jk}\|_{\bar{\Omega}_\mu} < \frac{1}{2^{\nu+1}}, \quad 1 \leq \mu \leq \nu. \tag{4.3.29}$$

Thus, $\tilde{g}_{\nu+1}$ is determined. In this way, we take inductively $\tilde{g}_\nu$, $\nu = 2, 3, \ldots$, by (4.3.28) so that (4.3.29) is satisfied.

By approximation (4.3.29) the following function series converges absolutely and uniformly in Ω_ν:

$$b_{\nu k} = \sum_{\lambda=\nu}^{\infty} \sum_{j} (a_{\lambda+1j} - \tilde{a}_{\lambda+1j})\alpha_{(\lambda+1,\nu)jk} \in \mathscr{O}(\Omega_\nu). \tag{4.3.30}$$

In Ω_ν we set

$$G_\nu = \tilde{g}_\nu + \sum_{k=1}^{M_\nu} b_{\nu k}\sigma_{(\nu)k} \in \Gamma(\Omega_\nu, \mathscr{F}), \quad \nu = 1, 2, \ldots. \tag{4.3.31}$$

We claim that $G_{\nu+1}|_{\Omega_\nu} = G_\nu$. In what follows we consider the restrictions to a fixed Ω_ν, and so the restriction symbol "$|_{\Omega_\nu}$" is dropped for simplicity. By (4.3.30) and (4.3.31) we see that

$$\begin{aligned}
G_{\nu+1} &= \tilde{g}_{\nu+1} + \sum_{l=1}^{M_{\nu+1}} b_{\nu+1l}\sigma_{(\nu+1)l} \\
&= \tilde{g}_\nu + \tilde{g}_{\nu+1} - \tilde{g}_\nu + \sum_{l=1}^{M_{\nu+1}} b_{\nu+1l}\sigma_{(\nu+1)l} \\
&= \tilde{g}_\nu + \sum_k \sum_j (a_{\nu+1j} - \tilde{a}_{\nu+1j})\alpha_{(\nu+1,\nu)jk}\,\sigma_{(\nu)k} \\
&\quad + \sum_k \sum_{\lambda=\nu+1}^{\infty} \sum_j (a_{\lambda+1j} - \tilde{a}_{\lambda+1j})\alpha_{(\lambda+1,\nu)jk}\,\sigma_{(\nu)k} \\
&= \tilde{g}_\nu + \sum_k \sum_{\lambda=\nu}^{\infty} \sum_j (a_{\lambda+1j} - \tilde{a}_{\lambda+1j})\alpha_{(\lambda+1,\nu)jk}\,\sigma_{(\nu)k} \\
&= \tilde{g}_\nu + \sum_{k=1}^{M_\nu} b_{\nu k}\,\sigma_{(\nu)k} = G_\nu.
\end{aligned}$$

Therefore the limit $G = \lim_{\nu\to\infty} G_\nu \in C^0(\mathscr{U}, \mathscr{F})$ exists and $\delta G = f$. This finishes the proof of $[f] = 0$. □

Remark 4.3.32 In the case of (c) ($q = 1$) above we fixed a finite generator system of $\mathscr{F}$ on each $\bar{\Omega}_\nu$, and the limit was taken only for the coefficients functions to construct G_ν. It was not an argument such as a topology in the section space $\Gamma(\Omega_\nu, \mathscr{F})$ was introduced and the limit was taken. It is possible to introduce a topology of uniform convergence on compact subsets for the space of sections of $\mathscr{F}$. Interested readers may go to Sect. 8.1. If such a topology was introduced, the arguments to take limits would become simpler, but it requires some more preparations to do so.

Remark 4.3.33 In the vanishing of cohomologies, that of the first cohomology $H^1(*, \star) = 0$ is most important. This is nothing but the Cousin I Problem for coherent sheaves, which K. Oka solved in Oka VII: What Oka proved after obtaining his First Coherence Theorem was the existence of Oka's Syzygies (Lemmas 4.3.7 and 4.3.9). The purpose was an existence theorem, and its content was the Oka–

Cartan Fundamental Theorem 4.4.2 ($q = 1$) formulated by H. Cartan in terms of sheaf cohomology.

It was J.-P. Serre who proposed to deal with cohomologies $H^q(*, \star)$ of general degree $q \geq 1$ (Cartan [11], p. 51). Because of this reason it is called "the Oka–Cartan–Serre Theorem" in some references.

As seen in the above proof, it is natural and consistent for the cohomology theory to handle cohomologies of all degrees. There is, however, an essential difference between the vanishing of the first cohomology and that of cohomologies of degree 2 and above, because the first requires an approximation theorem of Runge type, but the latter does not. The cohomologies of degree 2 and above play here an intermediate role in understanding the first cohomology. In the solution of the Levi Problem (Hartogs' Inverse Problem) it suffices to use the first cohomology (cf. Chap. 7).

4.4 Oka–Cartan Fundamental Theorem

We begin with the definition of an analytic polyhedron.

Definition 4.4.1 Let $G \subset \mathbf{C}^n$ be a domain, and let $f_j \in \mathscr{O}(G)$, $1 \leq j \leq l$, be given. An *$\mathscr{O}(G)$-analytic polyhedron* is a relatively compact connected component or a union of finitely many relatively compact connected components of the open subset

$$\{z \in G;\ |f_j(z)| < 1,\ 1 \leq j \leq l\}.$$

If $f_j \in \mathscr{O}(G)$ is clear, we simply call it an *analytic polyhedron*.

We prove the following on holomorphically convex domains.

Theorem 4.4.2 (Oka–Cartan Fundamental Theorem) *Let $\Omega \subset \mathbf{C}^n$ be a holomorphically convex domain, and let $\mathscr{F} \to \Omega$ be a coherent sheaf. Then we have*

$$H^q(\Omega, \mathscr{F}) = 0, \qquad q \geq 1.$$

We show steps Sect. 4.3.1 (viii) and (ix). To prove those steps we replace open cubes Ω_ν with analytic polyhedra P_ν and the open covering by relatively compact open cubes with an open covering by relatively compact convex cylinder domains in the proof of Oka's Fundamental Lemma 4.3.15; we then repeat the same arguments as there. We now give the proof, comparing the necessary properties.

(a) In general, let $G \subset \mathbf{C}^n$ be a domain and let $X \subset G$ be a closed complex submanifold of dimension k. Let $\mathscr{O}_X$ denote the sheaf of germs of holomorphic functions over X. Locally about every $x \in X$ there is a neighborhood $U \subset G$ of x such that U is biholomorphic to a polydisk $\mathrm{P}\Delta_k \times \mathrm{P}\Delta_{n-k}$ ($\mathrm{P}\Delta_k \subset \mathbf{C}^k, \mathrm{P}\Delta_{n-k} \subset \mathbf{C}^{n-k}$) with center 0, and after identification,

$$X \cap (\mathrm{P}\Delta_k \times \mathrm{P}\Delta_{n-k}) = \mathrm{P}\Delta_k \times \{0\} \quad (0 \in \mathrm{P}\Delta_{n-k}).$$

The geometric ideal sheaf $\mathscr{I}\langle X\rangle$ of X is given by

$$\mathscr{I}\langle X\rangle = \{\underline{f_V}_x \in \mathscr{O}_G; x \in G, f_V|_{V\cap X} = 0\}. \tag{4.4.3}$$

Here V stands for a neighborhood of x where the holomorphic function f_V is defined.

Lemma 4.4.4 *Let the notation be as above, and let $x = (x_1, \ldots, x_k, x_{k+1}, \ldots, x_n) \in \mathrm{P}\Delta_k \times \mathrm{P}\Delta_{n-k}$ be the natural complex coordinate system. Then it follows that*

$$\mathscr{I}\langle X\rangle_x = \sum_{j=k+1}^{n} \mathscr{O}_{G,x} \cdot \underline{x_j}_x, \quad x \in \mathrm{P}\Delta_k \times \mathrm{P}\Delta_{n-k}. \tag{4.4.5}$$

In particular, $\mathscr{I}\langle X\rangle$ is locally finite and hence a coherent sheaf over G.

Proof It suffices to show (4.4.5) at $x = 0$. Let $f(x) = \sum_\alpha c_\alpha x^\alpha$ with $c_\alpha \in \mathbf{C}$ be a holomorphic function about 0, and write

$$f(x) = b_{n-1}(x_1, \ldots, x_{n-1}) + a_n(x_1, \ldots, x_n)x_n,$$

where the expression is unique. We write $b_{n-1}(x_1, \ldots, x_{n-1})$ similarly as

$$b_{n-1}(x_1, \ldots, x_{n-1}) = b_{n-2}(x_1, \ldots, x_{n-2}) + a_{n-1}(x_1, \ldots, x_{n-1})x_{n-1}.$$

By repeating this, we have

$$f(x) = b_k(x_1, \ldots, x_k) + \sum_{j=k+1}^{n} a_j(x_1, \ldots, x_j)x_j.$$

Therefore, the following equivalences hold:

$$\underline{f}_0 \in \mathscr{I}\langle X\rangle_0 \Longleftrightarrow b_k(x_1, \ldots, x_k) = 0 \Longleftrightarrow \underline{f}_0 \in \sum_{j=k+1}^{n} \mathscr{O}_{\mathrm{P}\Delta,0} \cdot \underline{x_j}_0.$$

Thus, $\mathscr{I}\langle X\rangle \subset \mathscr{O}_{\mathrm{P}\Delta}$ is locally finite, and by Proposition 2.4.7 (i) $\mathscr{I}\langle X\rangle$ is coherent. □

Since $\mathscr{O}_X|_{U\cap X} \cong \mathscr{O}_{\mathrm{P}\Delta_k}$, $\mathscr{O}_X$ is coherent as a sheaf over X. We define a *simple extension* $\widehat{\mathscr{O}}_X$ by extending the sheaf $\mathscr{O}_X$ to be 0 outside X over G. Taking account of an exact sequence

$$0 \to \mathscr{I}\langle X\rangle \to \mathscr{O}_G \to \mathscr{O}_G/\mathscr{I}\langle X\rangle \to 0,$$

we see that $\widehat{\mathscr{O}}_X \cong \mathscr{O}_G/\mathscr{I}\langle X\rangle$. It follows from Lemma 4.4.4 and Serre's Theorem 3.3.1 that $\widehat{\mathscr{O}}_X \cong \mathscr{O}_G/\mathscr{I}\langle X\rangle$ is a coherent sheaf over G.

In general, for a coherent sheaf $\mathscr{F}$ over X the *simple extension* $\widehat{\mathscr{F}}$ of $\mathscr{F}$ over G extended to be 0 outside X as above is coherent over G. The locally finite generation follows from the definition. Next, take finitely many $s_j \in \Gamma(U, \widehat{\mathscr{F}})$, $1 \leq j \leq l$, and consider the relation sheaf

$$\mathscr{R}(s_1, \ldots, s_l) = \left\{ \left(\underline{f_j}_x\right) \in \mathscr{O}_G^l;\ x \in G, \sum_{j=1}^{l} \underline{f_j}_x s_j(x) = 0 \right\}.$$

We would like to show that $\mathscr{R}(s_1, \ldots, s_l)$ is a locally finite sheaf of $\mathscr{O}_G$-modules; at $x \in G \setminus X, \mathscr{R}(s_1, \ldots, s_l)_x = \mathscr{O}_{G,x}^l$, and so it holds. In a neighborhood $V \subset G$ of $x \in X$ the assumption implies that $\mathscr{R}(s_1, \ldots, s_l) \otimes (\mathscr{O}_G / \mathscr{I}\langle X \rangle)$ is coherent over $\mathscr{O}_G / \mathscr{I}\langle X \rangle$ in V. Thus, if V is sufficiently small, it is finitely generated over $\mathscr{O}_G / \mathscr{I}\langle X \rangle$ in V. That is, it is finitely generated modulo $\mathscr{I}\langle X \rangle$. Moreover, by Lemma 4.4.4 we may assume that $\mathscr{I}\langle X \rangle$ is finitely generated over $\mathscr{O}_G$ in V. Therefore, $\mathscr{R}(s_1, \ldots, s_l)$ is finitely generated over $\mathscr{O}_G$ in V.

Since the facts above are used frequently, we summarize them:

Theorem 4.4.6 *Let X be a closed complex submanifold of a domain $G \subset \mathbf{C}^n$. Then the following holds.*

(i) *The geometric ideal sheaf $\mathscr{I}\langle X \rangle$ of X is a coherent sheaf over G.*[3]
(ii) *The simple extension $\widehat{\mathscr{O}_X}$ of $\mathscr{O}_X$ over G is coherent in G.*
(iii) *Let $\mathscr{F}$ be a coherent sheaf over X. The simple extension $\widehat{\mathscr{F}}$ of $\mathscr{F}$ over G is coherent in G.*

We then construct a sequence of analytic polyhedra of a holomorphically convex domain which approximates the domain from inside.

Lemma 4.4.7 *A holomorphically convex domain Ω always carries an increasing open covering by $\mathscr{O}(\Omega)$-analytic polyhedra P_ν ($\nu = 1, 2, \ldots$) of Ω such that*

$$P_\nu \Subset P_{\nu+1}, \quad \nu \in \mathbf{N},$$
$$\Omega = \bigcup_\nu P_\nu.$$

Proof Set the boundary distance function of Ω by

$$d(z; \partial\Omega) = \inf\{\|z - \zeta\|; \zeta \in \partial\Omega\}.$$

Fix any $a_0 \in \Omega$. For a sufficiently large $r_0 > 0$,

$$U_1 = \left\{ z \in \Omega;\ \|z\| < r_0,\ d(z, \partial\Omega) > \frac{1}{r_0} \right\} \ni a_0.$$

[3] More generally, Oka's Second Coherence Theorem 6.5.1 says that the geometric ideal sheaf of an analytic subset, which may be singular, is coherent.

Let V_1 be the connected component of this open set containing a_0. Let V_ν be the connected component of

$$U_\nu = \left\{ z \in \Omega;\ \|z\| < \nu r_0,\ d(z, \partial\Omega) > \frac{1}{\nu r_0} \right\} \quad (\nu = 1, 2, \ldots),$$

which contains a_0. By the definition, $U_\nu \Subset U_{\nu+1}$ and $\bigcup_{\nu=1}^{\infty} U_\nu = \Omega$. For any point $z \in \Omega$, we connect a_0 and z by a curve C. Since C is compact and $C \subset \bigcup_{\nu=1}^{\infty} U_\nu$, there is a number ν_0 such that $C \subset U_{\nu_0}$. The connectedness implies $C \subset V_{\nu_0}$, and $z \in V_{\nu_0}$. Thus, we have constructed an increasing sequence of relatively compact domains V_ν of Ω, which cover Ω:

$$V_\nu \Subset V_{\nu+1}, \quad \bigcup_{\nu=1}^{\infty} V_\nu = \Omega.$$

Let $\widehat{\bar{V}}_{1\Omega}$ be the holomorphic hull of $\bar{V}_1$. Since Ω is holomorphically convex, $\widehat{\bar{V}}_{1\Omega}$ is compact. Take a neighborhood W with $\widehat{\bar{V}}_{1\Omega} \Subset W \Subset \Omega$. For a boundary point $a \in \partial W$ of W, $a \notin \widehat{\bar{V}}_{1\Omega}$, and hence there is a holomorphic function $h \in \mathscr{O}(\Omega)$ such that

$$\max_{\widehat{\bar{V}}_{1\Omega}} |h| < |h(a)|.$$

Take θ so that

$$\max_{\widehat{\bar{V}}_{1\Omega}} |h| < \theta < |h(a)|.$$

There is a neighborhood $\omega(a) \subset \Omega$ of a such that

$$\max_{\widehat{\bar{V}}_{1\Omega}} |h| < \theta < |h(z)|, \quad z \in \omega(a).$$

Replacing h with h/θ, we have

$$\max_{\widehat{\bar{V}}_{1\Omega}} |h| < 1 < |h(z)|, \quad z \in \omega(a). \tag{4.4.8}$$

Since ∂W is compact, there exist a finite number of $h_j \in \mathscr{O}(\Omega)$ and $\omega(a_j)$ $(1 \le j \le l)$ such as (4.4.8), and the following holds:

$$\widehat{\bar{V}}_{1\Omega} \subset Q := \{z \in \Omega;\ |h_j(z)| < 1,\ 1 \le j \le l\}, \quad \bigcup_{j=1}^{l} \omega(a_j) \supset \partial W.$$

Because $Q \cap \partial W = \emptyset$, the connected component P_1 of Q containing V_1 is an $\mathscr{O}(\Omega)$-analytic polyhedron, and satisfies

$$V_1 \Subset P_1 \Subset W.$$

Take V_{ν_2} with $V_{\nu_2} \supset \bar{P}_1 \cup \bar{V}_2$. For $\bar{V}_{\nu_2}$ we apply the same argument as done for $\bar{V}_1$, and obtain an $\mathscr{O}(\Omega)$-analytic polyhedron P_2 such that $V_{\nu_2} \Subset P_2 \Subset \Omega$. Repeating this procedure, we obtain an increasing sequence of analytic polyhedra P_ν, $\nu = 1, 2, \ldots$, satisfying the requirements. □

Lemma 4.4.9 *Let $P \Subset G$ be an analytic polyhedron of a domain G in general. Then for every coherent sheaf $\mathscr{S}$ over P we have that*

$$H^q(P, \mathscr{S}) = 0, \quad q \geq 1.$$

Proof Take finitely many $h_i \in \mathscr{O}(G)$, $1 \leq i \leq m$, so that $P \Subset G$ is a finite union of connected components of

$$\{z \in G;\ |h_j(z)| < 1,\ 1 \leq j \leq m\}.$$

Since $P(\Subset \mathbf{C}^n)$ is bounded, there is a polydisk $\mathrm{P}\Delta'$ with $P \subset \mathrm{P}\Delta'$. We consider the following proper embedding as a closed complex submanifold:

$$\iota_P : z \in P \to (z, h_1(z), \ldots, h_m(z)) \in \mathrm{P}\Delta' \times \Delta(0, 1)^m.$$

We set polydisks, $\mathrm{P}\Delta'' = \Delta(0, 1)^m$, $\mathrm{P}\Delta = \mathrm{P}\Delta' \times \mathrm{P}\Delta''$. Since the above embedding is frequently used from now on, we call

(4.4.10) $$\iota_P : P \to \mathrm{P}\Delta' \times \mathrm{P}\Delta'' = \mathrm{P}\Delta$$

the *Oka map* of the analytic polyhedron P (cf. Fig. 4.6).

We identify P with the image $\iota_P(P)(\subset \mathrm{P}\Delta)$. Note that the simple extension $\widehat{\mathscr{S}}$ of $\mathscr{S}$ over $\mathrm{P}\Delta$ is coherent in $\mathrm{P}\Delta$ (Theorem 4.4.6). Since $\mathrm{P}\Delta$ is a convex cylinder domain, Oka's Fundamental Lemma 4.3.15 implies that

$$H^q(\mathrm{P}\Delta, \widehat{\mathscr{S}}) = 0, \quad q \geq 1.$$

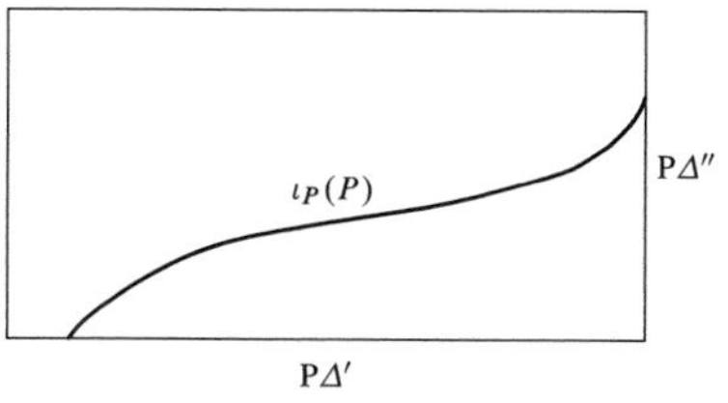

Fig. 4.6 Oka map

By the construction, $H^q(P, \mathscr{S}) \cong H^q(\mathrm{P}\Delta, \widehat{\mathscr{S}})$, and hence the claimed vanishing follows. □

In general, for an open subset $U \subset \mathbf{C}^n$ and finitely many holomorphic functions $f_j \in \mathscr{O}(U)$, $1 \leq j \leq L$, we call

$$Q = \{z \in U;\ |f_j(z)| < 1,\ 1 \leq j \leq L\}$$

a *semi-analytic polyhedron* of U.

Lemma 4.4.11 *Let Q be a semi-analytic polyhedron of a convex cylinder domain G and let $\mathscr{S} \to Q$ be any coherent sheaf. Then,*

$$H^q(Q, \mathscr{S}) = 0, \quad q \geq 1.$$

Proof Let Q be written with finitely many $f_j \in \mathscr{O}(G)$, $1 \leq j \leq m$ as

$$Q = \{z \in G;\ |f_j(z)| < 1, 1 \leq j \leq m\}.$$

By the same idea as Oka maps, the image of the holomorphic injection

$$\iota_Q : z \in Q \to (z, f_1(z), \ldots, f_m(z)) \in G \times \Delta(0; 1)^m$$

is a closed complex submanifold of $R = G \times \Delta(0; 1)^m$, which is identified with Q. Let $\widehat{\mathscr{S}}$ be the simple extension of $\mathscr{S} \to Q$ over the convex cylinder domain R. Since $\widehat{\mathscr{S}}$ is coherent in R (Theorem 4.4.6), Oka's Fundamental Lemma 4.3.15 implies that

$$H^q(Q, \mathscr{S}) \cong H^q(R, \widehat{\mathscr{S}}) = 0, \quad q \geq 1.$$ □

Let $\Omega \subset \mathbf{C}^n$ be a holomorphically convex domain. We take a locally finite open covering $\mathscr{U} = \{U_\alpha\}$ of Ω with convex cylinder domains $U_\alpha \Subset \Omega$. Note that the support $|\sigma|$ of every simplex $\sigma \in N_q(\mathscr{U})$ is a convex cylinder domain. By Oka's Fundamental Lemma 4.3.15 we have

$$H^q(|\sigma|, \mathscr{F}) = 0, \quad q \geq 1 \tag{4.4.12}$$

for any coherent sheaf $\mathscr{F} \to \Omega$, and so $\mathscr{U}$ is a Leray covering with respect to $\mathscr{F}$ (Theorem 3.4.40). Therefore,

$$H^q(\Omega, \mathscr{F}) \cong H^q(\mathscr{U}, \mathscr{F}), \quad q \geq 0. \tag{4.4.13}$$

By Lemma 4.4.7 we take an increasing covering $\{P_\nu\}$ with $\mathscr{O}(\Omega)$-analytic polyhedra P_ν. Consider the restriction $\mathscr{U}_\nu = \{P_\nu \cap U_\alpha\}$ of $\mathscr{U}$ to P_ν. If $U_\alpha \subset P_\nu$, $P_\nu \cap U_\alpha = U_\alpha$ is a convex cylinder domain; otherwise, it is a semi-analytic polyhedron of a convex cylinder domain. Therefore it follows from Lemma 4.4.11 that

$$H^q(|\tau|, \mathscr{F}) = 0, \quad \tau \in N_q(\mathscr{U}_\nu), q \geq 1.$$

We see that $\mathscr{U}_\nu$ is a Leray covering of P_ν with respect to $\mathscr{F}$, and by Lemma 4.4.9 that

$$H^q(\mathscr{U}_\nu, \mathscr{F}) \cong \begin{cases} H^0(P_\nu, \mathscr{F}), & q = 0, \\ H^q(P_\nu, \mathscr{F}) = 0, & q \geq 1. \end{cases} \tag{4.4.14}$$

(b) *Proof of Oka–Cartan Fundamental Theorem 4.4.2* for $q \geq 2$: With the preparation above, we can give a proof for the case of $q \geq 2$. We follow the proof of Oka's Fundamental Lemma 4.3.15, replacing Ω_ν with P_ν: In the same way as the proof (b) of the lemma, for every $f \in Z^q(\mathscr{U}, \mathscr{F})$, by (4.4.14) we find $\tilde{g}_\nu \in C^{q-1}(\mathscr{U}_\nu, \mathscr{F})$ satisfying

$$\delta \tilde{g}_\nu = f|_{\Omega_\nu},$$
$$\tilde{g}_\nu(\sigma) = \tilde{g}_{\nu-1}(\sigma), \quad \sigma \in N_{q-1}(\mathscr{U}), \ |\sigma| \subset \Omega_{\nu-1}.$$

Therefore, the limit $\tilde{g} = \lim_{\nu \to \infty} \tilde{g}_\nu \in C^{q-1}(\mathscr{U}, \mathscr{F})$ is defined and $\delta \tilde{g} = f$ holds. Thus, $[f] = 0 \in H^q(\mathscr{U}, \mathscr{F})$.

(c) *Proof of Oka–Cartan Fundamental Theorem 4.4.2* for $q = 1$: In this case, in addition to what was proved in (a), it is necessary to construct a finite generator system of $\mathscr{F}$ in a neighborhood of $\bar{P}_\nu$, and to approximate functions of $\mathscr{O}(P_\nu)$ by those of $\mathscr{O}(\Omega)$.

Lemma 4.4.15 *Let P be an $\mathscr{O}(G)$-analytic polyhedron of a domain $G \subset \mathbf{C}^n$, and let $\mathscr{S}$ be a coherent sheaf defined in a neighborhood of the closure $\bar{P}$. Then, $\mathscr{S}$ carries a finite generator system defined in a neighborhood of $\bar{P}$; i.e., there are a neighborhood $U \supset \bar{P}$, and finitely many $s_j \in \Gamma(U, \mathscr{S})$, $1 \leq j \leq l$, such that the following is exact:*

$$(\underline{f_j}_z) \in \mathscr{O}_U^l \to \sum_j \underline{f_j}_z s_j(z) \in \mathscr{S}|_U \to 0 \quad (z \in U).$$

Proof We take finitely many $h_i \in \mathscr{O}(G)$, $1 \leq i \leq m$, so that P ($\Subset G$) is a finite union of connected components of

$$\{z \in G; |h_j(z)| < 1, 1 \leq j \leq m\}.$$

With sufficiently small $\varepsilon > 0$, the union $\tilde{P}$ of the connected components of

$$\{z \in G; (1 - \varepsilon)|h_j(z)| < 1, 1 \leq j \leq m\}$$

which contains a point of P, is contained in the neighborhood where $\mathscr{S}$ is defined, and $\bar{P} \subset \tilde{P} \Subset G$. We write $\mathscr{S} = \mathscr{S}|_{\tilde{P}}$.

Let

$$\iota_{\tilde{P}} : \tilde{P} \to \mathrm{P}\Delta' \times \mathrm{P}\Delta'' = \mathrm{P}\Delta$$

be the Oka map of the $\mathscr{O}(G)$-analytic polyhedron $\tilde{P}$. Identify $\tilde{P}$ with the image $\iota_{\tilde{P}}(\tilde{P})$. The simple extension $\widehat{\mathscr{S}}$ of $\mathscr{S}$ over $\mathrm{P}\Delta$ is coherent (Theorem 4.4.6). By Riemann's Mapping Theorem, $\mathrm{P}\Delta$ may be assumed to be an open cube R. Take a closed cube E with $\bar{P} \subset E \Subset R$. By Lemma 4.3.7, $\widehat{\mathscr{S}}$ carries a finite generator system in a neighborhood V of E. Restricting it to $U = V \cap \tilde{P}$, one obtains a finite generator system of $\mathscr{S}$ in U. □

Lemma 4.4.16 *Let $P \Subset G$ and $\mathscr{S}$ be the same as in the previous lemma. Let $s_j \in \Gamma(P, \mathscr{S})$, $1 \le j \le l$, form a finite generator system of $\mathscr{S}$ in P. Then the following is exact (surjective):*

$$(f_j) \in \mathscr{O}(P)^l \to \sum_j f_j s_j \in \Gamma(P, \mathscr{S}) \to 0.$$

Proof By the assumption,

$$\phi : (\underline{f_j}_z) \in \mathscr{O}_P^l \to \sum_j \underline{f_j}_z s_j(z) \in \mathscr{S}|_P$$

is surjective. Oka's First Coherence Theorem 2.5.1 and Serre Theorem 3.3.1 imply the coherence of $\operatorname{Ker}\phi$. We obtain a short exact sequence of coherent sheaves:

$$0 \to \operatorname{Ker}\phi \to \mathscr{O}_P^l \to \mathscr{S}|_P \to 0.$$

From Theorem 3.4.17 the following long exact sequence follows:

$$H^0(P, \mathscr{O}_P^l) \to H^0(P, \mathscr{S}) \to H^1(P, \operatorname{Ker}\phi) \to \cdots .$$

By Lemma 4.4.9, $H^1(P, \operatorname{Ker}\phi) = 0$, and so

$$\phi_* : H^0(P, \mathscr{O}_P^l) \to H^0(P, \mathscr{S}) \to 0$$

is surjective. □

The following is the approximation theorem which plays an essential role in making the solutions obtained on analytic polyhedra P_ν converge.

Lemma 4.4.17 (Runge–Oka Approximation) *Let P be an analytic polyhedron of a domain G. Every holomorphic function in P is arbitrarily approximated uniformly on compact subsets of P by elements of $\mathscr{O}(G)$.*

Proof Let the Oka map of P be

$$\iota_P : z \in P \to (z, h(z)) = (z, w) \in \mathrm{P}\Delta' \times \mathrm{P}\Delta'' = \mathrm{P}\Delta,$$

where $h(z) = (h_j(z)), h_j \in \mathscr{O}(G)$. Identifying P with the image $\iota_P(P)$, we consider the geometric ideal sheaf $\mathscr{I}\langle P\rangle$ of $P \subset \mathrm{P}\Delta$. From Oka's First Coherence Theorem 2.5.1 we obtain a short exact sequence

$$0 \to \mathscr{I}\langle P\rangle \to \mathscr{O}_{\mathrm{P}\Delta} \to \widehat{\mathscr{O}}_P \to 0.$$

By Theorem 3.4.17 the long sequence

$$H^0(\mathrm{P}\Delta, \mathscr{O}_{\mathrm{P}\Delta}) \to H^0(\mathrm{P}\Delta, \widehat{\mathscr{O}}_P) \cong H^0(P, \mathscr{O}_P) \to H^1(\mathrm{P}\Delta, \mathscr{I}\langle P\rangle) \to \cdots$$

is exact. Since $\mathscr{I}\langle P\rangle$ is coherent, $H^1(\mathrm{P}\Delta, \mathscr{I}\langle P\rangle) = 0$ by Oka's Fundamental Lemma 4.3.15. Hence,

$$H^0(\mathrm{P}\Delta, \mathscr{O}_{\mathrm{P}\Delta}) \to H^0(P, \mathscr{O}_P) \to 0.$$

It follows that there is an element $F \in \mathscr{O}(\mathrm{P}\Delta)$ with $F|_P = f$. We expand $F(z, w)$ ($(z, w) \in \mathrm{P}\Delta' \times \mathrm{P}\Delta''$) in a power series:

$$F(z, w) = \sum_{\alpha,\beta} c_{\alpha\beta} z^\alpha w^\beta.$$

For every compact subset $K \Subset P (\subset \mathrm{P}\Delta)$ and every $\varepsilon > 0$ there is a number $N \in \mathbf{N}$ such that

$$\left| F(z, w) - \sum_{|\alpha|, |\beta| \leq N} c_{\alpha\beta} z^\alpha w^\beta \right| < \varepsilon, \quad (z, w) \in K \subset \mathrm{P}\Delta.$$

Substituting $w = h(z)$, one gets

$$\left| f(z) - \sum_{|\alpha|, |\beta| \leq N} c_{\alpha\beta} z^\alpha h^\beta(z) \right| < \varepsilon, \quad z \in K \subset P \subset G.$$

Since $\sum_{|\alpha|, |\beta| \leq N} c_{\alpha\beta} z^\alpha h^\beta(z) \in \mathscr{O}(G)$, the claimed statement is proved. □

Now we begin to prove the case of $q = 1$. With the preparation above, we follow after the proof of Oka's Fundamental Lemma 4.3.15 (c), replacing the open cubes Ω_ν with analytic polyhedra P_ν. For example, in place of Runge's Approximation we use Lemma 4.4.17 to approximate holomorphic functions in P_ν by elements of

$\mathscr{O}(\Omega)$ uniformly on $\bar{P}_{\nu-1}$. As a consequence we can inductively choose the following sequence (cf. (4.3.31)):

$$G_\nu \in C^0(\mathscr{U}_\nu, \mathscr{F}) \quad (\text{on } P_\nu), \quad \nu = 1, 2, \ldots, \tag{4.4.18}$$
$$\delta G_\nu = f|_{P_\nu}, \quad \nu = 1, 2, \ldots,$$

$$G_{\nu+1}|_{P_\nu} = G_\nu, \quad \nu = 1, 2, \ldots. \tag{4.4.19}$$

Thus, the limit $G = \lim_{\nu\to\infty} G_\nu \in C^0(\mathscr{U}, \mathscr{F})$ exists, and satisfies $\delta G = f$. Therefore, $[f] = 0$. This finishes the proof of Theorem 4.4.2. □

Here we give several easy but non-trivial applications of the Oka–Cartan Fundamental Theorem.

Firstly, we have the following from Dolbeault's Theorem 3.6.9.

Corollary 4.4.20 ($\bar\partial$-equation[4]) *Let Ω be a holomorphically convex domain. Then, for every $f \in \Gamma(\Omega, \mathscr{E}_\Omega^{(p,q)})$ with $q \geq 1$ and $\bar\partial f = 0$ there is an element $g \in \Gamma(\Omega, \mathscr{E}_\Omega^{(p,q-1)})$ such that $\bar\partial g = f$.*

Corollary 4.4.21 *Let $\Omega \subset \mathbf{C}^n$ be a domain.*

(i) (**Interpolation Theorem**) *For every discrete subset $X = \{x_\nu\}_{\nu=1}^\infty$ of Ω without accumulation point in Ω and for every sequence of complex numbers $\{\alpha_\nu\}_{\nu=1}^\infty$ there exists a holomorphic function $f \in \mathscr{O}(\Omega)$ with $f(x_\nu) = \alpha_\nu$, $\nu = 1, 2, \ldots$, if and only if Ω is holomorphically convex.*

(ii) *Let $\mathscr{F}$ be a coherent sheaf over a holomorphically convex domain Ω. At every point $a \in \Omega$ the stalk $\mathscr{F}_a$ is generated by a finite number of sections $\sigma_j \in \Gamma(\Omega, \mathscr{F})$, $1 \leq j \leq l$, over $\mathscr{O}_{\Omega,a}$; i.e.,*

$$\mathscr{F}_a = \sum_{j=1}^{l} \mathscr{O}_{\Omega,a} \cdot \sigma_j(a). \tag{4.4.22}$$

Proof (i) Assume that Ω is holomorphically convex. Regarding X as a 0-dimensional closed complex submanifold of Ω, the geometric ideal sheaf $\mathscr{I}\langle X\rangle$ is a coherent sheaf over Ω by Theorem 4.4.6. At each x_ν, $\mathscr{I}\langle X\rangle_{x_\nu}$ coincides with the maximal ideal $\mathfrak{m}_{x_\nu}$ of $\mathscr{O}_{\Omega,x_\nu}$. Therefore,

$$(\mathscr{O}_\Omega/\mathscr{I}\langle X\rangle)_{x_\nu} = \mathscr{O}_{\Omega,x_\nu}/\mathfrak{m}_{x_\nu} \cong \mathbf{C}.$$

Thus it suffices to show that the following natural sequence induced from the restriction map is exact:

---

[4] In this sense the work of Oka [62], I (1936) and II (1937), is the first to solve $\bar\partial$-equation globally on domains of holomorphy in n variables (equivalently, on holomorphically convex domains; cf. Theorem 5.3.1). Cf. Hörmander [33], Notes at the end of Chap. IV.

(4.4.23) $$f \in H^0(\Omega, \mathscr{O}_\Omega) \to f|_X \in H^0(\Omega, \mathscr{O}_\Omega/\mathscr{I}\langle X\rangle) \cong H^0(X, \mathbf{C}) \to 0.$$

From the short exact sequence of coherent sheaves

(4.4.24) $$0 \to \mathscr{I}\langle X\rangle \to \mathscr{O}_\Omega \to \mathscr{O}_\Omega/\mathscr{I}\langle X\rangle \to 0,$$

follows a long exact sequence

(4.4.25) $$H^0(\Omega, \mathscr{O}_\Omega) \to H^0(\Omega, \mathscr{O}_\Omega/\mathscr{I}\langle X\rangle) \to H^1(\Omega, \mathscr{I}\langle X\rangle) \to \cdots.$$

By the Oka–Cartan Fundamental Theorem 4.4.2, $H^1(\Omega, \mathscr{I}\langle X\rangle) = 0$. Therefore, we have (4.4.23).

Conversely, suppose that Ω is not holomorphically convex. Then there is a compact subset $K \Subset \Omega$ with $\hat{K}_\Omega \not\Subset \Omega$. We may take a discrete sequence of points $x_\nu \in \hat{K}_\Omega \subset \Omega$, $\nu = 1, 2, \ldots$, without accumulation point in Ω. Let $\{\alpha_\nu\}_\nu$ be any sequence of numbers such that $|\alpha_\nu| \nearrow \infty$. Then there is no $f \in \mathscr{O}(\Omega)$ satisfying $f(x_\nu) = \alpha_\nu$, $\nu = 1, 2, \ldots$. For by definition,

$$|\alpha_\nu| = |f(x_\nu)| \le \sup_K |f| < \infty, \quad {}^\forall f \in \mathscr{O}(\Omega),\ \nu = 1, 2, \ldots.$$

(ii) Next we consider the case where $X = \{a\}$ and $\mathscr{I}\langle X\rangle = \mathfrak{m}_a$. From (4.4.24), Theorems 3.2.5 and 3.3.6, one obtains a short exact sequence of coherent sheaves,

$$\mathfrak{m}_a \otimes \mathscr{F} \to \mathscr{F} \to (\mathscr{O}_\Omega/\mathfrak{m}_a) \otimes \mathscr{F} \to 0.$$

Let $\mathscr{K}$ denote the kernel of the homomorphism $\mathscr{F} \to (\mathscr{O}_{\Omega,a}/\mathfrak{m}_a) \otimes \mathscr{F} \cong \mathscr{F}/\mathfrak{m}_a\mathscr{F}$. Serre's Theorem 3.3.1 implies the coherence of $\mathscr{K}$, and the short sequence

$$0 \to \mathscr{K} \to \mathscr{F} \to \mathscr{F}/(\mathfrak{m}_a\mathscr{F}) \to 0$$

is exact. By arguments similar to those of (i), $H^1(\Omega, \mathscr{K}) = 0$, and so follows a surjection

(4.4.26) $$H^0(\Omega, \mathscr{F}) \to H^0(\Omega, \mathscr{F}/(\mathfrak{m}_a\mathscr{F})) \cong \mathscr{F}_a/(\mathfrak{m}_a\mathscr{F}_a) \to 0.$$

Take a finite number of generators α_j, $1 \le j \le l$ of $\mathscr{F}_a$ over $\mathscr{O}_{\Omega,a}$. By (4.4.26) there are $\sigma_j \in H^0(\Omega, \mathscr{F})$, $1 \le j \le l$, such that

$$\sigma_j(a) \equiv \alpha_j \pmod{\mathfrak{m}_a\mathscr{F}_a}.$$

That is, there are elements $\underline{h_{jk}}_a \in \mathfrak{m}_a$ such that

$$\sigma_j(a) = \alpha_j + \sum_{k=1}^{l} \underline{h_{jk}}_a \cdot \alpha_k.$$

With Kronecker's symbol δ_{jk} we have

$$\sum_{k=1}^{l} \left(\delta_{jk} + \underline{h_{jk}}_a\right)\alpha_k = \sigma_j(a), \quad 1 \le j \le l.$$

Since $h_{jk}(a) = 0$, $\det\left(\delta_{jk} + h_{jk}(a)\right) = \det(\delta_{jk}) = 1$, and the inverse

$$\left(\delta_{jk} + \underline{h_{jk}}_a\right)^{-1} = (\gamma_{jk}), \quad \gamma_{jk} \in \mathscr{O}_{\Omega,a},$$

exists. One gets $\alpha_j = \sum_k \gamma_{jk}\sigma_k(a)$. Therefore, $\{\alpha_j(a)\}_j$ generates $\mathscr{F}_a$ over $\mathscr{O}_{\Omega,a}$. □

Remark 4.4.27 (i) Readers should notice that this Interpolation Theorem is very non-trivial even in the simplest case such as $\Omega = \mathbf{C}^n$ or $\Omega = B(0; R)$ (open ball).

(ii) H. Cartan named statement (ii) above "Théorème A"; on the other hand, he named the Oka–Cartan Fundamental Theorem "Théorème B". In view of their contents, statement (ii) is an immediate corollary of the Fundamental Theorem 4.4.2. By Proposition 2.4.6 (point-local generation), (4.4.22) holds in a neighborhood $U \ni a$. In this sense, one may say for coherent sheaves that in the converse of the point-local generation, even the "global-point generation" holds over holomorphically convex domains.

We next deal with the analytic de Rham Theorem. The analytic de Rham complex (3.5.21) leads to the *analytic de Rham cohomology*

$$(4.4.28)\quad H^q_{\mathrm{ADR}}(\Omega, \mathbf{C}) := \mathscr{H}^q\left(\Omega, \left\{\mathscr{O}^{(p)}_{\Omega}\right\}_{p\ge 0}\right)$$
$$= \left\{f \in \Gamma\left(\Omega, \mathscr{O}^{(q)}_{\Omega}\right); df = 0\right\} / d\Gamma\left(\Omega, \mathscr{O}^{(q-1)}_{\Omega}\right), \quad q \ge 0.$$

Here $\mathscr{O}^{(-1)}_{\Omega} := 0$.

Corollary 4.4.29 (Analytic de Rham Theorem) *If $\Omega \subset \mathbf{C}^n$ is holomorphically convex,*

$$H^q(\Omega, \mathbf{C}) \cong H^q_{\mathrm{ADR}}(\Omega, \mathbf{C}), \quad q \ge 0.$$

In particular, $H^q(\Omega, \mathbf{C}) = 0$, $q \ge n+1$.

Proof By Corollary 3.5.20 there is a resolution of **C**:

$$0 \to \mathbf{C} \to \mathscr{O}_\Omega = \mathscr{O}_\Omega^{(0)} \to \mathscr{O}_\Omega^{(1)} \to \cdots \to \mathscr{O}_\Omega^{(n)} \to 0.$$

Since all $\mathscr{O}_\Omega^{(p)}$ are coherent sheaves, the Oka–Cartan Fundamental Theorem 4.4.2 implies that $H^q(\Omega, \mathscr{O}_\Omega^{(p)}) = 0$, $q \geq 1$, $p \geq 0$. The required isomorphism is deduced from Theorem 3.4.30. □

Cousin I, II Problems. Since the Oka–Cartan Fundamental Theorem 4.4.2 was proved, the Cousin I and II Problems (Sect. 3.7) have been solved over holomorphically convex domains. The problems were, however, originally proposed over domains of holomorphy. In the next chapter we will show the equivalence between those two kinds of domains. After it we will discuss the Cousin Problems (Sect. 5.5).

4.5 Oka–Cartan Fundamental Theorem on Stein Manifolds

The concept of a Stein manifold is an abstraction of the properties which holomorphically convex domains carry. We begin with the definition of complex manifolds.

4.5.1 Complex Manifolds

4.5.1.1 Differentiable Manifolds

We give the definition and the elementary properties of differentiable manifolds. Cf., e.g., Murakami [44] or Matsushima [41] for more details.

Definition 4.5.1 (*Differentiable manifold*) A connected Hausdorff topological space M is called a *differentiable manifold* or C^∞*-manifold* if the following conditions are satisfied:

(i) There are an open covering $M = \bigcup_{\alpha \in \Gamma} U_\alpha$ and a homeomorphism $\phi_\alpha : U_\alpha \to \Omega_\alpha$ from every U_α to an open subset Ω_α of $\mathbf{R}^n$.
(ii) For every $U_\alpha \cap U_\beta \neq \emptyset$ the restriction map

$$\phi_\beta \circ \phi_\alpha^{-1}|_{\phi_\alpha(U_\alpha \cap U_\beta)} : \phi_\alpha(U_\alpha \cap U_\beta) \to \phi_\beta(U_\alpha \cap U_\beta)$$

is differentiable (of C^∞-class).

It is not necessary to assume the connectedness of M, but it suffices to deal with each connected component: Henceforth, M is assumed to be connected. We call the above n the (real) *dimension* of M, and write $n = \dim M$ ($\dim_{\mathbf{R}} M$). The triple $(U_\alpha, \phi_\alpha, \Omega_\alpha)$ is called a *local chart* of M. For $x \in U_\alpha, \phi_\alpha(x) = (x_\alpha^1, \ldots, x_\alpha^n)$ is called a *local coordinate system*. When U_α is considered as a neighborhood of a point $x \in U_\alpha$, U_α or $(U_\alpha, \phi_\alpha, \Omega_\alpha)$ is called a *local coordinate neighborhood*. A function $f : U \to \mathbf{C}$ defined on an open subset U of M is said to be differentiable or of C^∞-class if for every $U_\alpha \cap U \neq \emptyset, f \circ \phi_\alpha^{-1}(x_\alpha^1, \ldots, x_\alpha^n)$ is a differentiable function (of C^∞-class) in $(x_\alpha^1, \ldots, x_\alpha^n) \in \phi_\alpha(U \cap U_\alpha)$. This property is independent of the choice of U_α, and it is well-defined. Similarly, a continuous map $F : N \to M$ between differentiable manifolds is differentiable if for every $x \in N$ and a local chart $(U_\alpha, \phi_\alpha, \Omega_\alpha)$ of M with $F(x) \in U_\alpha$,

$$\phi_\alpha \circ F|_{F^{-1}U_\alpha} = (F_\alpha^1, \ldots, F_\alpha^n),$$

holds with differentiable functions F_α^j.

In general, a topological space X is locally compact if every point $x \in X$ carries a fundamental neighborhood system $\{V_\alpha\}$ such that the closure $\bar{V}_\alpha$ is compact. A differentiable manifold is locally compact.

If X carries a countable base of open sets, X is said to satisfy the second countability axiom. We say X to be *σ-compact* if X carries an increasing open covering $\{U_j\}_{j=1}^\infty$ such that $\bar{U}_j$ is compact and $\bar{U}_j \Subset U_{j+1}, j = 1, 2, \ldots$. The following fact is known:

Theorem 4.5.2 *Let X be a locally compact topological space. Then the following three conditions are equivalent:*

(i) *X satisfies the second countability axiom.*
(ii) *X is paracompact (Definition 3.4.15).*
(iii) *X is σ-compact.*

The following is elementary.

Theorem 4.5.3 (Partition of unity) *Let M be a paracompact differentiable manifold and $\{U_\alpha\}_{\alpha \in \Gamma}$ be a locally finite open covering of M. Then there is a family $\{c_\alpha\}_{\alpha \in \Gamma}$ of C^∞ functions $c_\alpha : U_\alpha \to \mathbf{R}^+$ satisfying the following three conditions:*

(i) $0 \le c_\alpha \le 1$ *for every* α.
(ii) $\operatorname{Supp} c_\alpha \subset U_\alpha$ *for every* α.
(iii) $\sum_\alpha c_\alpha = 1$.

Remark 4.5.4 Similarly to Example 3.4.34, the sheaf $\mathscr{E}_M$ of germs of C^∞-functions over a paracompact differentiable manifold M is a fine sheaf, and hence every sheaf $\mathscr{S}$ of $\mathscr{E}_M$-modules over M is fine. Therefore it follows from Theorem 3.4.35 that

$$H^q(M, \mathscr{S}) = 0, \quad q \ge 1.$$

4.5.2 Complex Manifolds

Complex manifolds are defined by replacing differentiable maps with holomorphic maps as follows.

Definition 4.5.5 (*Complex manifold*) A connected Hausdorff topological space M is a *complex manifold* if the following conditions are satisfied:

(i) There are an open covering $M = \bigcup_{\alpha \in \Gamma} U_\alpha$ and a homeomorphism $\phi_\alpha : U_\alpha \to \Omega_\alpha$ from every U_α onto an open subset Ω_α of $\mathbf{C}^n$.
(ii) For every $U_\alpha \cap U_\beta \neq \emptyset$, the restriction

$$\phi_\beta \circ \phi_\alpha^{-1}|_{\phi_\alpha(U_\alpha \cap U_\beta)} : \phi_\alpha(U_\alpha \cap U_\beta) \to \phi_\beta(U_\alpha \cap U_\beta)$$

is biholomorphic.

The above n is called the (complex) *dimension* of M, and is denoted by $n = \dim M$ ($\dim_{\mathbf{C}} M$); in particular, in the case of $n = 1$, M is called a *Riemann surface*.

Similarly to the case of differentiable manifolds, the triple $(U_\alpha, \phi_\alpha, \Omega_\alpha)$ is called a *local chart* of M. For $x \in U_\alpha$, $\phi_\alpha(x) = (x_\alpha^1, \ldots, x_\alpha^n)$ is called a *holomorphic local coordinate system* of M, and U_α is called a *holomorphic local coordinate neighborhood*.

By making use of holomorphic local charts, holomorphic functions on an open subset of M, and holomorphic or biholomorphic maps between two complex manifolds are defined as in the case of differentiable manifolds above.

We denote by $\mathscr{O}_M$ the sheaf of germs of holomorphic functions over a complex manifold M, and by $\mathscr{O}(M)$ $(= \Gamma(M, \mathscr{O}_M) = H^0(M, \mathscr{O}_M))$ the space of all holomorphic functions on M.

Let $\pi : M \to N$ be a holomorphic map between complex manifolds. If $\pi : M \to N$ is a topological covering, i.e., π is surjective and for every point $y \in N$ there is a connected neighborhood V of y such that for each connected component U of $\pi^{-1}V$, the restriction $\pi|_U : U \to V$ is biholomorphic, $\pi : M \to N$ is called an *unramified covering* over N. Here, if $\pi : U \to V$ is finite (cf. Convention), $\pi : M \to N$ is called a *ramified covering* over N.

Definition 4.5.6[5] We say that $\pi : M \to N$ is an *unramified covering domain* over N if for every $x \in M$ there are neighborhoods $U \ni x$ and $V \ni \pi(x)$ such that $\pi|_U : U \to V$ is biholomorphic. With this property we say that π is *locally biholomorphic*.

In general, if $\pi|_U : U \to V$ is not necessarily biholomorphic but a finite map, $\pi : M \to N$ is called a *ramified covering domain* over N. In particular, if $\pi : M \to N$ is an unramified covering domain and π is injective, we call it a *univalent domain*; in this case, M is regarded as a subdomain of N.

[5]The notion dealt with here should be defined over complex spaces with singularities in the final form (cf. Definition 6.9.11).

4.5.3 Stein Manifolds

Definition 4.5.7 A complex manifold M is called a *Stein manifold* if

(0) M satisfies the second countability axiom,

and the following *Stein conditions* (i)–(iii) are satisfied:

(i) (holomorphic separability) For distinct two points $x, y \in M$ there is a holomorphic function $f \in \mathscr{O}(M)$ with $f(x) \neq f(y)$.
(ii) (holomorphic local coordinate system) For every point $x \in M$ there exist elements $f_j \in \mathscr{O}(M)$, $1 \leq j \leq n$, with $n = \dim M$ such that in a neighborhood of x, $(f_j)_{1 \leq j \leq n}$ gives rise to a holomorphic local coordinate system.
(iii) (holomorphic convexity) For every compact subset $K \Subset M$ its holomorphic hull defined by

$$\hat{K}_M = \left\{ x \in M;\ |f(x)| \leq \max_K |f|,\ {}^\forall f \in \mathscr{O}(M) \right\}$$

is again compact.

An $\mathscr{O}(M)$-analytic polyhedron $P \Subset M$ is defined in the same way as in Definition 4.4.1. Let $f_j \in \mathscr{O}(M)$ be those used to define P. Assume that M is a Stein manifold. By making use of Stein conditions (i) and (ii), we may add some more to f_j to form f_j, $1 \leq j \leq l$ so that the map

$$\iota_P = (f_j) : P \to \mathrm{P}\Delta$$

is a proper embedding of P into a polydisk $\mathrm{P}\Delta$ of $\mathbf{C}^l$ as a complex submanifold. This is an *Oka map* in the case of a Stein manifold.

Every point of a complex manifold M carries a basis of neighborhoods consisting of open subsets, biholomorphic to polydisks. Therefore, M carries a locally finite open covering $\mathscr{U} = \{U_\alpha\}$ such that every U_α is biholomorphic to a holomorphically convex domain. Since a connected component of a finite intersection of U_α's is a holomorphically convex domain, $\mathscr{U}$ is a Leray covering with respect to any coherent sheaf.

Again by making use of Stein condition (iii) (Definition 4.5.7), we can construct a sequence of increasing analytic polyhedra

$$P_\nu \Subset P_{\nu+1}, \quad \nu \in \mathbf{N},$$
$$M = \bigcup_\nu P_\nu,$$

in the same way as in Lemma 4.4.7.

Now, we see that all lemmas on coherent sheaves over analytic polyhedra and holomorphically convex domains shown in the previous section hold for coherent sheaves over analytic polyhedra of a Stein manifold M and for those over M. Therefore, we deduce the following.

Theorem 4.5.8 (Oka–Cartan Fundamental Theorem) *For a Stein manifold M and a coherent sheaf $\mathscr{F} \to M$, we have that*

$$H^q(M, \mathscr{F}) = 0, \quad q \geq 1.$$

Similarly to Corollary 4.4.20 we have:

Corollary 4.5.9 ($\bar{\partial}$-equation) *Let M be a Stein manifold. Then, for every $f \in \Gamma(M, \mathscr{E}_M^{(p,q)})$ with $\bar{\partial} f = 0$ there is an element $g \in \Gamma(M, \mathscr{E}_M^{(p,q-1)})$ such that $\bar{\partial} g = f$.*

Similarly to Corollary 4.4.29 we immediately get:

Corollary 4.5.10 (Analytic de Rham Theorem) *Let M be a Stein manifold. Then the following isomorphism holds:*

$$\begin{aligned} H^q(M, \mathbf{C}) &\cong \mathscr{H}^q\left(M, \{\mathscr{O}_\Omega^{(p)}\}_{p \geq 0}\right) \\ &= \left\{f \in \Gamma\left(M, \mathscr{O}_M^{(q)}\right); df = 0\right\} / d\Gamma\left(M, \mathscr{O}_M^{(q-1)}\right), \quad q \geq 0. \end{aligned}$$

In particular, $H^q(M, \mathbf{C}) = 0$, $q \geq n + 1$.

N.B. Although we avoid more repetition, Corollary 4.4.21 is valid for Stein manifolds.

Theorem 4.5.11 (General interpolation; cf. Sect. 6.5) *Let M be a Stein manifold and let Y be an analytic subset of M. Then, for any holomorphic function g on Y there exists a holomorphic function f on M with $f|_Y = g$.*

Proof With the geometric ideal sheaf $\mathscr{I}\langle Y \rangle$ of Y we have a short exact sequence of coherent sheaves (see Theorem 6.5.1)

$$0 \to \mathscr{I}\langle Y \rangle \to \mathscr{O}_M \to \mathscr{O}_M / \mathscr{I}\langle Y \rangle \to 0.$$

It follows from this short exact sequence and the Oka–Cartan Fundamental Theorem 4.5.8 that (cf. Definition 6.5.16)

$$H^0(M, \mathscr{O}_M) \to H^0(M, \mathscr{O}_M / \mathscr{I}\langle Y \rangle)(\cong H^0(Y, \mathscr{O}_Y)) \to H^1(M, \mathscr{I}\langle Y \rangle) = 0.$$

Hence, for $g \in H^0(Y, \mathscr{O}_Y)$ there is an element $f \in H^0(M, \mathscr{O}_M)$ such that $f|_Y = g$. □

Remark 4.5.12 (i) It is known that *every open Riemann surface is a Stein manifold* (cf. Behnke–Stein [5], and [55] for another proof which relies only on the contents presented in this book).

(ii) After Theorem 4.5.11 it is a natural problem to look for a *solution f with estimate*, which has been explored by $\bar{\partial}$–L^2 method (cf., e.g., Ohsawa [59, 60]).

4.5.4 Influence on Other Fields

We have already mentioned that coherent sheaves and the Oka–Cartan Fundamental Theorem provide bases of various fields of modern Mathematics. They are not only complex analysis and complex geometry, but also algebraic geometry, theory of differential equations, Sato's hyperfunction theory, theory of *D*-modules, representation theory, etc. (cf. Chap. 9).

After the success of analytic coherent sheaves, J.-P. Serre [72] introduced the notion of coherence in algebraic geometry in 1955.

In the theory of differential equations, e.g., in K. Aomoto and M. Kita [3], hypergeometric function theory is developed assuming the Analytic de Rham Theorem 4.4.29.

In Sato's hyperfunction theory and the related fields, besides A. Kaneko [34], one finds the following comment in the introduction of M. Kashiwara et al. [36]: After referring to Oka's "theory of ideals of undetermined domains",

> The following theorems are crucial. In particular, Theorem 1.2.2 seems to be one of the most profound results in the field of analysis in this century.

Theorem 1.2.2 above is the Oka–Cartan Fundamental Theorem, referred to without proof.

In books such as M. Kashiwara [35] and T. Tanisaki and R. Hotta [75], the analytic parts are described assuming the coherence and the Oka–Cartan Fundamental Theorem to develop the theory.

Exercises

1 (Related to the Interpolation Theorem). Construct an example for Hartogs' domain $\Omega_{\mathrm{H}}(a;\gamma)$ for which the interpolation problem is insolvable.

2 Let $a_{\nu\mu} \in \mathbf{C}^2$, $(\nu,\mu) \in \mathbf{Z}^2$, be distinct points, and let $\alpha_{\nu\mu} \in \mathbf{C}$, $(\nu,\mu) \in \mathbf{Z}^2$, be complex numbers satisfying

$$\sum_{(\nu,\mu)\in\mathbf{Z}^2} \frac{|\alpha_{\nu\mu}|}{|\nu|^k|\mu|^l} < \infty$$

for some $(k,l) \in \mathbf{N}^2$. Show that the series

$$f(z,w) = (\sin \pi z)^k (\sin \pi w)^l \sum_{(\nu,\mu)\in\mathbf{Z}^2} \frac{(-1)^{\nu k+\mu l}\alpha_{\nu\mu}}{\pi^k \pi^l (z-\nu)^k (w-\mu)^l}$$

converges normally to an entire function $f(z,w)$ satisfying $f(a_{\nu\mu}) = \alpha_{\nu\mu}$ for all $(\nu,\mu) \in \mathbf{Z}^2$.

3 Let $\{a_\nu\}_{\nu=1}^{\infty}$ be a discrete subset of a holomorphically convex domain Ω of $\mathbf{C}^n$ without accumulation point in Ω. For each a_ν a polynomial $P_\nu(z - a_\nu)$ of degree d_ν is assigned. Show that there is a holomorphic function $f \in \mathscr{O}(\Omega)$ satisfying

$$f(z) - P_\nu(z - a_\nu) = O(\|z - a_\nu\|^{d_\nu+1}) \quad (z \to a_\nu), \quad {}^\forall \nu \geq 1.$$

4 For a compact subset $K \subset \mathbf{C}^n$ the *polynomially convex hull* $\hat{K}_P$ is defined by

$$\hat{K}_P = \left\{ z \in \mathbf{C}^n;\ |P(z)| \leq \sup_K |P|,\ {}^\forall P, \text{ polynomial} \right\}.$$

If $\hat{K}_P = K$, K is said to be *polynomially convex.*
Show that a holomorphic function in a neighborhood of a polynomially convex compact subset K of $\mathbf{C}^n$ can be approximated uniformly on K by polynomials on $\mathbf{C}^n$.

5 Let $\mathscr{O}_B(\Omega)$ denote the set of all bounded holomorphic functions on a domain Ω of $\mathbf{C}^n$. For a compact subset $K \subset \Omega$ we define the holomorphically convex $\mathscr{O}_B$-hull $\hat{K}_B$ by

$$\hat{K}_B = \left\{ z \in \Omega;\ |f(z)| \leq \sup_K |f|;\ {}^\forall f \in \mathscr{O}_B(\Omega) \right\}.$$

Assume that $K = \hat{K}_B$. Then, show that a holomorphic function in a neighborhood of K can be approximated uniformly on K by bounded holomorphic functions on Ω.

6 Let $X \subset \mathbf{C}^n$ be a convex domain, and let $Y \subset X$ be the intersection of a complex hyperplane of $\mathbf{C}^n$ with X. Show that for any holomorphic function $g \in \mathscr{O}(Y)$ there is a holomorphic function $f \in \mathscr{O}(X)$ with $f|_Y = g$.

7 Show that any complex submanifold of a Stein manifold is Stein.

8 Let M be a Stein manifold (or a holomorphically convex domain of $\mathbf{C}^n$). Let $f_j \in \mathscr{O}(M)$, $1 \leq j \leq N$, be a finite number of holomorphic functions such that $\{f_1 = \cdots = f_N = 0\} = \emptyset$. Show that there exist holomorphic functions $g_j \in \mathscr{O}(M)$, $1 \leq j \leq N$, such that

$$g_1 f_1 + \cdots + g_N f_N = 1.$$

9 Let M be a complex manifold, satisfying the second countability axiom. Assume that $H^1(M, \mathscr{I}) = 0$ for any coherent ideal sheaf $\mathscr{I}$ of $\mathscr{O}_M$. Then, show that M is Stein.

Chapter 5
Domains of Holomorphy

A domain of holomorphy is defined as a domain with no boundary point b such that there is a neighborhood V of b and all holomorphic functions on the doamin analytically extend over V (no Hartogs' phenomenon happens at any boundary point). From the viewpoint of an existence domain of holomorphic functions this idea is more natural than the holomorphic convexity, and the notion of a domain of holomorphy is historically older than that of a holomorphically convex domain.

Cousin Problems that K. Oka solved had been asked on domains of holomorphy. The Cartan–Thullen Theorem claims equivalence between the two notions. Therefore, the Oka–Cartan Fundamental Theorem holds on domains of holomorphy.

5.1 Envelope of Holomorphy

The notion of a domain of holomorphy was given in Definition 1.2.30. We begin with the definition of the envelope of holomorphy.

Definition 5.1.1 (i) Let $\Omega \subset \Omega'$ be domains of $\mathbf{C}^n$ (or more generally, a complex manifold). If every $f \in \mathscr{O}(\Omega)$ analytically extends on Ω', Ω' is called an *extension of holomorphy* of Ω (Hartogs' phenomenon).

(ii) A maximal domain among the extensions of holomorphy of Ω is called the *envelope of holomorphy* of Ω.

Therefore, Ω is a domain of holomorphy, if and only if Ω is the envelope of holomorphy of itself.

We have seen that Hartogs' phenomenon takes place on Hartogs' domains $\Omega_{\mathrm{H}}(a; \gamma)$ in Sect. 1.2.4. In fact, the polydisk $\mathrm{P}\Delta(a; \gamma)$ is the envelope of holomorphy of $\Omega_{\mathrm{H}}(a; \gamma)$. We here discuss this kind of properties in detail.

We first investigate the support of the solution of Dolbeault's Lemma 3.6.4 in the case of $p = 0$ and $q = 1$:

J. Noguchi, *Analytic Function Theory of Several Variables*,
DOI 10.1007/978-981-10-0291-5_5

Lemma 5.1.2 *Let* $\eta = \sum_{j=1}^{n} \eta_j d\bar{z}_j \in \mathscr{E}^{(0,1)}(\mathbf{C}^n)$ *with* $n \geq 2$, *and assume that*

$$\mathrm{Supp}\,\eta \Subset \mathbf{C}^n,$$
$$\bar{\partial}\eta = \sum_{1 \leq j < k \leq n} \left(-\frac{\partial \eta_j}{\partial \bar{z}_h} + \frac{\partial \eta_k}{\partial \bar{z}_j} \right) d\bar{z}_j \wedge d\bar{z}_h = 0.$$

Then there exists an element $\psi \in \mathscr{E}^{(0,0)}(\mathbf{C}^n)$ *such that* $\bar{\partial}\psi = \eta$ *and* $\mathrm{Supp}\,\psi \Subset \mathbf{C}^n$.

Proof We set

$$\psi(z_1, \ldots, z_n) = \frac{1}{2\pi i} \int_{\mathbf{C}} \frac{\eta_1(\zeta_1, z_2, \ldots, z_n)}{\zeta_1 - z_1} d\zeta_1 \wedge d\bar{\zeta}_1.$$

A variable transformation by $\zeta_1 - z_1 = \xi$ yields

$$\psi(z_1, \ldots, z_n) = \frac{1}{2\pi i} \int_{\mathbf{C}} \frac{\eta_1(z_1 + \xi, z_2, \ldots, z_n)}{\xi} d\xi \wedge d\bar{\xi}.$$

By this expression, $\psi \in \mathscr{E}^{(0,0)}(\mathbf{C}^n)$. If $\|z'\| = \|(z_2, \ldots, z_n)\| \gg 1$, then for every $\xi \in \mathbf{C}$, $\eta_1(\xi, z') \equiv 0$. Therefore,

$$\psi(z_1, \ldots, z_n) = 0, \quad \|z'\| \gg 1. \tag{5.1.3}$$

We see by Lemma 3.6.3 that

$$\frac{\partial \psi}{\partial \bar{z}_1} = \eta_1(z_1, \ldots, z_n).$$

It follows from the condition $\bar{\partial}\eta = 0$ that for $2 \leq h \leq n$,

$$\begin{aligned}
\frac{\partial \psi}{\partial \bar{z}_h} &= \frac{1}{2\pi i} \int_{\mathbf{C}} \frac{\frac{\partial \eta_1}{\partial \bar{z}_h}(z_1 + \xi, z_2, \ldots, z_n)}{\xi} d\xi \wedge d\bar{\xi} \\
&= \frac{1}{2\pi i} \int_{\mathbf{C}} \frac{\frac{\partial \eta_h}{\partial \bar{z}_1}(z_1 + \xi, z_2, \ldots, z_n)}{\xi} d\xi \wedge d\bar{\xi} \\
&= \frac{\partial}{\partial \bar{z}_1} \frac{1}{2\pi i} \int_{\mathbf{C}} \frac{\eta_h(\xi, z_2, \ldots, z_n)}{\xi - z_1} d\xi \wedge d\bar{\xi} \\
&= \eta_h(z_1, z_2, \ldots, z_n).
\end{aligned}$$

Thus we see that $\bar{\partial}\psi = \eta$, and hence

$$\bar{\partial}\psi = 0 \quad \text{in } \mathbf{C}^n \setminus \mathrm{Supp}\,\eta.$$

Therefore ψ is holomorphic in $\mathbf{C}^n \setminus \mathrm{Supp}\,\eta$. By (5.1.3), $\psi = 0$ for $\|z'\| \gg 1$. It follows that $\psi \equiv 0$ in a connected component of $\mathbf{C}^n \setminus \mathrm{Supp}\,\eta$ containing $\|z'\| \gg 1$. Hence, $\mathrm{Supp}\,\psi \Subset \mathbf{C}^n$. □

Theorem 5.1.4 (Hartogs) *Let $\Omega \subset \mathbf{C}^n$ be a domain, let $n \geq 2$, and let $K \Subset \Omega$ be a compact subset. Assume that $\Omega \setminus K$ is connected (important). Then every $f \in \mathscr{O}(\Omega \setminus K)$ analytically continued over Ω. That is, Ω is an extension of holomorphy of $\Omega \setminus K$.*

Proof We take $\varphi \in C_0^\infty(\Omega)$ so that $\varphi \equiv 1$ on a neighborhood of K. Set

$$u_0 = (1 - \varphi)f \in C^\infty(\Omega).$$

We obtain $\eta := \bar{\partial} u_0 = -\bar{\partial}\varphi \cdot f = -f\bar{\partial}\varphi$. Then,

$$\eta \in \mathscr{E}^{(0,1)}(\mathbf{C}^n), \quad \mathrm{Supp}\,\eta \Subset \mathbf{C}^n, \quad \bar{\partial}\eta = 0.$$

By Lemma 5.1.2 there exists an element $g \in \mathscr{E}_0^{(0,0)}(\mathbf{C}^n)$ such that

$$\bar{\partial} g = \eta, \quad \mathrm{Supp}\,g \Subset \mathbf{C}^n.$$

We infer from the assumption that $g = 0$ on $V \cap \Omega$ with a neighborhood V of $\partial\Omega$ (cf. Fig. 5.1).

Setting $\tilde{f} = u_0 - g$, we have

$$\bar{\partial}\tilde{f} = \bar{\partial} u_0 - \bar{\partial} g = \eta - \eta = 0.$$

Therefore, $\tilde{f} \in \mathscr{O}(\Omega)$. Taking V smaller if necessary, we have that $u_0 = f$ on $V \cap \Omega$, and then

$$f(z) = \tilde{f}(z), \quad z \in V \cap \Omega.$$

Because of the connectedness of $\Omega \setminus K$ and the uniqueness of analytic continuation, $f(z) = \tilde{f}(z)$ on $\Omega \setminus K$. □

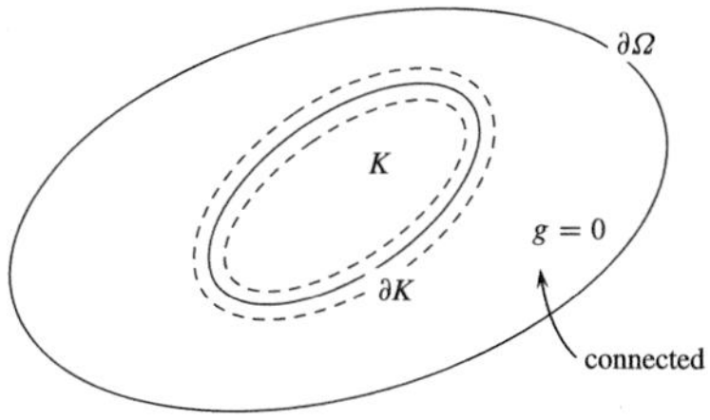

Fig. 5.1 Hartogs' phenomenon

So far the given examples of extensions of holomorphy are all univalent domains, but here we give an example such that an extension of holomorphy of a univalent domain $\Omega \subset \mathbf{C}^n$ is not necessarily univalent.

Example 5.1.5 Let $\Omega \subset \mathbf{C}^2 (\ni (z, w))$ be a domain defined as follows. We fix a branch of $\arg z$ with $\arg 1 = 0$:

$$\frac{1}{2} < |z| < 1,$$
$$-\frac{\pi}{2} + \arg z < |w| < \frac{\pi}{2} + \arg z.$$

When a point z of the annulus moves one round in the anti-clockwise direction, then $\arg z$ increases by 2π, and the defining domain of w with $(z, w) \in \Omega$ does not overlap. Therefore, Ω is univalent (a subdomain of $\mathbf{C}^n$) (cf. Fig. 5.2).

Let $f(z, w) \in \mathscr{O}(\Omega)$. With a point in $\frac{1}{2} < |z| < 1$ given, w forms an annulus such as (b) and (c), where we have a Laurent expansion

$$\text{(5.1.6)} \qquad f(z, w) = \sum_{k=-\infty}^{\infty} a_k(z) w^k, \quad -\frac{\pi}{2} + \arg z < |w| < \frac{\pi}{2} + \arg z,$$
$$a_k(z) = \frac{1}{2\pi i} \int_{|w|=r} \frac{f(z, w)}{w^{k+1}} dw, \quad \text{holomorphic in } z.$$

By the analytic continuation, as z moves to the range $0 < \arg z < \frac{\pi}{2}, f(z, w)$ is holomorphic in $|w| < \frac{\pi}{2} + \arg z$, and hence $a_k(z) \equiv 0,\ k < 0$. By the uniqueness of analytic continuation, $a_k(z) \equiv 0,\ k < 0$ in the series (5.1.6). Thus, it follows from (5.1.6) that

$$f(z, w) = \sum_{k=0}^{\infty} a_k(z) w^k, \quad |w| < \frac{\pi}{2} + \arg z,$$

and $f(z, w)$ is holomorphic in the whole $\{|w| < \frac{\pi}{2} + \arg z\}$. That is, $f(z, w) \in \mathscr{O}(\tilde{\Omega})$, where $\tilde{\Omega}$ is a multivalent domain over $\mathbf{C}^n$ with infinitely many sheets (cf. Definition 4.5.6), defined by:

$$\tilde{\Omega} : \frac{1}{2} < |z| < 1, \quad \arg 1 = 0,$$
$$|w| < \frac{\pi}{2} + \arg z,$$
$$p : (z, w) \in \tilde{\Omega} \to (z, w) \in \mathbf{C}^2 \quad \text{(with infinitely many sheets)}.$$

A domain of this kind is called a Riemann domain and will be discussed in more detail later (cf. Sect. 7.5.1).

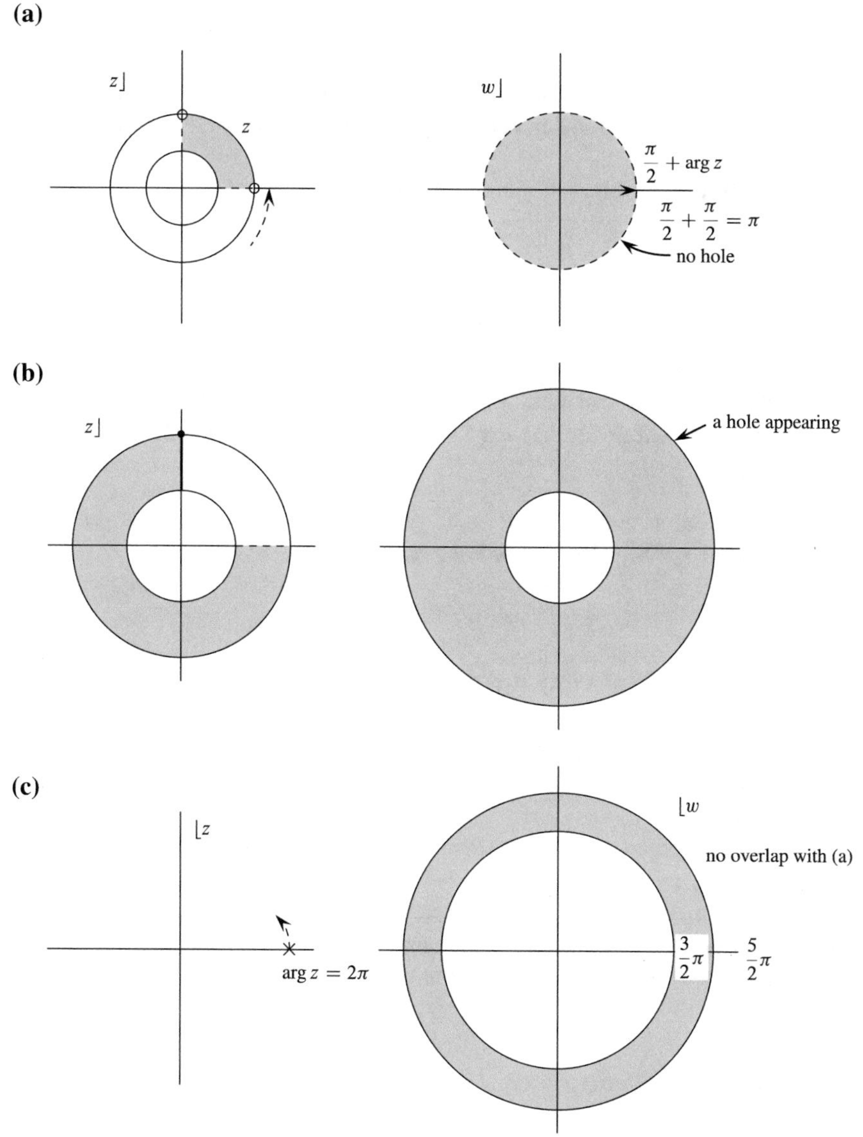

Fig. 5.2 Univalent domain with an infinitely multivalent extension of holomorphy: (**a**) $z : -\frac{\pi}{2} < \arg z < \frac{\pi}{2}$; (**b**) $z : \frac{\pi}{2} \leq \arg z < 2\pi$; (**c**) $z : \arg z \geq 2\pi$

5.2 Reinhardt Domains

Here we explain Reinhardt domains which describe the convergence domains of power series. The shape of a domain of holomorphy and the holomorphic convexity appears clearly.

We first study elementary properties of power series. Let

$$f(z) = \sum_{|\alpha| \geq 0} a_\alpha z^\alpha \quad (z \in \mathbf{C}^n) \tag{5.2.1}$$

be a power series in n variables.

Lemma 5.2.2 *Assume that there are $M \geq 0$ and $w \in \mathbf{C}^n$ with $|a_\alpha w^\alpha| \leq M$ for all $\alpha \in (\mathbf{Z}^+)^n$. Then, the series $f(z)$ in (5.2.1) normally converges in $\mathrm{P}\Delta = \{z; |z_j| < |w_j|, 1 \leq j \leq n\}$.*

Proof We take an arbitrary $0 < \theta < 1$. We may assume that every $w_j \neq 0$. Let $z \in \mathrm{P}\Delta$ with $|z_j| \leq \theta |w_j|$, $1 \leq j \leq n$. We have

$$|a_\alpha z^\alpha| = |a_\alpha| \theta^{|\alpha|} |w^\alpha| \leq M\theta^{|\alpha|}.$$

Thus, the convergence of a majorant series follows:

$$\sum_\alpha |a_\alpha z^\alpha| \leq \sum_{\alpha_1, \ldots, \alpha_n \geq 0} M\theta^{\alpha_1 + \cdots + \alpha_n} = M \left(\frac{1}{1-\theta} \right)^n. \qquad \square$$

To discuss the convergence domain of a power series in more than one variable, we may have the convergence of the power series by making one variable zero, even if it contains some divergent subseries; the power series may converge on a subset without interior points. It does not make sense to discuss the convergence on such a set. Therefore we define the *convergence domain* $\Omega(f)$ of f as follows:

$$\begin{aligned} &\Omega(f)^* = \{w \in (\mathbf{C}^*)^n; \ {}^\exists M > 0, \ |a_\alpha w^\alpha| \leq M, \ {}^\forall \alpha \in (\mathbf{Z}^+)^n\}^\circ, \\ &\Omega(f) = \{z = (z_j) \in \mathbf{C}^n; \ {}^\exists (w_j) \in \Omega(f)^*, \ |z_j| < |w_j|, \ 1 \leq {}^\forall j \leq n\}. \end{aligned} \tag{5.2.3}$$

Here, $\{ \ \}^\circ$ denotes the set of interior points. Since a line segment connecting every point of $\Omega(f)$ and the origin is contained in $\Omega(f)$, $\Omega(f)$ is a domain.

Lemma 5.2.4 *We have that $\Omega(f)^* = \Omega(f) \cap (\mathbf{C}^*)^n$; in particular, $\Omega(f)^*$ is a domain.*

Proof For every point $z \in \Omega(f) \cap (\mathbf{C}^*)^n$, there are by the definition a point $w \in \Omega(f)^*$ and a constant $M > 0$ such that

$$|z_j| < |w_j|, \quad |a_\alpha w^\alpha| \le M.$$

Therefore, $z \in \Omega(f)^*$.

Conversely, we take any $z \in \Omega(f)^*$. For a sufficiently small $\varepsilon > 0$, $(|z_j| + \varepsilon) \in \Omega(f)^*$. Therefore, there is an $M > 0$ such that

$$|a_\alpha|(|z_j| + \varepsilon)^\alpha \le M.$$

With $w_j = |z_j| + \varepsilon$ $(1 \le j \le n)$, $w = (w_j) \in \Omega(f)^*$, and hence $z \in \Omega(f) \cap (\mathbf{C}^*)^n$ follows. □

For every $\theta = (\theta_1, \ldots, \theta_n) \in \mathbf{R}^n$ we have

$$z = (z_1, \ldots, z_n) \in \Omega(f) \Rightarrow e^{i\theta} \cdot z := (e^{i\theta_1} z_1, \ldots, e^{i\theta_n} z_n) \in \Omega(f).$$

(Here, note that "$e^{i\theta}$" is not defined as a number, but just a symbol.) We call the action $e^{i\theta} \cdot z$ a *plurirotation*.

Definition 5.2.5 A domain $\Omega \subset \mathbf{C}^n$ is called a *Reinhardt domain* if Ω is invariant under plurirotations; i.e., for every $\theta = (\theta_1, \ldots, \theta_n) \in \mathbf{R}^n$

$$z \in \Omega \Longrightarrow e^{i\theta} \cdot z \in \Omega.$$

In this case, the boundary $\partial\Omega$ is also invariant under plurirotations.

Theorem 5.2.6 *Let Ω be a Reinhardt domain with $\Omega \ni 0$ and let $f \in \mathscr{O}(\Omega)$. Then $f(z)$ is represented by a unique power series,*

$$f(z) = \sum a_\alpha z^\alpha,$$

and the convergence is normal in Ω.

Proof The uniqueness is determined in a neighborhood of 0 (the Identity Theorem 1.2.14).

For a given $\varepsilon > 0$ we let Ω_ε be the connected component of an open set

$$\{z \in \Omega;\ d(z; \partial\Omega) > \varepsilon\|z\|\},$$

containing the origin 0. Here, $d(z; \partial\Omega)$ denotes the boundary distance function with respect to the Euclidean metric. Since every point $z \in \Omega$ is connected in Ω to 0 by a polygonal line C (which is compact), there is an $\varepsilon > 0$ such that

$$d(C; \partial\Omega) > \varepsilon\|z\|, \quad {}^\forall z \in C.$$

By the definition, $C \subset \Omega_\varepsilon$. Thus, we have $z \in \Omega_\varepsilon$, and

$$\Omega = \bigcup_{\varepsilon > 0} \Omega_\varepsilon.$$

We take an arbitrary point $z \in \Omega_\varepsilon$ and set $\xi_i = (1+\varepsilon)e^{i\theta_j}$ with $\theta = (\theta_j) \in \mathbf{R}^n$. For every $a \in \partial\Omega$,

$$\begin{aligned}\|(\xi_j z_j) - a\| &= \|e^{i\theta} \cdot z - a + \varepsilon e^{i\theta} \cdot z\| \\ &\geq \|e^{i\theta} \cdot z - a\| - \varepsilon \|e^{i\theta} \cdot z\| \\ &\geq d(e^{i\theta} \cdot z, \partial\Omega) - \varepsilon\|z\| = d(z, \partial\Omega) - \varepsilon\|z\| \\ &> \varepsilon\|z\| - \varepsilon\|z\| = 0.\end{aligned}$$

Therefore, $(\xi_i z_j) \in \Omega$ follows. Set

$$g(z_1, \ldots, z_n) = \left(\frac{1}{2\pi i}\right)^n \int \cdots \int_{|\xi_j| = 1+\varepsilon} \frac{f(\xi_1 z_1, \ldots, \xi_n z_n)}{\prod_j (\xi_j - 1)} d\xi_1 \cdots d\xi_n.$$

Then $g \in \mathcal{O}(\Omega_\varepsilon)$. If $\|z\|$ is sufficiently small, the Cauchy integral formula implies that $g(z) = f(z)$. It follows from the Identity Theorem 1.2.14 that for $z \in \Omega_\varepsilon$

$$\begin{aligned}(5.2.7)\quad f(z) &= \left(\frac{1}{2\pi i}\right)^n \int \cdots \int_{|\xi_j|=1+\varepsilon} \frac{f(\ldots, \xi_j z_j, \ldots)}{\prod_j (\xi_j - 1)} d\xi_1 \cdots d\xi_n \\ &= \left(\frac{1}{2\pi i}\right)^n \int \cdots \int_{|\xi_j|=1+\varepsilon} \frac{f(\ldots, \xi_j z_j, \ldots)}{\prod_j \xi_j \left(1 - \frac{1}{\xi_j}\right)} d\xi_1 \cdots d\xi_n \\ &= \left(\frac{1}{2\pi i}\right)^n \int \cdots \int_{|\xi_j|=1+\varepsilon} f(\ldots, \xi_j z_j, \ldots) \prod_{j=1}^n \left(\sum_{\alpha_j=0}^\infty \frac{1}{\xi_j^{\alpha_j+1}}\right) d\xi_1 \cdots d\xi_n \\ &= \sum_{|\alpha| \geq 0} \left(\frac{1}{2\pi i}\right)^n \int \cdots \int_{|\xi_j|=1+\varepsilon} \frac{f(\xi_1 z_1, \ldots, \xi_n z_n)}{\xi_1^{\alpha_1+1} \cdots \xi_n^{\alpha_n+1}} d\xi_1 \cdots d\xi_n.\end{aligned}$$

It is noticed that the above last series is absolutely convergent for any $z \in \Omega_\varepsilon$. Let $\|z\|$ be small so that $(\xi_j z_j)$ belongs to a polydisk about the origin where f is expanded to a power series, and set

$$(5.2.8)\qquad f(\xi_1 z_1, \ldots, \xi_u z_u) = \sum_\beta c_\beta \xi_1^{\beta_1} z_1^{\beta_1} \cdots \xi_n^{\beta_n} z_n^{\beta_n}.$$

For each α we get

$$(5.2.9)\qquad \left(\frac{1}{2\pi i}\right)^n \int\cdots\int_{|\xi_j|=1+\varepsilon} \left(\sum_\beta \frac{c_\beta \xi_1^{\beta_1} z_1^{\beta_1}\cdots \xi_n^{\beta_n} z_n^{\beta_n}}{\xi_1^{\alpha_1+1}\cdots \xi_n^{\alpha_n+1}}\right) d\xi_1\cdots d\xi_n$$

$$= \left(\frac{1}{2\pi i}\right)^n \sum_\beta \left(\int\cdots\int_{|\xi_j|=1+\varepsilon} \xi_1^{\beta_1-\alpha_1-1}\cdots \xi_n^{\beta_n-\alpha_n-1} d\xi_1\cdots d\xi_n\right) c_\beta z^\beta$$

$$= c_\alpha z^\alpha.$$

Therefore we infer from (5.2.7)–(5.2.9) that

$$f(z) = \sum_{|\alpha|\geq 0} c_\alpha z^\alpha, \quad z \in \Omega_\varepsilon.$$

Since $\varepsilon > 0$ is arbitrary, this formula holds in Ω. □

For a point $a = (a_j) \in (\mathbf{C}^*)^n$ we set

$$(5.2.10)\qquad \log a^* = (\log|a_1|, \ldots, \log|a_n|) \in \mathbf{R}^n.$$

For a plurirotation invariant subset $A \subset \mathbf{C}^n$ we define

$$(5.2.11)\qquad \begin{aligned} A^* &= A \cap (\mathbf{C}^*)^n, \\ \log A^* &= \{(\lambda_1, \ldots, \lambda_u) \in \mathbf{R}^n;\ (e^{\lambda_1}, \ldots, e^{\lambda_n}) \in A^*\} \subset \mathbf{R}^n. \end{aligned}$$

If A is an open subset, $\log A^*$ is open. We denote by $\mathrm{co}(\log A^*)$ the convex hull of $\log A^*$ in $\mathbf{R}^n$. If $\log A^* = \mathrm{co}(\log A^*)$, we say that A is *logarithmically convex.*

A Reinhardt domain Ω is said to be *complete* if

$$\{z = (z_j);\ |z_j| < |w_j|,\ 1 \leq j \leq n\} \subset \Omega$$

for all $w = (w_j) \in \Omega$.

Theorem 5.2.12 *The domain $\Omega(f)$ satisfies the following:*

(i) *$\Omega(f)$ is a complete Reinhardt domain.*
(ii) *It is necessary and sufficient for a point $z \in \mathbf{C}^n$ to belong to $\Omega(f)$ that $|z_j| < e^{\lambda_j}$, $1 \leq j \leq n$, hold for some $\lambda = (\lambda_1, \ldots, \lambda_n) \in \log \Omega(f)^*$.*
(iii) *$\Omega(f)$ is a logarithmically convex open set.*

Proof (i) By Lemma 5.2.2, $\prod \Delta(0; |w_j|) \subset \Omega(f)$ for $(w_j) \in \Omega(f)^*$. Thus, $\Omega(f)$ is complete.

(ii) A point $z = (z_j) \in \mathbf{C}^n$ belongs to $\Omega(f)$ if and only if there exists a point $(z'_j) \in \Omega(f)$ with $|z_j| < |z'_j|$, $1 \le j \le n$. Putting $\lambda_j = \log |z'_j|$, $1 \le j \le n$, we get

$$|z_j| < e^{\lambda_j}, \quad 1 \le j \le n.$$

Conversely, suppose that $|z_j| < e^{\lambda_j}$, $1 \le j \le n$, for a point $(\lambda_j) \in \Omega(f)^*$. It follows from (i) that $(z_j) \in \Omega(f)$.

(iii) Since $\Omega(f)^*$ is an open set, $\log \Omega(f)^*$ is open, too. For arbitrary $\lambda, \lambda' \in \log \Omega(f)^*$, we may take a sufficiently small $\varepsilon > 0$ such that $\lambda + (\varepsilon, \ldots, \varepsilon)$ and $\lambda' + (\varepsilon, \ldots, \varepsilon) \in \log \Omega(f)^*$. By Lemma 5.2.2 there is an $M > 0$ such that

$$|a_\alpha| e^{\sum \alpha_j (\lambda_j + \varepsilon)} \le M, \qquad |a_\alpha| e^{\sum \alpha_j (\lambda'_j + \varepsilon)} \le M.$$

For every $0 \le t \le 1$ we have

$$\begin{aligned} |a_\alpha| e^{\sum \alpha_j (t\lambda_j + (1-t)\lambda'_j + \varepsilon)} &= |a_\alpha| e^{\sum \alpha_j (t(\lambda_j + \varepsilon) + (1-t)(\lambda'_j + \varepsilon))} \\ &= \left(|a_\alpha| e^{\sum \alpha_j (\lambda_j + \varepsilon)} \right)^t \left(|a_\alpha| e^{\sum \alpha_j (\lambda'_j + \varepsilon)} \right)^{1-t} \\ &\le M^t \cdot M^{1-t} = M. \end{aligned}$$

We infer from Lemma 5.2.2 that

$$(e^{t\lambda_1 + (1-t)\lambda'_j}, \ldots, e^{t\lambda_n + (1-t)\lambda'_n}) \in \Omega(f)^*.$$

Therefore, $t\lambda + (1-t)\lambda' \in \log \Omega(f)^*$. □

In general, for a Reinhardt domain Ω we set

$$\begin{aligned} \widehat{\Omega} = \big\{ (z_1, \ldots, z_n) \in \mathbf{C}^n; {}^{\exists}(\lambda_j) \in \mathrm{co}(\log \Omega^*), |z_j| < e^{\lambda_j}, \\ 1 \le j \le n \big\} \ (\supset \Omega). \end{aligned}$$

By definition, $\widehat{\Omega}$ is a complete Reinhardt domain.

Theorem 5.2.13 *Let Ω be a Reinhardt domain containing the origin 0. Then, $\widehat{\Omega}$ is an extension of holomorphy of Ω; i.e., for every $f \in \mathscr{O}(\Omega)$ there is an element $\hat{f} \in \mathscr{O}(\widehat{\Omega})$ with $\hat{f}|_\Omega = f$.*

Proof Let $f \in \mathscr{O}(\Omega)$ be any element. By Theorem 5.2.6, $f(z)$ is expressed in Ω by a normally convergent power series,

$$f(z) = \sum_\alpha a_\alpha z^\alpha.$$

We consider the domain $\Omega(f)$ of the convergence of this series. It is immediate that $f \in \mathscr{O}(\Omega(f))$, and $\log \Omega^* \subset \log \Omega(f)^*$. By Theorem 5.2.12, $\log \Omega(f)^*$ is convex.

Therefore,

$$\text{co}(\log \Omega^*) \subset \log \Omega(f)^*.$$

From this and Lemma 5.2.4, $\widehat{\Omega} \subset \Omega(f)$ follows. Therefore, we may regard $f \in \mathscr{O}(\widehat{\Omega})$. □

We will see later that a logarithmically convex complete Reinhardt domain is a domain of holomorphy (Theorem 5.2.18).

Example 5.2.14 We consider again a Hartogs domain $\Omega_{\mathrm{H}}(\subset \mathbf{C}^2)$ defined by

$$\begin{aligned} &0 < r,\ s < 1, \\ \Omega_{\mathrm{H}} &= \{(z, w) \in \mathbf{C}^2;\ |z| < 1,\ |w| < s\} \\ &\quad \cup \{(z, w) \in \mathbf{C}^2;\ r < |z| < 1,\ |w| < 1\}. \end{aligned}$$

This is a Reinhardt domain (Fig. 5.3). The domain $\widehat{\Omega}_{\mathrm{H}} = (\Delta(0; 1))^2$ is the envelope of holomorphy of Ω such as Fig. 5.4; in fact, $\widehat{\Omega}_{\mathrm{H}} \supsetneqq \Omega_{\mathrm{H}}$.

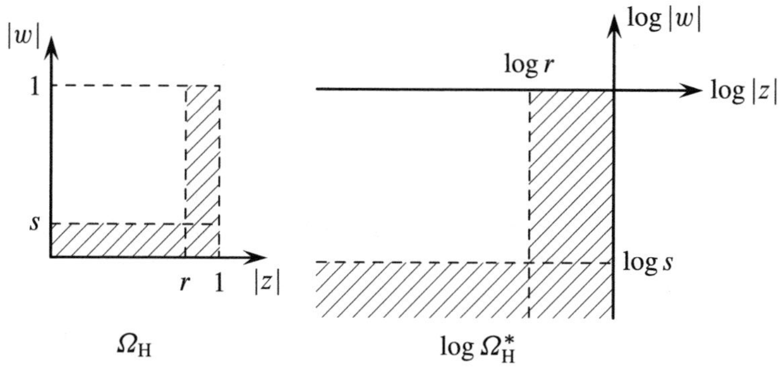

Fig. 5.3 Hartogs' domain

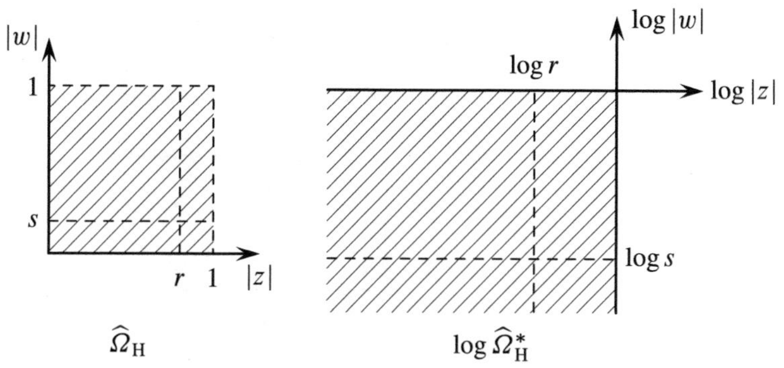

Fig. 5.4 Envelope of holomorphy of Hartogs' domain

Example 5.2.15 Let

$$f(z) = \sum_{\alpha} c_\alpha z^\alpha, \quad z = (z_1, z_2)$$

be a power series, and assume that it converges at $\left(\frac{1}{2}, 1\right)$ and $\left(1, \frac{1}{2}\right)$. Then $\Omega(f)$ contains at least the domain described by Fig. 5.5, in which the following are satisfied:

$$\log|z_i| < 0, \quad i = 1, 2,$$
$$\log|z_1| + \log|z_2| < -\log 2.$$

In particular, $\left(\frac{1}{2}, \frac{1}{2}\right)$ is a point of $\Omega(f)$.

Let Ω be a Reinhardt domain, and let

$$I \sqcup J = \{1, 2, \ldots, n\}, \quad I \cap J = \emptyset,$$
$$|I| = k, \quad |J| = n - k$$

be a decomposition of the index set of the coordinate system $z = (z_1, \ldots, z_n) \in \mathbf{C}^n$. We then set

(5.2.16)
$$\Omega_I = \Omega \cap \{(z_j) \in \mathbf{C}^n;\ z_j = 0,\ {}^\forall j \in J\}$$
$$\subset \{(z_j) \in \mathbf{C}^n;\ z_j = 0,\ {}^\forall j \in J\} \cong \mathbf{C}^k.$$

Lemma 5.2.17 *Let $\Omega \subset \mathbf{C}^n$ be a logarithmically convex complete Reinhardt domain. For every decomposition $I \sqcup J$ of the coordinate index set, Ω_I is a Reinhardt domain in $\mathbf{C}^k$ ($|I| = k$), which is logarithmically convex and complete.*

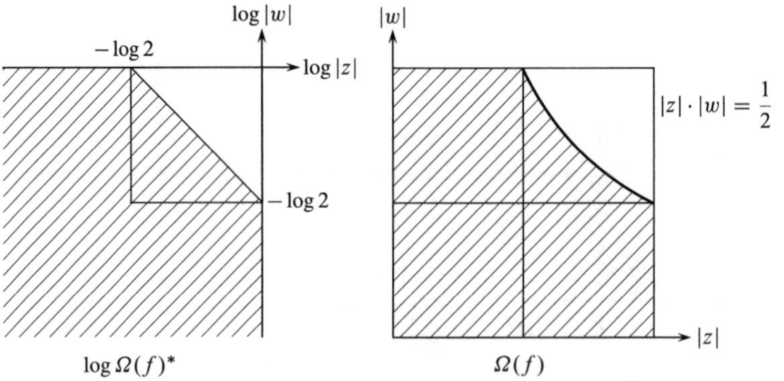

Fig. 5.5 Logarithmically convex domain

Proof Changing the order of indices, we may assume that $I = \{1, \dots, k\}$ and $J = \{k+1, \dots, n\}$. We write $z = (z', z'') \in \mathbf{C}^k \times \mathbf{C}^{n-k}$. It is immediate by the definition that Ω_I is plurirotationally invariant and complete. We show the logarithmic convexity. We take arbitrarily two points $z', w' \in \Omega_I^*$. Because $(z', 0), (w', 0) \in \Omega$, $(z', z''), (w', w'') \in \Omega^*$ for sufficiently small $z'', w'' \in (\mathbf{C}^*)^{n-k}$. For every $0 \le t \le 1$ we see that

$$\begin{aligned} & t(\dots, \log|z_i'|, \dots, \log|z_j''|, \dots) \\ & + (1-t)(\dots, \log|w_i'|, \dots, \log|w_j''|, \dots) \in \log \Omega^*. \end{aligned}$$

The completeness of Ω implies that

$$t(\dots, \log|z_i'|, \dots) + (1-t)(\dots, \log|w_i'|, \dots) \in \log \Omega_I^*, \quad 0 \le {}^{\forall}t \le 1.$$

Therefore, Ω_I is logarithmically convex. □

Theorem 5.2.18 *Let Ω be a Reinhardt domain containing* 0. *Then the following conditions are equivalent:*

(i) *Ω is a holomorphically convex domain.*
(ii) *Ω is a domain of holomorphy.*
(iii) *There is a power series $f(z) = \sum_{|\alpha| \ge 0} a_\alpha z^\alpha$ with $\Omega = \Omega(f)$.*
(iv) *Ω is complete and logarithmically convex.*

Proof (i)⇒(ii): This follows from Theorem 5.3.1 (i), which will be proved in the next section. (The result of the present theorem will not be used, of course, in the proof of Theorem 5.3.1.)

(ii)⇒(iii): This follows from Theorem 5.3.1 (ii) and Theorem 5.2.6.

(iii)⇒(iv): We have that $\Omega = \widehat{\Omega}$. By Theorem 5.2.13, Ω is logarithmically convex complete Reinhardt domain.

(iv)⇒(i): For a point $w \in (\mathbf{C}^*)^n$ we set

$$\mathrm{P}\Delta^w = \{(z_1, \dots, z_n) \in \mathbf{C}^n;\ |z_j| < |w_j|, 1 \le j \le n\}.$$

The completeness implies

$$\Omega = \bigcup_{w \in \Omega^*} \mathrm{P}\Delta^w \quad \text{(an open covering)}.$$

Let $K \Subset \Omega$ be any compact subset. We are going to show that $\widehat{K}_\Omega \Subset \Omega$ (Ω is holomorphically convex). There are finitely many $w_1, \dots, w_N \in \Omega^*$ such that

$$K \subset \bigcup_{\nu=1}^{N} \mathrm{P}\Delta^{w_\nu} \subset \bigcup_{\nu=1}^{N} \overline{\mathrm{P}\Delta}^{w_\nu} \Subset \Omega.$$

Put $L = \bigcup_{\nu=1}^{N} \overline{\mathrm{P}\Delta}^{w_\nu}$. It follows that $\widehat{K}_\Omega \subset \widehat{L}_\Omega \subset \widehat{L}_{\mathbf{C}^n}$, and $\widehat{L}_{\mathbf{C}^n}$ is compact. Thus, it suffices to show that $\widehat{L}_{\mathbf{C}^n} \subset \Omega$. Note that $\widehat{L}_{\mathbf{C}^n}$ is plurirotationally invariant.

(a) We show first that $\widehat{L}_{\mathbf{C}^n} \cap (\mathbf{C}^*)^n \subset \Omega$. Take any $\zeta = (\zeta_j) \in \widehat{L}_{\mathbf{C}^n} \cap (\mathbf{C}^*)^n$.

Lemma 5.2.19 $\mathrm{co}(\log L^*) \supset \log(\widehat{L}_{\mathbf{C}^n})^*$ *holds.*

Proof For arbitrary $\alpha_j \in \mathbf{Z}^+$, $1 \le j \le n$, with $|\alpha| > 0$ we consider the holomorphic functions $z^\alpha = z_1^{\alpha_1} \cdots z_n^{\alpha_n}$, and infer that

$$\max_L |z^\alpha| \le \max_{1 \le \nu \le N} |w_\nu^\alpha|.$$

Therefore, $\zeta = (\zeta_j)$ satisfies

$$|\zeta^\alpha| \le \max_{1 \le \nu \le N} |w_\nu^\alpha|. \tag{5.2.20}$$

Taking the logarithm, we get

$$\sum_{j=1}^{n} \alpha_j \log |\zeta_j| \le \max_\nu \sum_{j=1}^{n} \alpha_j \log |w_{\nu j}|.$$

Dividing this by $\alpha_1 + \cdots + \alpha_u = |\alpha|$ and setting $\lambda_j = \frac{\alpha_j}{|\alpha|} \in \mathbf{Q} \ge 0$, we obtain

$$\sum_{j=1}^{n} \lambda_j \log |\zeta_j| \le \max_\nu \sum_{j=1}^{n} \lambda_j \log |w_{\nu j}|.$$

This inequality is valid for arbitrary n numbers $\lambda_j \in \mathbf{Q}$, $\lambda_j \ge 0$ with $\sum_{j=1}^n \lambda_j = 1$. Hence for arbitrary $\lambda_j \in \mathbf{R}$, $\lambda_j \ge 0$ with $\sum_{j=1}^n \lambda_j = 1$,

$$\sum_{j=1}^{n} \lambda_j \log |\zeta_j| \le \max_\nu \sum_{j=1}^{n} \lambda_j \log |w_{\nu j}|. \tag{5.2.21}$$

We set

$$M = \Big\{(\eta_1, \ldots, \eta_n) \in \mathbf{R}^n;\ {}^\forall\lambda_j \in \mathbf{R},\ \lambda_j \ge 0,\ \sum_{j=1}^{n} \lambda_j = 1,$$
$$\sum_{j=1}^{n} \lambda_j \eta_j \le \max_\nu \sum_{j=1}^{n} \lambda_j \log |w_{\nu j}|\Big\}. \tag{5.2.22}$$

It follows that $\log \zeta^* \in M$.

Claim $\mathrm{co}(\log L^*) = M$.

$\because$) Set $\log w_\nu^* = (\log |w_{\nu 1}|, \ldots, \log |w_{\nu n}|) \in \mathbf{R}^n$. The set $\log(\mathrm{P}\Delta^{w_\nu})^*$ is a "quadrant" with a vertex $\log w_\nu^*$. Therefore, taking any $(\eta_1, \ldots, \eta_n) \in \mathrm{co}(\log L^*)$, we have

$$\sum_{j=1}^{n} \lambda_j \eta_j \le \max_{\nu} \sum_{j=1}^{n} \lambda_j \log |w_{\nu j}|$$

for all $\lambda_j \in \mathbb{R}$, $\lambda_j \ge 0$ with $\sum_{j=1}^{n} \lambda_j = 1$. Thus we see that $\mathrm{co}(\log L^*) \subset M$.

On the other hand, suppose that $\eta = (\eta_1, \ldots, \eta_n) \notin \mathrm{co}(\log L^*)$. There is an affine linear function F such that

$$F(\eta) = \sum_{j=1}^{n} l_j \eta_j + C,$$
$$F(\eta) > 0, \qquad \log L^* \subset \{F < 0\}.$$

In $\log L^*$, we can let every $\eta_j \to -\infty$, so that all $l_j \ge 0$. We may suppose that $\sum_{j=1}^{n} l_j = 1$. The definition of M implies that $\eta \notin M$; hence, $\mathrm{co}(\log L^*) \supset M$. Thus, $\mathrm{co}(\log L^*) = M$. This finishes the proof of Lemma 5.2.19. △

Since Ω is assumed to be logarithmically convex and complete, $\log \zeta^* \in M \subset \mathrm{co}(\log \Omega^*) = \log \Omega^*$. Therefore we see that $\zeta \in \Omega$.

(b) We consider the case where some coordinates of $\zeta = (\zeta_1, \ldots, \zeta_n) \in \widehat{L}_{\mathbf{C}^n}$ are 0. Let I (resp. J) denote the set of indices with $\zeta_i \ne 0$ (resp. $\zeta_j = 0$). Put $k = |I|$. Then, since

$$L_I = \bigcup_{\nu=1}^{N} \overline{\mathrm{P}\Delta_I^{w_\nu}} \Subset \Omega_I \subset \mathbf{C}^n{}_I \cong \mathbf{C}^k.$$

we infer that $\zeta \in \widehat{L_I}_{\mathbf{C}^k} \cap (\mathbf{C}^*)^k$. By Lemma 5.2.17, Ω_I is a logarithmically convex complete Reinhardt domain of $\mathbf{C}^k$. We now apply the arguments of (a) to conclude that $\zeta \in \Omega_I \subset \Omega$. This completes the proof of Theorem 5.2.18. □

5.3 Domains of Holomorphy and Holomorphically Convex Domains

We discussed holomorphically convex domains in Chap. 4. We prove in the present section that these are equivalent to domains of holomorphy.

Theorem 5.3.1 (Cartan–Thullen) *For a domain $\Omega \subset \mathbf{C}^n$ the following three conditions are equivalent:*

(i) *Ω is a domain of holomorphy.*
(ii) *There is a function $f \in \mathcal{O}(\Omega)$ such that Ω is the domain of existence of f.*
(iii) *Ω is holomorphically convex.*

For a moment we prepare for the proof. We fix a polydisk $\mathrm{P}\Delta = \mathrm{P}\Delta(0; r)$ with a polyradius $r = (r_1, \ldots, r_n)$ $(r_j > 0)$. Let $\Omega \subset \mathbf{C}^n$ be a domain, and set

$$\delta_{\mathrm{P}\Delta}(z, \partial\Omega) = \sup\{s > 0;\ z + s\mathrm{P}\Delta \subset \Omega\}(> 0), \quad z \in \Omega, \tag{5.3.2}$$

$$\|z - z'\|_{\mathrm{P}\Delta} = \inf\{s \geq 0; \quad z - z' \in s\mathrm{P}\Delta\} \geq 0, \quad z, z' \in \mathbf{C}^n. \tag{5.3.3}$$

We call $\delta_{\mathrm{P}\Delta}(z, \partial\Omega)$ the *boundary distance function* of Ω with respect to $\mathrm{P}\Delta$. An easy calculation yields

$$|\delta_{\mathrm{P}\Delta}(z, \partial\Omega) - \delta_{\mathrm{P}\Delta}(z', \partial\Omega)| \leq \|z - z'\|_{\mathrm{P}\Delta}, \quad z, z' \in \Omega. \tag{5.3.4}$$

Let $\|z - z'\|$ be the standard Euclidean norm. Then there is a constant $C > 0$ such that

$$C^{-1}\|z - z'\| \leq \|z - z'\|_{\mathrm{P}\Delta} \leq C\|z - z'\|,$$

and hence by (5.3.4), $\delta_{\mathrm{P}\Delta}(z, \partial\Omega)$ is a continuous function.

Lemma 5.3.5 *Assume that there are a holomorphic function $f \in \mathcal{O}(\Omega)$ and a compact subset $K \Subset \Omega$ satisfying*

$$|f(z)| \leq \delta_{\mathrm{P}\Delta}(z, \partial\Omega), \quad z \in K.$$

For an arbitrary element $u \in \mathcal{O}(\Omega)$ we expand it to a power series about a point $\xi \in \widehat{K}_\Omega$,

$$u(z) = \sum_{\alpha} \frac{\partial^\alpha u(\xi)}{\alpha!}(z - \xi)^\alpha. \tag{5.3.6}$$

Then, this converges at every $z \in \xi + |f(\xi)|\mathrm{P}\Delta$.

Proof For $0 < t < 1$ we set

$$\begin{aligned}\Omega_t &= \{(z_j);\ {}^\exists w \in K, |z_j - w_j| \leq tr_j|f(w)|,\ 1 \leq j \leq n\} \\ &\subset \bigcup_{w \in K}\{(z_j);\ (z_j) \in (w_j) + t\delta_{\mathrm{P}\Delta}(w, \partial\Omega)\,\overline{\mathrm{P}\Delta}\} \subset \Omega.\end{aligned}$$

Then Ω_t is compact in Ω. There is an $M > 0$ with $|u(z)| \leq M$ for $z \in \Omega_t$, by which we now estimate the partial differentiation. Let $w \in K$ and let $\rho_j > 0$ be sufficiently small so that we get

$$u(z) = \left(\frac{1}{2\pi i}\right)^n \int_{|\xi_j - w_j| = \rho_j} \cdots \int \frac{u(\xi)}{\prod_j (\xi_j - z_j)} d\xi_1 \cdots d\xi_n,$$

$$\partial^\alpha u(z) = \left(\frac{1}{2\pi i}\right)^n \alpha! \int_{|\xi_j - w_j| = \rho_j} \cdots \int \frac{u(\xi)}{(\xi - z)^{\alpha + (1, \ldots, 1)}} d\xi_1 \cdots d\xi_n.$$

Suppose that $f(w) \neq 0$, and set

$$z = w, \quad \rho_j = tr_j |f(w)|.$$

It follows that

$$\begin{aligned} |\partial^\alpha u(w)| &\leq \alpha! M \cdot \frac{1}{(t|f(w)|r)^\alpha} \\ &= \alpha! M \frac{1}{t^{|\alpha|} |f(w)|^{|\alpha|} r^\alpha}. \end{aligned}$$

Therefore,

$$\frac{|\partial^\alpha u(w)| t^{|\alpha|} \, |f(w)|^{|\alpha|} r^\alpha}{\alpha!} \leq M, \quad w \in K.$$

This estimate is valid trivially when $f(w) = 0$. After a transfer we obtain

$$|f(w)^{|\alpha|} \partial^\alpha u(w)| \leq \frac{\alpha! \cdot M}{t^{|\alpha|} r^\alpha}, \quad w \in K.$$

Since $f(w)^{|\alpha|} \partial^\alpha u(w) \in \mathscr{O}(\Omega)$, the definition of $\widehat{K}_\Omega$ implies that

$$|f(w)^{|\alpha|} \partial^\alpha u(w)| \leq \frac{\alpha! M}{t^{|\alpha|} r^\alpha}, \quad w \in \widehat{K}_\Omega.$$

With $w = \xi \in \widehat{K}_\Omega$, (5.3.6) converges for $z \in \xi + |f(\xi)| t \, \mathrm{P}\Delta$ by Lemma 5.2.2. Letting $t \nearrow 1$, we infer that (5.3.6) converges at $z \in \xi + |f(\xi)| \, \mathrm{P}\Delta$. □

Lemma 5.3.7 *Let $\Omega \subset \mathbf{C}^n$ be a domain of holomorphy. Let $f \in \mathscr{O}(\Omega)$ and let $K \Subset \Omega$ be compact. If*

$$|f(z)| \leq \delta_{\mathrm{P}\Delta}(z, \partial\Omega), \quad z \in K,$$

then

$$|f(z)| \leq \delta_{\mathrm{P}\Delta}(z, \partial\Omega), \quad z \in \widehat{K}_\Omega.$$

In particular, with constant f, we have that

(5.3.8) $$\inf_{z \in K} \delta_{\mathrm{P}\Delta}(z, \partial\Omega) = \inf_{z \in \widehat{K}_\Omega} \delta_{\mathrm{P}\Delta}(z, \partial\Omega).$$

Proof It follows from Lemma 5.3.5 that for $u \in \mathscr{O}(\Omega)$ and $z \in \widehat{K}_\Omega$, u is holomorphic in $z + |f(z)|\,\mathrm{P}\Delta$. Since Ω is a domain of holomorphy, $z + |f(z)|\,\mathrm{P}\Delta \subset \Omega$ must hold. Thus,

$$|f(z)| \le \delta_{\mathrm{P}\Delta}(z, \partial\Omega), \quad z \in \widehat{K}_\Omega.$$

In particular, with $f \equiv C = \min\{\delta_{\mathrm{P}\Delta}(z, \partial\Omega); z \in K\}$, we see that

$$C \le \delta_{\mathrm{P}\Delta}(z, \partial\Omega), \quad z \in \widehat{K}_\Omega.$$

Therefore,

$$\inf_{z \in K} \delta_{\mathrm{P}\Delta}(z, \partial\Omega) \le \inf_{z \in \widehat{K}_\Omega} \delta_{\mathrm{P}\Delta}(z, \partial\Omega).$$

The converse of this inequality is immediate from the inclusion relation, $K \subset \widehat{K}_\Omega$. Therefore (5.3.8) follows. □

Proof of Theorem 5.3.1

(i) ⇒ (iii): Let $K \Subset \Omega$ be any compact subset. Then $\hat{K}_\Omega$ is bounded (Proposition 4.1.3) and closed in Ω. Since Ω is a domain of holomorphy, (5.3.8) holds:

$$\inf_{z \in K} \delta_{\mathrm{P}\Delta}(z, \partial\Omega) = \inf_{z \in \widehat{K}_\Omega} \delta_{\mathrm{P}\Delta}(z, \partial\Omega).$$

Because of the continuity of $\delta_{\mathrm{P}\Delta}(z, \partial\Omega)$,

$$\inf_{z \in K} \delta_{\mathrm{P}\Delta}(z, \partial\Omega) = \min_{z \in K} \delta_{\mathrm{P}\Delta}(z, \partial\Omega) > 0.$$

Therefore, $\inf_{z \in \widehat{K}_\Omega} \delta_{\mathrm{P}\Delta}(z, \partial\Omega) > 0$, and $\widehat{K}_\Omega \Subset \Omega$.

(iii) ⇒ (ii): This easily follows from the interpolation Theorem 4.4.21 (i), which is an application of the Oka–Cartan Fundamental Theorem 4.4.2 in Chap. 4. In fact, we take a discrete sequence $\{b_j\}_{j=1}^\infty$ of Ω, similar to the one in Lemma 1.1.10; i.e., for any connected open set V with $V \not\subset \Omega$ and a connected component W of $\Omega \cap V$, there is a subsequence of $\{b_j\}$ contained in W, which converges to a point of $\partial\Omega \cap \partial W \cap V$.

Then, by Theorem 4.4.21 (i) there is an element $f \in \mathscr{O}(\Omega)$ with $f(b_j) = j$. For any V and W as above, there is a subsequence $\{b_{j_\mu}\}_\mu$ in W, converging to a point $b \in \partial\Omega \cap \partial W \cap V$. Since,

$$f(a_{j_\mu}) = j_\mu \to \infty \quad (\mu \to \infty),$$

f cannot be analytically continued over V through W. Thus, Ω is the domain of existence of f.

This proof, however, is due to the great theorem of Oka and Cartan, and it is possible to give a more elementary one as below; historically, this part was known much earlier than the Oka–Cartan Fundamental Theorem.

Let $\{a_j\}_{j=1}^{\infty}$ be a discrete subset of Ω as above. Put

$$D_j = a_j + \delta_{\mathrm{P}\Delta}(a_j, \partial\Omega)\mathrm{P}\Delta \subset \Omega.$$

Let $K_j, j = 1, 2, \ldots$ be increasing compact subsets of Ω such that the interior point sets K_j° satisfy

$$K_j \Subset K_{j+1}^{\circ}, \quad \bigcup_{j=1}^{\infty} K_j^{\circ} = \Omega.$$

By choice, $D_j \cap (\Omega \setminus \widehat{K}_{j\Omega}) \neq \emptyset$ for all $j \geq 1$. For a point $z_j \in D_j \setminus \widehat{K}_{j\Omega}$, there is an element $f_j \in \mathscr{O}(\Omega)$ satisfying

$$\max_{K_j} |f_j| < |f_j(z_j)|.$$

Dividing f_j by $f_j(z_j)$, we have $f_j(z_j) = 1$ and

$$\max_{K_j} |f_j| < |f_j(z_j)| = 1.$$

Taking a power f_j^{ν} with a sufficiently large ν, we rewrite f_j for it, and then we may assume that

$$\max_{K_j} |f_j| < \frac{1}{2^j}, \quad f_j(z_j) = 1.$$

Since $\sum_j \frac{j}{2^j} < \infty$, the infinite product

$$f(z) = \prod_{j=1}^{\infty} (1 - f_j(z))^j$$

converges uniformly on every compact subset of Ω to $f(z) \not\equiv 0$. (cf., e.g., Noguchi [52], Chap. 2, Sect. 6).

We show that Ω is the domain of existence of f. Otherwise, there is a point $b \in \partial\Omega$ over which $f(z)$ is analytically continued to a holomorphic function in a polydisk neighborhood of b. Take a subsequence $\{a_{j_\nu}\}$ converging to b. Since $\delta_{\mathrm{P}\Delta}(a_{j_\nu}, \partial\Omega) \to 0$ $(\nu \to \infty)$, the sequence $\{z_{j_\nu}\}$ also converges to b. $f(z)$ has zero of order j_ν at $z = z_{j_\nu}$. That is, the partial differential $\partial^{\alpha} f(z)$ with $|\alpha| \leq j_\nu$ satisfies

$$\partial^{\alpha} f(z_{j_\nu}) = 0.$$

Therefore, for every fixed ∂^{α}, we see that $\partial^{\alpha} f(z_{j_\nu}) = 0$ for all $\nu \gg 1$, and that

$$\partial^{\alpha} f(z_{j_\nu}) \to \partial^{\alpha} f(b), \quad \nu \to \infty.$$

It follows that

$$\partial^\alpha f(b) = 0, \quad {}^\forall \alpha.$$

By the Identity Theorem 1.2.14, $f(z) \equiv 0$, which is a contradiction.

(ii) $\Rightarrow$ (i): Since f cannot be analytically continued over a properly larger domain than Ω, Ω is the maximal domain of extension. Therefore, Ω is a domain of holomorphy.

Thus, the proof of Theorem 5.3.1 is completed. □

For two subsets $E, F \subset \mathbf{C}^n$ we set

$$\delta_{\mathrm{P}\Delta}(E, F) = \inf\{\|z - w\|_{\mathrm{P}\Delta};\ z \in E,\ w \in F\}.$$

Corollary 5.3.9 *Let $\{\Omega_\gamma\}_{\gamma \in \Gamma}$ be an arbitrary family of domains of holomorphy. Then a connected component Ω of the interior point set of $\bigcap_{\gamma \in \Gamma} \Omega_\gamma$ is a domain of holomorphy.*

Proof Let $K \Subset \Omega$ be any compact subset. Then, $K \subset \widehat{K}_\Omega \subset \widehat{K}_{\Omega_\gamma}$. Since Ω_γ are domains of holomorphy, it follows from (5.3.8) that

$$\begin{aligned}\delta_0 &:= \delta_{\mathrm{P}\Delta}(K, \partial\Omega) \le \delta_{\mathrm{P}\Delta}(K, \partial\Omega_\gamma) \\ &= \delta_{\mathrm{P}\Delta}(\widehat{K}_{\Omega_\gamma}, \partial\Omega_\gamma).\end{aligned}$$

The inclusion relation implies that

$$\delta_{\mathrm{P}\Delta}(K, \partial\Omega_\gamma) \ge \delta_{\mathrm{P}\Delta}(\widehat{K}_\Omega, \partial\Omega_\gamma) \ge \delta_{\mathrm{P}\Delta}(\widehat{K}_{\Omega_\gamma}, \partial\Omega_\gamma).$$

Therefore, $\delta_{\mathrm{P}\Delta}(\widehat{K}_\Omega, \partial\Omega_\gamma) = \delta_{\mathrm{P}\Delta}(K, \partial\Omega_\gamma) \ge \delta_0 > 0$. We deduce that for all $a \in \widehat{K}_\Omega$,

$$a + \delta_0 \mathrm{P}\Delta \subset \Omega_\gamma, \quad {}^\forall \gamma \in \Gamma.$$

Since $a + \delta_0 \mathrm{P}\Delta$ is connected, $a + \delta_0 \mathrm{P}\Delta \subset \Omega$. It follows that $\widehat{K}_\Omega \Subset \Omega$. Hence, Ω is holomorphically convex, and by Theorem 5.3.1 it is a domain of holomorphy. □

Corollary 5.3.10 *Let Ω be a domain of holomorphy (or equivalently, a holomorphically convex domain), let $f \in \mathscr{O}(\Omega)$ and let $c > 0$ be a constant. Then, every connected component Ω' of $\{z \in \Omega;\ |f(z)| < c\}$ is a domain of holomorphy. In particular, for a polydisk $\mathrm{P}\Delta(a; r)$, every connected component of $\Omega \cap \mathrm{P}\Delta(a; r)$ is a domain of holomorphy.*

Proof Take any compact subset $K \Subset \Omega'$. Set

$$\theta := \sup_K |f| < c.$$

It follows that $\sup_{\widehat{K}_{\Omega'}} |f| = \theta < c$. Since $\widehat{K}_{\Omega'} \subset \widehat{K}_\Omega \Subset \Omega$, and $\widehat{K}_{\Omega'} \Subset \{z \in \Omega;\ |f(z)| \le \theta\}$, it is deduced that $\widehat{K}_{\Omega'} \Subset \Omega'$. □

For notational simplicity we write

$$\mathrm{P}\Delta((\rho)) = \mathrm{P}\Delta(0; (\rho, \ldots, \rho)) \subset \mathbf{C}^n, \quad \rho > 0. \tag{5.3.11}$$

Corollary 5.3.12 *Let Ω be a domain of holomorphy (or equivalently, a holomorphically convex domain).*

(i) *For every $\varepsilon > 0$, any connected component of $\Omega_\varepsilon = \{z \in \Omega; \delta_{\mathrm{P}\Delta}(z, \partial\Omega) > \varepsilon\}$ is a domain of holomorphy.*
(ii) *Any connected component of $\Omega_\varepsilon \cap \mathrm{P}\Delta(0; r)(\Subset \Omega)$ is a domain of holomorphy.*
(iii) *In particular, with a given point $a_0 \in \Omega$, the connected components Ω_ν of $\Omega_{1/\nu} \cap \mathrm{P}\Delta((\nu))$ containing a_0 ($\nu = 1, 2, \ldots$) are domains of holomorphy, and satisfy*

$$\Omega_\nu \Subset \Omega_{\nu+1}, \quad \bigcup_{\nu=1}^{\infty} \Omega_\nu = \Omega.$$

Proof (i) Let Ω'_ε be any connected component of Ω_ε. For a compact subset $K \Subset \Omega'_\varepsilon$ we set $\varepsilon' = \inf_K \delta_{\mathrm{P}\Delta}(z, \partial\Omega)\ (>\varepsilon)$. Since $\widehat{K}_\Omega$ is bounded and closed in Ω, Lemma 5.3.7 implies that

$$\inf_{\widehat{K}_\Omega} \delta_{\mathrm{P}\Delta}(z, \partial\Omega) = \varepsilon' > \varepsilon.$$

Therefore, $\widehat{K}_\Omega \Subset \Omega_\varepsilon$, and then $\widehat{K}_{\Omega'_\varepsilon} \Subset \Omega'_\varepsilon$. Thus, Ω'_ε is holomorphically convex. By Theorem 5.3.1, Ω'_ε is a domain of holomorphy.

(ii) It follows from Corollary 5.3.10 that every connected component of

$$\Omega'_\varepsilon \cap \mathrm{P}\Delta(0; r) = \{z \in (z_j) \in \Omega'_\varepsilon;\ |z_j| < r_j\}$$

is a domain of holomorphy.

(iii) This follows from (ii). □

5.4 Domains of Holomorphy and Exhaustion Sequences

As a consequence of the previous section, we saw the equivalence of the notion of "domain of holomorphy" and that of "holomorphically convex domain". Therefore we obtain the following important theorem.

Theorem 5.4.1 (Oka–Cartan Fundamental Theorem) *Let $\Omega \subset \mathbf{C}^n$ be a domain of holomorphy. Then for every coherent sheaf $\mathscr{F} \to \Omega$ we have*

$$H^q(\Omega, \mathscr{F}) = 0, \quad q \geq 1.$$

Remark 5.4.2 Although we avoid repetitions, we see by this theorem that Corollary 4.4.20 on $\bar{\partial}$-equation and the Analytic de Rham Theorem (Corollary 4.4.29) hold for domains of holomorphy.

In this section we discuss the approximation of holomorphic functions by making use of the Oka–Cartan Fundamental Theorem 5.4.1, and further show that the limit of an increasing sequence of domains of holomorphy is again a domain of holomorphy; in the course, analytic polyhedra (Definition 4.4.1) and Oka's Jôku-Ikô play essential roles.

Theorem 5.4.3 (Runge–Oka Approximation) *Let $\Omega \subset \mathbf{C}^n$ be a domain and let $K = \widehat{K}_\Omega \Subset \Omega$.*

(i) *K carries a fundamental system of neighborhoods by $\mathscr{O}(\Omega)$-analytic polyhedra.*
(ii) *Every holomorphic function in a neighborhood of K can be uniformly approximated on K by elements of $\mathscr{O}(\Omega)$.*

Proof (i) Take a neighborhood U of K such that $K \Subset U \Subset \Omega$. For every $\xi \in \partial U$ there is an $f \in \mathscr{O}(\Omega)$ satisfying

$$\sup_K |f| < |f(\xi)|.$$

Then, there are a neighborhood $V_\xi \ni \xi$ and a constant $\theta_\xi > 0$ such that

$$\sup_K |f| < \theta_\xi < |f(z)|, \quad {}^\forall z \in V_\xi.$$

Since ∂U is compact, there are finitely many $\xi_1, \ldots, \xi_N \in \partial U$ with $\bigcup_{j=1}^N V_{\xi_j} \supset \partial U$, $f_j \in \mathscr{O}(\Omega)$ such that

$$\sup_K |f_j| < \theta_{\xi_j} < |f_j(z)|, \quad z \in V_{\xi_j}.$$

Replacing f_j/θ_{ξ_j} by f_j, we get

$$\sup_K |f_j| < 1 < |f_j(z)|, \quad z \in V_{\xi_j}, \quad 1 \le j \le N,$$

and $K \subset \{z \in \Omega;\ |f_j(z)| < 1\}$. Let P be the finite union of the connected components of $\{z \in \Omega;\ |f_j(z)| < 1\}$ covering K. Then, $K \Subset P \Subset U$, and P is an $\mathscr{O}(\Omega)$-analytic polyhedron.

(ii) Let g be a holomorphic function in a neighborhood U of K. By (i), we take an $\mathscr{O}(\Omega)$-analytic polyhedron P with $K \Subset P \Subset U$. Naturally, $g|_P \in \mathscr{O}(P)$. It follows from Lemma 4.4.17 that the restriction $g|_K$ can be uniformly approximated on K by elements of $\mathscr{O}(\Omega)$. □

Definition 5.4.4 A couple of domains $\Omega_1 \subset \Omega_2$ is called a *Runge pair* if all $f \in \mathscr{O}(\Omega_1)$ can be approximated uniformly on every compact subset of Ω_1 by elements of $\mathscr{O}(\Omega_2)$.

Theorem 5.4.5 *For a couple of domains of holomorphy, $\Omega_1 \subset \Omega_2$, the following conditions are equivalent:*

(i) *$\Omega_1 \subset \Omega_2$ is a Runge pair.*
(ii) *For every compact subset $K \Subset \Omega_1$,*

$$\widehat{K}_{\Omega_1} = \widehat{K}_{\Omega_2}.$$

(iii) *For every compact subset $K \Subset \Omega_1$,*

$$\widehat{K}_{\Omega_2} \Subset \Omega_1.$$

Proof (i) $\Rightarrow$ (ii): We write $K_1 = \widehat{K}_{\Omega_1}$. Then, $K_1 \Subset \Omega_1$. It suffices to show:
Claim $\widehat{K}_{\Omega_2} = K_1$.

$\because$) By definition, $\widehat{K}_{\Omega_2} \supset K_1$. Suppose that $\widehat{K}_{\Omega_2} \neq K_1$. Take a point $\xi \in \widehat{K}_{\Omega_2} \setminus K_1$ and put

$$K_2 = K_1 \cup \{\xi\} \ (\Subset \Omega_2).$$

It follows from the assumption that $\widehat{K_2}_{\Omega_2} \Subset \Omega_2$. By Theorem 5.4.3 we take an $\mathscr{O}(\Omega_2)$-analytic polyhedron P_2 such that

$$\widehat{K_2}_{\Omega_2} \Subset P_2 \Subset \Omega_2.$$

We take a neighborhood U of K_1 such that

$$K_1 \Subset U \Subset P_2 \cap \Omega_1, \quad \xi \notin \bar{U},$$

and an $\mathscr{O}(\Omega_1)$-analytic polyhedron P_1 with

$$K_1 \Subset P_1 \Subset U$$

(cf. Fig. 5.6). There are finitely many elements $f_1, \ldots, f_N \in \mathscr{O}(\Omega_1)$ such that P_1 is a finite union of connected components of

$$\{z \in \Omega_1;\ |f_j(z)| < 1,\ 1 \leq j \leq N\}.$$

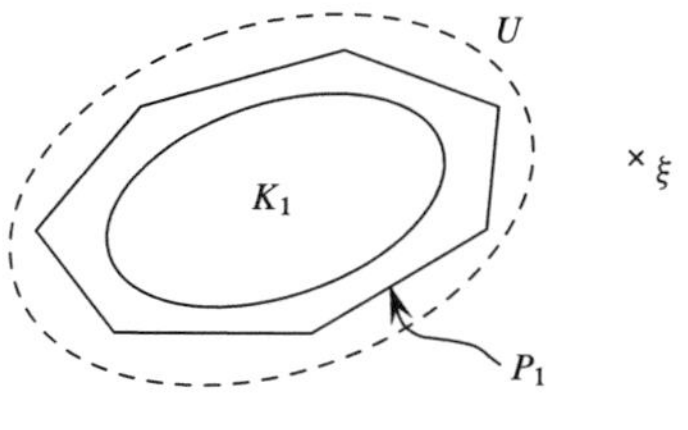

Fig. 5.6 Analytic polyhedron

By the assumption, each f_j is uniformly approximated on $\bar{U}$ by $g_j \in \mathscr{O}(\Omega_2)$. Approximating f_j on $\bar{U}$ by $g_j \in \mathscr{O}(\Omega_2)$ sufficiently, we have an $\mathscr{O}(\Omega_2)$-analytic polyhedron Q', consisting of finitely many connected components of

$$Q = \{z \in P_2;\ |g_j(z)| < 1,\ 1 \le j \le N\},$$

such that $K_1 \Subset Q' \Subset U$.

By the choice of ξ, $|g_j(\xi)| < 1$, $1 \le j \le N$, so that $\xi \in Q$. Let Q'' be the connected component of Q containing ξ. Then, $\xi \in Q'' \Subset \Omega_2$, and $Q' \cap Q'' = \emptyset$. Now, $Q_0 = Q' \cup Q''$ is an $\mathscr{O}(\Omega_2)$-analytic polyhedron with

$$Q_0 \Supset K_2. \tag{5.4.6}$$

We put

$$F(z) = \begin{cases} 0, & z \in Q'\ (\ni K_1), \\ 1, & z \in Q''\ (\ni \xi). \end{cases}$$

Since $F \in \mathscr{O}(Q_0)$ and $K_2 \Subset Q_0$, by Theorem 5.4.3, F can be approximated uniformly on K_2 by elements of $\mathscr{O}(\Omega_2)$. Therefore, there is an element $G \in \mathscr{O}(\Omega_2)$ such that

$$\sup_{K_1} |G| < \frac{1}{2}, \quad |G(\xi)| > \frac{1}{2}.$$

Thus we obtain $\xi \notin \widehat{K}_{\Omega_2}$; this is a contradiction.

(ii) $\Rightarrow$ (iii): This is clear.

(iii) $\Rightarrow$ (i): Take a compact subset $K \Subset \Omega_1$. Since $\widehat{K}_{\Omega_2} \Subset \Omega_1$, by Theorem 5.4.3 (i) there is an $\mathscr{O}(\Omega_2)$-analytic polyhedron P such that

$$\widehat{K}_{\Omega_2} \Subset P \Subset \Omega_1.$$

For every $f \in \mathscr{O}(\Omega_1)$, $f|_P \in \mathscr{O}(P)$ and so by Theorem 5.4.3 (ii), f is approximated uniformly on K by elements of $\mathscr{O}(\Omega_2)$. That is, $\Omega_1 \subset \Omega_2$ is a Runge pair. □

Proposition 5.4.7 *Let*

$$\Omega_1 \subset \Omega_2 \subset \cdots \subset \Omega_\nu \subset \Omega_{\nu+1} \subset \cdots$$

be an increasing sequence of domains of holomorphy with the limit domain $\Omega = \bigcup_{\nu=1}^{\infty} \Omega_\nu$. *Assume that all* $\Omega_\nu \subset \Omega_{\nu+1}$ $(\nu = 1, 2, \ldots)$ *are Runge pairs. Then the following hold:*

(i) $\Omega_\nu \subset \Omega$ $(\nu = 1, 2, 3, \ldots)$ *are Runge pairs.*
(ii) Ω *is a domain of holomorphy.*

Proof (i) Let $K \Subset \Omega_\nu \Subset \Omega_{\nu+1} \Subset \Omega_{\nu+2}$ be a compact subset, and let $f \in \mathscr{O}(\Omega_\nu)$ be any element. For every $\varepsilon > 0$, we take $f_k \in \mathscr{O}(\Omega_{\nu+k})$ successively as follows:

$$\begin{aligned} &f_1 \in \mathscr{O}(\Omega_{\nu+1}), \quad \|f - f_1\|_K < \varepsilon, \\ &f_2 \in \mathscr{O}(\Omega_{\nu+2}), \quad \|f_2 - f_1\|_{\overline{\Omega}_\nu} < \frac{\varepsilon}{2}, \\ &\vdots \\ &f_k \in \mathscr{O}(\Omega_{\nu+k}), \quad \|f_k - f_{k-1}\|_{\overline{\Omega}_{\nu+k-2}} < \frac{\varepsilon}{2^{k-1}}, \\ &\vdots \end{aligned}$$

We define $F \in \mathscr{O}(\Omega)$ by the series

$$F = f_1 + \sum_{k=1}^{\infty}(f_{k+1} - f_k) = f_\mu + \sum_{k=\mu}^{\infty}(f_{k+1} - f_k).$$

On every $\overline{\Omega}_{\nu+\mu}$ we have a majorant convergence:

$$\sum_{k=\mu+1}^{\infty} \|f_{k+1} - f_k\|_{\overline{\Omega}_{\nu+\mu}} \leq \sum_{k=\mu+1}^{\infty} \frac{\varepsilon}{2^k} = \frac{\varepsilon}{2^\mu}.$$

Therefore, $F \in \mathscr{O}(\Omega_{\nu+\mu})$, and so $F \in \mathscr{O}(\Omega)$. It follows that

$$\|f - F\|_K \leq \|f - f_1\|_K + \sum_{k=1}^{\infty} \|f_{k+1} - f_k\|_K \leq \sum_{k=0}^{\infty} \frac{\varepsilon}{2^k} = 2\varepsilon.$$

Since $\varepsilon > 0$ is arbitrary, $\Omega_\nu \subset \Omega$ is a Runge pair.

(ii) For any compact subset $K \Subset \Omega$, we take Ω_ν with $K \Subset \Omega_\nu$, and then we have $\widehat{K}_{\Omega_\nu} \Subset \Omega_\nu$. It suffices to show that $\widehat{K}_\Omega = \widehat{K}_{\Omega_\nu}$. Suppose that $\widehat{K}_\Omega \supsetneqq \widehat{K}_{\Omega_\nu}$. There is a point $z_0 \in \widehat{K}_\Omega \setminus \widehat{K}_{\Omega_\nu}$. Take Ω_μ so that $K_1 := \widehat{K}_{\Omega_\nu} \cup \{z_0\} \Subset \Omega_\mu$. It follows from Theorem 5.4.5 that

$$\widehat{K}_{\Omega_\nu} = \widehat{K}_{\Omega_\mu} \not\ni z_0.$$

Hence there is an element $f \in \mathscr{O}(\Omega_\mu)$ with

$$\max_K |f| < |f(z_0)|.$$

As a consequence of (i), this f can be approximated uniformly on K_1 by $\tilde{f} \in \mathscr{O}(\Omega)$, and then

$$\max_K |\tilde{f}| < |\tilde{f}(z_0)|.$$

This contradicts the choice of $z_0 \in \widehat{K}_\Omega$. □

Proposition 5.4.7 (ii) without condition of Runge pair for $\Omega_\nu \subset \Omega_{\nu+1}$ is the Behnke–Stein Theorem, which we are going to show. For that purpose we prepare two lemmas; we shall use the distance function $\delta_{\mathrm{P}\Delta}(\cdot, \cdot)$ with respect to $\mathrm{P}\Delta$ used in (5.3.2).

Lemma 5.4.8 *Let $D_1 \Subset D_2 \Subset D_3 \Subset \mathbf{C}^n$ be domains. Assume that D_3 is a domain of holomorphy, and that*

$$\delta_{\mathrm{P}\Delta}(\partial D_1, \partial D_3) > \max_{z_2 \in \partial D_2} \delta_{\mathrm{P}\Delta}(z_2, \partial D_3).$$

Then there is an $\mathscr{O}(D_3)$-analytic polyhedron P such that

$$D_1 \Subset P \Subset D_2.$$

Proof Note that $K = \bar{D}_1$ is compact. Set $K_1 = \widehat{K}_{D_3} (\Subset D_3)$. It follows from the assumption and (5.3.8) that

$$\begin{aligned}\max_{z_2 \in \partial D_2} \delta_{\mathrm{P}\Delta}(z_2, \partial D_3) &< \delta_{\mathrm{P}\Delta}(\partial D_1, \partial D_3) \\ &= \inf_{z_1 \in K} \delta_{\mathrm{P}\Delta}(z_1, \partial D_3) = \inf_{z_1 \in K_1} \delta_{\mathrm{P}\Delta}(z_1, \partial D_3).\end{aligned}$$

Therefore, $K_1 \Subset D_2$. Since $K_1 = \widehat{K}_{D_3}$, Theorem 5.4.3 (i) implies the existence of an $\mathscr{O}(D_3)$-analytic polyhedron P with $K_1 \Subset P \Subset D_2$. □

Lemma 5.4.9 *Let $\Omega \subset \mathbf{C}^n$ be a domain, and let $\Omega_{(r)}$ denote an arbitrary connected component of $\Omega \cap \mathrm{P}\Delta((r))$ for $r > 0$. Then, Ω is a domain of holomorphy if and only if $\Omega_{(r)}$ is a domain of holomorphy for every $r > 0$.*

Proof If Ω is a domain of holomorphy, by Corollary 5.3.10, every $\Omega_{(r)}$ is a domain of holomorphy.

We show the converse. By Proposition 5.4.7, it suffices to prove that every pair $\Omega_{(r)} \subset \Omega_{(R)}$ for arbitrary $R > r > 0$ is a Runge pair. Let $K \Subset \Omega_{(r)}$ be a compact subset. Since $\Omega_{(R)}$ is a domain of holomorphy, $K_1 := \widehat{K}_{\Omega_{(R)}} \Subset \Omega_{(R)}$. On the other hand, $K \subset \mathrm{P}\Delta((r))$, so that for $s < r$ sufficiently close to r, $K_1 \subset \mathrm{P}\Delta((s))$; thus, $K_1 \Subset \Omega_{(r)}$. We infer from Theorem 5.4.5 that $\Omega_{(r)} \subset \Omega_{(R)}$ is a Runge pair. □

Theorem 5.4.10 (Behnke–Stein) *Let $\Omega_j \subset \mathbf{C}^n$, $j = 1, 2, \ldots$, be an increasing sequence ($\Omega_j \subset \Omega_{j+1}$) of domains of holomorphy. Then $\Omega = \bigcup_{j=1}^{\infty} \Omega_j$ is also a domain of holomorphy.*

Proof By Lemma 5.4.9, it suffices to show that for every $r > 0$ an arbitrary connected component of

$$\Omega \cap \mathrm{P}\Delta((r)) = \bigcup_{j=1}^{\infty} \Omega_j \cap \mathrm{P}\Delta((r))$$

is a domain of holomorphy. Thus, we may assume that $\Omega \Subset \mathbf{C}^n$. Moreover, by Corollary 5.3.12 one may assume that

$$\Omega_j \Subset \Omega_{j+1}, \quad j = 1, 2, \ldots.$$

Let $\mathrm{P}\Delta$ be a polydisk with center at the origin, and set

$$M_j = \max_{z \in \partial \Omega_j} \delta_{\mathrm{P}\Delta}(z, \partial \Omega) \quad \searrow 0 \quad (j \nearrow \infty),$$
$$m_j = \min_{z \in \partial \Omega_j} \delta_{\mathrm{P}\Delta}(z, \partial \Omega) \quad \searrow 0 \quad (j \nearrow \infty).$$

Of course, $m_j \leq M_j$, the monotonicities follow from the maximum principle (Theorem 1.2.17). In the sequel, for $j_1 < j_2$ we set

$$M_{j_1 j_2} = \max_{z \in \partial \Omega_{j_1}} \delta_{\mathrm{P}\Delta}(z, \partial \Omega_{j_2}),$$
$$m_{j_1 j_2} = \min_{z \in \partial \Omega_{j_1}} \delta_{\mathrm{P}\Delta}(z, \partial \Omega_{j_2}).$$

As $j_2 \nearrow \infty$, the monotone convergences,

$$M_{j_1 j_2} \nearrow M_{j_1}, \quad m_{j_1 j_2} \nearrow m_{j_1}$$

hold.

We are going to choose a sequence, $\nu_1 < \nu_2 < \cdots$, inductively such that

(5.4.11)
$$\text{(i)} \quad m_{\nu_{q-1}} > M_{\nu_q}, \quad q = 2, 3, \ldots$$
$$\text{(ii)} \quad m_{\nu_{q-2}\nu_q} > M_{\nu_{q-1}\nu_q}, \quad q = 3, 4, \ldots.$$

We firstly let $\nu_1 = 1$. Since $m_\nu > 0$ and $M_\nu \searrow 0$ ($\nu \to \infty$), there is some $\nu_2 > \nu_1$ with $m_{\nu_1} > M_{\nu_2}$. As $M_{\nu_2 \nu} \nearrow M_{\nu_2}$ ($\nu \to \infty$), we may take a number $\nu_3 > \nu_2$ so that

$$m_{\nu_2} > M_{\nu_3}$$

and

$$m_{\nu_1 \nu_3} > M_{\nu_2 \nu_3}$$

are satisfied.

Suppose that up to ν_q, $q \geq 3$, are determined. Since $m_{\nu_{q-1}} > M_{\nu_q}$ $(\geq m_{\nu_q} > 0)$, with a sufficiently large $\nu_{q+1} > \nu_q$ we have

$$m_{\nu_q} > M_{\nu_{q+1}},$$
$$m_{\nu_{q-1}\nu_{g+1}} > M_{\nu_q \nu_{q+1}}.$$

The sequence of three domains thus defined inductively,

$$\Omega_{\nu_{q-1}} \Subset \Omega_{\nu_q} \Subset \Omega_{\nu_{q+1}}, \quad q = 2, 3, \ldots$$

satisfies the condition of Lemma 5.4.8, by which there is an $\mathscr{O}(\Omega_{\nu_{q+1}})$-analytic polyhedron P_{q-1} such that

$$\Omega_{\nu_{q-1}} \Subset P_{q-1} \Subset \Omega_{\nu_q}, \quad q = 2, 3, \ldots.$$

Since

$$P_{q-1} \Subset \Omega_{\nu_q} \Subset P_q \Subset \Omega_{\nu_{q+1}},$$

P_{q-1} is also an $\mathscr{O}(P_q)$-analytic polyhedron. By Lemma 4.4.17, $P_{q-1} \Subset P_q$ is a Runge pair. Since $\Omega = \bigcup_{q=1}^{\infty} P_q$, Proposition 5.4.7 implies that Ω is a domain of holomorphy. □

Remark 5.4.12 The Behnke–Stein Theorem 5.4.10 remains valid for Riemann domains over $\mathbf{C}^n$, but not for general complex manifolds (cf. Nishino [49], Remark after Theorem 8.8, T. Ueda [77], and J.E. Fornæss [16]).

Here we confirm the following theorem, which will be used in the proof of Oka's Theorem in Chap. 7, solving the Levi Problem (Hartogs' Inverse Problem).

We first generalize the notion of a Runge pair, Definition 5.4.4, over a general complex manifold.

Theorem 5.4.13 *Let M be a complex manifold with a sequence of subdomains,*

$$\Omega_1 \subset \Omega_2 \subset \cdots \subset \Omega_\nu \subset \Omega_{\nu+1} \subset \cdots,$$
$$M = \bigcup_{\nu=1}^{\infty} \Omega_\nu.$$

If all Ω_ν are Stein, and $\Omega_\nu \subset \Omega_{\nu+1}$ ($\nu = 1, 2, \ldots$) are Runge pairs, then we have:

(i) *$\Omega_\nu \subset M$ ($\nu = 1, 2, 3, \ldots$) is a Runge pair.*
(ii) *M is Stein.*

Proof (i) The proof is similar to that of Proposition 5.4.7.

(ii) The proof of Stein condition (iii) (Definition 4.5.7) is similar to that of Proposition 5.4.7. To prove Stein condition (i), we take two distinct points $a, b \in M$, and $\Omega_\nu \supset \{a, b\}$. Since Ω_ν is Stein, there is an element $f \in \mathscr{O}(\Omega_\nu)$ with $f(a) \neq f(b)$. Approximating f on $\{a, b\}$ by $g \in \mathscr{O}(M)$, we have that $g(a) \neq g(b)$.

Now we show Stein condition (iii). Take arbitrarily a point $a \in M$. Fix one $\Omega_\nu \ni a$. By the assumption there are $x_j \in \mathscr{O}(\Omega_\nu)$, $1 \le j \le n$ ($n = \dim M$) such that $(x_j)_{1 \le j \le n}$ forms a holomorphic local coordinate system in a neighborhood $U \Subset \Omega_\nu$ of a. It follows from (i) that x_j are approximated uniformly on U by $g_j \in \mathscr{O}(M)$. ($g_j \to x_j$ ($1 \le j \le n$)). With a neighborhood $V \Subset U$ of a fixed, the partial differentials of g_j approximate those of x_j uniformly on V. Therefore, the Jacobian $\frac{\partial(g_k)}{\partial(x_j)} \to 1$ uniformly on V. By the inverse function Theorem 1.2.43, the functions $g_j \in \mathscr{O}(M)$ which approximate x_j sufficiently on U, give a holomorphic local coordinate system in V. □

5.5 Cousin Problems and Oka Principle

5.5.1 *Cousin I Problem*

We recall Mittag-Leffler's Theorem of one variable. Let $\Omega \subset \mathbf{C}$ be a domain, and let $\{\zeta_\nu\}_{\nu \in \mathbf{N}} \subset \Omega$ be a discrete subset of Ω without accumulation point in Ω. At each point ζ_ν a so-called main part

$$Q_\nu(z) = \sum_{j>0,\ \text{finite}} \frac{a_{\nu j}}{(z - \zeta_\nu)^j}, \quad a_{\nu j} \in \mathbf{C}$$

is given.

Theorem 5.5.1 (Mittag-Leffler) *Let Ω, $\{\zeta_\nu\}$, $\{Q_\nu(z)\}$ be as above. Then there is a meromorphic function f in Ω such that in a neighborhood of every ζ_ν,*

$$f(z) - Q_\nu(z)$$

is holomorphic.

We take a neighborhood U_ν of ζ_ν which does not contain other ζ_μ ($\mu \neq \zeta_\nu$), and set $U_0 = \Omega \setminus \{\zeta_\nu\}$. Then we obtain an open covering $\mathscr{U} = \{U_\nu\}_{\nu=0}^{\infty}$ of Ω. We take $Q_0 = 0$ on U_0. With $c_{\nu\mu} = Q_\mu - Q_\nu \in \mathscr{O}(U_\nu \cap U_\mu)$, we obtain a 1-cocycle

$$(c_{\nu\mu}) \in Z^1(\mathscr{U}, \mathscr{O}_\Omega).$$

Taking a solution f of Theorem 5.5.1, we have that

$$b_\nu = Q_\nu - f \in \mathscr{O}(U_\nu).$$

By definition, $\delta(b_\nu) = (c_{\nu\mu})$. That is, $[(c_{\nu\mu})] = 0 \in H^1(\mathscr{U}, \mathscr{O}_\Omega)$ as a cohomology class. The converse is easily confirmed.

We consider a version of this existence theorem of meromorphic function of several variables. Let $\mathscr{M}_\Omega$ denote the sheaf of germs of meromorphic functions over a domain Ω ($\subset \mathbf{C}^n$) (Definition 1.3.14).

5.5.2 (**Cousin I Problem**) Let Ω be a domain of holomorphy with an open covering $\Omega = \bigcup_{\alpha\in\Gamma} U_\alpha$ and let $f_\alpha \in \Gamma(U_\alpha, \mathscr{M}_\Omega)$, $\alpha \in \Gamma$, be given so that

$$f_\alpha - f_\beta \in \mathscr{O}(U_\alpha \cap U_\beta).$$

The pair $(\{U_\alpha\}_{\alpha\in\Gamma}, \{f_\alpha\}_{\alpha\in\Gamma})$ is called a *Cousin I distribution*. With this data, is there a meromorphic function $F \in \Gamma(\Omega, \mathscr{M}_\Omega)$ such that in each U_α,

$$F - f_\alpha \in \mathscr{O}(U_\alpha)?$$

This problem was a leading problem of the Analytic Function Theory of Several Variables at the dawn of the theory in the 1930s. Oka I, II (1936, '37) solved this affirmatively. We are going to give a proof of this problem by making use of the Oka–Cartan Fundamental Theorem 5.4.1, but in the original fact of the development, the study to understand Oka's solution of the Cousin I Problem (Oka I, II) furthermore led to the invention of the concept of "Coherence" ("Ideal of undetermined domains with finite pseudobase" as termed by Oka himself) and Oka's Coherence Theorems were proved (Oka VII, VIII). As seen already in the proof, the Fundamental Theorem 5.4.1 is equivalent to the existence of Oka's Syzygies of arbitrary lengths. From this sense the Fundamental Theorem 5.4.1 may be called simply "Oka's Fundamental Theorem", but the formulation of the theorem by means of sheaves and cohomologies due to the work of H. Cartan, which is easier for comprehension, cannot be disregarded. It is, however, noticed that the proof by sheaves and cohomologies did not replace the proof of Oka; rather, it follows after the proof of Oka by making use of sheaves and cohomologies.

Theorem 5.5.3 (Oka) *The Cousin I Problem is solvable on a domain of holomorphy.*

Proof We use the notation given above. Put $\mathscr{U} = \{U_\alpha\}$. With setting $g_{\alpha\beta} = f_\alpha - f_\beta$, we consider $g = (g_{\alpha\beta}) \in C^1(\mathscr{U}, \mathscr{O}_\Omega)$, which satisfies

$$\begin{aligned}(\delta g)_{\alpha\beta\gamma} &= g_{\beta\gamma} - g_{\alpha\gamma} + g_{\alpha\beta}\\ &= (f_\beta - f_\gamma) - (f_\alpha - f_\gamma) + (f_\alpha - f_\beta) = 0.\end{aligned}$$

Thus, $g \in Z^1(\mathscr{U}, \mathscr{O}_\Omega)$, which determines a cohomology class $[g] \in H^1(\mathscr{U}, \mathscr{O}_\Omega)$. By Proposition 3.4.11,

$$H^1(\mathscr{U}, \mathscr{O}_\Omega) \hookrightarrow H^1(\Omega, \mathscr{O}_\Omega).$$

Since Ω is a domain of holomorphy, by the Oka–Cartan Fundamental Theorem 5.4.1, $H^1(\Omega, \mathscr{O}_\Omega) = 0$. Therefore, $[g] \in H^1(\mathscr{U}, \mathscr{O}_\Omega) = 0$. There exist sections $h_\alpha \in \mathscr{O}(U_\alpha)$, $\alpha \in \Gamma$, such that

$$\begin{aligned} h_\beta - h_\alpha &= g_{\alpha\beta} = f_\alpha - f_\beta, \\ f_\alpha + h_\alpha &= f_\beta + h_\beta \quad (\text{in } U_\alpha \cap U_\beta). \end{aligned}$$

Defining $F \in \Gamma(\Omega, \mathscr{M}_\Omega)$ by

$$F|_{U_\alpha} = f_\alpha + h_\alpha,$$

we obtain the required solution. □

5.5.2 *Cousin II Problem*

In one variable, the following Weierstrass' Theorem is known.

Theorem 5.5.4 *Let $\Omega \subset \mathbf{C}$ be a domain, and let $Z \subset \Omega$ be a discrete subset without accumulation point in Ω. For each $\zeta \in Z$, an integer $\nu_\zeta \in \mathbf{Z}\setminus\{0\}$ is given. Then there is a meromorphic function $f(z)$ in Ω such that in a neighborhood U of every ζ there is an element $h \in \mathscr{O}(U)$ with*

$$f(z) = (z - \zeta)^{\nu_\zeta} \cdot h(z), \quad h(\zeta) \neq 0.$$

Moreover, $f(z)$ has neither zero nor pole in $\Omega\setminus Z$.

As in the case of Mittag-Leffler's Theorem we consider a version of this theorem in several variables. Let $\Omega \subset \mathbf{C}^n$ be a domain, and consider the following sheaves of abelian groups with respect to multiplications:

$$\begin{aligned} \mathscr{O}^*_{\Omega,z} &= \left\{ \underline{f}_z \in \mathscr{O}_{\Omega,z};\quad f(z) \neq 0 \right\}, \\ \mathscr{O}^*_\Omega &= \bigcup_{z\in\Omega} \mathscr{O}^*_{\Omega,z}, \\ \mathscr{M}^*_{\Omega,z} &= \left\{ \underline{g}_z \in \mathscr{M}_{\Omega,z};\quad \underline{g}_z \neq 0 \right\}, \\ \mathscr{M}^*_\Omega &= \bigcup_{z\in\Omega} \mathscr{M}^*_{\Omega,z}. \end{aligned}$$

Since $\mathscr{O}^*_\Omega$ is a subsheaf of abelian groups of $\mathscr{M}^*_\Omega$, we may define the quotient sheaf,

$$\mathscr{D}_\Omega = \mathscr{M}^*_\Omega / \mathscr{O}^*_\Omega.$$

This is called the *sheaf of divisor groups* over Ω. The following short exact sequence is obtained:

$$0 \longrightarrow \mathscr{O}^*_\Omega \longrightarrow \mathscr{M}^*_\Omega \longrightarrow \mathscr{D}_\Omega \longrightarrow 0. \tag{5.5.5}$$

We call $H^0(\Omega, \mathscr{D}_\Omega)(= \Gamma(\Omega, \mathscr{D}_\Omega))$ the *divisor group* of Ω, and call its element a *divisor* on Ω.

We consider what a divisor $\varphi \in H^0(\Omega, \mathscr{D}_\Omega)$ is. For every point $a \in \Omega$ there are a connected neighborhood U_a of a and $f_{U_a}, g_{U_a} \in \mathscr{O}(U_a)$ such that $f_{U_a} \neq 0$, $g|_{U_a} \neq 0$, and

$$\varphi(z) = \frac{\underline{f_{U_a}}_z}{\underline{g_{U_a}}_z} \cdot \mathscr{O}^*_{\Omega,z}, \quad z \in U_a.$$

Let U_b $(b \in \Omega)$, f_{U_b}, and g_{U_b} be similar to the above. If $U_a \cap U_b \neq \emptyset$, then

$$\frac{\underline{f_{U_a}}_z}{\underline{g_{U_a}}_z} \cdot \frac{\underline{g_{U_b}}_z}{\underline{f_{U_b}}_z} \in \mathscr{O}^*_{\Omega,z}, \quad z \in U_a \cap U_b.$$

Therefore, there are an open covering $\Omega = \bigcup U_\alpha$ and $f_\alpha, g_\alpha \in \mathscr{O}(U_\alpha)$ with $f_\alpha \neq 0$, $g_\alpha \neq 0$, such that

$$\varphi_\alpha(z) = \frac{\underline{f_\alpha}_z}{\underline{g_\alpha}_z}, \quad z \in U_\alpha,$$

$$\varphi_\alpha(z)\varphi_\beta^{-1}(z) \in \mathscr{O}^*_{\Omega,z}, \qquad z \in U_\alpha \cap U_\beta. \tag{5.5.6}$$

Definition 5.5.7 Let an open covering $\mathscr{U} = \{U_\alpha\}_{\alpha\in\Gamma}$ of Ω and sections $\varphi_\alpha \in \Gamma(U_\alpha, \mathscr{M}^*_\Omega)$ satisfying (5.5.6) be given. The pair $(\{U_\alpha\}_{\alpha\in\Gamma}, \{\varphi_\alpha\}_{\alpha\in\Gamma})$ is called a *Cousin II distribution* on Ω.

5.5.8 (**Cousin II Problem**) Let Ω be a domain of holomorphy, and let a Cousin II distribution $(\{U_\alpha\}_{\alpha\in\Gamma}, \{\varphi_\alpha\}_{\alpha\in\Gamma})$ be given on Ω. Is there a meromorphic function $F \in \Gamma(\Omega, \mathscr{M}^*_\Omega)$ such that

$$F|_{U_\alpha} \cdot \varphi_\alpha^{-1} \in \Gamma(U_\alpha, \mathscr{O}^*_\Omega), \quad {}^\forall\alpha \in \Gamma\ ?$$

Assuming the existence of a solution F of this Cousin II Problem, we set

$$\psi_\alpha = F|_{U_\alpha} \cdot \varphi_\alpha^{-1} \in \Gamma(U_\alpha, \mathscr{O}^*_\Omega).$$

On $U_\alpha \cap U_\beta$,

$$\psi_\alpha \cdot \psi_\beta^{-1} = F|_{U_\alpha} \cdot \varphi_\alpha^{-1} \cdot (F|_{U_\beta} \cdot \varphi_\beta^{-1})^{-1} = \varphi_\beta \cdot \varphi_\alpha^{-1}.$$

Thus,

$$\psi_\alpha = \frac{\varphi_\beta}{\varphi_\alpha} \cdot \psi_\beta, \quad \text{in } U_\alpha \cap U_\beta. \tag{5.5.9}$$

Conversely, assuming the existence of $\psi_\alpha \in \Gamma(U_\alpha, \mathscr{O}_\Omega^*)$ satisfying (5.5.9), we define F in U_α by

$$F|_{U_\alpha} = \varphi_\alpha \psi_\alpha.$$

Then, $F \in \Gamma(\Omega, \mathscr{M}_\Omega^*)$, which gives rise to a solution of the Cousin II Problem.

Definition 5.5.10 A Cousin II Problem 5.5.8 is said to be *topologically solvable* if there exist nowhere vanishing continuous functions (complex valued) $c_\alpha, \alpha \in \Gamma$ satisfying

$$\frac{c_\beta(z)}{c_\alpha(z)} = \frac{\varphi_\beta(z)}{\varphi_\alpha(z)}, \quad z \in U_\alpha \cap U_\beta. \tag{5.5.11}$$

We consider the sheaf $\mathscr{C}_\Omega$ of germs of continuous functions over Ω and the sheaf $\mathscr{C}_\Omega^*$ of germs of zero-free (nowhere vanishing) continuous functions on Ω. For $\underline{c}_z \in \mathscr{C}_{\Omega,z}$ we put

$$\mathbf{e}(\underline{c}_z) = \underline{\exp(2\pi i\, c)}_z \in \mathscr{C}_{\Omega,z}^*.$$

Then we obtain the following short exact sequence over Ω:

$$0 \longrightarrow \mathbf{Z} \longrightarrow \mathscr{C}_\Omega \xrightarrow{\mathbf{e}} \mathscr{C}_\Omega^* \longrightarrow 0. \tag{5.5.12}$$

Setting

$$\xi_{\alpha\beta} = \varphi_\beta / \varphi_\alpha, \tag{5.5.13}$$

we obtain

$$(\xi_{\alpha\beta}) \in Z^1(\mathscr{U}, \mathscr{O}_\Omega^*) \subset Z^1(\mathscr{U}, \mathscr{C}_\Omega^*),$$

where $\mathscr{U} = \{U_\alpha\}$. With the coboundary morphism

$$\delta : H^0(\Omega, \mathscr{D}_\Omega) \to H^1(\Omega, \mathscr{O}_\Omega^*)$$

induced from the short exact sequence (5.5.5), we have that

$$\iota : [(\xi_{\alpha\beta})] = \delta[(\varphi_\alpha)] \in H^1(\Omega, \mathscr{O}^*_\Omega) \to \iota([(\xi_{\alpha\beta})]) \in H^1(\Omega, \mathscr{C}^*_\Omega).$$

Since

$$H^1(\mathscr{U}, \mathscr{C}^*_\Omega) \longrightarrow H^1(\Omega, \mathscr{C}^*_\Omega)$$

is injective (Proposition 3.4.11), (5.5.11) is equivalent to

$$\iota([(\xi_{\alpha\beta})]) = 0 \in H^1(\Omega, \mathscr{C}^*_\Omega).$$

Since $\mathscr{C}_\Omega$ is a fine sheaf, $H^q(\Omega, \mathscr{C}_\Omega) = 0$, $q \geq 1$ (Theorem 3.4.35). It follows from (5.5.12) that

$$H^1(\Omega, \mathscr{C}_\Omega) = 0 \to H^1(\Omega, \mathscr{C}^*_\Omega) \xrightarrow{\delta} H^2(\Omega, \mathbf{Z}) \to H^2(\Omega, \mathscr{C}_\Omega) = 0. \tag{5.5.14}$$

Therefore, we have that

$$H^1(\Omega, \mathscr{C}^*_\Omega) \cong H^2(\Omega, \mathbf{Z}), \tag{5.5.15}$$
$$c_1(\varphi) := -\delta(\iota([(\xi_{\alpha\beta})])) \in H^2(\Omega, \mathbf{Z}).$$

Here, $c_1(\varphi)$ is called the *first Chern class* of the divisor φ. Summarizing the above, we see the following.

Proposition 5.5.16 *A Cousin II Problem 5.5.8 is topologically solvable if and only if* $c_1(\varphi) = 0$.

Now, we prove a fundamental theorem on the original Cousin II Problem on analytic functions that is called the *Oka Principle*.

Theorem 5.5.17 (Oka Principle, Oka III (1939)) *On a domain of holomorphy, a Cousin II Problem 5.5.8 is solvable if and only if it is topologically solvable.*

Proof Let Ω be a domain of holomorphy. Assume that the Cousin II Problem 5.5.8 for a divisor $\varphi = (\varphi_\alpha)$ on Ω is topologically solvable. It follows from Proposition 5.5.16 that

$$c_1(\varphi) = 0 \in H^2(\Omega, \mathbf{Z}).$$

Similarly to (5.5.12) we have the following short exact sequence:

$$0 \to \mathbf{Z} \to \mathscr{O}_\Omega \xrightarrow{\mathbf{e}} \mathscr{O}^*_\Omega \to 0. \tag{5.5.18}$$

By the Oka–Cartan Fundamental Theorem 5.4.1, $H^q(\Omega, \mathscr{O}_\Omega) = 0$, $q \geq 1$. As in (5.5.14)–(5.5.15), we infer that

$$
\begin{array}{ccc}
\delta : H^1(\Omega, \mathscr{O}^*_\Omega) & \stackrel{\cong}{\longrightarrow} & H^2(\Omega, \mathbf{Z}). \\
\Psi & & \Psi \\
[(\xi_{\alpha\beta})] & \mapsto & c_1(\varphi)
\end{array}
\tag{5.5.19}
$$

Here, as in (5.5.13), we put $\xi_{\alpha\beta} = \frac{\varphi_\beta}{\varphi_\alpha}$. Since $c_1(\varphi) = 0$, $[(\xi_{\alpha\beta})] = 0 \in H^1(\Omega, \mathscr{O}^*_\Omega)$. By Proposition 3.4.11, for every α there is an element $\psi_\alpha \in \Gamma(U_\alpha, \mathscr{O}^*_\Omega)$ such that on $U_\alpha \cap U_\beta$,

$$\xi_{\alpha\beta} = \psi_\beta \cdot \psi_\alpha^{-1}.$$

Because of the choice of $\xi_{\alpha\beta}$,

$$\varphi_\alpha \cdot \psi_\alpha^{-1} = \varphi_\beta \cdot \psi_\beta^{-1} \quad \text{on } U_\alpha \cap U_\beta.$$

Therefore, with $F|_{U_\alpha} = \varphi_\alpha \cdot \psi_\alpha^{-1}$ on U_α, $F \in \Gamma(\Omega, \mathscr{M}^*_\Omega)$ is well-defined, and gives rise to the required solution.

The converse is clear. □

5.5.3 *Oka Principle*

Theorem 5.5.17 was a striking result in the sense that the existence of an analytic solution is completely characterized by a purely topological condition. Henceforth in general, it is called the *Oka Principle* that the solvability of an analytic problem is characterized by a topological condition.

In general, over a complex manifold M a *holomorphic vector bundle* or simply a *vector bundle* E is a complex manifold defined as follows:

Definition 5.5.20 (i) E is a complex manifold with a holomorphic surjection $p : E \to M$.

(ii) A fiber $E_a = p^{-1}\{a\}$ of a point $a \in M$ has a structure of r-dimensional complex vector space.

(iii) At every point $a \in M$ there are a holomorphic local coordinate system $U(x^1, \ldots, x^m)$ and a biholomorphic map

$$\Phi_U : p^{-1}U \longrightarrow U \times \mathbf{C}^r, \tag{5.5.21}$$

such that with projection $q : U \times \mathbf{C}^r \ni (x, v) \to x \in U$,

(a) $q \circ \Phi_U(w) = p(w), w \in p^{-1}U$.
(b) For each $x \in U$,

$$\Phi_U|_{E_x} : E_x \longrightarrow \{x\} \times \mathbf{C}^r \cong \mathbf{C}^r$$

is a linear isomorphism.

We call Φ_U in (5.5.21) a *local trivialization* and r the *rank* of E. When $r = 1$, E is called a *holomorphic line bundle* (or simply a line bundle).

Let $p : E \to M$ and $q : F \to M$ be two vector bundles over M. If a holomorphic map $\Phi : E \to F$ satisfies $p = q \circ \Phi$ and if

$$\Phi_x = \Phi|_{E_x} : E_x \longrightarrow F_x, \quad {}^\forall x \in M$$

is linear, Φ is called a *homomorphism (of vector bundles)* from E to F. If the inverse $\Phi^{-1} : F \to E$ exists, Φ is called an *isomorphism*, and we say that E and F are *isomorphic*, written as $E \cong F$.

If Φ is simply a continuous map, it is called a topological homomorphism (or isomorphism), and E is said to be topological isomorphic to F.

The product $F = M \times \mathbf{C}^r$ with the first projection $q : F \to M$ is a vector bundle over M; this is called a *trivial vector bundle*. If a vector bundle E is (resp. topologically) isomorphic to a trivial vector bundle, E is said to be (resp. topologically) trivial.

Let $p : E \to M$ be a holomorphic vector bundle over M as above with rank r. Let $V \subset M$ be an open subset. A *section* σ of E on V is a map $\sigma : V \to E$ such that $p \circ \sigma = \mathrm{id}_V$ (the identity map of V). If σ is holomorphic (resp. continuous), we call it a holomorphic (resp. continuous) section.

A family $\{\sigma_j : V \to E, 1 \le j \le r\}$ of r holomorphic sections of E on an open subset $V \subset M$ is called a *frame* if at every $x \in V$, $\{\sigma_1(x), \ldots, \sigma_r(x)\}$ forms a basis of the vector space E_x; in this case, any $w_x \in E_x (x \in V)$ is uniquely written as

$$w_x = \sum_{j=1}^{r} w_j(x)\sigma_j(x), \quad w_j(x) \in \mathbf{C},$$

and the following surjection is defined:

$$\begin{array}{rcll} \psi : p^{-1}V \ni w_x & \longrightarrow & (x, (w_j(x))) & \in V \times \mathbf{C}^n \\ \downarrow p & & \downarrow & \\ V \ni x & = & x \in V, & \\ \psi|_{E_x} : E_x & \longrightarrow & \{x\} \times \mathbf{C}^r \cong \mathbf{C}^r, & \text{liner isomorphism.} \end{array}$$

Here, ψ is biholomorphic. If the frame $\{\sigma_j\}$ consists of just continuous sections, ψ is a homeomorphism.

Let $\mathscr{O}(E)$ denote the sheaf of germs of holomorphic sections of the vector bundle E over M. Then $\mathscr{O}(E)$ is a locally free sheaf of $\mathscr{O}_M$-modules with rank r, and therefore, *coherent*. Assume that M is Stein. Then the Oka–Cartan Fundamental Theorem 4.5.8 implies

$$H^q(M, \mathscr{O}(E)\otimes\mathscr{F}) = 0, \quad q \geq 1 \tag{5.5.22}$$

for any coherent sheaf $\mathscr{F} \to M$. In particular, putting $\mathscr{F} = \mathscr{O}_M/\mathfrak{m}_{M,a}$ with $a \in M$ and $\mathscr{F} = \mathscr{O}_M/\mathscr{I}\langle\{a, b\}\rangle$ with distinct $a, b \in M$, we have the following.

Proposition 5.5.23 *Let E be a holomorphic vector bundle of rank r over a Stein manifold M.*

(i) *For every $v \in E_a$ with $a \in M$ there is a holomorphic section $\sigma \in \Gamma(M, \mathscr{O}(E))$ with $\sigma(a) = v$; in particular there are holomorphic sections $\sigma_j \in \Gamma(M, \mathscr{O}(E))$, $1 \leq j \leq r$, such that $\{\sigma_j\}$ forms a holomorphic frame in a neighborhood of a.*
(ii) *Let $a, b \in M$ be two distinct points and let $v \in E_a$, $w \in E_b$ be any vectors. Then, there exists a holomorphic section $\sigma \in \Gamma(M, \mathscr{O}(E))$ such that $\sigma(a) = v$ and $\sigma(b) = w$.*

For a moment in the sequel, we consider a holomorphic line bundle $L \stackrel{p}{\to} M$. By Definition 5.5.20, there are an open covering $\mathscr{U} = \{U_\alpha\}$ of M and local trivializations such as (5.5.21),

$$\varphi_\alpha : p^{-1}U_\alpha \to U_\alpha \times \mathbf{C}.$$

For every $w_x \in p^{-1}\{x\}$ with $x \in U_\alpha$, there corresponded $\varphi_\alpha(w_x) = (x, \xi_\alpha) \in U_\alpha \times \mathbf{C}$. If there is another U_β with $x \in U_\beta$, then one may write

$$\varphi_\beta(w_x) = (x, \xi_\beta) \in U_\beta \times \mathbf{C}.$$

Then, $\varphi_\alpha \circ \varphi_\beta^{-1}(x, \xi_\beta) = (x, \xi_\alpha)$ and it is linear in variable ξ_β. Therefore, there are zero-free holomorphic functions in $U_\alpha \cap U_\rho$, $\varphi_{\alpha\beta} \in \mathscr{O}^*(U_\alpha \cap U_\beta)$ such that

$$\xi_\alpha = \varphi_{\alpha\beta}(x)\xi_\beta, \qquad \text{on } U_\alpha \cap U_\beta. \tag{5.5.24}$$

We call $\{\varphi_{\alpha\beta}\}$ the *system of transition functions* associated with the open covering $\mathscr{U} = \{U_\alpha\}$ of local trivializations.

The family $\{\varphi_{\alpha\beta}\}$ satisfies the so-called cocycle condition:

$$\begin{aligned} &\text{(i)} \quad \varphi_{\alpha\beta} \cdot \varphi_{\beta\alpha} = 1, && \text{on } U_\alpha \cap U_\beta, \\ &\text{(ii)} \quad \varphi_{\alpha\beta} \cdot \varphi_{\beta\gamma} \cdot \varphi_{\gamma\alpha} = 1, && \text{on } U_\alpha \cap U_\beta \cap U_\gamma. \end{aligned} \tag{5.5.25}$$

Conversely, let an open covering $\mathscr{U} = \{U_\alpha\}$ of M with holomorphic functions $\varphi_{\alpha\beta} \in \mathscr{O}^*(U_\alpha \cap U_\beta)$ be given, so that $\{\varphi_{\alpha\beta}\}$ satisfies cocycle condition (5.5.25). Then we may construct a holomorphic line bundle $L \longrightarrow M$ with the system of transition functions $\{\varphi_{\alpha\beta}\}$, as follows.

We take a disjoint union

$$\mathscr{L} = \bigsqcup_{\alpha} (U_\alpha \times \mathbf{C}).$$

Let $U_\alpha \times \mathbf{C}$ be endowed the product topology. For two elements $(x_\alpha, \zeta_\alpha) \in U_\alpha \times \mathbf{C}$ and $(x_\beta, \zeta_\beta) \in U_\beta \times \mathbf{C}$ of $\mathscr{L}$ we define a relation "$\sim$" by

$$(x_\alpha, \zeta_\alpha) \sim (x_\beta, \zeta_\beta) \Longleftrightarrow \begin{cases} \text{(i)} & x_\alpha = x_\beta, \\ \text{(ii)} & \zeta_\alpha = \varphi_{\alpha\beta}(x_\beta)\zeta_\beta. \end{cases}$$

In fact, by cocycle condition (5.5.25), this is an equivalence relation. The quotient topological space $L = \mathscr{L}/\sim$ with the natural projection

$$p : [(x_\alpha, \zeta_\alpha)] \in L \longrightarrow x_\alpha \in M$$

is defined, and L gives rise to a holomorphic line bundle over Ω. We denote this line bundle by $L(\{\varphi_{\alpha\beta}\})$. We consider a divisor $\varphi = (\varphi_\alpha) \in H^0(M, \mathscr{D}_M)$ as discussed in Sect. 5.5.2. It follows from (5.5.13) that

$$\delta(\varphi_\alpha) = [(\xi_{\alpha\beta})] \in H^1(M, \mathscr{O}_M^*).$$

The family $\{\xi_{\alpha\beta}\}$ satisfies cocycle condition (5.5.25). We define a line bundle determined by the divisor φ by

$$L(\varphi) = L\left(\left\{\frac{1}{\xi_{\alpha\beta}}\right\}\right).$$

By the definition, (φ_α) is naturally a section of $L(\varphi)$.

Conversely, let $\varphi_{\alpha\beta} \in H^1(M, \mathscr{O}_M^*)$ be elements satisfying cocycle condition (5.5.25). Then we have a line bundle $L(\{\varphi_{\alpha\beta}\})$. Applying Proposition 5.5.23, we have:

Proposition 5.5.26 *Let M be a Stein manifold and $\mathscr{U} = \{U_\alpha\}$ be an open covering. For any $\varphi_{\alpha\beta} \in H^1(\mathscr{U}, \mathscr{O}_M^*)$ satisfying cocycle condition* (5.5.25)*, there exists a Cousin II distribution $\{\varphi_\alpha\}$ with holomorphic functions $\varphi_\alpha \in \mathscr{O}(U_\alpha)$ ($\varphi_\alpha \neq 0$) such that*

$$\varphi_{\alpha\beta} = \frac{\varphi_\alpha}{\varphi_\beta} \quad \text{in } U_\alpha \cap U_\beta.$$

We have by the construction:

Proposition 5.5.27 *Let M be a complex manifold, let $\mathscr{U} = \{U_\alpha\}$ be an open covering of M, and let L be a line bundle with the system $\{\xi_{\alpha\beta}\}$ of transition functions. Then the following are equivalent:*

(i) *L is trivial.*
(ii) *There is a frame of L over M; i.e., there is a holomorphic section* $\sigma : M \to L$, $p \circ \sigma = \mathrm{id}_M$ *with*

$$\sigma(x) \neq 0, \quad {}^\forall x \in M.$$

(iii) *There is a holomorphic function* $\phi_\alpha : U_\alpha \to \mathbf{C}^*$ *on every* U_α *such that*

$$\xi_{\alpha\beta} = \phi_\beta \cdot \phi_\alpha^{-1} \qquad on\ U_\alpha \cap U_\beta.$$

Proof (i) $\Rightarrow$ (ii): There is an isomorphism $\Phi : L \to M \times \mathbf{C}$. With $x \in M \to \sigma(x) = \Phi^{-1}(x, 1)$, $\sigma \in \Gamma(M, L)$ gives rise to a frame of L over M.

(ii) $\Rightarrow$ (iii): On each U_α there is a trivialization,

$$L|_{U_\alpha} \cong U_\alpha \times \mathbf{C}.$$

A frame σ may be written as a holomorphic section of $U_\alpha \times \mathbf{C}$ with

$$\sigma|_{U_\alpha} : x \in U_\alpha \to (x, \sigma_\alpha(x)) \in U_\alpha \times \mathbf{C},$$

which satisfies $\sigma_\alpha(x) \neq 0$ (${}^\forall x \in U_\alpha$). It suffices to take $\phi_\alpha = \sigma_\alpha$.

(iii)$\Rightarrow$(i): We define $\Phi : L \to M \times \mathbf{C}$ by

$$\Phi|_{L|_{U_\alpha}} : L|_{U_\alpha} \cong U_\alpha \times \mathbf{C} \ni (x, \zeta_\alpha) \mapsto (x, \zeta_\alpha \phi_\alpha(x)) \in U_\alpha \times \mathbf{C}$$

over each U_α. Since for $x \in U_\alpha \cap U_\beta$,

$$\zeta_\alpha \phi_\alpha(x) = \xi_{\alpha\beta}(x)\zeta_\beta \frac{\phi_\alpha(x)}{\phi_\beta(x)} \phi_\beta(x) = \zeta_\beta \phi_\beta(x).$$

$\Phi : L \to M \times \mathbf{C}$ is well-defined. Because $\phi_\alpha(x) \neq 0$, Φ is an isomorphism. □

Let $L_j, j = 1, 2$, be two given holomorphic line bundles over M. We may assume that there is a common open covering $\mathscr{U} = \{U_\alpha\}$, with respect to which L_j is given by the system $\{\varphi_{j\alpha\beta}\}$ of transition functions. Then, the product $\varphi_{1\alpha\beta} \cdot \varphi_{2\alpha\beta}$ yields a holomorphic line bundle denoted by $L_1 \otimes L_2$. Let $\mathrm{Pic}(M)$ denote the set of all isomorphism classes of holomorphic line bundles over M. We call $\mathrm{Pic}(M)$ the *Picard group* of M. This forms an abelian group and as a consequence of the above arguments, the following homomorphisms are obtained:

(5.5.28) $$H^0(M, \mathscr{D}_M) \longrightarrow H^1(M, \mathscr{O}_M^*) \xrightarrow{\cong} \mathrm{Pic}(M).$$

We regard the system $\{\varphi_{\alpha\beta}\}$ of transition functions of L as an element of $H^1(M, \mathscr{O}_M^*)$. With the coboundary morphism (cf. (5.5.19))

$$\delta : H^1(M, \mathscr{O}_M^*) \longrightarrow H^2(M, \mathbf{Z}),$$

we define

$$c_1(L) = \delta(\{\varphi_{\alpha\beta}\}) \in H^2(M, \mathbf{Z}), \tag{5.5.29}$$

which is called the *first Chern class* of L. If L is determined by a divisor φ, we have

$$c_1(L(\varphi)) = c_1(\varphi) \tag{5.5.30}$$

(cf. (3.7.6)).

Theorem 5.5.31 (Oka Principle) *Let $L \to M$ be a holomorphic line bundle over a Stein manifold M. Then, L is trivial if and only if $c_1(L) = 0$.*

Remark 5.5.32 This theorem is equivalent to Theorem 5.5.17 which solves the Cousin II Problem.

Proof Since M is Stein, it follows from (5.5.19) that

$$H^1(M, \mathscr{O}_M^*) \cong H^2(M, \mathbf{Z}).$$

Thus, $c_1(L) = 0$ is equivalent to $\{\varphi_{\alpha\beta}\} = 0 \in H^1(M, \mathscr{O}_M^*)$, where $\{\varphi_{\alpha\beta}\}$ is the system of transition functions of L. In $H^1(M, \mathscr{O}_M^*)$, $\{\varphi_{\alpha\beta}\} = 0$ means that there are holomorphic functions $\phi_\alpha : U_\alpha \longrightarrow \mathbf{C}^*$ with

$$\varphi_{\alpha\beta} = \phi_\beta \cdot \phi_\alpha^{-1},$$

so that by Proposition 5.5.27, this is equivalent to the triviality of L. □

H. Grauert extended the Oka Principle as follows. We are to be content only to state the result, since the proof exceeds the level of this book.

Theorem 5.5.33 (Grauert's Oka Principle) *Let $E \to M$ be a holomorphic vector bundle over a Stein manifold. If E is topologically trivial, then E is holomorphically trivial.*

5.5.4 Hermitian Holomorphic Line Bundles

In general, it is not easy to study the first Chern class $c_1(L) \in H^2(M, \mathbf{Z})$ of a given holomorphic line bundle $p : L \to M$ over a complex manifold M. By making use of the homomorphism

$$\iota : \lambda \in H^2(M, \mathbf{Z}) \to \lambda_{\mathbf{R}} \in H^2(M, \mathbf{R}) \subset H^2(M, \mathbf{C}),$$
$$\iota(c_1(L)) = c_1(L)_{\mathbf{R}}$$

induced from $\iota : \mathbf{Z} \hookrightarrow \mathbf{C}$, we can calculate it easier since an element of $H^2(M, \mathbf{C})$ is represented by a closed 2-form by de Rham Theorem 3.5.11.

Let $\mathscr{U} = \{U_\alpha\}$ be an open covering of local trivializations of L, and let $\{\varphi_{\alpha\beta}\}$ be the system of transition functions.

Definition 5.5.34 A family $h = \{h_\alpha\}$ of positively valued functions h_α of class C^∞ on U_α is a *Hermitian metric* in L if on every $U_\alpha \cap U_\beta$,

$$h_\alpha = |\varphi_{\alpha\beta}|^2 h_\beta. \tag{5.5.35}$$

A line bundle endowed with a hermitian metric is called a *hermitian line bundle* and denoted by (L, h).

In this case, for two elements $L|_{U_\alpha} \cong U_\alpha \times \mathbf{C} \ni v = (x, v_\alpha), w = (x, w_\alpha)$, an inner product

$$\langle v, w \rangle = \frac{v_\alpha \bar{w}_\alpha}{h_\alpha(x)} \in \mathbf{C}$$

is well-defined; thus, a hermitian form of C^∞-class in z

$$(v, w) \in L \times_M L \to \langle v, w \rangle \in \mathbf{C} \tag{5.5.36}$$

is obtained, where

$$L \times_M L = \{(v, w) \in L \times L;\ p(v) = p(w)\}.$$

A C^∞ map (5.5.36) is also called a hermitian metric in L. By (5.5.35), $\partial\bar{\partial} \log |\varphi_{\alpha\beta}|^2 = 0$, and so a real closed $(1, 1)$-form

$$\omega_h = \frac{i}{2\pi} \partial\bar{\partial} \log h_\alpha \tag{5.5.37}$$

is well-defined on M. We call ω_h the *curvature form* or the *Chern form* of the hermitian line bundle. Since ω is d-closed, it defines a second cohomology class $[\omega_h] \in H^2_{\mathrm{DR}}(M, \mathbf{R}) \subset H^2_{\mathrm{DR}}(M, \mathbf{C})$.

Theorem 5.5.38 *Let M be a complex manifold with the second countability axiom.*

(i) *Every holomorphic line bundle over M carries a hermitian metric.*
(ii) *Let $L \to M$ be a hermitian line bundle with hermitian metric $h = \{h_\alpha\}$. Then we have*

$$c_1(L)_{\mathbf{R}} = [\omega_h] \in H^2_{\mathrm{DR}}(M, \mathbf{R}).$$

Proof (i) Take the system $\{\varphi_{\alpha\beta}\}$ of transition functions associated with an open covering $\mathscr{U} = \{U_\alpha\}$ of local trivializations of L. Let $\{c_\alpha\}$ be a partition of unity subordinated to $\mathscr{U}$. On U_α we set

$$\rho_\alpha = \sum_\gamma c_\gamma \log |\varphi_{\alpha\gamma}|^2.$$

On $U_\alpha \cap U_\beta$ we have

$$\begin{aligned}\rho_\alpha - \rho_\beta &= \sum_\gamma \left(c_\gamma \log |\varphi_{\alpha\gamma}|^2 - c_\gamma \log |\varphi_{\beta\gamma}|^2\right)\\ &= \sum_\gamma \left(c_\gamma \log |\varphi_{\alpha\gamma}|^2 + c_\gamma \log |\varphi_{\gamma\beta}|^2\right)\\ &= \sum_\gamma c_\gamma \log |\varphi_{\alpha\beta}|^2 = \log |\varphi_{\alpha\beta}|^2.\end{aligned}$$

Then the family of $h_\alpha = e^{\rho_\alpha}$ is a required hermitian metric.

(ii) We choose the covering taken in (i) such that that all $\mathrm{Supp}\,\sigma_q$ of $\sigma_q \in N_q(\mathscr{U})$ ($q \geq 0$) are diffeomorphic to $\mathbf{C}^n$ ($n = \dim M$).[1] Then, $\mathscr{U}$ is a Leray covering for the constant sheaf $\mathbf{C}$ over M, so that by Theorem 3.4.40

$$H^q(\mathscr{U}, \mathbf{C}) \cong H^q(M, \mathbf{C}), \quad q \geq 0.$$

Next, we have the de Rham resolution (3.5.5),

$$0 \to \mathbf{C} \to \mathscr{E}_M^{(0)} \xrightarrow{d} \mathscr{E}_M^{(1)} \xrightarrow{d} \mathscr{E}_M^{(2)} \xrightarrow{d} \cdots.$$

We apply diagram (3.4.44) for this resolution; we only need up to the second degree for the diagram chasing:

[1]If M is a domain of $\mathbf{C}^n$, it suffices to take all U_α convex; then, every $\mathrm{Supp}\,\sigma_q$ is again convex and diffeomorphic to $\mathbf{C}^n$. If M satisfies the second countability axiom, then M admits a Riemannian metric; then, we may take a neighborhood $U(a)$ of every $a \in M$, and then take a covering $\mathscr{U}$ of M by such $U(a)$'s. Then, every $\mathrm{Supp}\,\sigma_q$, if it is not empty, is geodesically convex, too, and hence diffeomorphic to $\mathbf{C}^n$. For the proof, cf. T. Sakai [70], Chap. IV, Sect. 5.

(5.5.39)

$$
\begin{array}{ccccccccc}
 & 0 & & 0 & & 0 & & 0 & \\
 & \downarrow & & \downarrow & & \downarrow & & \downarrow & \\
0 \to & \Gamma(M, \mathbf{C}) & \to & \Gamma(M, \mathscr{E}^{(0)}) & \longrightarrow & \Gamma(M, \mathscr{E}^{(1)}) & \longrightarrow & \Gamma(M, \mathscr{E}^{(2)}) & \to \\
 & & & & & & & \rotatebox{90}{$\in$} & \\
 & & & & & & & (\frac{1}{2\pi i}\bar{\partial}\partial \log h_\alpha) & \\
 & \downarrow & & \downarrow & & \downarrow & & \downarrow & \\
0 \to & C^0(\mathscr{U}, \mathbf{C}) & \to & C^0(\mathscr{U}, \mathscr{E}^{(0)}) & \longrightarrow & C^0(\mathscr{U}, \mathscr{E}^{(1)}) & \xrightarrow{d} & C^0(\mathscr{U}, \mathscr{E}^{(2)}) & \to \\
 & & & & & \rotatebox{90}{$\in$} & & \rotatebox{90}{$\in$} & \\
 & & & & & (\frac{1}{2\pi i}\partial \log h_\alpha) & & (\frac{1}{2\pi i} d\partial \log h_\alpha) & \\
 & \downarrow & & \downarrow & & \downarrow \delta & & \downarrow & \\
0 \to & C^1(\mathscr{U}, \mathbf{C}) & \to & C^1(\mathscr{U}, \mathscr{E}^{(0)}) & \xrightarrow{-d} & C^1(\mathscr{U}, \mathscr{E}^{(1)}) & \longrightarrow & C^1(\mathscr{U}, \mathscr{E}^{(2)}) & \to \\
 & & & \rotatebox{90}{$\in$} & & \rotatebox{90}{$\in$} & & & \\
 & & & (\frac{1}{2\pi i}\log \varphi_{\alpha\beta}) & & (-d\frac{1}{2\pi i}\log \varphi_{\alpha\beta}) & & & \\
 & \downarrow & & \downarrow \delta & & \downarrow & & \downarrow & \\
0 \to & C^2(\mathscr{U}, \mathbf{C}) & \to & C^2(\mathscr{U}, \mathscr{E}^{(0)}) & \longrightarrow & C^2(\mathscr{U}, \mathscr{E}^{(1)}) & \longrightarrow & C^2(\mathscr{U}, \mathscr{E}^{(2)}) & \to \\
 & \rotatebox{90}{$\in$} & & \rotatebox{90}{$\in$} & & & & & \\
 & c_1(L)_{\mathbf{R}} & & \delta(\frac{1}{2\pi i}\log \varphi_{\alpha\beta}) & & & & & \\
 & \downarrow & & \downarrow & & \downarrow & & \downarrow &
\end{array}
$$

By making use of (5.5.39), we compute a representative of $c_1(L)_{\mathbf{R}} \in H^2(\mathscr{U}, \mathbf{C})$. We choose a branch of $\log \varphi_{\alpha\beta}$ in $U_\alpha \cap U_\beta$. Then we have

$$
\begin{aligned}
\left(\frac{1}{2\pi i}\log \varphi_{\alpha\beta}\right) &\in C^1(\mathscr{U}, \mathscr{E}^{(0)}), \\
\delta\left(\frac{1}{2\pi i}\log \varphi_{\alpha\beta}\right) &\in Z^2(\mathscr{U}, \mathbf{C}) \subset C^2(\mathscr{U}, \mathscr{E}^{(0)}), \\
\left[\delta\left(\frac{1}{2\pi i}\log \varphi_{\alpha\beta}\right)\right] &= c_1(L)_{\mathbf{R}} \in H^2(\mathscr{U}, \mathbf{C}).
\end{aligned}
$$

The hermitian metric $h = \{h_\alpha\}$ satisfies that $h_\alpha = |\varphi_{\alpha\beta}|^2 h_\beta$. Taking "log" and applying ∂, we obtain $\partial \log \bar{\varphi}_{\alpha\beta} = \bar{\partial} \log \varphi_{\alpha\beta} = 0$, and then

$$
\partial \log h_\alpha = \partial \log \varphi_{\alpha\beta} + \partial \log h_\beta = d \log \varphi_{\alpha\beta} + \partial \log h_\beta.
$$

Thus,

$$
\begin{aligned}
d \log \varphi_{\alpha\beta} &= \partial \log h_\alpha - \partial \log h_\beta, \\
\left(\frac{1}{2\pi i}\partial \log h_\alpha\right) &\in C^0(\mathscr{U}, \mathscr{E}^{(1)}), \\
\left(-d\frac{1}{2\pi i}\log \varphi_{\alpha\beta}\right) &= \delta\left(\frac{1}{2\pi i}\partial \log h_\alpha\right) \in C^1(\mathscr{U}, \mathscr{E}^{(1)}).
\end{aligned}
$$

And then we have

$$d\left(\frac{1}{2\pi i}\partial \log h_\alpha\right) \in \Gamma(M, d\mathscr{E}^{(1)}) \subset \Gamma(M, \mathscr{E}^{(2)}),$$

$$c_1(L)_{\mathrm{DR}} = \left[\left(\frac{1}{2\pi i} d\partial \log h_\alpha\right)\right] = \left[\left(\frac{i}{2\pi}\partial\bar{\partial}\log h_\alpha\right)\right].$$

Note that $\frac{i}{2\pi}\partial\bar{\partial}\log h_\alpha$ is a real form. □

5.5.5 *Stein's Example of Non-solvable Cousin II Distribution*

Here we discuss an interesting example of a non-solvable Cousin II distribution due to K. Stein [73] (1941). We consider the case where M is a domain $\Omega \subset \mathbf{C}^n$. We look for a domain Ω of holomorphy with $H^2(\Omega, \mathbf{Z}) \neq 0$ as simple as possible.

(a) If $\Omega = \mathbf{C}^n$, $H^q(\Omega, \mathbf{Z}) = 0$ $(q \geq 1)$.

(b) If $\Omega = \mathbf{C}^* \times \mathbf{C}^{n-1}$,

$$H^1(\Omega, \mathbf{Z}) \cong \mathbf{Z}, \qquad H^q(\Omega, \mathbf{Z}) = 0 \quad (q \geq 2).$$

(c) If $\Omega = (\mathbf{C}^*)^2 \times \mathbf{C}^{n-2}$,

$$H^1(\Omega, \mathbf{Z}) \cong \mathbf{Z} \oplus \mathbf{Z}, \quad H^2(\Omega, \mathbf{Z}) \cong \mathbf{Z}.$$

Now, we firstly have $H^2(\Omega, \mathbf{Z}) \neq 0$. For simplicity, we let $\Omega = (\mathbf{C}^*)^2$, on which there should be a non-solvable Cousin II distribution by (5.5.19) and Proposition 5.5.26. Let $(z, w) \in (\mathbf{C}^*)^2$ be the coordinate system, and consider the following divisors given by multi-valued holomorphic functions:

$$D^+ : \ w = z^i = e^{i \log z},$$
$$D^- : \ w = z^{-i} = e^{-i \log z}.$$

We write $L(D^\pm)$ for the line bundles determined by $D^\pm$. There is a generator

$$e_2 = \{|z| = 1\} \times \{|w| = 1\} \subset (\mathbf{C}^*)^2$$

of the homology space $H_2((\mathbf{C}^*)^2, \mathbf{R})$ as a dual of $H^2((\mathbf{C}^*)^2, \mathbf{R})$.

Theorem 5.5.40 *Let the notation be as above, and let* $c_1(L(D^\pm))_{\mathrm{DR}}$ *be the first Chern classes of* $L(D^\pm)$. *Then*

$$(5.5.41) \qquad \langle c_1(L(D^+))_{\mathrm{DR}}, e_2\rangle = 1, \quad \langle c_1(L(D^-))_{\mathrm{DR}}, e_2\rangle = -1.$$

In particular, the Cousin II Problem for $D^{\pm}$ is non-solvable, but since

$$c_1(L(D^+ + D^-)) = 0,$$

the Cousin II Problem for $D^+ + D^-$ is solvable.

Proof As an analytic representation K. Stein gave an infinite product expression of the divisor D^+:

$$F^+(z, w) = \exp\left(\frac{(\log z)^2}{4\pi} + \frac{\log z}{1 - i}\right) \times \prod_{\nu=0}^{\infty}\left(1 - \frac{w}{e^{i\log z + 2\nu\pi}}\right) \times \prod_{\nu=1}^{\infty}\left(1 - \frac{1}{we^{-i\log z + 2\nu\pi}}\right). \tag{5.5.42}$$

This infinite product converges absolutely and uniformly on compact subsets. As z moves around the origin anti-clockwise, it causes a change:

$$i \log z \longrightarrow i \log z - 2\pi.$$

Then, $F^+(z, w)$ is transformed as

$$F^+(z, w) \longrightarrow wF^+(z, w). \tag{5.5.43}$$

It is one-valued in w, and

$$D^+ = \{F^+ = 0\}.$$

We set

$$F^-(z, w) = F^+\left(z, \frac{1}{w}\right). \tag{5.5.44}$$

As z moves around the origin in anti-clockwise, we get from (5.5.43)

$$F^-(z, w) \longrightarrow w^{-1}F^-(z, w). \tag{5.5.45}$$

Therefore,

$$D^- = \{F^-(z, w) = 0\}. \tag{5.5.46}$$

The holomorphic function $F^+(z, w) \cdot F^-(z, w)$ is one-valued in $(\mathbf{C}^*)^2$, and

$$D^+ + D^- = \{F^+(z, w) \cdot F^{-2}(z, w) = 0\}. \tag{5.5.47}$$

Therefore, $L(D^+ + D^-)$ is trivial, and $c_1(D^+ + D^-) = c_1(L(D^+ + D^-)) = 0$.

For a hermitian metric in $L(D^+)$ we consider a multi-valued positive function

$$h(z,w) = |w|^{(\arg z)/\pi}.$$

In fact,

$$\frac{|F^+(z,w)|^2}{|w|^{(\arg z)/\pi}}$$

is one-valued. Therefore the Chern form ω of h is given by

$$\omega = \frac{i}{2\pi}\partial\bar{\partial}\log|w|^{(\arg z)/\pi} = dd^c \log|w|^{(\arg z)/\pi},$$

where d^c is defined by (1.2.1). Writing $z = r_1 e^{i\theta_1}$ and $w = r_2 e^{i\theta_2}$ in polar coordinates, we have

$$d = \sum_j \frac{\partial}{\partial r_j} dr_j + \frac{\partial}{\partial \theta_j} d\theta_j,$$
$$d^c = \frac{1}{4\pi}\sum_{j=1}^{2}\left(r_j \frac{\partial}{\partial r_j} d\theta_j - \frac{1}{r_j}\frac{\partial}{\partial \theta_j} dr_j \right).$$

A direct computation yields

$$\begin{aligned}\omega &= dd^c \log|w|^{\frac{1}{\pi}\arg z}\\ &= d\left\{ \frac{1}{4\pi}\sum_{j=1}^{2}\left(r_j \frac{\partial}{\partial r_j} d\theta_j - \frac{1}{r_j}\frac{\partial}{\partial \theta_j} dr_j \right)\frac{\theta_1}{\pi}\log r_2 \right\}\\ &= \frac{1}{4\pi^2}\left(d\log r_1 \wedge d\log r_2 + d\theta_1 \wedge d\theta_2 \right).\end{aligned}$$

The first term in the last equation is $d((\log r_1) d\log r_2)$, and so d-exact. Hence, as a cohomology class we get

$$[\omega] = \left[\frac{1}{4\pi^2} d\theta_1 \wedge d\theta_2 \right] \in H^2((\mathbf{C}^*)^2, \mathbf{R}).$$

It follows that

$$\langle [\omega], e_2 \rangle = \int_0^{2\pi}\int_0^{2\pi} \frac{1}{4\pi^2} d\theta_1 \wedge d\theta_2 = 1.$$

It is similar for D^-. □

See Abe–Hamano–Noguchi [1] for more topics. Cf. F. Forstnerič [18] for the advancements of the Oka Principle in more general settings.

Historical Supplements

The content of this chapter is mainly due to Cartan–Thullen[2] and Oka I–III. Starting with Oka VII and tracing backwards, we have finished Oka I. "Oka's Jôku-Ikô" is due to Oka I. Readers may understand, hopefully, how finely "Oka's Jôku-Ikô" matches "Oka's First Coherence Theorem", which was invented twelve years later from Oka I.

K. Oka has written in a number of his essays about the strong emotion he felt when he found the "Jôku-Ikô" in various ways. In one of them he stated "When he found the "Jôku-Ikô", he felt as if the Universe lined up in straight with center himself". Shigeo Nakano made a "Choka" (Japanese long verse) for a "Homage to Kiyoshi Oka", writing of the deep emotion of K. Oka as

> ...... directly heard myself from
> the great Oka, recalling the time with telling
> "When the first paper was made,
> the Universe lined up in straight
> with center myself"; the deep emotion ....
> ......

Exercises

1. Let $\Omega \subset \mathbf{C}^n$ $(n \geq 2)$ be a domain, and let $f \in \mathscr{O}(\Omega)$. Show that if $\{z \in \Omega; f(z) = 0\} \Subset \Omega$, then f has no zeros in Ω.
2. Let $f(z_1, z_2) = \sum_{\nu=0}^{\infty} z_1^{\nu} z_2^{\nu}$. Describe the figures of $\Omega(f)$ and $\log \Omega(f)^*$.
3. Show that $\|\cdot\|_{\mathrm{P}\Delta}$ defined by (5.3.2) satisfies the axioms of a norm.
4. Let $B = \{z \in \mathbf{C}^n;\ \|z\| < 1\}$ be the open unit ball in $\mathbf{C}^n$ with center at the origin and let $\Omega \subset \mathbf{C}^n$ be a domain. Define $\delta_B(z, \partial\Omega)$ as in (5.3.2) by making use of B in place of $\mathrm{P}\Delta$.

   a. Show that $|\delta_B(z, \partial\Omega) - \delta_B(w, \partial\Omega) \leq \|z - w\|$ for all $z, w \in \Omega$.
   b. Let $\mathrm{P}\Delta \subset B$ be a polydisk with center at the origin. Show that
   $$\delta_B(z, \partial\Omega) \leq \delta_{\mathrm{P}\Delta}(z, \partial\Omega), \quad {}^{\forall}z \in \Omega.$$

5. Show Lemma 5.3.5 with $\delta_B(z, \partial\Omega)$ in place of $\delta_{\mathrm{P}\Delta}(z, \partial\Omega)$.
6. Show Lemma 5.3.7 with $\delta_B(z, \partial\Omega)$ in place of $\delta_{\mathrm{P}\Delta}(z, \partial\Omega)$.
7. Let $\Omega \subset \mathbf{C}^n$ be a domain of holomorphy, and let $f_j \in \mathscr{O}(\Omega)$, $1 \leq j \leq k (< \infty)$. Show that any connected component of $\{z \in \Omega;\ |f_j(z)| < 1, 1 \leq j \leq k\}$ is a domain of holomorphy.

---

[2] H. Cartan und P. Thullen, Regularitäts- und Konvergenzbereiche, Math. Ann. **106** (1932), 617–647.

Chapter 6
Analytic Sets and Complex Spaces

If one considers a holomorphic function f in a domain Ω of $\mathbf{C}^n$ or more generally of a complex manifold, one is naturally led to investigate a constant hypersurface $\{f = c\}$ or the zero-hypersurface $\{f = 0\}$. More generally, one has to study an intersection X of locally finite zero-hypersurfaces of holomorphic functions. Thus the defined X is called an "analytic set". If $n \geq 2$, X carries singularities in general, and has further more involved geometric structures than complex submanifolds.

The central theme of the first half of this chapter is "Oka's Second Coherence Theorem", claiming the coherence of a geometric ideal sheaf (the ideal sheaf of an analytic set). By making use of it, the subset of singular points of an analytic set is proved to be an analytic subset of lower dimension. In the latter half, the notion of a complex space is introduced. Oka's normalization theorem, which reduces a singular point to a normal one with better property, and "Oka's Third Coherence Theorem" claiming the coherence of the normalization sheaf are proved. It is proved that the codimension of a normal complex space is at least two, and that on a normal complex space the analyticity of a function is characterized only on the space itself without information from the local ambient (locally embedded) space. Finally, the Oka–Cartan Fundamental Theorem is obtained on Stein spaces.

6.1 Preparations

6.1.1 Algebraic Sets

An *algebraic subset* X of $\mathbf{C}^n$ is defined as a subset

$$X = \{z \in \mathbf{C}^n;\ P_j(z) = 0,\ 1 \leq j \leq l\}$$

with finitely many polynomials $P_1(z), \ldots, P_l(z)$. Conversely, for an arbitrary subset $A \subset \mathbf{C}^n$

J. Noguchi, *Analytic Function Theory of Several Variables*,
DOI 10.1007/978-981-10-0291-5_6

$$I\langle A\rangle = \{P \in \mathbf{C}[z_1, \ldots, z_n];\ P|_A \equiv 0\}$$

forms an ideal of the polynomial ring $\mathbf{C}[z_1, \ldots, z_n]$, and by the Noetherian property it is finitely generated.

Let $I \subset \mathbf{C}[z_1, \ldots, z_n]$ be an ideal. Since I is finitely generated, we take its generators, Q_j, $1 \le j \le m$ $(< \infty)$. The zero set of I defined by

$$\begin{aligned} V(I) &= \{z \in \mathbf{C}^n;\ P(z) = 0,\ {}^\forall P \in I\} \\ &= \{z \in \mathbf{C}^n;\ Q_j(z) = 0,\ 1 \le j \le m\} \end{aligned}$$

is an algebraic subset. With an arbitrarily given family of algebraic subsets X_α, $\alpha \in \Gamma$, the intersection $\bigcap_{\alpha \in \Gamma} X_\alpha$ is algebraic, too.

We define an action of the multiplicative group $\mathbf{C}^* = \mathbf{C}\setminus\{0\}$ on $\mathbf{C}^{n+1}\setminus\{0\}$ by

$$(\lambda, z) \in \mathbf{C}^* \times (\mathbf{C}^{n+1}\setminus\{0\}) \to \lambda z \in \mathbf{C}^{n+1}\setminus\{0\}.$$

The quotient space by this action is written as $\mathbf{P}^n(\mathbf{C}) = (\mathbf{C}^{n+1}\setminus\{0\})/\mathbf{C}^*$, which is called n-dimensional *complex projective space*; $\mathbf{P}^n(\mathbf{C})$ is a compact Hausdorff space. By definition the projection

$$\pi : \mathbf{C}^{n+1}\setminus\{0\} \to \mathbf{P}^n(\mathbf{C}) \tag{6.1.1}$$

is continuous and called the *Hopf map*. For $z = (z_0, \ldots, z_n) \in \mathbf{C}^{n+1}\setminus\{0\}$, we write $\pi(z) = [z] = [z_0, \ldots, z_n] \in \mathbf{P}^n(\mathbf{C})$, and $[z_0, \ldots, z_n]$ is called a *homogeneous coordinate system*. The open subsets

$$U_i = \{[z_0, \ldots, z_n];\ z_i \ne 0\},\ 0 \le i \le n$$

form a covering of $\mathbf{P}^n(\mathbf{C})$, $\mathbf{P}^n(\mathbf{C}) = \bigcup_{i=0}^n U_i$. For a point $[z_0, \ldots, z_n]$ of U_i, the representative element

$$\left(\frac{z_0}{z_i}, \ldots, \overset{i\text{th}}{1}, \ldots, \frac{z_n}{z_r}\right)$$

is uniquely determined. We define a coordinate system in U_i by

$$(z_{i0}, \ldots, \check{z}_{ii}, \ldots, z_{in}) = \left(\frac{z_0}{z_i}, \ldots, \overset{i\text{th}}{\check{1}}, \ldots, \frac{z_n}{z_i}\right) \in \mathbf{C}^n, \quad z_{ik} = z_k/z_i,$$

where $\check{\bullet}$ stands for the deletion. By this coordinate system $(z_{ik})_{0 \le k \ne i \le n}$ we may regard $U_i = \mathbf{C}^n$. On $U_i \cap U_j$ $(i \ne j)$ there are transitions of coordinates such that

$$z_{ik} = z_{jk}/z_{ji}.$$

With this $\mathbf{P}^n(\mathbf{C})$ carries the structure of an n-dimensional complex manifold. The Hopf map $\pi : \mathbf{C}^{n+1}\setminus\{0\} \to \mathbf{P}^n(\mathbf{C})$ (cf. (6.1.1)) is holomorphic.

A subset $X \subset \mathbf{P}^n(\mathbf{C})$ is said to be *(projective) algebraic* if $X \cap U_i \subset \mathbf{C}^n$ is an algebraic subset for every i. Set

$$X \cap U_i = \left\{ P_{ij}\left(\frac{z_0}{z_i}, \ldots, \frac{\check{z}_i}{z_i}, \ldots, \frac{z_n}{z_i}\right) = 0, \quad 1 \le j \le l_i \right\}.$$

Assuming $X \cap U_i \neq \phi$, we put

$$P_{ij}(z_0, \ldots, z_i, \ldots, z_n) = z_i^{\deg P_{ij}} P_{ij}\left(\frac{z_0}{z_i}, \ldots, \frac{\check{z}_i}{z_i}, \ldots, \frac{z_n}{z_i}\right).$$

Then P_{ij} are homogeneous polynomials. Because of the homogeneity of P_{ij} we may write

$$X = \bigcup_{i=0}^{n} \{[z_0, \ldots, z_n] \in \mathbf{P}^n(\mathbf{C});\ P_{ij}(z_0, \ldots, z_n) = 0, \quad 1 \le j \le l_i\}.$$

6.1.2 Analytic Sets

Let $\Omega \subset \mathbf{C}^n$ be an open subset. In Definition 2.3.1 the notion of analytic subsets was given.

Proposition 6.1.2 *For a subset $X \subset \Omega$ the following are equivalent:*

(i) *X is an analytic subset.*
(ii) *X is closed in Ω, and for every $a \in X$ there are a neighborhood $U \subset \Omega$ of a and finitely many holomorphic functions, $f_\nu \in \mathscr{O}(U)$, $1 \le \nu \le l$, such that*

$$U \cap X = \{z \in U;\ f_\nu(z) = 0, \quad 1 \le \nu \le l\}.$$

N.B. In the case where the closedness condition in (ii) is not assumed, X is called a *locally closed analytic subset* of Ω; an open subset is one of locally closed analytic subsets.

Definition 6.1.3 Let X be an analytic subset of Ω (the case of $X = \Omega$ be included). A subset $A \subset X$ is called a *thin set* in X if A is closed and for every $a \in A$ there are a neighborhood $U \subset \Omega$ and an analytic subset $S \subset U \cap X$ such that S contains no non-empty open subset of $X \cap U$ and $A \cap U \subset S \subsetneqq X \cap U$. When X is given clearly, A is simply called a *thin set*.

Theorem 6.1.4 *Let $\Omega \subset \mathbf{C}^n$ be a domain, and let $A \subset \Omega$ be a thin set.*

(i) *$\Omega\setminus A$ is dense in Ω.*
(ii) *$\Omega\setminus A$ is connected.*

(iii) *For every $a \in A$ there is a curve (continuous, but in fact analytic) φ satisfying the following:*

$$\begin{aligned}&\varphi : [0, 1] \longrightarrow \Omega,\\&\varphi(0) = a,\\&\varphi(t) \notin A, \quad 0 < t \le 1.\end{aligned}$$

Proof (i) It follows from Theorem 2.3.3 that A contains no interior point in Ω, and hence $\Omega \setminus A$ is dense in Ω.

(ii) If $\Omega \setminus A$ were disconnected, there would exist non-empty open subsets $U_i \subset \Omega$, $i = 1, 2$, such that $\Omega \setminus A = U_1 \cup U_2$, and $U_1 \cap U_2 = \emptyset$. Take any point $a \in A$ and a connected neighborhood $V \ni a$ with an analytic subset $S \subsetneqq V$ such that $S \supset A \cap V$. We have

$$\begin{aligned}V \setminus S \subset V \setminus A &= (V \cap U_1) \cup (V \cap U_2),\\V \setminus S &= ((V \cap U_1) \setminus S) \cup ((V \cap U_2) \setminus S).\end{aligned}$$

By Theorem 2.3.3, $(V \cap U_i) \setminus S \neq \emptyset$, $i = 1, 2$, are open and disjoint to each other, whereas $V \setminus S$ is connected by Theorem 2.3.5; this is absurd.

(iii) For a point $a \in A$ we take a polydisk $\mathrm{P}\Delta(a; r)$ and $f \in \mathscr{O}(\mathrm{P}\Delta(a; r))$ so that $f \not\equiv 0$, $A \cap \mathrm{P}\Delta(a; r) \subset \{f = 0\}$. We may assume that $\mathrm{P}\Delta(a; r) = \mathrm{P}\Delta_{n-1} \times \Delta(a_n; r_n)$ is a standard polydisk of f, and $a = 0$ by translation. Since $\{z_n \in \Delta(0; r_n) : f(0, z_n) = 0\} = \{0\}$, with $0 < s_n < r_n$ we set $\varphi(t) = (0, ts_n)$, $0 \le t \le 1$. Then

$$f(\varphi(t)) = f(0, ts_n) \neq 0, \quad t > 0.$$

Therefore, $\varphi(t) \notin A$, $t > 0$. □

6.1.3 Regular Points and Singular Points

Let $\Omega \subset \mathbf{C}^n$ be an open subset, and let $X \subset \Omega$ be an analytic subset.

Definition 6.1.5 A point $a \in X$ is called a *regular* or *non-singular point* if there is a neighborhood $U (\subset \Omega)$ of a such that $X \cap U$ is a complex submanifold of U. This is equivalent to the existence of finitely many $f_1, \ldots, f_q \in \mathscr{O}(U)$, $0 \le q \le n$, such that

$$\begin{aligned}&\{f_1 = \cdots = f_q = 0\} = X \cap U,\\&df_1(a) \wedge \cdots \wedge df_q(a) \neq 0\end{aligned} \tag{6.1.6}$$

(cf. Theorem 1.2.41). A point of X that is not regular is called a *singular point*, and we denote by $\Sigma(X)$ the set of all singular points of X. When $\Sigma(X) = \emptyset$, X is said to be *non-singular*; in this case, X is a complex submanifold of Ω.

By definition $X \setminus \Sigma(X)$ is open in X, and so $\Sigma(X)$ is a closed subset. It is, however, noticed that expression (6.1.6) with

$$df_1(a) \wedge \cdots \wedge df_q(a) = 0$$

does not necessarily imply a being singular (cf. Remark 6.5.14).

6.1.4 Finite Maps

Let X and Y be topological spaces. A continuous map $f : X \to Y$ is called an *open map* (resp. *closed map*) if the image of every open (resp. closed) subset of X by f is open (resp. closed).

Definition 6.1.7 (*Finite map*) A continuous map $f : X \to Y$ is called a *finite map* if f is proper and for every $y \in f(Y)$ the inverse image $f^{-1}y$ is finite.

Proposition 6.1.8 *Let X, Y be Hausdorff, locally countable and locally compact, and let $f : X \to Y$ be a continuous map.*

(i) *A proper map $f : X \to Y$ is closed.*
(ii) *Let $f : X \to Y$ be a finite map. For every point $y \in f(X)$ and every neighborhood U of $f^{-1}y$, there is a neighborhood V of y with $f^{-1}V \subset U$.*
(iii) *Let $f : X \to Y$ be finite and surjective. For a point $y \in Y$, we write $f^{-1}y = \{x_i\}_{i=1}^{l}$. Then, there are a neighborhood $V \ni y$ and mutually disjoint neighborhoods $U_i \ni x_i$ such that*
  (a) $f^{-1}V = \bigcup_{i=1}^{l} U_i$,
  (b) *every restriction $f|_{U_i} : U_i \to V (1 \leq i \leq l)$ is finite.*

Proof (i) Let f be proper. Let $E \subset X$ be a closed subset. If $b \in Y$ is an accumulation point of $f(E)$, there is a sequence $\{x_\nu\} \subset E$ with $\lim f(x_\nu) = b$. Since the set $\{f(x_\nu)\}_\nu \cup \{b\}$ is compact, so is $f^{-1}(\{f(x_\nu)\}_\nu \cup \{b\})$. Therefore, $\{x_\nu\}$ contains a convergent subsequence $\{x_{\nu_\mu}\}_\mu$. With $a = \lim_\mu x_{\nu_\mu}$, $a \in E$ because of the closedness of E, and $f(a) = b$. Hence, $f(E)$ is closed.

(ii) Since $X \setminus U$ is a closed subset, it follows from (i) that $f(X \setminus U)$ is closed. Since $y \notin f(X \setminus U)$, there is a neighborhood $V \ni y$ with $V \cap f(X \setminus U) = \emptyset$. Therefore $f^{-1}V \subset U$.

(iii) Take mutually disjoint neighborhoods $W_i \ni x_i$. Then, $W = \bigcup W_i$ is a neighborhood $f^{-1}y$. It follows from (ii) that there is a neighborhood V of y with $f^{-1}V \subset W$. Set $U_i = W_i \cap f^{-1}V (\ni x_i)$. For each i we show that

$$f|_{U_i} : U_i \to V$$

is proper and hence a finite map. We take, e.g., $i = 1$. For any compact subset $K \Subset V$, $f^{-1}K = f^{-1}K \cap (\bigcup_i U_i) = \bigcup_i f^{-1}K \cap U_i$ is compact. We take an arbitrary open covering of $f^{-1}K \cap U_1$,

$$f^{-1}K \cap U_1 \subset \bigcup_{\lambda \in \Lambda} O_\lambda.$$

We may assume that $\bigcup_\lambda O_\lambda \subset U_1$. It follows that $\{O_\lambda, U_i; \lambda \in \Lambda, 2 \le i \le l\}$ is an open covering of the compact subset $f^{-1}K$. Therefore there are finitely many O_{λ_j}, $1 \le j \le m$, such that $f^{-1}K \cap U_1 \subset \bigcup_{j=1}^m O_{\lambda_j}$. Thus we see that $f^{-1}K \cap U_1$ is compact. □

6.2 Germs of Analytic Sets

For a point $a \in \mathbf{C}^n$ we consider the set $\mathfrak{A}_a$ of all analytic subsets $A_U (\subset U)$ in neighborhoods U of a. For two elements A_U and $A_V \in \mathfrak{A}_a$ we write $A_U \sim A_V$ if there is a neighborhood $W \subset U \cap V$ of a with $A_U \cap W = A_V \cap W$; this is an equivalence relation in $\mathfrak{A}_a$. We denote the quotient set by $\underline{\mathfrak{A}}_a = \mathfrak{A}_a / \sim$. The equivalence class $\underline{A_U}_a \in \underline{\mathfrak{A}}_a$ of A_U is called a *germ of analytic subset* A_U. For two elements $\underline{A}_a, \underline{B}_a \in \underline{\mathfrak{A}}_a$, we write $\underline{A}_a \supset \underline{B}_a$ if there are a common neighborhood $W \ni a$ and their representatives A_W, $B_W \subset W$ with $A_W \supset B_W$.

We say that a germ $\underline{f}_a \in \mathscr{O}_{n,a} = \mathscr{O}_{\mathbf{C}^n,a}$ vanishes or takes value 0 on $\underline{A}_a$ if their representatives f and A respectively in a neighborhood $U \ni a$ satisfy $f|_A \equiv 0$. In this case we write $\underline{f}_a|_{\underline{A}_a} = 0$. The set

$$\mathscr{I}\langle \underline{A}_a \rangle = \left\{ \underline{f}_a \in \mathscr{O}_{n,a};\ \underline{f}_a|_{\underline{A}_a} = 0 \right\} \tag{6.2.1}$$

is an ideal of $\mathscr{O}_{n,a}$, which is called a *geometric ideal*.

We define the *radical* of an ideal $\mathfrak{a} \subset \mathscr{O}_{n,a}$ by

$$\sqrt{\mathfrak{a}} = \left\{ \underline{f}_a \in \mathscr{O}_{n,a};\ {}^\exists N \in \mathbf{N},\ \underline{f}_a^N \in \mathfrak{a} \right\}; \tag{6.2.2}$$

if $\mathfrak{a} = \sqrt{\mathfrak{a}}$, $\mathfrak{a}$ is called a *radical*.

Proposition 6.2.3 *The geometric ideal $\mathscr{I}\langle \underline{A}_a \rangle$ is a radical.*

This is immediate from definition (6.2.1).

Let $\mathfrak{a} \subset \mathscr{O}_{n,a}$ be an ideal. By the Noetherian property of $\mathscr{O}_{n,a}$ (Theorem 2.2.20) there are finitely many elements $f_j \in \mathscr{O}_{n,a}$, $1 \le j \le l$, such that

$$\mathfrak{a} = \sum \mathscr{O}_{n,a} \cdot \underline{f_j}_a.$$

Let U be a neighborhood of a with the representatives $f_j \in \mathscr{O}(U)$ of $\underline{f_j}_a$, and set

$$\mathscr{V}(\mathfrak{a}) = \underline{\{z \in U;\ f_j(z) = 0,\ 1 \le j \le l\}}_a \in \underline{\mathfrak{A}}_a.$$

Proposition 6.2.4 (i) *If* $\underline{A_1}_a \subset \underline{A_2}_a$, *then* $\mathscr{I}\langle \underline{A_1}_a \rangle \supset \mathscr{I}\langle \underline{A_2}_a \rangle$.
(ii) *If* $\mathfrak{a}_1 \subset \mathfrak{a}_2 \subset \mathscr{O}_{n,a}$ *are ideals, then* $\mathscr{V}(\mathfrak{a}_1) \supset \mathscr{V}(\mathfrak{a}_2)$.
(iii) $\mathscr{I}\langle \mathscr{V}(\mathfrak{a})\rangle \supset \mathfrak{a}$.
(iv) $\mathscr{V}(\mathscr{I}\langle \underline{A}_a \rangle) = \underline{A}_a$ *for* $\underline{A}_a \in \underline{\mathfrak{A}}_a$.
(v) *For ideals* $\mathfrak{a}_1, \ldots, \mathfrak{a}_l \subset \mathscr{O}_{n,a}$ $(l \le \infty)$,

$$\mathscr{V}\left(\sum_{j=1}^{l} \mathfrak{a}_j\right) = \bigcap_{j=1}^{l} \mathscr{V}(\mathfrak{a}_j) \in \underline{\mathfrak{A}}_a.$$

(vi) *In* (v), *if* $l < \infty$, *then*

$$\mathscr{V}\left(\bigcap_{j=1}^{l} \mathfrak{a}_j\right) = \bigcup_{j=1}^{l} \mathscr{V}(\mathfrak{a}_j).$$

(vii) *For finitely many* $\underline{A_1}_a, \ldots, \underline{A_l}_a \in \underline{\mathfrak{A}}_a$,

$$\mathscr{I}\left\langle \bigcup_{j=1}^{l} \underline{A_j}_a \right\rangle = \bigcap_{j=1}^{l} \mathscr{I}\left\langle \underline{A_j}_a \right\rangle.$$

Proof The proofs of (i)–(iii) are easy and left to the reader.

(iv) Let $\underline{f_j}_a \in \mathscr{I}\langle \underline{A}_a \rangle$, $1 \le j \le l < \infty$, and let $\underline{A}_a = \underline{\{f_1 = \cdots = f_l = 0\}}_a$. Then,

$$\mathscr{V}(\mathscr{I}\langle \underline{A}_a \rangle) \subset \underline{\{f_j = 0,\ 1 \le j \le l\}}_a = \underline{A}_a.$$

On the other hand, by definition, $\underline{f}_a \in \mathscr{I}\langle \underline{A}_a \rangle$, so that $\underline{f}_a|_{\underline{A}_a} = 0$, and then $\underline{\{f = 0\}}_a \supset \underline{A}_a$. Therefore,

$$\mathscr{V}(\mathscr{I}\langle \underline{A}_a \rangle) \supset \underline{A}_a.$$

It follows that $\mathscr{V}(\mathscr{I}\langle \underline{A}_a \rangle) = \underline{A}_a$.

(v) Since $\sum_{j=1}^{l} \mathfrak{a}_j \supset \mathfrak{a}_j$,

$$\mathscr{V}\left(\sum_{j=1}^{l} \mathfrak{a}_j\right) \subset \mathscr{V}(\mathfrak{a}_j).$$

Therefore,

$$\mathscr{V}\left(\sum_{j=1}^{l}\mathfrak{a}_j\right) \subset \bigcap_{j=1}^{l}\mathscr{V}(\mathfrak{a}_j).$$

By the Noetherian property, there is a number $\nu_0 \in \mathbf{N}$ such that $\sum_{j=1}^{\nu_0}\mathfrak{a}_j = \sum_{j=1}^{\nu_0+1}\mathfrak{a}_j = \cdots = \sum_{j=1}^{l}\mathfrak{a}_j$, from which follows

$$\mathscr{V}\left(\sum_{j=1}^{\nu_0}\mathfrak{a}_j\right) = \bigcap_{j=1}^{\nu_0}\mathscr{V}(\mathfrak{a}_j) \supset \bigcap_{j=1}^{l}\mathscr{V}(\mathfrak{a}_j).$$

Thus we see that

$$\mathscr{V}\left(\sum_{j=1}^{l}\mathfrak{a}_j\right) = \bigcap_{j=1}^{l}\mathscr{V}(\mathfrak{a}_j).$$

(vi) Since $l < \infty$, we take a neighborhood $U \ni a$ where all of the generators $\underline{f_{jk}}_a$ of every $\mathfrak{a}_j$ have the representatives $f_{jk} \in \mathscr{O}(U)$. It follows that

$$A_j = \{x \in U;\ f_{jk}(x) = 0,\ {}^\forall k\} \subset U, \quad \underline{A_j}_a = \mathscr{V}(\mathfrak{a}_j).$$

Taking U smaller if necessary, we may assume that $\bigcap_j \mathfrak{a}_j$ is generated by germs of finitely many elements $\{g_i\}$ of $\mathscr{O}(U)$. Then,

$$B = \{x \in U;\ g_i(x) = 0\} \subset U, \quad \underline{B}_a = \mathscr{V}\left(\bigcap_j \mathfrak{a}_j\right).$$

With k_j taken arbitrarily for each j, $\prod_j \underline{f_{jk_j}}_a \in \bigcap_j \mathfrak{a}_j$; since there are only finitely many such $\prod_j \underline{f_{jk_j}}_a$, we may have in U that

$$\prod_j f_{jk_j} = \sum_i h_i g_i, \quad h_i \in \mathscr{O}(U). \tag{6.2.5}$$

Suppose that $x \in U \setminus \bigcup_j A_j$. For every j there is some f_{jk_j} with $f_{jk_j}(x) \neq 0$. Hence, $\prod_{j=1}^{l} f_{jk_j}(x) \neq 0$. We infer from (6.2.5) that some $g_i(x) \neq 0$; thus, $x \notin B$. That is, $U \setminus \bigcup_j A_j \subset U \setminus B$. We see that

$$\bigcup_j A_j \supset B, \quad \bigcup_{j=1}^{l} \mathscr{V}(\mathfrak{a}_j) \supset \mathscr{V}\left(\bigcap_{j=1}^{l} \mathfrak{a}_j\right). \tag{6.2.6}$$

On the other hand, it follows from $\mathfrak{a}_i \supset \bigcap_j \mathfrak{a}_j$ and (ii) that $\mathscr{V}(\mathfrak{a}_i) \subset \mathscr{V}\left(\bigcap_j \mathfrak{a}_j\right)$, and that

$$\bigcup_{j=1}^{l} \mathscr{V}(\mathfrak{a}_j) \subset \mathscr{V}\left(\bigcap_{j=1}^{l} \mathfrak{a}_j\right).$$

Combining this with (6.2.6), we obtain the equality.

(vii) This follows from the definition. □

Proposition 6.2.7 *Let $\Omega \subset \mathbf{C}^n$ be an open subset, and let $A_\alpha \subset \Omega$, $\alpha \in \Gamma$, be any family of analytic subsets. Then $\bigcap_{\alpha \in \Gamma} A_\alpha$ is an analytic subset of Ω.*

Proof Firstly, $\bigcap_\alpha A_\alpha$ is closed in Ω. At a point $a \in \bigcap_{\alpha \in \Gamma} A_\alpha$ we consider the geometric ideal $\mathscr{I}\langle \underline{A_\alpha}_a \rangle \subset \mathscr{O}_{n,a}$ together with their sum

$$\mathfrak{a} = \sum_{\alpha \in \Gamma} \mathscr{I}\langle \underline{A_\alpha}_a \rangle \subset \mathscr{O}_{n,a}.$$

Since $\mathfrak{a}$ is finitely generated by the Noetherian property, there are a neighborhood $U \ni a$ and $f_j \in \mathscr{O}(U)$, $1 \leq j \leq l < \infty$, such that

$$\mathfrak{a} = \sum \mathscr{O}_{n,a}\, \underline{f_j}_a = \sum_{\alpha:\text{finitely many}} \mathscr{I}\langle \underline{A_\alpha}_a \rangle,$$

$$\underline{\bigcap_{\alpha \in \Gamma} A_\alpha}_a = \bigcap_{\alpha:\text{finitely many}} \underline{A_\alpha}_a = \underline{\{f_1 = \cdots = f_l = 0\}}_a.$$

Therefore, $\bigcap A_\alpha$ is an analytic subset of Ω. □

Definition 6.2.8 A germ of analytic subset $\underline{X}_a \in \underline{\mathfrak{A}}_a$ is said to be *reducible* if there are germs $\underline{X_i}_a \in \underline{\mathfrak{A}}_a$, $i = 1, 2$, satisfying

$$\underline{X}_a = \underline{X_1}_a \cup \underline{X_2}_a,$$
$$\underline{X}_a \neq \underline{X_i}_a, \quad i = 1, 2.$$

In this case we call "$\underline{X}_a = \underline{X_1}_a \cup \underline{X_2}_a$" a *proper decomposition*. If $\underline{X}_a$ is not reducible, it is said to be *irreducible*.

Theorem 6.2.9 (i) *A germ* $\underline{X}_a \in \mathfrak{A}_a$ *is irreducible if and only if* $\mathscr{I}\langle \underline{X}_a \rangle$ *is a prime ideal.*

(ii) *A germ* $\underline{X}_a \in \mathfrak{A}_a$ *decomposes uniquely up to the order into irreducible elements* $\underline{X_j}_a \in \mathfrak{A}_a$, $1 \le j \le l$, *with* $\underline{X}_a = \bigcup_{j=1}^{l} \underline{X_j}_a$, *such that there is no inclusion relation among* $\underline{X_j}_a$*'s.*

Proof (i) If $\mathscr{I}\langle \underline{X}_a \rangle$ is not a prime ideal, there exist two elements $\underline{f}_a, \underline{g}_a \in \mathscr{O}_{n,a} \setminus \mathscr{I}\langle \underline{X}_a \rangle$ such that $\underline{f}_a \cdot \underline{g}_a \in \mathscr{I}\langle \underline{X}_a \rangle$. It follows that

$$\begin{aligned} \underline{X}_a &= \underline{X}_a \cap \underline{\{f \cdot g = 0\}}_a \\ &= \left(\underline{X}_a \cap \underline{\{f = 0\}}_a \right) \cup \left(\underline{X}_a \cap \underline{\{g = 0\}}_a \right), \end{aligned}$$

but the choice of $\underline{f}_a$ and $\underline{g}_a$ implies that

$$\underline{X}_a \cap \underline{\{f = 0\}}_a \subsetneqq \underline{X}_a, \quad \underline{X}_a \cap \underline{\{g = 0\}}_a \subsetneqq \underline{X}_a.$$

Therefore, $\underline{X}_a$ is reducible.

Conversely, assume that $\underline{X}_a$ is reducible. Then there is a proper decomposition, $\underline{X}_a = \underline{X_1}_a \cup \underline{X_2}_a$, so that

$$\begin{aligned} &\mathscr{I}\langle \underline{X}_a \rangle = \mathscr{I}\langle \underline{X_1}_a \rangle \cap \mathscr{I}\langle \underline{X_2}_a \rangle, \\ &\mathscr{I}\langle \underline{X_i}_a \rangle \supsetneqq \mathscr{I}\langle \underline{X}_a \rangle, \quad i = 1, 2. \end{aligned}$$

It follows that there are elements $\underline{f_i}_a \in \mathscr{I}\langle \underline{X_i}_a \rangle \setminus \mathscr{I}\langle \underline{X}_a \rangle$, $i = 1, 2$, with $\underline{f_1}_a \cdot \underline{f_2}_a \in \mathscr{I}\langle \underline{X}_a \rangle$. Thus, $\mathscr{I}\langle \underline{X}_a \rangle$ is not prime.

(ii) If $\underline{X}_a$ is reducible, there is a proper decomposition $\underline{X}_a = \underline{X_1}_a \cup \underline{X_2}_a$. It follows that $\mathscr{I}\langle \underline{X}_a \rangle \subsetneqq \mathscr{I}\langle \underline{X_i}_a \rangle$, $i = 1, 2$. Moreover, if $\underline{X_i}_a$ is not irreducible, again there is a proper decomposition, $\underline{X_i}_a = \underline{X_{i1}}_a \cup \underline{X_{i2}}_a$. Then it follows that

$$\mathscr{I}\langle \underline{X}_a \rangle \subsetneqq \mathscr{I}\langle \underline{X_i}_a \rangle \subsetneqq \mathscr{I}\langle \underline{X_{ij}}_a \rangle, \qquad i, j = 1, 2.$$

We repeat this process, but since $\mathscr{O}_{n,a}$ is Noetherian, it terminates at a finite number of times.

We show the uniqueness. Suppose that there are two such decompositions,

$$\underline{X}_a = \bigcup \underline{X_i}_a = \bigcup \underline{\tilde{X}_j}_a.$$

For every $\underline{X_i}_a$, we have that $\underline{X_i}_a = \bigcup_j (\underline{X_i}_a \cap \underline{\tilde{X}_j}_a)$. Since $\underline{X_i}_a$ is irreducible, there is some $\underline{\tilde{X}_{j(i)}}_a$ with

$$\underline{X_i}_a \subset \underline{\tilde{X}_{j(i)}}_a.$$

Similarly, we have $\underline{\tilde{X}_{j(i)}}_a \subset \underline{X_{i(j(i))}}_a$. The members of the family $\left\{\underline{X_i}_a\right\}$ to which $\underline{X_{i(j(i))}}_a$ belongs are distinct and have no inclusion relation to each other; we deduce that

$$\underline{X_i}_a = \underline{\tilde{X}_{j(i)}}_a = \underline{X_{i(j(i))}}_a.$$

Therefore, $\left\{\underline{X_i}_a\right\}$ and $\left\{\underline{\tilde{X}_j}_a\right\}$ are the same up to the orders. □

Each $\underline{X_j}_a$ of Theorem 6.2.9 is called an *irreducible component* of $\underline{X}_a$ at a.

Remark 6.2.10 The relations above between the germs of analytic subsets and the ideals of $\mathcal{O}_{n,a}$ hold in a similar manner for the algebraic subsets of $\mathbf{C}^n$ and the ideals of the polynomial ring $\mathbf{C}[z_1, \ldots, z_n]$. For instance, an algebraic subset $A \subset \mathbf{C}^n$ is irreducible if and only if the corresponding ideal of $\mathbf{C}[z_1, \ldots, z_n]$ is prime.

Example 6.2.11 An irreducible algebraic subset is not necessarily an irreducible analytic subset. For example,

$$P(x, y) = x^2 - x^3 - y^2 \in \mathbf{C}[x, y]$$

is an irreducible polynomial. With

$$X = \{(x, y) \in \mathbf{C}^2;\ P(x, y) = 0\} \subset \mathbf{C}^2,$$

X is algebraically irreducible.

In a neighborhood of 0, we have analytically

$$\begin{aligned} x^2 - x^3 - y^2 &= x^2(1 - x) - y^2 \\ &= (x\sqrt{1 - x})^2 - y^2 \\ &= (x\sqrt{1 - x} - y)(x\sqrt{1 - x} + y), \end{aligned}$$

where a branch of $\sqrt{1 - x}$ is chosen. Then,

$$\underline{X}_0 = \underline{\{x\sqrt{1 - x} - y = 0\}}_0 \bigcup \underline{\{x\sqrt{1 - x} + y = 0\}}_0,$$

which is reducible.

Example 6.2.12 Even if $\underline{X}_a$ is irreducible, at a nearby point b of a, $\underline{X}_b$ is not necessarily irreducible. We define X in $\mathbf{C}^3$ by

$$x^2 - zy^2 = 0.$$

Then, $\underline{X}_0$ is irreducible. At any near point of 0 with $z \neq 0$, at $(0, 0, z)$ X decomposes into two irreducible components: $x = \pm\sqrt{z} \cdot y$.

6.3 Prerequisite from Algebra

In this section we prepare some elementary facts from algebra: For general references, cf., e.g., Nagata [45], Morita [43], Lang [38].

Let A be an integral domain and assume that $A \ni 1 \neq 0$ (cf. Convention (xvi)). Let $A[X]$ denote the polynomial ring of a variable X with coefficients in A. For an ideal $\mathfrak{a} \subset A$ we define the *radical* $\sqrt{\mathfrak{a}}$ similarly to (6.2.2) by

$$\sqrt{\mathfrak{a}} = \{a \in A; {}^{\exists}N \in \mathbf{N}, a^N \in \mathfrak{a}\}.$$

If $\mathfrak{a}$ is a prime ideal, $\sqrt{\mathfrak{a}} = \mathfrak{a}$. An ideal $\mathfrak{a}$ with prime $\sqrt{\mathfrak{a}}$ is called a primary ideal.

Theorem 6.3.1 (Primary decomposition) *Let A be Noetherian. For an ideal $\mathfrak{a} \subset A$ there are primary ideals $\mathfrak{p}_1, \ldots, \mathfrak{p}_l$ $(l < \infty)$ such that*

$$\mathfrak{a} = \bigcap_{i=1}^{l} \mathfrak{p}_i,$$

$$\mathfrak{a} \subsetneqq \bigcap_{i \in I} \mathfrak{p}_i, \quad {}^{\forall}I \subsetneqq \{1, \ldots, l\}.$$

With these conditions, $\{\mathfrak{p}_1, \ldots, \mathfrak{p}_l\}$ is unique up to the order.

Let k be a field with characteristic, $\operatorname{char} k = 0$.

Theorem 6.3.2 (Primitive element) *Let $K = k(u_1, \ldots, u_r)$ be a finite extension of k, and let $S \subset k$ be any infinite subset. Then there are elements $c_1, \ldots, c_r \in S$ such that*

$$K = k\left(\sum_{j=1}^{r} c_j u_j\right).$$

Definition 6.3.3 A homomorphism $\eta : R \to A$ between two rings is called a *finite homomorphism* if through η, A is a finitely generated module over R; that is, there are finitely many elements $v_1, \ldots, v_n \in A$ such that $A = \sum_{j=1}^{n} \eta(R) \cdot v_j$. Moreover, if $\eta : R \to A$ is injective, with regarding $R \subset A$, $A \supset R$ (or $R \subset A$) is called a *finite ring extension.*

Theorem 6.3.4 *Let $\widetilde{B} \supset B$ be an extension of modules over a ring R. Assume that $\widetilde{B}$ has a multiplicative structure (R-algebra) and B is finitely generated over R. Then for an element $f \in \widetilde{B}$ with $f \cdot B \subset B$, there is a monic polynomial in f with coefficients in R,*

$$D(f) = f^n + \sum_{\nu=1}^{n} c_\nu f^{n-\nu} \quad (c_\nu \in R),$$

such that

$$D(f) \cdot B = 0.$$

In particular, if B is a ring, then $D(f) = 0$, and hence f is integral over R.

Proof Let $\{v_1, \ldots, v_n\} \subset B$ be a finite generator system of B over R. We set $f \cdot v_j = \sum(-c_{jk}) \cdot v_k$ with $c_{jk} \in R$. Then,

$$\begin{pmatrix} f + c_{11} & c_{12} & \cdots & c_{1n} \\ c_{21} & f + c_{22} & \cdots & c_{2n} \\ \vdots & \vdots & \ddots & \vdots \\ c_{n1} & c_{n2} & \cdots & f + c_{nn} \end{pmatrix} \begin{pmatrix} v_1 \\ \vdots \\ \vdots \\ v_n \end{pmatrix} = \begin{pmatrix} 0 \\ \vdots \\ \vdots \\ 0 \end{pmatrix}.$$

Let $D(f) = f^n + \sum_{\nu=1}^{n} c_\nu f^{n-\nu}$ denote the determinant of the matrix in the left-hand side above. It follows from Cramer's formula that

$$D(f)v_j = 0, \quad 1 \leq j \leq n.$$

Therefore, $D(f) \cdot B = 0$.

If B is a ring, $B \ni 1$, so that $D(f) = 0$. □

Corollary 6.3.5 *In Theorem 6.3.4 we assume that A is an integral domain, and denote by L (resp. K) the quotient field of A (resp. R). Then the minimal polynomial of every $f \in A$ over K is written as*

$$P(f) = f^d + \sum_{\nu=1}^{d} a_\nu f^{d-\nu} = 0, \quad a_\nu \in R.$$

Proof By Theorem 6.3.4, f is integral over R, so that its minimal polynomial is integral over A (cf. Nagata [45], Lang [38], p. 240). □

Theorem 6.3.6 *Let R be an integrally closed integral domain. Let $A \supset R$ be a finite ring extension, and let L/K be the extension of the quotient fields of them. Let $L = K(\alpha)$ with $\alpha \in A$, and let $P(X) \in R[X]$ be the minimal polynomial of α.*

Then, for every $f \in A$ we have

$$P'(\alpha) \cdot f = Q_1(\alpha) \in R[\alpha], \quad \deg Q_1 < \deg P.$$

Here, P' is a formal derivative of the polynomial P. In particular, with the resultant $\Delta (\in R)$ of $P(X)$, we obtain

$$\Delta \cdot f = Q_2(\alpha) \in R[\alpha], \quad \deg Q_2 < \deg P.$$

Cf. Nagata [45], Theorem 3.9.2.

Proof Let $\alpha = \alpha^{(1)}, \ldots, \alpha^{(d)}$ be α and all of its conjugates. We consider the Galois extension $\tilde{L} = L(\alpha^{(1)}, \ldots, \alpha^{(d)})$ of K.

The element f can be written as $f = Q(\alpha)$ with a rational function of α. Since f is integral over R, $Q(\alpha^{(j)})$ is also integral over R. We consider the following equation:

$$
\begin{aligned}
(6.3.7) \qquad & \sum_{j=1}^{d} \frac{P(X)}{X-\alpha^{(j)}} Q(\alpha^{(j)}) \\
&= \sum_{j=1}^{d} \frac{P(X)-P(\alpha^{(j)})}{X-\alpha^{(j)}} Q(\alpha^{(j)}) \\
&= \sum_{j=1}^{d} \frac{X^d - (\alpha^{(j)})^d + \sum_{\nu=1}^{d-1} a_\nu (X^{d-\nu} - (\alpha^{(j)})^{d-\nu})}{X-\alpha^{(j)}} Q(\alpha^{(j)}) \\
&= \sum_{j=1}^{d} \left\{ \sum_{\nu=0}^{d-1} X^{d-\nu-1} (\alpha^{(j)})^{\nu} Q(\alpha^{(j)}) \right. \\
&\qquad \left. + \sum_{\nu=1}^{d-1} a_\nu \sum_{\mu=0}^{d-\nu-1} X^{d-\nu-\mu-1} (\alpha^{(j)})^{\mu} Q(\alpha^{(j)}) \right\} \\
&= \sum_{\nu=0}^{d-1} b_j X^{d-1-\nu}.
\end{aligned}
$$

We denote the last polynomial by $Q_1(X)$. Then, $\deg Q_1 < d = \deg P$. Since $b_j \in \tilde{L}$ is integral over R, and invariant with respect to the Galois group $\mathrm{Gal}(\tilde{L}/K)$, $b_j \in K$. Since R is integrally closed, $b_j \in R$. With substituting $X = \alpha$ in (6.3.7), we get

$$\text{the left-hand side} = P'(\alpha)Q(\alpha) = P'(\alpha) \cdot f = Q_1(\alpha) \in A[\alpha].$$

By Theorem 2.2.11 there are polynomials $\alpha(X), \beta(X) \in R[X]$ such that

$$(6.3.8) \qquad \alpha(X)P(X) + \beta(X)P'(X) = \Delta.$$

With $X = \alpha$ we obtain

$$\Delta \cdot f = \beta(\alpha)P'(\alpha)f = \beta(\alpha)Q_1(\alpha).$$

After dividing the last polynomial $\beta(\alpha)Q_1(\alpha)$ by $P(\alpha)$ we denote the remainder by $Q_2(\alpha) \in R[\alpha]$; we get

$$\Delta \cdot f = Q_2(\alpha), \quad \deg Q_2 < \deg P. \qquad \square$$

6.4 Ideals of Local Rings

Let $\mathfrak{a}$ be an ideal of the local ring $\mathscr{O}_{\mathbf{C}^n,0} = \mathscr{O}_{n,0}$. Unless otherwise mentioned, we set:

6.4.1 (Assumption) $\{0\} \subsetneqq \mathfrak{a} \subsetneqq \mathscr{O}_{n,0}$.

Let $z_1, \ldots, z_n$ be the standard coordinate system of $\mathbf{C}^n$, and set

$$\begin{array}{l} \mathscr{O}_{n,0} = \mathbf{C}\{z_1, \ldots, z_n\} \\ \cup \\ \mathscr{O}_{p,0} = \mathbf{C}\{z_1, \ldots, z_p\}, \quad 0 \le p \le n, \\ \mathscr{O}_{p,0} \hookrightarrow \mathscr{O}_{n,0} \\ \qquad {}_{\eta}\searrow \quad \downarrow \\ \qquad\quad \mathscr{O}_{n,0}/\mathfrak{a} = A \,. \end{array}$$

N.B. In what follows, to avoid the complication of notation, we sometimes use the same notation, e.g., f for a holomorphic function in a neighborhood of 0 and for its germ at 0, unless there occurs confusion; e.g., z_n is used also in the sense of $\underline{z_n}_0$.

It may be also better to use different notation for 0 of $\mathbf{C}^n$ and for 0 of $\mathbf{C}^p$, but both are the origin, so we use the same 0; there will be no confusion.

Proposition 6.4.2 *After a linear transform of $z_1, \ldots, z_n$, there is a number $0 \le p < n$ such that*

$$\eta : \mathscr{O}_{p,0} \longrightarrow A$$

is injective and A is a finite module over $\mathscr{O}_{p,0}$ through η. In particular, A is integral over $\mathscr{O}_{p,0}$.

Proof By Assumption 6.4.1 there is an element $\underline{f_n}_0 \in \mathfrak{a}$ such that $\underline{f_n}_0 \neq 0$. Take a standard coordinate system $(z_1, \ldots, z_n)$ of $\underline{f_n}_0$. Note that it is free to apply linear transformations to $z_1, \ldots, z_{n-1}$. Let

$$f_n(z', z_n) = u_n(z) W_n(z', z_n), \quad \deg W_n = d_n$$

be the Weierstrass decomposition of $f_n(z', z_n)$ at 0. If $\mathfrak{a} \cap \mathscr{O}_{n-1,0} = \{0\}$, we take $p = n-1$. If $\mathfrak{a} \cap \mathscr{O}_{n-1,0} \neq \{0\}$, we take an element $\underline{f_{n-1}}_0 \in \mathfrak{a} \cap \mathscr{O}_{n-1,0} \setminus \{0\}$, and a standard coordinate system $(z_1, \ldots, z_{n-2}, z_{n-1}) = (z'', z_{n-1})$ of $\underline{f_{n-1}}_0$. Then we have the Weierstrass decomposition:

$$f_{n-1}(z'', z_{n-1}) = u_{n-1}(z'', z_{n-1}) W_{n-1}(z'', z_{n-1}), \quad \deg W_{n-1} = d_{n-1},$$

where $W_{n-1}(z'', z_{n-1}) \in \mathfrak{a} \cap \mathscr{O}_{n-1,0}$. We repeat this procedure until $\mathfrak{a} \cap \mathscr{O}_{p,0} = \{0\}$ and obtain

$$W_\nu(z_1, \ldots, z_{\nu-1}, z_\nu) \in \mathfrak{a} \cap \mathscr{O}_{\nu,0}, \quad p+1 \le \nu \le n,$$
$$W_\nu(z_1, \ldots, z_{\nu-1}, z_\nu) \in \mathscr{O}_{\nu-1,0}[z_\nu], \quad \text{a Weierstrass polynomial},$$
$$\deg W_\nu = d_\nu.$$

Since $\mathfrak{a} \cap \mathscr{O}_{p,0} = \{0\}$, η is injective.

Claim 6.4.3 *A is finitely generated as $\mathscr{O}_{p,0}$-module.*

∵) To show this we take an arbitrary element $\underline{f}_0 \in \mathscr{O}_{n,0}$. By Weierstrass' Preparation Theorem 2.1.3 we write

$$\begin{aligned} f &= a_n W_n + \sum_{\nu=0}^{d_n-1} b_\nu(z_1, \ldots z_{n-1})\, z_n^\nu \\ &\equiv \sum_{\nu=0}^{d_n-1} b_\nu(z_1, \ldots z_{n-1})\, z_n^\nu \pmod{\mathfrak{a}}. \end{aligned}$$

Further, by division we have

$$b_\nu(z_1, \ldots, z_{n-1}) = a_{\nu n-1} W_{n-1} + \sum_{\mu=0}^{d_{n-1}-1} c_{\nu\mu}(z_1, \ldots, z_{n-2}) z_{n-1}^\mu,$$

so that

$$f \equiv \sum_{\nu=0}^{d_n-1} \sum_{\mu=0}^{d_{n-1}-1} c_{\nu\mu}(z_1, \ldots, z_{n-2}) z_{n-1}^\mu z_n^\nu \pmod{\mathfrak{a}}.$$

Repeating this, we obtain

$$f \equiv \sum_{\nu_n=0}^{d_n-1} \cdots \sum_{\nu_{p+1}=0}^{d_{p+1}-1} c_{\nu_{p+1}\cdots\nu_n}(z_1, \ldots, z_p) z_{p+1}^{\nu_{p+1}} \cdots z_n^{\nu_n} \pmod{\mathfrak{a}}.$$

Therefore,

$$A = \sum_{\substack{0 \le \nu_j \le d_j - 1 \\ p+1 \le j \le n}} \mathscr{O}_{p,0} \cdot z_{p+1}^{\nu_{p+1}} \cdots z_n^{\nu_n},$$

and A is finitely generated over $\mathscr{O}_{p,0}$. □

Since η is injective, we regard $\mathscr{O}_{p,0}$ as a subring of A.

Corollary 6.4.4 *Under the conditions of Proposition 6.4.2, we assume that $\mathfrak{a}$ is a prime ideal. Let K (resp. L) be the quotient field of $\mathscr{O}_{p,0}$ (resp. A). Then L/K is a finite field extension, and*

$$L = K(z_{p+1}, \ldots, z_n).$$

Taking $\lambda_j \in \mathbf{C}$, $p+1 \le j \le n$, with $\lambda_{p+1} \ne 0$, we set

$$z'_{p+1} = \sum_{j=p+1}^{n} \lambda_j z_j.$$

By Theorem 6.3.2 there is such a vector (λ_j) with $L = K(z'_{p+1})$. Denoting again z'_{p+1} by z_{p+1}, we have

(6.4.5) $$L = K(z_{p+1}).$$

Let $z' = (z_1, \ldots, z_p)$ and let the minimal polynomial of z_{p+1} be

(6.4.6) $$P_{p+1}(z', z_{p+1}) = z_{p+1}^{d_{p+1}} + \sum_{\nu=1}^{d_{p+1}} b_\nu(z') z_{p+1}^{d_{p+1}-\nu} = 0.$$

Since z_{p+1} is integral over $\mathscr{O}_{p,0}$, $b_\nu \in \mathscr{O}_{p,0}$ by Corollary 6.3.5.

Lemma 6.4.7 *$P_{p+1}(z', z_{p+1})$ is a Weierstrass polynomial.*

Proof If $b_\nu(0) \ne 0$ for one of $1 \le \nu \le d_{p+1}$, we denote the maximum of such ν by ν_0. Then we may write

$$P_{p+1}(0, z_{p+1}) = z_{p+1}^{d_{p+1}-\nu_0}(b_{\nu_0}(0) + O(z_{p+1})).$$

Let

$$P_{p+1}(0, z_{p+1}) = u(z', z_{p+1}) Q(z', z_{p+1}), \quad \deg Q = d_{p+1} - \nu_0 < d_{p+1}.$$

be the Weierstrass decomposition at 0. This, however, contradicts $P_{p+1}(z', z_{p+1})$ being the minimal polynomial of z_{p+1}. □

Remark 6.4.8 For each of z_j, $p+1 < j \le n$, the minimal polynomial $P_j(z', z_j)$ of z_j over K is a Weierstrass polynomial; therefore the root of $P_j(0, z_j) = 0$, $p+1 \le j \le n$ consists only of $z_j = 0$, $p+1 \le j \le n$.

Remark 6.4.9 If $p = 0$, then there is a number $N \in \mathbf{N}$ with $z_j^N = 0$ (in A), $1 \le j \le n$; i.e., $z_j^N \in \mathfrak{a}$ for all j, and since $\mathfrak{a}$ is prime, $z_j \in \mathfrak{a}$, $\mathscr{V}(\mathfrak{a}) = \underline{\{0\}}_0$. Conversely, for $\mathscr{I}\langle \underline{\{0\}}_0 \rangle$, $z_j \in \mathscr{I}\langle \underline{\{0\}}_0 \rangle$, $1 \le j \le n$, and so $p = 0$.

Below, we assume that $\mathfrak{a}$ is prime, and assume the conditions of Proposition 6.4.2 and (6.4.5). Set

$$P(z', z_{p+1}) = P_{p+1}(z', z_{p+1}), \quad d = d_{p+1},$$

and let $\delta(z')$ denote the discriminant of $P(z', z_{p+1})$. By Theorems 2.2.11 and 2.2.5 there are elements $\alpha(z_{p+1}), \beta(z_{p+1}) \in \mathscr{O}_{p,0}[z_{p+1}]$ of $\deg \alpha(z_{p+1}) < d-1$, $\deg \beta(z_{p+1}) < d$, respectively, such that

$$\alpha(z_{p+1})P_{p+1}(z_{p+1}) + \beta(z_{p+1})P'_{p+1}(z_{p+1}) = \delta(z'). \tag{6.4.10}$$

Lemma 6.4.11 *For every $f \in \mathscr{O}_{n,0}$, there exists an element $R_f(z_{p+1}) \in \mathscr{O}_{p,0}[z_{p+1}]$ of* $\deg R_f < d$ *such that*

$$\delta f - R_f(z_{p+1}) \in \mathfrak{a}.$$

Proof We apply Theorem 6.3.6 with $R = \mathscr{O}_{p,0}$, $A = \mathscr{O}_{n,0}/\mathfrak{a}$ and $\alpha = z_{p+1}$. Since $\mathscr{O}_{p,0}$ is a unique factorization domain, it is integrally closed, and then there is some $R_f(z_{p+1}) \in \mathscr{O}_{p,0}[z_{p+1}]$, $\deg R_f < d$, satisfying

$$\delta f - R_f(z_{p+1}) \in \mathfrak{a}.$$

□

The lemma above applied for $f = z_j,\ p+2 \le j \le n$, yields $Q_j(z_{p+1}) \in \mathscr{O}_{p,0}[z_{p+1}]$ such that

$$\delta z_j - Q_j(z_{p+1}) \in \mathfrak{a}, \quad p+2 \le j \le n, \quad \deg Q_j < d. \tag{6.4.12}$$

Lemma 6.4.13 *For every $f \in \mathscr{O}_{n,0}$, there are elements $g \in \mathscr{O}_{n,0} \backslash \mathfrak{a}$ and $h \in \mathscr{O}_{p,0}$ such that*

$$gf - h \in \mathfrak{a}.$$

Proof If $f \in \mathfrak{a}$, it suffices to take $g = 1$ and $h = 0$. Suppose that $f \notin \mathfrak{a}$, Since f is integral over $\mathscr{O}_{p,0}$, we may take a polynomial in f,

$$P(z', f) = f^m + \sum_{\nu=0}^{m-1} a_\nu(z') f^\nu \in \mathfrak{a}, \quad a_\nu \in \mathscr{O}_{p,0},$$

where m is the minimal among such degrees. Then this is also the minimal polynomial of f, and then by the same reason as in Lemma 6.4.7, $P(z', f)$ is a Weierstrass polynomial in f. Setting

$$g = f^{m-1} + \sum_{\nu=1}^{m-1} a_\nu(z') f^{\nu-1},$$

$$h = -a_0,$$

we have that $g \notin \mathfrak{a},\ h \in \mathscr{O}_{p,0}$, and $gf - h \in \mathfrak{a}$. □

We summarize the results obtained in Proposition 6.4.2, (6.4.5), (6.4.6), Lemma 6.4.7, Remark 6.4.8, Lemma 6.4.11, and (6.4.12).

Theorem 6.4.14 (Setting) *For a prime ideal, $\{0\} \subsetneqq \mathfrak{a} \subsetneqq \mathscr{O}_{n,0}$, we have the following coordinate system after a suitable linear change,*

$$z = (z_1, \ldots, z_p, z_{p+1}, \ldots, z_m), \qquad z' = (z_1, \ldots, z_p)$$

such that:

(i) *The natural homomorphism $\mathscr{O}_{p,0} \to \mathscr{O}_{n,0}/\mathfrak{a} = A$ is injective and finite, and A is integral over $\mathscr{O}_{p,0}$.*
(ii) *With the quotient fields K of $\mathscr{O}_{p,0}$ and L of A,*

$$L = K(z_{p+1}).$$

(iii) *For every index, $p+1 \leq j \leq n$, the minimal polynomial $P_j(z', z_j) \in \mathscr{O}_{p,0}[z_j]$ of z_j over K is a Weierstrass polynomial.*
(iv) *Set $d = \deg P_{p+1}$, and let $\delta(z')$ be the discriminant of $P_{p+1}(z', z_{p+1})$. For every index, $p+2 \leq j \leq n$, there is an element $Q_j(z_{p+1}) \in \mathscr{O}_{p,0}[z_{p+1}]$ of $\deg Q_j < d$ such that in A,*

$$\delta(z')z_j - Q_j(z_{p+1}) = 0, \quad p+2 \leq j \leq n. \tag{6.4.15}$$

Henceforth we assume the setting given in Theorem 6.4.14 for ideal $\mathfrak{a}$ and coordinates.

With the finite generators $\underline{f_1}_0, \ldots, \underline{f_m}_0$ of $\mathfrak{a}$ we set

$$S = \{f_1 = \cdots = f_m = 0\}$$

in a suitable neighborhood of 0. Then $\underline{S}_0 = \mathscr{V}(\mathfrak{a})$ is an irreducible germ of an analytic subset.

Theorem 6.4.16 *Let S be as above. Then there is a neighborhood basis $\{U\}$ of $0 \in \mathbf{C}^n$ such that with setting*

$$0 \in U = U' \times U'' \subset \mathbf{C}^p \times \mathbf{C}^{n-p},$$
$$\pi : U \cap S \to U' \quad \text{(projection)},$$

the following are satisfied:

(i) *π is finite, and $\pi^{-1}0 = 0$.*
(ii)

$$\begin{aligned} S \cap U \cap \{z = (z', z'') \in U' \times U''; \ \delta(z') \neq 0\} \\ = \{z \in U; \delta(z') \neq 0, \ P_{p+1}(z', z_{p+1}) = 0, \\ \delta(z')z_j - Q_j(z_{p+1}) = 0, \ p+2 \leq j \leq n\}. \end{aligned}$$

(iii) *If* $z = (z', z'') \in U' \times \mathbf{C}^{n-p}$ *satisfies* $\delta(z') \neq 0$ *and*

$$P_{p+1}(z', z_{p+1}) = \delta(z')z_j - Q_j(z_{p+1}) = 0, \quad p+2 \leq j \leq n,$$

then $z \in U' \times U'' = U$.

(iv) $\Sigma(S) \cap U \subset \pi^{-1}\{\delta(z') = 0\}$, *and*

$$\pi|_{S\cap U\setminus\pi^{-1}\{\delta=0\}} : S \cap U\setminus\pi^{-1}\{\delta = 0\} \to U'\setminus\{\delta = 0\}$$

is an unramified covering such that $S \cap U\setminus\pi^{-1}\{\delta = 0\}$ *is a connected open subset of* S.

(v) $\pi : S \cap U \to U'$ *is surjective.*

Proof We take a neighborhood V of 0 where all the germs of functions in Theorem 6.4.14, as well as $\underline{f_1}_0, \ldots, \underline{f_{m}}_0$ above, have the representatives, and consider $\mathrm{P}\Delta_p(0; r') \times \mathrm{P}\Delta_{n-p}(0; r'') \subset V$. Put

$$|z'| = \max\{|z_i|; 1 \leq i \leq p\}.$$

Since $P_j(z', z_j)$, $p+1 \leq j \leq n$, are all Weierstrass polynomials, for every $\rho > 0$ there is a number $\sigma > 0$ such that all roots of $P_j(z', z_j) = 0$ with $|z'| < \sigma$ are contained in $|z_j| \leq \rho/2$. As $\rho \searrow 0$, $\sigma \searrow 0$: Thus, with such ρ and σ we set

(6.4.17)
$$\begin{aligned} U'_\sigma &= \mathrm{P}\Delta_p(0; (\sigma)), \quad (\sigma) = (\sigma, \ldots, \sigma) \ (p\text{-vector}), \\ U''_\rho &= \mathrm{P}\Delta_{n-p}(0; (\rho)), \quad (\rho) = (\rho, \ldots, \rho) \ ((n-p)\text{-vector}), \\ U_{\sigma,\rho} &= U'_\sigma \times U''_\rho. \end{aligned}$$

(i) If $z = (z', z'') \in S \cap U_{\sigma,\rho}$, then

$$P_j(z', z_j) = 0, \quad p+1 \leq j \leq n,$$

and so $|z_j| \leq \rho/2$, $p+1 \leq j \leq n$. Therefore, $\pi : S \cap U_{\sigma,\rho} \to U'_\sigma$ is proper and $\pi^{-1}z'$ is a finite set. If $z' = 0$, then by Remark 6.4.8 we have $z'' = 0$.

(ii) On $U_{\sigma,\rho}$ we apply Weierstrass' Preparation Theorem 2.1.3 for f_i divided by $P_n(z', z_n)$:

$$\begin{aligned} f_i &= a_{in} P_n(z', z_n) + \sum_{\nu=1}^{d_n} b_{in\nu} z_n^{d_n-\nu}, \qquad a_{in} \in \mathscr{O}(U_{\sigma,\rho}), \\ b_{in\nu} &= b_{in\nu}(z', z_{p+1}, \ldots, z_{n-1}) \in \mathscr{O}(\mathrm{P}\Delta_p(0, (\sigma)) \times \mathrm{P}\Delta_{n-1-p}(0, (\rho))). \end{aligned}$$

Similarly, the division of $b_{in\nu}$ by $P_{n-1}(z', z_{n-1})$ yields

$$b_{in\nu} = a_{in-1\nu} P_{n-1}(z', z_{n-1}) + \sum_{\mu=1}^{d_{n-1}} b_{in-1\mu} z_{n-1}^{d_{n-1}-\mu}.$$

Repeating this up to $P_{p+1}(z', z_{p+1})$, we obtain

$$f_i \equiv \sum_{0 \le \nu_j < d_j} c_{i\nu_{p+1}\cdots\nu_n}(z') z_{p+1}^{\nu_{p+1}} \cdots z_n^{\nu_n} \pmod{P_{p+1}, \ldots, P_n}.$$

Next, taking a sufficiently large $N > 0$, we get

$$\delta^N f_i \equiv \sum_{0 \le \nu_j < d_j} c'_{i\nu_{p+1}\cdots\nu_n}(z') z_{p+1}^{\nu_{p+1}} (\delta z_{p+2})^{\nu_{p+2}} \cdots (\delta z_n)^{\nu_n} \pmod{P_{p+1}, \ldots, P_n}.$$

Since for $p + 2 \le j \le n$

$$\delta(z') z_j - Q_j(z_{p+1}) \in \mathfrak{a},$$

we deduce that

$$\delta^N f_i \equiv \sum_{\text{finite}} d_{i\nu_{p+1}}(z') z_{p+1}^{\nu_{p+1}} \pmod{\delta z_j - Q_j,\ p + 2 \le j \le n,\ P_{p+1}, \ldots, P_n}.$$

After dividing the right-hand side by $P_{p+1}(z', z_{p+1})$, we have

$$\begin{aligned}\delta^N f_i &\equiv R_i(z', z_{p+1}) \pmod{P_{p+1}, \ldots, P_n, \delta z_{p+2} - Q_{p+2}, \ldots, \delta z_n - Q_n},\\ &\deg R_i < d = \deg P_{p+1}.\end{aligned}$$

Since

$$\begin{aligned}&\underline{\delta_0^N}\, \underline{f_i}_0 \in \mathfrak{a}, \quad \underline{P_{p+1}}_0, \ldots, \underline{P_n}_0 \in \mathfrak{a},\\ &\underline{\delta_0 z_j}_0 - \underline{Q_j(z_{p+1})}_0 \in \mathfrak{a}, \quad p + 2 \le j \le n,\end{aligned}$$

it follows that $R_i(z', z_{p+1}) \in \mathfrak{a}$. Since P_{p+1} is the minimal polynomial of z_{p+1}, the comparison of degrees implies that $R_i(z', z_{p+1}) = 0$.

For an index, $p + 2 \le j \le n$, there is a large number M such that

$$\delta^M P_j(z', z_j) \equiv A_j(z', z_{p+1}) \pmod{P_{p+1}(z', z_{p+1}), \delta z_j - Q_j(z_{p+1})}.$$

Here, note that $\deg A_j < d$ and $\underline{A_j(z', z_{p+1})}_0 \in \mathfrak{a}$. Since P_{p+1} is the minimal polynomial, we deduce that $A_j = 0$. Thus, with N taken larger if necessary, we have

$$\begin{aligned}\delta^N f_i &\equiv 0 \pmod{P_{p+1}(z', z_{p+1}),\ \delta z_j - Q_j(z_{p+1}),\ p + 2 \le j \le n},\\ &1 \le i \le m.\end{aligned}$$

Therefore we see that for $z = (z', z'') \in U'_\sigma \times U''_\rho$ with $\delta(z') \neq 0$,

$$f_i(z) = 0, \ 1 \le i \le m \Longleftrightarrow \begin{cases} P_{p+1}(z', z_{p+1}) = 0, \\ \delta(z')z_j - Q_j(z_{p+1}) = 0, \ p+2 \le j \le n. \end{cases}$$

(iii) The roots of $P_{p+1}(z', z_{p+1}) = 0$ with $z' \in U_\sigma$ satisfy $|z_{p+1}| \le \rho/2$. Since $\delta(z') \neq 0$, each $z_j = Q_j(z_{p+1})/\delta(z')$, $p+2 \le j \le n$, satisfies $P_j(z', z_j) = 0$, and hence $|z_j| \le \rho/2$. Consequently, $(z', z_{p+1}, \ldots, z_n) \in U_{\sigma,\rho}$.

(iv) Suppose that $P_{p+1}(z', z_{p+1}) = 0$ and $\delta(z') \neq 0$. Since $\frac{\partial P_{p+1}}{\partial z_{p+1}}(z', z_{p+1}) \neq 0$, by the implicit function Theorem 1.2.41, z_{p+1} is locally written as a holomorphic function, $z_{p+1} = \varphi_{p+1}(z')$ in z'. Setting

$$z_j = Q_j(\varphi_{p+1}(z'))/\delta(z'), \quad p+2 \le j \le n,$$

S gives rise to a complex submanifold about $(z', z_{p+1}, \ldots, z_n)$, and $z' = (z_1, \ldots, z_p)$ provides a holomorphic local coordinate system.

By Theorem 6.1.4, $U'_\sigma \backslash \{\delta = 0\}$ is a dense connected open set in U'_σ. Let $a_1(z'), \ldots, a_{d'}(z')$ $(d' \le d)$ be the elementary symmetric polynomials formed by the branches of the analytic continuation of $z_{p+1} = \varphi_{p+1}(z')$ over $z' \in U'_\sigma \backslash \{\delta = 0\}$. Notice that $a_j \in \mathscr{O}(U'_\sigma \backslash \{\delta = 0\})$ and it is bounded. Riemann's extension Theorem 2.3.4 implies that $a_j \in \mathscr{O}(U'_\sigma)$. From the construction we obtain

$$P(z', z_{p+1}) = z_{p+1}^{d'} + a_1 z_{p+1}^{d'-1} + \cdots + a_{d'} = 0.$$

Since d is the degree of the minimal polynomial, we see after all that $d' = d$ and $P(z', z_{p+1}) = P_{p+1}(z', z_{p+1})$. Therefore, the local solutions of $P_{p+1}(z', z_{p+1}) = 0$, $z_{p+1} = \varphi_{p+1}(z')$, are connected by analytic continuation; this implies that $S \cap U_{\sigma,\rho} \backslash \pi^{-1}\{\delta = 0\}$ is connected.

(v) It follows that $\pi : S \cap U_{\sigma,\rho} \to U'_\sigma$ is finite and the image is closed. On the other hand, $\pi(S \cap U_{\sigma,\rho}) \supset \{\delta \neq 0\}$, and $\{\delta \neq 0\} \subset U'_\sigma$ is dense; hence, $\pi(S \cap U_{\sigma,\rho}) = U'_\sigma$. □

Definition 6.4.18 We call $U = U' \times U'' = U_{\sigma,\rho} = \mathrm{P}\Delta_p \times \mathrm{P}\Delta_{n-p}$ in Theorem 6.4.16 a *standard polydisk* of S about 0.

Lemma 6.4.19 *We keep the notation in Theorem 6.4.16 and (6.4.17). If a function $f \in \mathscr{O}(U_{\sigma,\rho})$ satisfies $f(z) = 0$ for every $z = (z', z'') \in S \cap U_{\sigma,\rho}$ with $z' \in U'_\sigma \backslash \{\delta = 0\}$, then $\underline{f}_0 \in \mathfrak{a}$.*

Proof Suppose that $\underline{f}_0 \notin \mathfrak{a}$. By Lemma 6.4.13, there are elements $g \in \mathscr{O}_{n,0} \backslash \mathfrak{a}$ and $h \in \mathscr{O}_{p,0}$ such that

$$\underline{g}_0 \underline{f}_0 - \underline{h}_0 \in \mathfrak{a}.$$

Since $\mathfrak{a}$ is prime, $\underline{g}_0 \underline{f}_0 \notin \mathfrak{a}$. Therefore, $\underline{h}_0 \notin \mathfrak{a}$. Take U'_σ smaller if necessary, so that $h \in \mathcal{O}(U'_\sigma)$. For every $z' \in U'_\sigma \setminus \{\delta = 0\}$ there is a point $z = (z', z'') \in S \cap U_{\sigma,\rho}$ with $f(z', z'') = 0$, and hence $h(z') = 0$; that is, $h|_{U'_\sigma \setminus \{\delta=0\}} \equiv 0$ implying $h = 0$. This contradicts $\underline{h}_0 \notin \mathfrak{a}$. □

The following theorem is sometimes called Hilbert's Nullstellen Satz after the algebraic case.

Theorem 6.4.20 (Rückert's Nullstellen Satz) *For every ideal* $\mathfrak{a} \subset \mathcal{O}_{n,0}$,

$$\mathscr{I}(\mathscr{V}(\mathfrak{a})) = \sqrt{\mathfrak{a}}.$$

Proof If $\mathfrak{a} = \{0\}$, then $\mathscr{V}(\mathfrak{a}) = \underline{\mathbf{C}^n}_0$, so that $\mathscr{I}(\underline{\mathbf{C}^n}_0) = \{0\}$.

If $\mathfrak{a} = \mathcal{O}_{n,0}$, then $\mathscr{V}(\mathfrak{a}) = \phi$, so that $\mathscr{I}(\mathscr{V}(\mathcal{O}_{n,0})) = \mathcal{O}_{n,0}$.

In the sequel, we assume that $\{0\} \subsetneqq \mathfrak{a} \subsetneqq \mathcal{O}_{n,0}$, and $\underline{S}_0 = \mathscr{V}(\mathfrak{a})$.

(1) The case of prime $\mathfrak{a}$: Let $\underline{f}_0 \in \mathscr{I}(\underline{S}_0)$. By Lemma 6.4.19, $\underline{f}_0 \in \mathfrak{a}$, and so $\mathscr{I}(\mathscr{V}(\mathfrak{a})) = \mathfrak{a}$.

(2) The case of general $\mathfrak{a}$: It follows from Theorem 6.3.1 that there is a primary decomposition $\mathfrak{p}_1, \ldots, \mathfrak{p}_l$ $(l < \infty)$ of $\mathfrak{a}$:

$$\mathfrak{a} = \mathfrak{p}_1 \cap \cdots \cap \mathfrak{p}_l.$$

Since $\mathscr{V}(\mathfrak{a}) = \bigcup_{\nu=1}^{l} \mathscr{V}(\mathfrak{p}_\nu)$,

$$\mathscr{I}(\mathscr{V}(\mathfrak{a})) = \mathscr{I}\left(\bigcup_{\nu=1}^{l} \mathscr{V}(\mathfrak{p}_\nu)\right) = \bigcap_{\nu=1}^{l} \mathscr{I}(\mathscr{V}(\mathfrak{p}_\nu)). \tag{6.4.21}$$

Now, $\mathscr{V}(\mathfrak{p}_\nu) = \mathscr{V}(\sqrt{\mathfrak{p}_\nu})$, and $\sqrt{\mathfrak{p}_\nu}$ are prime. Therefore it follows from the result of (1) that $\mathscr{I}(\mathscr{V}(\sqrt{\mathfrak{p}_\nu})) = \sqrt{\mathfrak{p}_\nu}$. Combining this with (6.4.21), we get

$$\mathscr{I}(\mathscr{V}(\mathfrak{a})) = \bigcap_{\nu=1}^{l} \sqrt{\mathfrak{p}_\nu} = \sqrt{\mathfrak{a}}.$$

□

The following is immediate by the above theorem:

Corollary 6.4.22 (i) *An ideal* $\mathfrak{a}$ *is prime if and only if* $\mathscr{V}(\mathfrak{a})$ *is irreducible.*
(ii) *Let* $\mathfrak{a} = \bigcap_{\nu=1}^{l} \mathfrak{p}_\nu$ *be the primary decomposition. Then*

$$\mathscr{V}(\mathfrak{a}) = \bigcup_{\nu} \mathscr{V}(\mathfrak{p}_\nu) = \bigcup_{\nu} \mathscr{V}(\sqrt{\mathfrak{p}_\nu})$$

is the irreducible decomposition of $\mathscr{V}(\mathfrak{a})$.

Lemma 6.4.23 *Let* $(z, w) \in \mathbf{C}^2$, $|z| < 1$, *and* $d \in \mathbf{N}$. *We consider a polynomial equation with coefficients* $a_j \in \mathscr{O}(\Delta(0; 1))$, $1 \le j \le d$:

$$P(z, w) = w^d + \sum_{\nu=1}^{d} a_\nu(z) w^{d-\nu} = 0.$$

Let $z_0 \in \Delta(0; 1)$ *and let* $w = w_0$ *be a root of the above equation with* $z = z_0$. *Then for every* z_1 *sufficiently close to* z_0, *there is a curve* $\gamma : t \in [0, 1] \to (z(t), w(t)) \in \{P(z, w) = 0\}$ *such that* $\gamma(0) = (z_0, w_0)$, $z(t) = (1 - t)z_0 + tz_1$.

Proof Without loss of generality we may assume that $z_0 = w_0 = 0$. By Weierstrass' Preparation Theorem 2.1.3 we may assume that $P(z, w)$ is an irreducible Weierstrass polynomial. We denote by $\delta(z)$ the discriminant of $P(z, w)$, so that $\underline{\delta_0} \neq 0$.

We apply Theorem 6.4.16 for the ideal $\mathfrak{a} = \mathscr{O}_{2,0} \cdot \underline{P(z, w)_0}$. With a sufficiently small $\sigma > 0$, $\delta(z) \neq 0$ for $0 < |z| < \sigma$. Set $S = \{P(z, w) = 0\}$. Since

$$\pi|_{S \cap U_{\sigma,\rho}} : S \cap U_{\sigma,\rho} \to U'_\sigma$$

is surjective, for every $z_1 \in \Delta^*(0; \sigma)$ there is a point $(z_1, w_1) \in S \cap U_{\sigma,\rho}$. Since

$$\pi|_{S \cap U_{\sigma,\rho} \setminus \{z=0\}} : S \cap U_{\sigma,\rho} \setminus \{z = 0\} \to U'_\sigma \setminus \{0\}$$

is an unramified covering, there exists a continuous lift $\gamma : (0, 1] \to S \cap U_{\sigma,\rho} \setminus \{z = 0\}$ of $z(t) : (0, 1] \in t \to tz_1 \in U'_\sigma \setminus \{0\}$. Writing $\gamma(t) = (z(t), w(t))$, we have

$$(w(t))^d + \sum_{\nu=1}^{d} a_\nu(z(t))(w(t))^{d-\nu} = 0.$$

Noting that $\lim_{t\to 0} z(t) = 0$ and $\lim_{t\to 0} a_\nu(z(t)) = a_\nu(0) = 0$, we see that $\lim_{t\to 0} w(t) = 0$. That is, with setting $\gamma(0) = (0, 0)$, $\gamma : [0, 1] \to S \cap U_{\sigma,\rho}$ is a continuous curve satisfying the required property. □

Theorem 6.4.24 *Let* $\mathrm{P}\Delta_n = \mathrm{P}\Delta_p \times \mathrm{P}\Delta_{n-p} \subset \mathbf{C}^n$ *be a polydisk with center at the origin* 0. *Let* $\phi \in \mathscr{O}(\mathrm{P}\Delta_p)$, $\phi \neq 0$, *and let* $X \subset (\mathrm{P}\Delta_p \setminus \{\phi = 0\}) \times \mathrm{P}\Delta_{n-p}$ *be an analytic subset. Assume that the projection*

$$\pi : (z', z'') \in X \to z' \in \mathrm{P}\Delta_p \setminus \{\phi = 0\}$$

is finite and unramified. Moreover, we assume that with the (topological) closure $\bar{X} \subset \mathrm{P}\Delta_p \times \mathrm{P}\Delta_{n-p}$, π *extends to a proper map*

$$\bar{\pi} : \bar{X} \to \mathrm{P}\Delta_p.$$

Then, $\bar{X}$ *is an analytic subset.*

Proof We set $I(X) = \{f \in \mathscr{O}(\mathrm{P}\Delta_n);\ f|_X \equiv 0)\}$, and set the analytic closure of X by

$$\bar{X}^{\mathrm{an}} = \{z \in \mathrm{P}\Delta_n;\ f(z) = 0,\ {}^{\forall} f \in I(X)\}.$$

By Proposition 6.2.7, $\bar{X}^{\mathrm{an}}$ is an analytic subset of $\mathrm{P}\Delta_n$. Since $\bar{X}^{\mathrm{an}} \supset X$, $\bar{X} \subset \bar{X}^{\mathrm{an}}$. For $z' \in \mathrm{P}\Delta_p$ we set

$$\bar{X}^{\mathrm{an}}_{z'} = \bar{X}^{\mathrm{an}} \cap (\{z'\} \times \mathrm{P}\Delta_{n-p}), \quad \bar{X}_{z'} = \bar{X} \cap (\{z'\} \times \mathrm{P}\Delta_{n-p}).$$

It suffices to show:
Claim *For every* $z' \in \mathrm{P}\Delta_p$, $\bar{X}^{\mathrm{an}}_{z'} = \bar{X}_{z'}$.

∵) (a) Suppose that there is a point $z'_0 \in \mathrm{P}\Delta_p$ with $\phi(z'_0) \neq 0$, satisfying $\bar{X}^{\mathrm{an}}_{z'_0} \supsetneqq \bar{X}_{z'_0}$. Take a point $z_0 \in \bar{X}^{\mathrm{an}}_{z'_0} \setminus \bar{X}_{z'_0}$. Since $\bar{X}_{z'_0}$ is a finite set, there is a bounded function $f \in \mathscr{O}(\mathrm{P}\Delta_n)$ such that

$$f(z_0) \notin f(\bar{X}_{z'_0}).$$

For a point $z' \in \mathrm{P}\Delta_p \setminus \{\phi = 0\}$, we write $X_{z'} = \bar{X}_{z'_0} = \{z^{(1)}, \ldots, z^{(d)}\}$, and take the elementary symmetric polynomial of degree ν in $f(z^{(1)}), \ldots, f(z^{(d)})$,

$$a_\nu(z') = (-1)^\nu \sum_{1 \le i_1 < \cdots < i_\nu \le d} f(z^{(i_1)}) \cdots f(z^{(i_\nu)}).$$

Since $\bar{\pi}$ is assumed to be proper, $a_\nu(z')$ is a bounded holomorphic function in $z' \in \mathrm{P}\Delta_p \setminus \{\phi = 0\}$, and then by Riemann's Extension Theorem 2.3.4, $a_\nu \in \mathscr{O}(\mathrm{P}\Delta_p)$. We set

$$\text{(6.4.25)} \qquad P_f(z) = P_f(z', z'') = \prod_{i=1}^{d} (f(z) - f(z^{(i)}))$$
$$= f(z)^d + \sum_{\nu=1}^{d} a_\nu(z') f(z)^{d-\nu} \in \mathscr{O}(\mathrm{P}\Delta_n).$$

From the construction, $P_f|_X \equiv 0$; i.e., $P_f \in \mathscr{I}\langle X\rangle$. On the other hand,

$$P_f(z_0) = \prod_{i=1}^{d} (f(z_0) - f(z^{(i)})) \neq 0.$$

Therefore, $z_0 \notin \bar{X}^{\mathrm{an}}$; this is absurd.

(b) There remains the case of $\phi(z'_0) = 0$. It suffices to see that for every bounded $f \in \mathscr{O}(\mathrm{P}\Delta_n)$,

$$f(\bar{X}^{\mathrm{an}}_{z'_0}) = f(\bar{X}_{z'_0}).$$

Take any $\alpha \in f(\bar{X}^{\mathrm{an}}_{z'_0})$. For the above f we form $P_f(z)$ as in (6.4.25). Since $P_f|_X \equiv 0$, there is a point $z_0 \in \bar{X}^{\mathrm{an}}_{z'_0}$ with $\alpha = f(z_0)$. Since $P_f|_{\bar{X}^{\mathrm{an}}} \equiv 0$,

$$\alpha^d + \sum_{\nu=1}^{d} a_\nu(z'_0)\alpha^{d-\nu} = 0.$$

Let $z' = (z_1, \ldots, z_{p-1}, z_p)$ be a standard coordinate system of ϕ at $z'_0 = (z_{01}, \ldots, z_{0p})$:

$$\phi(z_{01}, \ldots, z_{0p-1}, z_p) = (z_p - z_{0p})^e h(z_p), \quad h(z_{0p}) \neq 0.$$

Then there is a number $\varepsilon > 0$ such that for $0 < \left| z_p - z_{0p} \right| < \varepsilon$,

$$\phi(z_{01}, \ldots, z_{0p-1}, z_p) \neq 0.$$

With $\hat{a}_\nu(z_p) = a_\nu(z_{01}, \ldots, z_{0p-1}, z_p)$ we consider a polynomial equation in ξ,

$$\xi^d + \sum_{\nu=1}^{d} \hat{a}_\nu(z_p)\xi^{d-\nu} = 0. \tag{6.4.26}$$

Note that $\xi = \alpha$ is a root of (6.4.26) at $z_p = z_{0p}$. It follows from Lemma 6.4.23 that there are continuous functions, $z_p(t)$, $\xi(t)$, $0 \le t \le 1$ such that for $0 < t \le 1$, $z_p(t) \neq z_{0p}$ and $\phi(z_{01}, \ldots, z_{0p-1}, z_p(t)) \neq 0$. Notice that $\xi(t)$ is a root of (6.4.26) at $z_p = z_p(t)$ and satisfies

$$\lim_{t \to 0} z_p(t) = z_{0p}, \quad \lim_{t \to 0} \xi(t) = \alpha.$$

With $z'(t) = (z_{01}, \ldots, z_{0p}, z_p(t))$, $\xi(t) \in f(X_{z'(t)})$. Since $\pi : \bar{X} \to \mathrm{P}\Delta_p$ is proper, with letting $t \searrow 0$ we deduce that $\alpha \in f(\bar{X}_{z'_0})$. □

Theorem 6.4.27 *Under the assumptions of Theorem 6.4.16 and (6.4.17), the following hold:*

(i) $S \backslash \pi^{-1}\{\delta = 0\}$ *is dense in* S.
(ii) $S \cap U_{\sigma,\rho}$ *is arc-wise connected.*
(iii) $S \backslash \Sigma(S)$ *is connected and dense in* S. *In particular,* $\Sigma(S)$ *is thin in* S.

Proof (i) Set $S' = S \cap U_{\sigma,\rho} \backslash \pi^{-1}\{\delta = 0\}$. By Theorem 6.4.16,

$$\pi|_{S'} : S' \to U_\sigma \backslash \{\delta = 0\}$$

is a connected unramified covering. Since $S \supset \bar{S}'$,

$$\pi|_{\bar{S}'} : \bar{S}' \to U_\sigma$$

is proper. It follows from Theorem 6.4.24 that $\bar{S}'$ is an analytic subset.

Suppose that $\underline{\bar{S}'}_0 \subsetneqq \underline{S}_0$, Since $\bar{S}'$ and S are the same over $\pi^{-1}\{\delta \neq 0\}$, in a sufficiently small $U_{\sigma,\rho}$ we have

$$S = \bar{S}' \cup (S \cap \{\delta = 0\}) = \bar{S}' \cup S'',$$
$$\underline{S}_0 \neq \underline{\bar{S}'}_0, \quad \underline{S}_0 \neq \underline{S''}_0,$$

and then $\underline{S}_0$ would not be irreducible. On the other hand, $\mathfrak{a}$ is a prime ideal, and hence by Corollary 6.4.22 $\underline{S}_0$ is irreducible; this is a contradiction. After $U_{\sigma,\rho}$ taken smaller if necessary, we have $\bar{S}' = S$.

(ii) Note that S' is arc-wise connected. By Theorem 6.4.16 (ii) it suffices to show that a point of S' and $0 \in S$ is connected by a curve. Let $z' = (z_1, \ldots, z_{p+1}, z_p)$ be a standard coordinate system of $\delta(z')$. Then, with a small $\varepsilon > 0$

$$\delta(0, \ldots, 0, z_p) = z_p^e h(z_p), \quad h(0) \neq 0,$$
$$\delta(0, \ldots, 0, z_p) \neq 0, \qquad 0 < |z_p| < \varepsilon.$$

By making use of Lemma 6.4.23, every point $z \in S \cap U_{\sigma,\rho}$ with $z = (0, \ldots, 0, z_p, z'')$ $(0 < |z_p| < \varepsilon)$ is connected to $0 \in S$ by a curve.

(iii) Let S' be as above. Then, $S' \subset S \setminus \Sigma(S) \subset S$. Since S' is dense, so is $S \setminus \Sigma(S)$. Since $\Sigma(S) \subset S \setminus S'$, $\Sigma(S)$ is thin. Suppose that $S \setminus \Sigma(S)$ is not connected. Let $S \setminus \Sigma(S) = T_1 \cup T_2$, where $T_i \subset S \setminus \Sigma(S)$ are open subsets, $T_i \neq \emptyset$ and $T_1 \cap T_2 = \emptyset$. Since S' is dense in S, $S' \cap T_i \neq \emptyset$, $i = 1, 2$, and then $S' = (S' \cap T_1) \cup (S' \cap T_2)$; this contradicts the connectedness of S'. □

Let $\Omega \subset \mathbf{C}^n$ be an open set.

Definition 6.4.28 (*Dimension*) (i) In general, let $X \subset \Omega$ be an analytic subset and let $a \in X$. If $\underline{X}_a$ is irreducible, by Theorem 6.4.27 there is a neighborhood base $\{\mathrm{P}\Delta(a)\}$ of a such that $(X \setminus \Sigma(X)) \cap \mathrm{P}\Delta(a)$ is a connected complex submanifold. In this case, we define the (local) *dimension* of X at a by

$$\dim_a X = \dim (X \setminus \Sigma(X)) \cap \mathrm{P}\Delta(a).$$

If $\underline{X}_a$ is reducible, after the irreducible decomposition, $\underline{X}_a = \bigcup_{(\text{finite})} \underline{X_j}_a$, we define the (local) *dimension* of X at a by

$$\dim_a X = \max_j \dim_a X_j.$$

The dimension of X is defined by

$$\dim X = \max_{a \in X} \dim_a X.$$

(ii) If $\dim X = \dim_a X$ at every point $a \in X$, X is said to be *pure dimensional*.
(iii) For an analytic subset $Y \subset X$, the *codimension* of Y in X at $b \in Y$ is defined by

$$\mathrm{codim}_{X,b} Y = \dim_b X - \dim_b Y,$$

and the *codimension* of Y in X by

$$\mathrm{codim}_X Y = \min_{b \in Y} \mathrm{codim}_{X,b} Y.$$

By the definition of dimension we immediately have:

Proposition 6.4.29 (Semi-continuity of dimension) *Let $X \subset \Omega$ be an analytic subset. For every point $a \in X$ there is a neighborhood $U \ni a$ such that*

$$\dim_z X \le \dim_a X, \quad {}^{\forall} z \in U.$$

Proposition 6.4.30 *Let $X, Y \subset \Omega$ be two analytic subsets, and let $a \in X \cap Y$. If for every irreducible component $\underline{X_\alpha}_a$ of $\underline{X}_a$*

$$\underline{Y \cap X_\alpha}_a \ne \underline{X_\alpha}_a, \tag{6.4.31}$$

then,

$$\dim_a X \cap Y \le \dim_a X - 1.$$

In particular, if $Y \subset X$ and Y is thin in X, then $\dim Y \le \dim X - 1$.

Proof It suffices to prove the assertion with assuming that $\underline{X}_a$ and $\underline{Y}_a$ are irreducible. We may assume $a = 0$ by a translation. Let $\mathfrak{a} = \mathscr{I}\langle \underline{X}_0 \rangle$ be the geometric ideal of $\underline{X}_0$ in $\mathscr{O}_{n,0} (= \mathscr{O}_{\mathbf{C}^n, o})$. Then, $\dim_0 X$ is identical to the p determined by Proposition 6.4.2. With $\mathfrak{b} = \mathscr{I}\langle \underline{Y}_0 \rangle$, $\mathfrak{b} + \mathfrak{a} \ne \mathfrak{a}$ and hence, there is an element $g \in \mathfrak{b} \setminus \mathfrak{a}$. Note that g is integral over $\mathscr{O}_{p,0}$ mod $\mathfrak{a}$. Therefore, there are elements $c_\nu \in \mathscr{O}_{p,0}$ such that

$$g^d + \sum_{\nu=1}^{d} c_\nu g^{d-\nu} \equiv 0 \pmod{\mathfrak{a}}.$$

Let d be the minimum among such degrees. Then $c_d \ne 0$, and we obtain

$$g \left(g^{d-1} + \sum_{\nu=1}^{d-1} c_\nu g^{d-1-\nu} \right) = -c_d \in \mathscr{O}_{p,0} \cap \mathfrak{b}.$$

Therefore, one can further perform the inductive procedure in the proof of Proposition 6.4.2 for $\mathfrak{b} + \mathfrak{a}$; with the resulting non-negative number q, $q < p$ holds. Thus, $\dim_0 Y \cap X < \dim_0 X$ is shown.

If $Y \subset X$ is a thin subset, (6.4.31) holds at every point $a \in Y$, and hence $\dim Y \le \dim X - 1$. □

Theorem 6.4.32 *For an analytic subset $X \subset \Omega$, $\Sigma(X)$ is thin in X.*

Proof We consider it in a neighborhood of an arbitrary point $a \in X$.

(1) If $\underline{X}_a$ is irreducible, by Theorem 6.4.27 (iii), there is a neighborhood $U \subset \Omega$ of a such that $\Sigma(X) \cap U$ is thin in $X \cap U$.

(2) If $\underline{X}_a$ is not irreducible, we take the irreducible decomposition, $\underline{X}_a = \bigcup \underline{X_i}_a$. By (1) above, there is a small neighborhood $U \subset \Omega$ of a such that every $\Sigma(X_i) \cap U$ is thin in $X_i \cap U$. It is noticed that if $x \in X \setminus \Sigma(X)$, there is only one irreducible component of X through x. Therefore we obtain

$$\Sigma(X) \cap U = \bigcup_i (\Sigma(X_i) \cap U) \bigcup \left(\bigcup_{i \neq j} X_i \cap X_j \right). \tag{6.4.33}$$

We infer from Proposition 6.4.30 that $X_i \cap X_j$ is thin in $X \cap U$. Thus, $\Sigma(X) \cap U$ is thin in $X \cap U$. □

6.5 Oka's Second Coherence Theorem

6.5.1 Geometric Ideal Sheaves

Let $\Omega \subset \mathbf{C}^n$ be an open set, and let $X \subset \Omega$ be an analytic subset. At each point $a \in \Omega$ the ideal $\mathscr{I}\langle \underline{X}_a \rangle \subset \mathscr{O}_{\Omega,a}$ is determined, and the ideal sheaf of $\mathscr{O}_\Omega$,

$$\mathscr{I}\langle X \rangle = \bigcup_{a \in \Omega} \mathscr{I}\langle \underline{X}_a \rangle$$

is obtained and called the geometric ideal sheaf (cf. Definition 2.3.6). If X is non-singular, it follows from Theorem 4.4.6 that $\mathscr{I}\langle X \rangle$ is coherent. It is the aim of this section to prove the coherence of the geometric ideal sheaf with no assumption (cf. Oka VII, VIII, Cartan [10], and Chap. 9):

Theorem 6.5.1 (Oka's Second Coherence Theorem, Oka [62] VII, VIII, H. Cartan [10]) *Every geometric ideal sheaf $\mathscr{I}\langle X \rangle$ over an open set $\Omega \subset \mathbf{C}^n$ is coherent.*

Proof Since $\mathscr{I}\langle X \rangle \subset \mathscr{O}_\Omega$ and $\mathscr{O}_\Omega$ is coherent by Oka's First Coherence Theorem 2.5.1, it suffices to prove the local finiteness of $\mathscr{I}\langle X \rangle$ under the assumption,

$\emptyset \neq X \subsetneqq \Omega$. The problem is local, and so we consider it in a neighborhood of $a \in X$, and after a translation, we may assume that $a = 0$.

Let $\underline{X}_0 = \bigcup_{j=1}^{l} \underline{X_{j}}_0$ ($l < \infty$) be the decomposition into irreducible components $\underline{X_{j}}_0$. With a sufficiently small polydisk neighborhood $\mathrm{P}\Delta$ about 0 we have

$$\mathscr{I}\langle X \cap \mathrm{P}\Delta\rangle = \bigcap_{j=1}^{l} \mathscr{I}\langle X_j \cap \mathrm{P}\Delta\rangle.$$

If $\mathscr{I}\langle X_j \cap \mathrm{P}\Delta\rangle$ are all coherent, the coherence of $\mathscr{I}\langle X \cap \mathrm{P}\Delta\rangle$ follows from Proposition 2.4.9. Therefore we may assume that $\underline{X}_0$ is irreducible.

Now, $\mathscr{I}\langle \underline{X}_0\rangle$ is a prime ideal, and $0 \subsetneqq \mathscr{I}\langle \underline{X}_0\rangle \subsetneqq \mathscr{O}_{n,0} (= \mathscr{O}_{\mathbf{C}^n,0})$. With $\mathfrak{a} = \mathscr{I}\langle \underline{X}_0\rangle$ we take a standard polydisk $\mathrm{P}\Delta = U = U' \times U''$ given in Theorem 6.4.16, and functions, $P_j(z', z_j)$, $p+1 \leq j \leq n$, $\delta(z')$, $Q_j(z', z_{p+1})$, $p+2 \leq j \leq n$, moreover, a finite number of generators $\underline{f_1}_0, \ldots, \underline{f_L}_0$ of $\mathfrak{a}$ with $f_k \in \mathscr{O}(U)$, $1 \leq k \leq L$. Taking U smaller if necessary, we have

$$X \cap U = \{f_1 = \cdots = f_L = 0\},$$

and

$$\begin{aligned} & d = \max_j \deg P_j = \deg P_{p+1}, \\ & P_j(z', z_j) \in \Gamma(U, \mathscr{I}\langle X\rangle), \quad p+1 \leq j \leq n, \\ & \delta(z')z_j - Q_j(z', z_{p+1}) \in \Gamma(U, \mathscr{I}\langle X\rangle), \quad p+2 \leq j \leq n. \end{aligned}$$

It follows that

$$\begin{aligned} (6.5.2) \qquad & X \cap \{(z', z'') \in U;\ \delta(z') \neq 0\} \\ & \quad = \{(z', z'') \in U;\ \delta(z') \neq 0,\ P_{j+1}(z', z_{p+1}) = 0, \\ & \qquad \delta(z')z_j - Q_j(z', z_{p+1}) = 0,\ p+2 \leq j \leq n\}, \\ (6.5.3) \qquad & X \cap \{\delta(z') \neq 0\} \text{ is dense in } X \cap U. \end{aligned}$$

We define an ideal sheaf $\mathscr{J}$ of $\mathscr{O}_U$ by

$$(6.5.4) \qquad \mathscr{J}_z = \sum_{j=p+1}^{n} \mathscr{O}_{n,z} \cdot \underline{P_j}_z + \sum_{j=p+2}^{n} \mathscr{O}_{n,z} \cdot \underline{(\delta z_j - Q_j)}_z + \sum_{h=1}^{L} \mathscr{O}_{n,z} \cdot \underline{f_h}_z, \quad z \in U.$$

We denote the generators in the right-hand side above in order by $\underline{F_h}_z$ ($1 \leq h \leq H$), and we write $\mathscr{J}_z = \sum_{h=1}^{H} \mathscr{O}_{n,z} \underline{F_h}_z$ for it. For a number $N \in \mathbf{N}$ we consider an ideal sheaf $\mathscr{B}^{(N)} \subset \mathscr{O}_U$ defined by

$$\mathscr{B}_z^{(N)} = \left\{ \underline{f}_z \in \mathscr{O}_{n,z};\ \underline{\delta}_z^N \underline{f}_z \in \mathscr{J}_z \right\}, \quad z \in U.$$

By (6.5.4), $\underline{f}_z \in \mathscr{B}_z^{(N)}$ for $\underline{f}_z \in \mathscr{O}_{n,z}$ if and only if there are $\underline{f_h}_z \in \mathscr{O}_{n,z}$, $1 \le h \le H$, satisfying

$$\underline{f}_z \cdot \underline{\delta}_z^N = \sum_{h=1}^{H} \underline{f_h}_z \cdot \underline{F_h}_z.$$

This is equivalent to

$$(6.5.5) \qquad \underline{f}_z \cdot (-\underline{\delta}_z^N) + \sum_{h=1}^{H} \underline{f_h}_z \cdot \underline{F_h}_z = 0$$
$$\iff \left(\underline{f}_z, \underline{f_1}_z, \ldots, \underline{f_H}_z \right) \in \mathscr{R}\left(-\delta^N, F_1, \ldots, F_H\right).$$

That is, $\underline{f}_z$ is the first element of the relation sheaf $\mathscr{R}\left(-\delta^N, F_1, \ldots, F_H\right)$. By Oka's First Coherence Theorem 2.5.1 $\mathscr{R}\left(-\delta^N, F_1, \ldots, F_H\right)$ is coherent, and so is locally finitely generated. Therefore, $\mathscr{B}^{(N)}$ consisting of the first elements has a locally finite generator system. Thus we see that[1]

$$(6.5.6) \qquad \mathscr{B}^{(N)} \text{ is coherent.}$$

Therefore, the following finishes the proof:

Lemma 6.5.7 *There is a number $N \in \mathbf{N}$ such that at every point $b \in U$,*

$$\mathscr{B}_b^{(N)} = \mathscr{I}\langle X \rangle_b.$$

Proof We first take any element $\underline{f}_b \in \mathscr{B}_b^{(N)}$. Since $\underline{\delta}_b^N \underline{f}_b \in \mathscr{J}_b$, there is a neighborhood $V \subset U$ of b such that

$$\delta(z')^N f(z', z'') = 0, \quad (z', z'') \in V \cap X.$$

Therefore, $f(z', z'') = 0$ for $(z', z'') \in V \cap X \setminus \{\delta = 0\}$. By (6.5.3) $V \cap X \setminus \{\delta = 0\}$ is dense in $V \cap X$, and so $f|_{V \cap X} = 0$. That is, $\underline{f}_b \in \mathscr{I}\langle X \rangle_b$, and hence $\mathscr{B}_b^{(N)} \subset \mathscr{I}\langle X \rangle_b$.

We show the converse. Take arbitrarily a point $b = (b', b'') \in U$. If $b \notin X \cap U$, then there is some $f_k(b) \neq 0$, and so $\mathscr{B}_b^{(N)} = \mathscr{O}_{n,b}$. Since $\mathscr{I}\langle X \rangle_b = \mathscr{O}_{n,b}$, $\mathscr{B}_b^{(N)} = \mathscr{I}\langle X \rangle_b$ holds.

Henceforth we assume that $b \in X \cap U$. We set

$$b = (b', b'') = (b_1, \ldots, b_p, b_{p+1}, \ldots, b_n).$$

[1] The way to formulate the ideal $\mathscr{B}^{(N)}$ in (6.5.6) and its coherence were already prepared and proved in Oka [62] VII, Sect. 3, 6; in particular, see Problème (K) there.

It follows that $P_j(b', b_j) = 0$ for $p+1 \le j \le n$. Since $P_j(b', z_j)$ are monic in z_j, $P_j(b', z_j) \not\equiv 0$ in z_j. Thanks to the Weierstrass decomposition we may write

$$P_j(z', z_j) = u_j \cdot A_j(z', z_j - b_j) \in \mathscr{O}_{p+1,\,(b',b_j)},$$

where u_j is a unit at (b', b_j) and $A_j(z', z_j - b_j)$ is a Weierstrass polynomial at (b', b_j). It follows from Lemma 2.2.15 that $u_j \in \mathscr{O}_{p,b}[z_j]$, $\deg A_j = e_j \le d_j$.

Take any $f \in \mathscr{O}_{n,b}$. Dividing f by $A_{p+1}, \ldots, A_n$ we infer from Weierstrass' Preparation Theorem 2.1.3 that

$$\underline{f}_b \equiv \sum_{0 \le \alpha_j < e_j} f_\alpha(z') z_{p+1}^{\alpha_{p+1}} \cdots z_n^{\alpha_n} \quad \left(\mathrm{mod} \sum_j \underline{A_j}_b \cdot \mathscr{O}_{n,b} \subset \mathscr{I}_b \right).$$

In the above equation, we consider the right-hand side as a germ at $b = (b', b'')$, but we abbreviate it to avoid the complication of notation; we use the similar abbreviation in what follows. We define N by

$$(e_{p+2} - 1) + \cdots + (e_n - 1) \le \sum_{j=p+2}^{n} (d_j - 1) = N(\le (n-p-1)(d-1)).$$

We then obtain

$$\begin{aligned} \delta_{b'}^N \underline{f}_b &\equiv \sum_{0 \le \alpha_j < e_j} g_\alpha(z') z_{p+1}^{\alpha_{p+1}} (\delta z_{p+2})^{\alpha_{p+2}} \ldots (\delta z_n)^{\alpha_n} \\ &\equiv Q(z', z_{p+1}) \pmod{\mathscr{I}_b}. \end{aligned}$$

Dividing $Q(z', z_{p+1}) (\in \mathscr{O}_{p,b'}[z_{p+1}])$ by $A_{p+1}(z_{p+1})$, we get

$$\begin{aligned} \delta_{b'}^N \underline{f}_b &\equiv R(z', z_{p+1}) \pmod{\mathscr{I}_b}, \\ \deg R &< e_{p+1}. \end{aligned} \tag{6.5.8}$$

With $b = (b', b'')$ we consider z' near b' such that $\delta(z') \ne 0$. The solution of $P_{p+1}(z', z_{p+1}) = 0$ consists of distinct d simple roots. Therefore, the solution of $A_{p+1}(z', z_{p+1} - b_{p+1}) = 0$ consists also of distinct e_{p+1} simple roots, denoted by $z_{p+1}^{(\nu)}$, $1 \le \nu \le e_{p+1}$.

Since $\delta(z') \ne 0$, we determine $z_j^{(\nu)}$ by $\delta(z') z_j^{(\nu)} = Q_j(z', z_{p+1}^{(\nu)})$ for $p+2 \le j \le n$, so that by (6.5.2), $z^{(\nu)} = (z', z_{p+1}^{(\nu)}, \ldots, z_n^{(\nu)}) \in X$. Since $A_j(z', z_j^{(\nu)} - b_j) = 0$ for $p+2 \le j \le n$ and $A_j(z', z_j - b_j)$ are all Weierstrass polynomials at (b', b_j), we see that $z_j^{(\nu)} \to b_j$ $(p+1 \le j \le n)$ as $z' \to b'$. We deduce from (6.5.8) that

$$\delta(z')^N f(z^{(\nu)}) = R\left(z', z_{p+1}^{(\nu)}\right), \quad 1 \le \nu \le e_{p+1}.$$

If $\underline{f}_b \in \mathscr{I}\langle X\rangle_b$, then $f(z^{(\nu)}) = 0$ for $1 \le \nu \le e_{p+1}$, and then $R\left(z', z^{(\nu)}_{p+1}\right) = 0$ for all $\nu = 1, \ldots, e_{p+1}$; while $\deg R < e_{p+1}$, follows $R = 0$. Therefore, $\underline{f}_b \in \mathscr{B}^{(N)}_b$, and hence $\mathscr{I}\langle X\rangle_b \subset \mathscr{B}^{(N)}_b$. □

Corollary 6.5.9 *Let $X \subset \Omega$ be an analytic subset, and let $a \in X$. Let $\{\underline{f_1}_a, \ldots, \underline{f_M}_a\}$ be a finite generator system of $\mathscr{I}\langle X\rangle_a$ over $\mathscr{O}_{n,a}$. Then there is a neighborhood $V(\subset \Omega)$ of a such that at every point $b \in V$*

$$\mathscr{I}\langle X\rangle_b = \sum_{i=1}^{M} \mathscr{O}_{n,b} \cdot \underline{f_i}_b;$$

i.e.,

$$\mathscr{I}\langle X\rangle|_V = \sum_{i=1}^{M} \mathscr{O}_V \cdot f_i.$$

Proof This follows from Oka's Second Coherence Theorem 6.5.1 and Proposition 2.4.6. □

6.5.2 *Singularity Sets*

The singularity set $\Sigma(X)$ of an analytic set X is a thin subset in X (Theorem 6.4.32). Here as an application of Oka's Second Coherence Theorem 6.5.1 we prove the analyticity of $\Sigma(X)$. Let $\Omega \subset \mathbf{C}^n$ be an open subset as before.

Theorem 6.5.10 *Let $X \subset \Omega$ be an analytic subset. Then the singularity set $\Sigma(X)$ is also analytic and the following hold:*

(6.5.11)
$$\dim_a \Sigma(X) \le \dim_a X - 1, \quad {}^{\forall}a \in \Sigma(X),$$
$$\dim \Sigma(X) \le \dim X - 1.$$

Proof Since $\Sigma(X)$ is closed, it suffices to show the analyticity in a neighborhood of each point $a \in \Sigma(X)$. By Oka's Second Coherence Theorem 6.5.1, there are a neighborhood $U(\subset\Omega)$ of a and finitely many $f_i \in \mathscr{O}(U)$, $1 \le j \le l\,(<\infty)$, such that

$$X \cap U = \{f_1 = \cdots = f_l = 0\},$$

$$\mathscr{I}\langle X\rangle_b = \sum_{j=1}^{l} \mathscr{O}_{n,b} \cdot \underline{f_j}_b, \quad {}^{\forall}b \in U.$$

(a) We first assume that $\underline{X}_a$ is irreducible, and take a standard polydisk neighborhood of X with center at a. Then, $X \cap U \backslash \Sigma(X)$ is a connected complex submanifold of $U \backslash \Sigma(X)$, whose dimension is denoted by m. In a small neighborhood $V (\subset U \backslash \Sigma(X))$ of an arbitrary point $b \in X \cap U \backslash \Sigma(X)$, there are holomorphic functions $g_1, \ldots, g_{n-m} \in \mathscr{O}(V)$ such that

$$\begin{aligned} & X \cap V = \{g_1 = \cdots = g_{n-m} = 0\}, \\ & dg_1 \wedge \cdots \wedge dg_{n-m}(z) \\ & = \sum_{1 \leq j_1 < \cdots < j_{n-m} \leq n} \begin{vmatrix} \frac{\partial g_1}{\partial z_{j_1}} & \cdots & \frac{\partial g_1}{\partial z_{j_{n-m}}} \\ \vdots & \vdots & \vdots \\ \frac{\partial g_{n-m}}{\partial z_{j_1}} & \cdots & \frac{\partial g_{n-m}}{\partial z_{j_{n-m}}} \end{vmatrix} dz_{j_1} \wedge \cdots \wedge dz_{j_{n-m}} \\ & \neq 0, \quad {}^{\forall} z \in X \cap V \qquad \text{(cf. Sect. 3.5).} \end{aligned}$$

Because of the choice of $\{f_j\}$, after taking V smaller if necessary, we have expressions over V, $g_k = \sum_{j=1}^{l} \alpha_{kj} f_j$ with $\alpha_{kj} \in \mathscr{O}(V)$. On $X \cap V$ we write

$$dg_1 \wedge \cdots \wedge dg_{n-m} = \sum_{1 \leq j_1 < \cdots < j_{n-m} \leq l} \beta_{j_1 \cdots j_{n-m}} df_{j_1} \wedge \cdots \wedge df_{j_{n-m}},$$
$$\beta_{j_1 \cdots j_{n-m}} \in \mathscr{O}(V).$$

There are indices $j_1 < \cdots < j_{n-m}$ such that

$$df_{j_1} \wedge \cdots \wedge df_{j_{n-m}}(z) \neq 0 \tag{6.5.12}$$

in V, where V may be taken smaller. Conversely, if (6.5.12) holds in some $V \subset U$, $V \cap \{f_{j_1} = \cdots = f_{j_{n-m}} = 0\}$ is an m-dimensional complex submanifold, containing $X \cap V$. Since $\dim X \cap V = m$, it follows from Proposition 6.4.30 that

$$X \cap V = \{f_{j_1} = \cdots = f_{j_{n-m}} = 0\} \cap V = \{f_1 = \cdots = f_l = 0\} \cap V,$$

so that $X \cap V$ consists of regular points of X. Thus, we have

$$\Sigma(X) \cap U = \{df_{j_1} \wedge \cdots \wedge df_{j_{n-m}} = 0;\ 1 \leq j_1 < \cdots < j_{n-m} \leq l\} \cap X, \tag{6.5.13}$$

and see that $\Sigma(X)$ is an analytic subset.

(b) If $\underline{X}_a$ is reducible, we infer from (a) and (6.4.33) that $\Sigma(X) \cap U$ is an analytic subset of U.

By Theorem 6.4.32, $\Sigma(X)$ is a thin analytic subset of X, and then from Proposition 6.4.30 the dimension inequalities follow. □

Remark 6.5.14 In the above proof, the coherence of $\mathscr{I}\langle X\rangle$ is essential. In (6.5.13) it is immediate from the definition and the implicit function theorem that

$$\Sigma(X)\cap U\subset\{df_{j_1}\wedge\cdots\wedge df_{j_{n-m}}=0;\ 1\le j_1<\cdots<j_{n-m}\le l\}\cap X.$$

Take arbitrarily a point z other than a in the right-hand side, It is unknown if there might be other defining equations, $h_j\in\mathscr{O}(W)$, $1\le j\le n-m$, in a neighborhood W of z satisfying

$$X\cap W=\{h_1=\cdots=h_{n-m}=0\},$$
$$dh_1\wedge\cdots\wedge dh_{n-m}(w)\ne 0,\quad w\in W.$$

The coherence of $\mathscr{I}\langle X\rangle$ guarantees that this does not happen.

6.5.3 Hartogs' Extension Theorem

For later use we extend Theorem 1.2.26 on Hartogs' phenomenon:

Theorem 6.5.15 (Hartogs' Extension) *Let $\Omega\subset\mathbf{C}^n$ be a domain and let $A\subset\Omega$ be an analytic subset. If* codim $A\ge 2$, *then every $f\in\mathscr{O}(\Omega\setminus A)$ extends to a unique element of $\mathscr{O}(\Omega)$.*

Proof The uniqueness follows from the Identity Theorem 1.2.14. It suffices to deal with the problem locally in a neighborhood of each point $a\in A$. In a neighborhood of a regular point $a\in A\setminus\Sigma(A)$, after choosing a suitable holomorphic local coordinate system, it is reduced to the case of Theorem 1.2.26. Therefore, f extends holomorphically over $\Omega\setminus\Sigma(A)$. By Theorem 6.5.10, $\Sigma(A)$ is an analytic subset of Ω, and $\dim\Sigma(A)<\dim A$. Repeating the same argument for $\Sigma(A)$ in place of A, we see inductively that f extends holomorphically over Ω. □

6.5.4 Coherent Sheaves over Analytic Sets

Let $\Omega\subset\mathbf{C}^n$ be an open subset. Let $X\subset\Omega$ be an analytic subset and let $\mathscr{I}\langle X\rangle$ be the geometric ideal of X.

Definition 6.5.16 We define the *sheaf of germs of holomorphic functions on* X by

$$\mathscr{O}_X=\mathscr{O}_\Omega/\mathscr{I}\langle X\rangle.$$

We call this the *structure sheaf of the analytic set* X.

At a point $a\in\Omega$, $\mathscr{O}_{n,a}=\mathscr{O}_{\Omega,a}$ is a local ring with maximal ideal $\mathfrak{m}_{n,a}\subset\mathscr{O}_{n,a}$, and therefore $\mathscr{O}_{X,a}$ is a local ring with maximal ideal $\mathfrak{m}_{X,a}$ $(=\mathfrak{m}_{n,a}/\mathscr{I}\langle X\rangle_a)$.

We call a section $f \in \Gamma(U, \mathscr{O}_X)$ over an open subset $U \subset X$ a *holomorphic function* on U; in this case, for every $a \in U$ there are a neighborhood $V \subset \Omega$ of a and $F \in \mathscr{O}(V)$ such that

$$f = F|_{V \cap X}.$$

Therefore, a continuous function $f : U \to \mathbf{C}$ is canonically determined, and is denoted by the same f unless confusion occurs. We write $\mathscr{O}_X(U) = \Gamma(U, \mathscr{O}_X)$.

The following theorem, which is a direct consequence of Oka's First and Second Coherence Theorem, is fundamental:

Theorem 6.5.17 (Oka) *Let X be an analytic subset of Ω. Then $\mathscr{O}_X$ is coherent.*

Proof By definition we have a short exact sequence of sheaves:

$$0 \to \mathscr{I}\langle X\rangle \to \mathscr{O}_\Omega \to \mathscr{O}_X \to 0.$$

By Oka's First and Second Coherence Theorems, $\mathscr{O}_\Omega$ and $\mathscr{I}\langle X\rangle$ are coherent. It follows from Serre's Theorem 3.3.1 that $\mathscr{O}_X$ is coherent. □

A continuous map determined by $f_j \in \Gamma(X, \mathscr{O}_X)$, $1 \le j \le m$,

$$f = (f_1, \ldots, f_m) : X \to \mathbf{C}^m$$

is called a *holomorphic map* from X to $\mathbf{C}^m$. If Y is an analytic subset of an open set of $\mathbf{C}^m$ with $f(X) \subset Y$, we call

$$f : X \to Y$$

a *holomorphic map* from X to Y. The map f naturally induces a homomorphism between local rings,

$$f^* : \mathscr{O}_{Y,f(x)} \to \mathscr{O}_{X,x}, \quad x \in X. \tag{6.5.18}$$

We call this the *pull-back* by f. Thus for a holomorphic function $\varphi \in \Gamma(Y, \mathscr{O}_Y)$ on Y we obtain the pull-back $f^*\varphi \in \Gamma(X, \mathscr{O}_X)$.

Let $X (\subset \Omega)$ be an analytic set. We consider a sheaf $\mathscr{S} \to X$ of $\mathscr{O}_X$-modules over X. We extend $\mathscr{S}$ as 0 over $\Omega \setminus X$, and obtain a sheaf of $\mathscr{O}_\Omega$-modules over Ω, which is denoted by $\widehat{\mathscr{S}} \to \Omega$. We call $\widehat{\mathscr{S}}$ the *simple extension* of $\mathscr{S}$ over Ω. Since the coherence of the geometric ideal sheaf $\mathscr{I}\langle X\rangle$ is already obtained, the following holds (cf. Theorem 4.4.6).

Proposition 6.5.19 *Let $X (\subset \Omega)$ be an analytic subset. A sheaf $\mathscr{S}$ of $\mathscr{O}_X$-modules is coherent over $\mathscr{O}_X$ if and only if the simple extension $\widehat{\mathscr{S}}$ is coherent over $\mathscr{O}_\Omega$.*

6.6 Irreducible Decompositions of Analytic Sets

Let $\Omega \subset \mathbf{C}^n$ be an open set, and let $X \subset \Omega$ be an analytic subset. At $a \in X$ we have the irreducible decomposition of $\underline{X}_a$ (Theorem 6.2.9 (iii)):

(6.6.1) $$\underline{X}_a = \bigcup_{j=1}^{l} \underline{X_j}_a \quad (l < \infty).$$

Each $\mathscr{I}\langle \underline{X_j}_a \rangle$ is a prime ideal. We apply Theorem 6.4.14 for $\mathfrak{a} = \mathscr{I}\langle \underline{X_j}_a \rangle$. Here it is noticed that the linearly transformed coordinate system $(z_1, \ldots, z_{p_j}, \ldots z_n)$ can be taken commonly for all j, although the number $p = p_j$ depends on j. Therefore there is a neighborhood base of a consisting of polydisks $\mathrm{P}\Delta(a)$ about a, where the representatives of $\underline{X_j}_a$ exist, and from Theorem 6.4.27 (iii) the following is deduced.

Lemma 6.6.2 *For every $j = 1, 2, \ldots, l$, $X_j \cap \mathrm{P}\Delta(a) \setminus \Sigma(X_j)$ is dense in X_j, and connected.*

Lemma 6.6.3 *Let $Y \subsetneqq X_j \cap \mathrm{P}\Delta(a) \setminus \Sigma(X_j)$ be an analytic subset. Then $X_j \cap \mathrm{P}\Delta(a) \setminus (\Sigma(X_j) \cup Y)$ is dense in X_j and connected.*

Proof With the help of the Identity Theorem this holds by the same reason as Theorem 6.1.4. □

Theorem 6.6.4 *Let $\underline{X}_a$ be a germ of analytic subsets at a. Then there is a neighborhood base $\{\mathrm{P}\Delta(a)\}$ of polydisks with center a such that with $X \cap \mathrm{P}\Delta(a) \setminus \Sigma(X) = \bigcup_\alpha X'_\alpha$ being a decomposition into connected components, for each X'_α there corresponds bijectively some X_j in (6.6.1), and*

$$\bar{X}'_\alpha = X_j,$$
$$X'_\alpha = X_j \setminus \Sigma(X).$$

In particular, $X \setminus \Sigma(X)$ is dense in X.

Proof Let $\{\mathrm{P}\Delta(a)\}$ be the neighborhood base of Lemma 6.6.2. Then, $X_j \cap \mathrm{P}\Delta(a) \setminus \Sigma(X_j)$ is connected. Set

(6.6.5) $$X'_j = X_j \cap \mathrm{P}\Delta(a) \setminus \Sigma(X) = X_j \cap \mathrm{P}\Delta(a) \setminus \left(\Sigma(X_j) \cup \bigcup_{k \neq j} X_k \right).$$

By Lemma 6.6.3, X'_j is connected, dense in $X_j \cap \mathrm{P}\Delta(a)$, and

$$X'_j \cap X'_h = \emptyset, \quad j \neq h,$$
$$X \setminus \Sigma(X) = \bigcup X'_j.$$

That is, $\{X'_j\}$ is the family of connected components of $X\backslash\Sigma(X)$, and coincides with $\{X'_\alpha\}$. Since X'_j is dense in X_j, $\bar{X}'_j = X_j$ follows. □

Each polydisk $\mathrm{P}\Delta(a)$ taken in Theorem 6.6.4 is called a *standard polydisk (neighborhood)* of X at a.

An analytic subset $X \subset \Omega$ decomposes locally at each point $a \in X$ into finitely many irreducible components (Theorem 6.6.4), and the elementary geometric property of each irreducible component X_j in a neighborhood of a is understood by Theorems 6.4.16 and 6.4.27.

We next study the global property of X in Ω.

Definition 6.6.6 Let $X \subset \Omega$ be an analytic subset. We say that X is *reducible*, if there are analytic subsets $X_i \subset \Omega$ $(i = 1, 2)$ satisfying $X_i \neq X$ $(i = 1, 2)$ and $X = X_1 \cup X_2$. Otherwise, X is said to be *irreducible*.

Theorem 6.6.7 *Let $X \subset \Omega$ be an irreducible analytic subset. Then $X' = X\backslash\Sigma(X)$ is connected and dense in X. Conversely, if $X\backslash\Sigma(X)$ is connected, then X is irreducible.*

Proof Let $\mathrm{P}\Delta(a)$ be a standard polydisk of X about $a \in X$. Then, $X' \cap \mathrm{P}\Delta(a) = X \cap \mathrm{P}\Delta(a)\backslash\Sigma(X)$ is dense in $X \cap \mathrm{P}\Delta(a)$, and hence X' is dense in X.

If X' were disconnected, there would be non-empty open subsets $Z_i \subset X', i = 1, 2$ with $X' = Z_1 \cup Z_2$. We take arbitrarily a point $a \in \bar{Z}_1 \cap X$. Let $\mathrm{P}\Delta(a)$ be a standard polydisk of X about a, and let $X' \cap \mathrm{P}\Delta(a) = \bigcup_j Y'_j$ be the decomposition into connected components. Then for Y'_j, either $Y'_j \cap Z_i = Y'_j$ or $Y'_j \cap Z_1 = \emptyset$. With $Y_i = \bigcup_{Y'_j \cap Z_1 = Y'_i} \bar{Y}'_j$, Y_i is analytic in $\mathrm{P}\Delta(a)$, and $\bar{Z}_i \cap \mathrm{P}\Delta(a) = Y_i$. Therefore $X_i = \bar{Z}_i$ is analytic in Ω. It follows that $X \neq X_i \neq \emptyset$ $(i = 1, 2)$ and $X = X_1 \cup X_2$; this contradicts the irreducibility of X.

On the other hand, supposing that X is not irreducible, we have a decomposition $X = X_1 \cup X_2$ with $X \neq X_i, i = 1, 2$. Set

$$W_1 = X_1\backslash\Sigma(X) = X_1\backslash\left(\Sigma(X_1) \cup X_2\right),$$
$$W_2 = X_2\backslash\Sigma(X) = X_2\backslash\left(\Sigma(X_2) \cup X_1\right).$$

Then, $W_i \neq \emptyset$ (Theorem 6.5.10), $W_1 \cap W_2 = \emptyset$, and $X\backslash\Sigma(X) = W_1 \cup W_2$; thus, $X\backslash\Sigma(X)$ is disconnected. □

Theorem 6.6.8 *An analytic subset $X \subset \Omega$ is decomposed, uniquely up to the order, into irreducible analytic subsets $X_\alpha \subset \Omega$ $(\alpha \in \Gamma)$, so that $X = \bigcup_{\alpha\in\Gamma} X_\alpha$.*

Each X_α corresponds bijectively to the closure of a connected component of $X\backslash\Sigma(X)$, and the family $\{X_\alpha\}_{\alpha\in\Gamma}$ is locally finite in Ω.

Proof Let X'_0 be a connected component of $X\backslash\Sigma(X)$, and let $a \in \bar{X}'_0$ $(\subset X)$. Take a standard polydisk $\mathrm{P}\Delta(a)$ of X about a, and let

$$\mathrm{P}\Delta(a) \cap X \setminus \Sigma(X) = \bigcup_h Z'_h \quad \text{(finite union)}$$

be the decomposition into connected components. Then, either, $Z'_h \subset X_0$ or $Z'_h \cap X_0 = \emptyset$. Hence, taking the union for $Z'_h \subset X_0$,

$$\bar{X}'_0 \cap \mathrm{P}\Delta(a) = \bigcup \bar{Z}'_h, \tag{6.6.9}$$

we obtain an analytic subset $\bar{X}'_0 \cap \mathrm{P}\Delta(a)$. It follows that

$$\bar{X}'_0 \supset \bar{X}'_0 \setminus \Sigma(\bar{X}'_0) \supset X_0.$$

Since X_0 is connected and dense in $\bar{X}'_0$, $\bar{X}'_0 \setminus \Sigma(\bar{X}'_0)$ is connected, and by Theorem 6.6.7, $\bar{X}'_0$ is irreducible.

Let $X \setminus \Sigma(X) = \bigcup_{\alpha \in \Gamma} X'_\alpha$ be the decomposition into connected components. It follows from (6.6.9) that the number of $\bar{X}'_\alpha$ with $\mathrm{P}\Delta \cap \bar{X}'_\alpha \neq \emptyset$ is finite. Therefore, $\{\bar{X}'_\alpha\}_{\alpha \in \Gamma}$ is locally finite, and hence Γ is at most countable. Each $\bar{X}'_\alpha$ is an irreducible analytic subset of Ω, and $X = \bigcup \bar{X}'_\alpha$.

Since the decomposition $X \setminus \Sigma(X) = \bigcup X'_\alpha$ into connected components is unique up to the order, so is the decomposition $X = \bigcup \bar{X}'_\alpha$. □

N.B. Theorem 6.6.8 holds with Ω replaced by a complex manifold except for the countability of Γ. If the complex manifold satisfies the second countability axiom, then Γ is at most countable.

Each X_α of Theorem 6.6.8 is called an *irreducible component* of X and $X = \bigcup_\alpha X_\alpha$ is called the *irreducible decomposition* of X or the *decomposition of X into irreducible components X_α*.

Even if X is irreducible, a germ $\underline{X}_a$ $(a \in X)$ is not necessarily irreducible.

Example 6.6.10 We define $X \subset \mathbf{C}^2$ by the following equation:

$$z^2 - w^2(1-w) = 0.$$

Although X is irreducible, the germ $\underline{X}_0$ decomposes into two irreducible components determined by

$$z = w\sqrt{1-w}, \quad z = -w\sqrt{1-w},$$

where a branch of $\sqrt{1-w}$ is chosen about $w = 0$.

Proposition 6.6.11 *Let X, Y be two analytic subsets of Ω, and assume that Y is irreducible. If there is a neighborhood U of a point $a \in Y$ satisfying that $Y \cap U \subset X$ and* $\dim_a Y = \dim_a X$, *then Y is an irreducible component of X.*

Proof First, note that if X and Y have no singular point, then the assertion follows from the Identity Theorem 1.2.14.

Set $\dim Y = k$. Let $X = \bigcup X_\alpha$ be the irreducible decomposition. Set

$$X' = \bigcup_{\dim X_\alpha \le k} X_\alpha, \quad X'' = \bigcup_{\dim X_\alpha > k} X_\alpha.$$

Since $a \in X' \backslash X''$ by the assumption, we may assume that $X = X'$. Set $Z = \Sigma(Y) \cup \Sigma(X)$. It follows from Theorem 6.5.10 that Z is an analytic subset of $\dim Z < k$. Since Y is irreducible, $Y \backslash \Sigma(Y)$ is connected by Theorem 6.6.8. By Theorem 2.3.5, $Y \backslash Z$ is connected, too. Let $X \backslash Z = \cup X'_j$ be the decomposition into connected components. By the assumption there is some X'_j such that $\dim X'_j = k$ and $X'_j \supset (Y \backslash Z) \cap U$. Note that $\bar{X}'_j$ is an irreducible component of X. Since the dimensions are the same,

$$X'_j = Y \backslash Z, \quad \bar{X}'_j = Y.$$

□

Theorem 6.6.12 (Maximum Principle) *Let $X \subset \Omega$ be an analytic subset and let $f \in \mathscr{O}(X)$. If $|f(z)|$ admits the maximum value at $a \in X$, then $f(z)$ is constant on the connected component of X containing a.*

Proof Let $\underline{X}_a = \bigcup \underline{X_\alpha}_a$ be the irreducible decomposition of X at a. It suffices to show that f is constant on each X_α. We assume that $\underline{X}_a$ is irreducible. By a translation we may assume that $a = 0$ and $p = \dim_a X$. Let $\mathrm{P}\Delta_p \times \mathrm{P}\Delta_{n-p}$ be a standard polydisk of X about 0. Henceforth, we restrict X to $\mathrm{P}\Delta_p \times \mathrm{P}\Delta_{n-p}$. With $\pi : X \to \mathrm{P}\Delta_p$ the projection, we apply Theorem 6.4.16. It follows that π is proper and $\pi^{-1}0 = 0$; moreover, there is a thin analytic subset $Z \subsetneqq \mathrm{P}\Delta_p$ such that with $X' = X \backslash \pi^{-1}Z$, X' is a dense open subset of X and

$$\pi|_{X'} : X' \to \mathrm{P}\Delta_p \backslash Z$$

is a finite unramified covering. Let the cardinality of $\pi^{-1}\zeta$ ($\zeta \in \mathrm{P}\Delta_p \backslash Z$) be k. Set

$$g(\zeta) = \sum_{z \in \pi^{-1}\zeta} f(z).$$

Since $g(\zeta)$ is a bounded holomorphic function, by Riemann's Extension Theorem 2.3.4 it may be regarded as a holomorphic function on $\mathrm{P}\Delta_p$. It follows that

$$|g(\zeta)| \le k|f(a)|, \quad \zeta \in \mathrm{P}\Delta_p,$$
$$|g(a)| = k|f(a)|.$$

By the Maximum Principle (Theorem 1.2.17), $g(\zeta)$ is constant. Since for every $\zeta \in \mathrm{P}\Delta_p \backslash Z$,

$$k|f(a)| = k|g(\zeta)| \le \sum_{z\in\pi^{-1}\zeta} |f(z)| \le k|f(a)|,$$

$$|f(z)| \le |f(a)|,$$

we see that $|f(z)| = |f(a)|$ for all $z \in \pi^{-1}\zeta$. Again by the Maximum Principle (Theorem 1.2.17), $f(z)$ is constant on X'. Since $X = \bar{X}'$ and f is continuous on X, so is $f(z)$ on X. □

Corollary 6.6.13 *A compact analytic subset* $X \subset \Omega$ $(\subset \mathbf{C}^n)$ *is a finite set.*

Proof The X is decomposed into finitely many connected components X_γ. Applying Theorem 6.6.12 for coordinate functions restricted on each connected component X_γ, we see that they are constants on X_γ; thus, X_γ consists of one point. Therefore, X is a finite set. □

6.7 Finite Holomorphic Maps

In this section we describe several important and useful properties of finite holomorphic maps (Definition 6.1.7). Let $U \subset \mathbf{C}^n$ and $V \subset \mathbf{C}^m$ be open sets and let $X \subset U$ and $Y \subset V$ be analytic subsets.

Proposition 6.7.1 *Let* $f : X \to Y$ *be a holomorphic map. Then* f *is finite if and only if* f *is proper.*

Proof By definition, it suffices to show that if f is proper, it is finite. For any $y \in Y$ the inverse image $f^{-1}y$ is compact. By Corollary 6.6.13, $f^{-1}y$ is a finite set. □

Theorem 6.7.2 (Finite map theorem) *Let* $f : X \to Y$ *be a holomorphic map. If* f *is finite, the image* $f(X)$ *is an analytic subset. Moreover, if* X *is irreducible, so is* $f(X)$.

This holds for proper holomorphic maps between general complex spaces (cf. the proper map Theorem 6.9.7).

Proof We first show the last statement. Suppose that $f(X)$ is reducible. Then, there is a decomposition, $f(X) = Y_1 \cup Y_2$ $(f(X) \neq Y_i, i = 1, 2)$. Since each $X_i = f^{-1}Y_i \neq X$ is an analytic subset and with $X = X_1 \cup X_2$, X is reducible.

Since f is proper, $f(X) \subset V$ is closed (Proposition 6.1.8). It suffices to prove the assertion with $Y = V$.

Take a point $b \in f(X)$. Set $f^{-1}b = \{a_i\}_{i=1}^{l}$ $(l \in \mathbf{N})$. It follows from Proposition 6.1.8 that for a sufficiently small neighborhood ω of b, there are neighborhoods $W_i \ni a_i$ satisfying that $W_i \cap W_j = \emptyset$ $(i \neq j)$, $f^{-1}\omega = \bigcup W_i$, and for all i

$$f|_{W_i} : W_i \to \omega$$

are finite. Therefore it suffices to show the analyticity of each $f(W_i) \subset \omega$. Thus the problem is reduced to a local one, and $f^{-1}b = \{a\}$ (one point) is assumed. Let $X \subset U$ (resp. $Y \subset V$) represent the one restricted to a neighborhood of a (resp. b) in what follows.

By translations, we may assume $a = b = 0$. We take the graph of f,

$$G = \{(x, f(x)) \in X \times V\} \subset X \times V \subset \mathbf{C}^n \times \mathbf{C}^m$$

and the projection $\pi : G \to V$. Note that $\pi^{-1}0 = \{0\}$. Now we see that it is sufficient to prove:

Claim 6.7.3 *If $Z \subset X \times V$ is an analytic subset, and the projection $\pi : Z \to V$ is finite, then the image $\pi(Z)$ is analytic.*

∵) With m arbitrary we use induction on n. For a moment we let $n \geq 1$ for preparations.

(a) We write $(x, y) = (x_1, \ldots, x_n, y_1, \ldots, y_m) \in \mathbf{C}^n \times \mathbf{C}^m$ by coordinates. Let $\mathscr{I}\langle \underline{Z}_0 \rangle \subset \mathscr{O}_{\mathbf{C}^n \times \mathbf{C}^m, 0} = \mathscr{O}_{n+m,0}$ be the geometric ideal of Z at $0 = (0, 0)$, and let $\{g_\mu\}_{\mu=1}^N$ be a finite generator system of $\mathscr{I}\langle \underline{Z}_0 \rangle$. Let $\mathfrak{a}$ be the ideal of $\mathscr{O}_{n+m,0}$ generated by $\mathscr{I}\langle \underline{Z}_0 \rangle$ and $\underline{y_j}_0$, $1 \leq j \leq m$. It follows from the definition and Rückert's Nullstellen Theorem 6.4.20 that

$$\sqrt{\mathfrak{a}} = \mathfrak{m}_{n+m,0}, \tag{6.7.4}$$

where $\mathfrak{m}_{n+m,0}$ denotes the maximum ideal of $\mathscr{O}_{n+m,0}$. There is a number $d \in \mathbf{N}$ such that $\underline{x_1^d}_0 \in \mathfrak{a}$; therefore, there are $\underline{a_\mu}_0, \underline{b_j}_0 \in \mathscr{O}_{n+m,0}$ satisfying

$$\underline{x_1^d}_0 = \sum_{\mu=1}^N \underline{a_\mu}_0 \cdot \underline{g_\mu}_0 + \sum_{j=1}^m \underline{b_j}_0 \cdot \underline{y_j}_0.$$

With the substitution of $(x, y) = (x_1, 0, \ldots, 0, 0, \ldots, 0)$ the following holds locally:

$$x_1^d = \sum_{\mu=1}^N a_\mu(x_1, 0, \ldots, 0) \cdot g_\mu(x_1, 0, \ldots, 0).$$

Therefore there is at least one g_μ with $g_\mu(x_1, 0, \ldots, 0, 0, \ldots, 0) \not\equiv 0$; e.g., $\mu = 1$. By Weierstrass' Preparation Theorem 2.1.3, $g_1(x_1, x', y)$ with $x' = (x_2, \ldots, x_n)$ may be assumed to be a Weierstrass polynomial in x_1 of degree d_1, and as for other g_μ, we divide them by g_1, and take the remainders: We may write

(6.7.5)
$$g_1(x_1, x', y) = x_1^{d_1} + \sum_{\nu=1}^{d_1} c_{1\nu}(x', y) x_1^{d_1-\nu}, \quad c_{1,\nu}(0,0) = 0,$$
$$g_\mu(x_1, x', y) = \sum_{\nu=1}^{d_1} c_{\mu\nu}(x', y) x_1^{d_1-\nu}, \quad 2 \le \mu \le N.$$

We take a small disk neighborhood Δ_1 about $x_1 = 0$. With a sufficiently small polydisk neighborhood $\mathrm{P}\Delta_{n-1} \times \mathrm{P}\Delta_m$ about $(0,0)$ all roots of $g_1(x_1, x', y) = 0$ belong to Δ_1 for every $(x', y) \in \mathrm{P}\Delta_{n-1} \times \mathrm{P}\Delta_m$. The projection $\pi_1 : (x_1, x', y) \in \{g_1 = 0\} \to (x', y) \in \mathrm{P}\Delta_{n-1} \times \mathrm{P}\Delta_m$ is finite, and with restriction to $Z' := Z \cap (\Delta_1 \times \mathrm{P}\Delta_{n-1} \times \mathrm{P}\Delta_m)$,

(6.7.6)
$$\pi_1 : Z' \to \mathrm{P}\Delta_{n-1} \times \mathrm{P}\Delta_m$$

is finite. The image $Z_1 = \pi_1(Z')$ consists of all points (x', y) such that

$$g_\mu(x_1, x', y) = 0, \quad 1 \le \mu \le N$$

have a common root $x_1 \in \Delta_1$.

Introducing a new variable t ($\in \mathbf{C}$), we set

$$G(x_1, x', y; t) = \sum_{\mu=2}^{N} g_\mu(x_1, x', y) t^{\mu-2}.$$

We develop the resultant of $g_1(x_1, x', y)$ and $G(x_1, x', y; t)$ in variable t (cf. (2.2.4)) as

$$R(x', y; t) = \sum_{\lambda=1}^{L} R_\lambda(x', y) t^\lambda, \quad {}^\exists L \in \mathbf{N}.$$

By Theorem 2.2.7 the following equivalences hold for (x', y):

$$g_\mu(x_1, x', y) = 0, {}^\exists x_1 \in \Delta_1, 1 \le {}^\forall\mu \le N \iff R(x', y; t) = 0, {}^\forall t \in \mathbf{C}$$
$$\iff R_\lambda(x', y) = 0, \ 1 \le {}^\forall\lambda \le L.$$

Therefore we have

$$Z_1 = \{(x', y) \in \mathrm{P}\Delta_{n-1} \times \mathrm{P}\Delta_m; \ R_\lambda(x', y) = 0, \ 1 \le {}^\forall\lambda \le L\},$$

and hence, Z_1 is an analytic subset. The case of $n = 1$ is finished.

(b) Let $n \ge 2$ and assume the case of $n-1$. It follows from (a) that $Z_1 = \pi_1(Z) \subset \mathrm{P}\Delta_{n-1} \times \mathrm{P}\Delta_m$ is analytic. Let $\pi_2 : (x', y) \in Z_1 \to y \in \mathrm{P}\Delta_m$ be the projection. Note that $\pi = \pi_2 \circ \pi_1$. Since π is finite, so is π_2. The induction hypothesis implies the analyticity of $\pi(Z) = \pi_2(Z_1) \subset \mathrm{P}\Delta_m$. □

Remark 6.7.7 (Inductive projection method) In the above proof, after taking the graph of a map f an assertion on the map f is transferred to a problem on an analytic subset in the product space $\mathbf{C}^n \times \mathbf{C}^m$ satisfying a certain property, and then by induction on n it is reduced to $n = 0$. We here call this procedure the *inductive projection method*. This method is comprehensive and useful, and will be used in the proof of the following important Theorems 6.7.8 and 6.7.16.

Theorem 6.7.8 *Let $f : X \to Y$ be a holomorphic map, and let $a \in X, b = f(a) \in Y$. Then the following are equivalent:*

(i) *Through the homomorphism $f^* : \mathscr{O}_{Y,b} \to \mathscr{O}_{X,a}$, $\mathscr{O}_{X,a}$ is finitely generated as a $\mathscr{O}_{Y,b}$-module.*
(ii) *$f^* : \mathscr{O}_{Y,b} \to \mathscr{O}_{X,a}$ is injective and the ring extension $\mathscr{O}_{X,a}$ over $\mathscr{O}_{Y,b}$ through f^* is integral.*
(iii) *There are neighborhoods $W \ni a$ and $\omega \ni b$ such that $f|_W : W \to \omega$ is finite.*
(iv) *The point $a \in f^{-1}b$ is isolated.*

Proof By translations we let $a = b = 0$.

(i) $\Rightarrow$ (ii): This immediately follows from Theorem 6.3.4.

(ii) $\Rightarrow$ (iii): Let $(x_1, \ldots, x_n)$ be the standard coordinate system of $\mathbf{C}^n$. Then, restricted to X in a neighborhood of 0, each x_i satisfies

$$x_i^{d_i} + \sum_{\nu=1}^{d_i} c_{i\nu}(y) x_i^{d_1-\nu} = 0, \quad 1 \le i \le n. \tag{6.7.9}$$

By Weierstrass' Preparation Theorem 2.1.3 we may assume all these to be Weierstrass polynomials in x_i at 0; thus, all $c_{i\nu}(0) = 0$. With a small polydisk neighborhood $0 \in \mathrm{P}\Delta_n \Subset U$ and with a sufficiently small neighborhood $0 \in \omega \Subset V$, the restriction

$$f|_{X \cap \mathrm{P}\Delta_n} : X \cap \mathrm{P}\Delta_n \to \omega$$

is finite.

(iii) $\Rightarrow$ (iv): This follows from the definition of the finiteness.

(iv) $\Rightarrow$ (i)[2]: We consider the graph $G = \{(x, f(x)) \in X \times Y;\ x \in X\}$ of f and the projection $\pi : (x, y) \in G \to y \in Y$, and then apply the inductive projection method as in Theorem 6.7.2 above. It suffices to prove:

Claim *Let $Z \subset U \times V$ be an analytic subset, and let $\pi : Z \to V (\subset \mathbf{C}^m)$ be the projection with $\pi(0) = 0$. If $0 \in \pi^{-1}0$ is isolated, through π^*, $\mathscr{O}_{Z,a}$ is finitely generated as $\mathscr{O}_{m,0}$-module.*

---

[2] As far as the author knows, the present proof may be new; it is the point to prove Theorem 6.7.2 in advance, which uses only the resultant.

∵) (a) Let $x = (x_1, \ldots, x_n)$, $y = (y_1, \ldots, y_m)$ be the coordinate system and write $x' = (x_2, \ldots, x_n)$. Let $\mathscr{I}\langle \underline{Z}_0 \rangle \subset \mathscr{O}_{n+m,0}$ be the geometric ideal of $\underline{Z}_0$. From the arguments in Theorem 6.7.2 (a) there is a Weierstrass polynomial $g_1(x_1, x', y)$ in x_1 belonging to $\mathscr{I}\langle \underline{Z}_0 \rangle$, the degree of which is denoted by d_1. By Weierstrass' Preparation Theorem every $\underline{f}_0 \in \mathscr{O}_{n+m,0}$ is expressed at 0 as

$$f(x_1, x', y) = u(x_1, x', y) g_1(x_1, x', y) + \sum_{\nu=1}^{d_1} c_\nu(x', y) x_1^{d_1 - \nu}.$$

Therefore, $\mathscr{O}_{Z,0}$ is generated by $\{x_1^\nu\}_{\nu=0}^{d_1-1}$ over $\mathscr{O}_{(n-1)+m,0,0}$. This finishes the case of $n = 1$.

(b) Let $n \geq 2$ and assume the case of $n - 1$. With a suitable polydisk neighborhood $\Delta_1 \times \mathrm{P}\Delta_{n-1} \times \mathrm{P}\Delta_m$ about $(0, 0, 0) \in \mathbf{C} \times \mathbf{C}^{n-1} \times \mathbf{C}^m$, the projection $\pi_1 : Z \to \mathrm{P}\Delta_{n-1} \times \mathrm{P}\Delta_m$ is finite. It follows from Theorem 6.7.2 that the image $Z_1 = \pi_1(Z) \subset \mathrm{P}\Delta_{n-1} \times \mathrm{P}\Delta_m$ is analytic; moreover, with projection $\pi_2 : Z_1 \to \mathrm{P}\Delta_m$, $\pi = \pi_2 \circ \pi_1$ and $(0, 0) \in \pi_2^{-1} 0$ is isolated. By the argument of (a) above, $\mathscr{O}_{Z,0}$ is finitely generated over $\mathscr{O}_{Z_1,(0,0)}$. By the induction hypothesis, $\mathscr{O}_{Z_1,(0,0)}$ is a finitely generated $\mathscr{O}_{m,0}$-module through π_2^*. Thus, $\mathscr{O}_{Z,0}$ is finitely generated over $\mathscr{O}_{m,0}$. □

Corollary 6.7.10 *Under the assumption of Theorem 6.7.8, if $a \in f^{-1}b$ is isolated, there is a neighborhood $W \subset X$ of a with*

$$\dim_z f^{-1} f(z) = 0, \qquad {}^\forall z \in W :$$

That is, $z \in f^{-1} f(z)$ is isolated.

Proof In a neighborhood U of a, (6.7.9) holds, and then $f^{-1} f(z)$ is a finite set for $z \in U$. □

Corollary 6.7.11 *Under the assumption of Theorem 6.7.2 we have*

$$\dim_a X = \dim_{f(a)} f(X) \leq \dim_{f(a)} Y, \quad {}^\forall a \in X.$$

In particular, $\dim X = \dim f(X)$.

Proof We may assume that $\underline{X}_a$ is irreducible. Henceforth, we deal with the restrictions to a neighborhood of a. Set $Z = f(X) \subset Y$ and take the graph, $G = \{(f(x), x);\ x \in X\} \subset V \times X$ of f. Let $\pi : G \to Z = \pi(G) \subset V$ be the first projection. Then, $\dim_a X = \dim_{(b,a)} G$, and $\pi^* : \mathscr{O}_{Z,b} \to \mathscr{O}_{G,(b,a)}$ is injective. Then $\underline{Z}_b$ is irreducible, and we take a standard polydisk $\mathrm{P}\Delta_p \times \mathrm{P}\Delta_{m-p}$ of Z at b. It follows that $p = \dim_b Z \leq \dim_{f(a)} Y$. With a small polydisk neighborhood $\mathrm{P}\Delta_n$ of a, $\mathrm{P}\Delta_p \times (\mathrm{P}\Delta_{m-p} \times \mathrm{P}\Delta_n)$ gives rise to a standard polydisk neighborhood of G at (b, a). Therefore, $p = \dim_{(b,a)} G = \dim_a X$. □

Theorem 6.7.12 (Semi-continuity of dimension) *Let $f : X \to Y$ be a holomorphic map. For every point $a \in X$ there is a neighborhood $W \subset X$ such that*

$$\dim_x f^{-1}f(x) \leq \dim_a f^{-1}f(a), \quad {}^{\forall}x \in W.$$

Proof We may assume that $Y = \mathbf{C}^m$, $a = 0$ and $f(a) = 0$. Put $p = \dim_0 f^{-1}f(0)$. We take a standard polydisk neighborhood, $z = (z', z'') \in \mathrm{P}\Delta_p \times \mathrm{P}\Delta_{n-p}$, of $f^{-1}f(0)$ at 0; in particular, 0 is isolated in $f^{-1}f(0) \cap \{z_1 = \cdots = z_p = 0\}$. Therefore, with setting

$$F : x \in X \to (f(x), z_1, \ldots, z_p) \in \mathbf{C}^m \times \mathbf{C}^p,$$

the origin $0 \in F^{-1}(0, 0)$ is isolated. It follows from Theorem 6.7.8 (iii) that there is a neighborhood $W \subset X$ of 0 such that for every $x \in W$, $x \in F^{-1}F(x)$ is isolated. Fixing $x \in W$ arbitrarily, we consider the projection

$$\pi : \zeta \in (f^{-1}f(x)) \cap W \to (z_1(\zeta), \ldots, z_p(\zeta)) \in \mathbf{C}^p.$$

Then, $\zeta \in \pi^{-1}\pi(\zeta)$ is isolated. We infer from Theorem 6.7.8 and Corollary 6.7.11 that $\dim_x f^{-1}f(x) \leq p$. □

Theorem 6.7.13 *Let $f : X \to Y$ be a holomorphic map, and let $a \in X$ and $b = f(a)$. Assume the following:*

(a) *$\underline{Y_b}$ is irreducible.*
(b) *$a \in f^{-1}b$ is an isolated point.*
(c) *$\dim_a X = \dim_b Y$.*

Then there are neighborhood bases $\{U'\}$ of a and $\{V'\}$ of b such that:

(i) *$f(U' \cap X) = V' \cap Y$;*
(ii) *$f|_{U' \cap X} : U' \cap X \to V' \cap Y$ is finite;*
(iii) *with X assumed to be pure dimensional, there is an analytic subset $R \subsetneqq Y$ such that $R \supset \Sigma(Y)$, $f^{-1}R \supset \Sigma(X)$, and*

$$f|_{X \setminus f^{-1}R} : X \setminus f^{-1}R \to Y \setminus R$$

is a finitely sheeted unramified covering.

Proof The statements (i) and (ii) follow from Theorem 6.7.8 and Corollary 6.7.11.

We show (iii). Let $\underline{X_a} = \bigcup_\alpha \underline{X_{\alpha_a}}$ be the irreducible decomposition. If suffices to show the theorem for each irreducible component X_α, and so we assume that $\underline{X_a}$ is irreducible.

(1) We first deal with the case of $Y = \mathbf{C}^p$. Let $f : X \to \mathbf{C}^p$ and let $a = b = 0$. Suppose that X is an analytic subset of a neighborhood Ω of 0. Let $\Gamma(f) = \{(f(z), z); z \in X\} \subset \mathbf{C}^p \times \Omega$ be the graph of f, and let $\pi : \Gamma(f) \ni (w, z) \mapsto w \in \mathbf{C}^p$ be the projection. Note that $0 \in f^{-1}0$ being isolated is equivalent to $0 \in \pi^{-1}0$ being isolated. Therefore, in general, it suffices to show the assertion for $X \subset \Omega \subset \mathbf{C}^n \ni z = (z_1, \ldots, z_p, \ldots, z_n)$ and the projection

$$\pi : z = (z_1, \ldots, z_p, z_{p+1}, \ldots, z_n) \in X \to (z_1, \ldots, z_p) \in \mathbf{C}^p.$$

We set

$$\mathfrak{a} = \mathscr{I}\langle \underline{X}_0 \rangle \subset \mathscr{O}_{n,0}.$$

By the assumption we have

$$\mathscr{V}\left(\mathfrak{a} + \sum_{j=1}^{p} \mathscr{O}_{n,0} \cdot \underline{z_j}_0\right) = \{0\},$$
$$\mathscr{I}\langle\{0\}\rangle = \mathfrak{m}_{n,0} = \sum_{j=1}^{n} \mathscr{O}_{n,0} \cdot \underline{z_j}_0.$$

It follows from Rückert's Nullstellen Theorem 6.4.20 that there is a number $d \in \mathbf{N}$ satisfying

$$\underline{z_k}_0^d \in \mathfrak{a} + \sum_{j=1}^{p} \mathscr{O}_{n,0} \cdot \underline{z_j}_0, \quad p < {}^{\forall}k \le n. \tag{6.7.14}$$

We repeat the procedure in the proof of Proposition 6.4.2: Let $\mathscr{O}_{n,0}, \mathscr{O}_{n-1,0}, \ldots, \mathscr{O}_{p,0}$ be the notation used there. Then, by (6.7.14) we see that $\mathfrak{a} \cap \mathscr{O}_{h,0} \ne \{0\}$ for $p < h \le n$. Since $\dim \underline{X}_0 = p$, Proposition 6.4.2 and Theorem 6.4.16 imply that a projection

$$\pi : X \cap (\mathrm{P}\Delta_p \times \mathrm{P}\Delta_{n-p}) \to \mathrm{P}\Delta_p$$

is surjective and finite. Thus, $\dim X = p$, and then with δ and $R = \{\delta = 0\} \subsetneqq \mathrm{P}\Delta_q$ in Theorem 6.4.16, the restriction

$$\pi|_{X \cap (\mathrm{P}\Delta_p \times \mathrm{P}\Delta_{n-p}) \setminus \pi^{-1}R} : X \cap (\mathrm{P}\Delta_p \cap \mathrm{P}\Delta_{n-p}) \setminus \pi^{-1}R \to \mathrm{P}\Delta_p \setminus R$$

is a finite unramified covering.

(2) In the case of general Y we take a standard disk neighborhood $\mathrm{P}\Delta_p \times \mathrm{P}\Delta_{m-p}$ of Y at 0 as in Theorem 6.4.16, and we restrict Y there: The projection

$$\pi : Y \to \mathrm{P}\Delta_p$$

is finite, and there is an analytic subset $S \subsetneqq \mathrm{P}\Delta_p$ such that $Y \setminus \pi^{-1}S \to \mathrm{P}\Delta_p \setminus S$ is a finite unramified covering. Then, the composed map $g = \pi \circ f : X \to \mathrm{P}\Delta_p$ satisfies the condition of the theorem. By the result of (1) above there is an analytic subset $T \subsetneqq \mathrm{P}\Delta_p$ such that

$$g|_{X \setminus g^{-1}T} : X \setminus T \to \mathrm{P}\Delta_p \setminus T$$

is a finite unramified covering. Set $R = \pi^{-1}(S \cup T) \subsetneqq Y$. Then, $X \setminus f^{-1}R$ and $Y \setminus R$ are p-dimensional complex manifolds, and

$$(\pi \circ f)|_{X \setminus f^{-1}R} : X \setminus f^{-1}R \to \mathrm{P}\Delta_p \setminus (S \cup T),$$
$$\pi|_{Y \setminus \pi^{-1}(S \cup T)} : Y \setminus \pi^{-1}(S \cup T) \to \mathrm{P}\Delta \setminus (S \cup T)$$

are both finite unramified coverings. Thus,

$$f|_{X\setminus f^{-1}R} : X\setminus f^{-1}R \to Y\setminus R$$

is a finite unramified covering. □

Corollary 6.7.15 *Under the conditions of Theorem 6.7.13, the homomorphism $f^* : \mathscr{O}_{Y,b} \to \mathscr{O}_{X,a}$ is injective, and $\mathscr{O}_{X,a}$ is a finitely generated module over $\mathscr{O}_{Y,b}$ through f^*.*

Proof The injectivity follows from Theorem 6.7.13. The finiteness is already proved in Theorem 6.7.8. □

Finally, we consider the direct images of a coherent sheaves (cf. Sect. 1.3.3 (8)). The following is a special case of Grauert's Ein Theorem [23] (cf. Theorem 6.9.8) for a proper holomorphic map between general complex spaces (cf. Sect. 6.9).

Theorem 6.7.16 (Direct Image Theorem) *Let $f : X \to Y$ be a finite holomorphic map, and let $\mathscr{S} \to X$ be a coherent sheaf. Then the direct image sheaf $f_*\mathscr{S} \to Y$ is coherent.*

Proof Since f is finite, it follows from Proposition 6.1.8 (iii) that

$$(f_*\mathscr{S})_y \cong \oplus_{x\in f^{-1}y}\mathscr{S}_x, \tag{6.7.17}$$

$$(f_*\mathscr{O}_X)_y \cong \oplus_{x\in f^{-1}y}\mathscr{O}_{X,x}. \tag{6.7.18}$$

Here, note that through $f^* : \mathscr{O}_{Y,y} \to \mathscr{O}_{X,x}$, $\mathscr{S}_x$ and $\mathscr{O}_{X,x}$ are regarded as $\mathscr{O}_{Y,y}$-modules. The coherence of $\mathscr{S}$ implies the following:

(i) $f_*\mathscr{S}$ is a finitely generated module over $f_*\mathscr{O}_X$.
(ii) A relation sheaf of $f_*\mathscr{S}$ as a $f_*\mathscr{O}_X$-module is locally finite over $f_*\mathscr{O}_X$.

Therefore it suffices to show the following: □

Lemma 6.7.19 *If a holomorphic map $f : X \to Y$ is finite, $f_*\mathscr{O}_X$ is locally finite over $\mathscr{O}_Y$.*

Proof The proof relies on the inductive projection method (cf. Remark 6.7.7). Since the problem is local, we may assume that $0 \in X$, $f(0) = 0$, and $X \subset U = \mathrm{P}\Delta_n$ (a polydisk of $\mathbf{C}^n$ about 0), $Y = V = \mathrm{P}\Delta_m$ (a polydisk of $\mathbf{C}^m$ about 0). We set the graph of f and the projection by

$$G = \{(x, f(x)) \in X \times Y; x \in X\},$$
$$\pi : (x, y) \in G \to y \in Y.$$

Now the following is sufficient for the proof:

Claim *Let* $Z \subset \mathrm{P}\Delta_n \times \mathrm{P}\Delta_m$ *be an analytic subset containing* $(0, 0)$, *and let* $\pi : Z \to \mathrm{P}\Delta_m$ *be the projection. If* π *is finite, then* $\pi_* \mathscr{O}_Z$ *is locally finite over* $\mathscr{O}_{\mathrm{P}\Delta_m}$.

∵) (a) Let $x = (x_1, \ldots, x_n)$ and $y = (y_1, \ldots, y_m)$ be the standard coordinate systems, and write $x' = (x_2, \ldots, x_n)$. As in (6.7.18), for a point $y \in \pi(Z)$ there is an isomorphism as $\mathscr{O}_{\mathrm{P}\Delta_m, y}$-modules:

$$(6.7.20) \qquad (\pi_* \mathscr{O}_Z)_y \cong \oplus_{(x,y) \in \pi^{-1} y} \mathscr{O}_{Z,(x,y)}.$$

It suffices to show it in a neighborhood of 0. Let $\mathscr{I}\langle Z \rangle_0 \subset \mathscr{O}_{\mathrm{P}\Delta_n \times \mathrm{P}\Delta_m, 0} = \mathscr{O}_{n+m,0}$ be the geometric ideal of $\underline{Z}_0$. We infer from the arguments in (a) of the proof of Claim 6.7.3 that there is a Weierstrass polynomial $W(x_1, x', y)$ in x_1 belonging to $\mathscr{I}\langle Z \rangle_0$. Let d be its degree. After taking $\mathrm{P}\Delta_n$ and $\mathrm{P}\Delta_m$ smaller if necessary, we may assume that $W(x_1, x', y) \in \mathscr{O}(\mathrm{P}\Delta_n \times \mathrm{P}\Delta_m)$. By means of Weierstrass' Preparation Theorem 2.1.3, every $\underline{f}_0 \in \mathscr{O}_{n+m,0}$ is expressed in a neighborhood of 0 as

$$f(x_1, x', y) = u(x_1, x', y) W(x_1, x', y) + \sum_{\nu=1}^{d} c_\nu(x', y) x_1^{d-\nu},$$
$$x = (x_1, x') \in \Delta_{(1)} \times \mathrm{P}\Delta_{n-1}, \quad y \in \mathrm{P}\Delta_m.$$

Hence we see that $\mathscr{O}_{Z,0}$ is generated by $\{1, x_1, x_1^2, \ldots, x_1^{d-1}\}$ as an $\mathscr{O}_{\mathrm{P}\Delta_{n-1} \times \mathrm{P}\Delta_m, 0}$-module. Since π is finite, the projection

$$\pi_1 : (x_1, x', y) \in Z \to (x', y) \in \mathrm{P}\Delta_{n-1} \times \mathrm{P}\Delta_m$$

is also finite. By Theorem 6.7.2, the image $Z_1 = \pi_1(X) \subset \mathrm{P}\Delta_{n-1} \times \mathrm{P}\Delta_m$ is an analytic subset. Take any point $(a, b) = (a_1, a', b) \in Z$. Then, $W(a, b) = 0$, and we let

$$W(x_1, x', y) = u(x, y) W_1(x_1 - a_1, x', y),$$
$$W_1(x_1 - a_1, x', y) = (x_1 - a_1)^{d_1} + \sum_{\nu=1}^{d_1} \beta_\nu(x', y)(x_1 - a_1)^{d_1 - \nu}, \ d_1 \le d$$

be the Weierstrass decomposition of W at (a, b) with respect to variable x_1. With the help of Weierstrass' Preparation Theorem 2.1.3 we deduce that $\mathscr{O}_{Z,(a,b)}$ is generated by $\{(x_1 - a_1)^\nu\}_{\nu=0}^{d_1 - 1}$ over $\mathscr{O}_{Z_1,(a',b)}$ through $\pi_1^* : \mathscr{O}_{Z_1,(a',b)} \to \mathscr{O}_{Z,(a,b)}$. Noting that $\{(x_1 - a_1)^\nu\}_{\nu=0}^{d_1 - 1}$ is generated by

$$1, x_1, x_1^2, \ldots, x_1^{d_1 - 1}, \ldots, x_1^{d-1},$$

we see by (6.7.20) that $\pi_{1*} \mathscr{O}_Z$ is finitely generated over $\mathscr{O}_{Z_1}$. The case of $n = 1$ is thus proved.

(b) Let $n \geq 2$ and assume that the case of $n - 1$ holds. It follows from the arguments of (a) that the projection $\pi_1 : Z \to \mathrm{P}\Delta_{n-1} \times \mathrm{P}\Delta_m$ is finite for a suitably chosen polydisk neighborhood $\Delta_{(1)} \times \mathrm{P}\Delta_{n-1} \times \mathrm{P}\Delta_m$ of $(0, 0, 0) \in \mathbf{C} \times \mathbf{C}^{n-1} \times \mathbf{C}^m$. By Theorem 6.7.2, the image $Z_1 = \pi_1(Z) \subset \mathrm{P}\Delta_{n-1} \times \mathrm{P}\Delta_m$ is an analytic subset, and $\pi_{1*}\mathscr{O}_Z$ is locally finite as a sheaf of modules over $\mathscr{O}_{Z_1}$. Moreover, with the projection $\pi_2 : Z_1 \to \mathrm{P}\Delta_m$, we have $\pi = \pi_2 \circ \pi_1$, and π_2 is also finite. The induction hypothesis implies that $\pi_{2*}\mathscr{O}_{Z_1}$ is locally finite over $\mathscr{O}_{\mathrm{P}\Delta_m}$. Therefore, $\pi_*\mathscr{O}_Z = \pi_{1*}(\pi_{2*}\mathscr{O}_Z)$ is locally finite over $\mathscr{O}_{\mathrm{P}\Delta_m}$. □

6.8 Continuation of Analytic Subsets

(a) We study the analytic continuation of analytic subsets as in the case of holomorphic functions. Let $\Omega \subset \mathbf{C}^n$ be a domain.

Theorem 6.8.1 (Remmert's Extension Theorem) *Let $Y \subset \Omega$ be an analytic subset, and let $X \subset \Omega \setminus Y$ be an analytic subset such that*

(6.8.2) $$\dim_a X > \dim Y, \quad {}^\forall a \in X.$$

Then the topological closure $\bar{X}$ of X in Ω is an analytic subset.

Proof If $X = \Omega \setminus Y$, then $\bar{X} = \Omega$ and so the assertion holds. Suppose that $X \neq \Omega \setminus Y$. Let $X = \bigcup X_\alpha$ be the irreducible decomposition. For a number m with $\dim Y < m < n$ we set $X^{(m)} = \bigcup_{\dim X_\alpha = m} X_\alpha$. Then $X = \bigcup_{m \geq 0} X^{(m)}$. If each closure $\bar{X}^{(m)}$ of $X^{(m)}$ is analytic, then $\bar{X}$ is analytic; thus X may be assume to be of pure dimension m. Take any point $a \in \bar{X} \cap Y$. We are going to show that $U \cap \bar{X}$ is analytic in a neighborhood U of a.

Claim 6.8.3 *If $a \in Y \setminus \Sigma(Y)$, then there is a neighborhood U of a such that $\bar{X} \cap U$ is analytic.*

It follows from this that $\bar{X} \cap (\Omega \setminus \Sigma(Y))$ is analytic. Since $\dim Y > \dim \Sigma(Y)$, the repetition of this leads the analyticity of $\bar{X}$.

Proof of Claim 6.8.3. Set $l = \dim_a Y$. We take a holomorphic local coordinate system $(z_1, \ldots, z_n)$ in a neighborhood U of a, so that $a = 0$, and

(6.8.4) $U \cap Y$ is the intersection of an l-dimensional linear subspace and U.

Let L_1 be a linear functional in $(z_1, \ldots, z_n)$ such that for every irreducible component X' of X, $\dim X' \cap \{L_1 = 0\} < m$. Repeating this, we take $L_1, \ldots, L_m$, so that

$$\dim X \cap \{L_1 = \cdots = L_m = 0\} = 0,$$
$$Y \cap \{L_1 = \cdots = L_m = 0\} = \{0\}.$$

We change the coordinate system $z = (z_1, \ldots, z_m, \ldots, z_n)$ with $z_j = L_j, 1 \leq j \leq m$. Then $X \cap \{z_1 = \cdots = z_m = 0\}$ is at most countable, and written as $X \cap \{z_1 = \cdots = z_m = 0\} = \{x^{(\nu)}\}_{\nu=1}^{\infty}$. We put

$$\varphi(z) = \max_{m+1 \leq k \leq n} |z_k|.$$

Then we may take an arbitrarily small number $\delta \notin \{\varphi(x^{(\nu)})\}$ such that

$$(X \cup Y) \cap U \cap \{z = (0, \ldots, 0, z_{m+1}, \ldots, z_n);\ \varphi(z) = \delta\} = \emptyset.$$

Since $(X \cup Y) \cap U$ is closed in U, and $\{z = (0, \ldots, 0, z_{m+1}, \ldots, z_m); \varphi(z) = \delta\}$ is compact, there is a number $\varepsilon > 0$ such that

$$\begin{aligned}(X \cup Y) \cap U \cap \{z = (z_1, \ldots, z_m, \ldots, z_n);\\ |z_j| \leq \varepsilon, 1 \leq j \leq m, \varphi(z) = \delta\} = \emptyset.\end{aligned}$$

Now we set

$$\begin{aligned}\mathrm{P}\Delta' &= \{(z_1, \ldots, z_m);\ |z_j| < \varepsilon\} \subset \mathbf{C}^m,\\ \mathrm{P}\Delta'' &= \{(z_{m+1}, \ldots, z_n); |z_j| < \delta\} \subset \mathbf{C}^{n-m},\\ \mathrm{P}\Delta &= \mathrm{P}\Delta' \times \mathrm{P}\Delta'' \Subset U.\end{aligned}$$

Then the projection

$$\pi : z = (z', z'') \in (X \cup Y) \cap \mathrm{P}\Delta \to z' \in \mathrm{P}\Delta'$$

is finite. Set $E' = \pi(Y \cap \mathrm{P}\Delta)$. Then it follows from (6.8.4) that there is an $l\ (< m)$-dimensional linear subspace E of $\mathbf{C}^m$ with $E' = \mathrm{P}\Delta' \cap E$. For every point $z' \in \mathrm{P}\Delta' \setminus E'$, $X \cap (\pi^{-1}z')$ is a compact analytic subset, and hence a finite set. By Theorem 6.7.13, $\pi(X)$ contains a neighborhood of z'. Therefore, $\pi((X \cup Y) \cap \mathrm{P}\Delta) = \mathrm{P}\Delta'$. Set

$$d = \max_{z' \in \mathrm{P}\Delta' \setminus E'} |X \cap (\pi^{-1}z')|\ (< \infty).$$

Let $a_{k1}(z'), \ldots, a_{kd}(z')$ be the elementary symmetric polynomials of the d values of $z_k\ (m+1 \leq k \leq n)$ on $X \cap (\pi^{-1}z')$. Then we have

$$z_k^d + \sum_{\nu=1}^{d} a_{k\nu}(z') z_k^{d-1} = 0, \quad m+1 \leq k \leq n. \tag{6.8.5}$$

Since $a_{k\nu}(z')$ are bounded holomorphic in $\mathrm{P}\Delta' \setminus E'$, they are holomorphic in $\mathrm{P}\Delta'$ by Riemann's Extension Theorem 2.3.4. Let Z be the analytic subset of $\mathrm{P}\Delta$ defined by (6.8.5). Then Z is of pure dimension m. In what follows we restrict everything to $\mathrm{P}\Delta$. We have

$$Z \supset \bar{X}.$$

Let Z'_α be the connected component of $Z \setminus \Sigma(Z)$, which contains a connected component X'_α of $X \setminus \Sigma(X)$. Then,

$$X'_\alpha = Z'_\alpha \setminus Y.$$

Therefore, $\bar{X}'_\alpha = \bar{Z}'_\alpha$ is an irreducible component of Z. Thus we see that $\bar{X} = \bigcup \bar{Z}'_\alpha$ is analytic. □

(b) We present a simple but important application of Remmert's Extension Theorem.

Theorem 6.8.6 (Chow) *An analytic subset $X \subset \mathbf{P}^n(\mathbf{C})$ is algebraic.*

Proof We take the Hopf map (6.1.1),

$$\pi : \mathbf{C}^{n+1} \setminus \{0\} \to \mathbf{P}^n(\mathbf{C}).$$

The inverse image $\widetilde{X} = \pi^{-1} X$ is an analytic subset of $\mathbf{C}^{n+1} \setminus \{0\}$. Since $\dim_z \widetilde{X} > 0$ at every $z \in \widetilde{X}$, and $Y = \{0\}$ is of 0-dimension, By Remmert's Extension Theorem 6.8.1, the closure $\widehat{X} = \bar{\widetilde{X}}$ of $\widetilde{X}$ is an analytic subset of $\mathbf{C}^{n+1}$.

Let $\underline{f_1}_0, \ldots, \underline{f_{l_0}}_0$ be generators of the ideal $\mathscr{I}\langle \widehat{X} \rangle_0$. Taking a polydisk $\mathrm{P}\Delta$ about 0 where all f_j are defined, we expand f_j to power series:

$$f_j(z_0, \ldots, z_n) = \sum_\alpha c_{j\alpha} z^\alpha = \sum_{\nu=1}^{\infty} P_{j\nu}(z), \quad z \in \mathrm{P}\Delta,$$
$$P_{j\nu}(z) = \sum_{|\alpha|=\nu} c_{j\alpha} z^\alpha,$$

where $P_{j\nu}(z)$ are homogeneous polynomial of degree ν. Let $z \in \widehat{X} \cap \mathrm{P}\Delta$ be any point. By the definition of $\widehat{X}$, $\zeta z \in \widehat{X}$ for all $\zeta \in \mathbf{C}$, and hence for $\zeta \in \mathbf{C}$ with $\zeta z \in \mathrm{P}\Delta \cap \widehat{X}$,

$$f_j(\zeta z) = \sum_{\nu=1}^{\infty} P_{j\nu}(z) \zeta^\nu = 0.$$

It follows that $P_{j\nu}(z) = 0$, $\nu = 1, 2, \ldots$, for every $z \in \widehat{X}$. Therefore,

$$\widehat{X} = \bigcap_{j,\nu} \{P_{j,\nu} = 0\}.$$

By the Noetherian property of a polynomial ring (Theorem 2.2.2), there are finitely many (j, ν) with which

$$\widehat{X} = \bigcap_{(j,\nu)} \{P_{j,\nu} = 0\}.$$

Thus, $\widehat{X}$ is algebraic, and so is X. $\square$

6.9 Complex Spaces

In general, a pair $(X, \mathscr{R}_X)$ of a topological space X and a sheaf of rings $\mathscr{R}_X$ over X is called a *ringed space*.

Definition 6.9.1 A ringed space $(X, \mathscr{O}_X)$ is called a *complex space* if the following conditions are satisfied:

(i) X is a Hausdorff topological space.
(ii) There are an open covering $X = \bigcup_{\alpha\in\Gamma} U_\alpha$ and for each $\alpha \in \Gamma$, a homeomorphism $\varphi_\alpha : U_\alpha \to A_\alpha$ onto an analytic subset A_α of an open set Ω_α of some $\mathbf{C}^{n_\alpha}$, and a sheaf isomorphism $\Phi_\alpha : \mathscr{O}_X|_{U_\alpha} \to \mathscr{O}_{A_\alpha}$ compatible with $\varphi_\alpha : U_\alpha \to A_\alpha$, i.e., the following is commutative:

$$\begin{array}{ccc} \mathscr{O}_X|_{U_\alpha} & \xrightarrow{\Phi_\alpha} & \mathscr{O}_{A_\alpha} \\ \downarrow & & \downarrow \\ U_\alpha & \xrightarrow{\varphi_\alpha} & A_\alpha . \end{array}$$

We call $\mathscr{O}_X$ the *structure sheaf* of the complex space $(X, \mathscr{O}_X)$, and the triple $(U_\alpha, \varphi_\alpha, A_\alpha)$ a *local chart*. In the case where U_α is considered to be a neighborhood of a point $x \in U_\alpha$, U_α or the triple $(U_\alpha, \varphi_\alpha, A_\alpha)$ is called a *local chart neighborhood* of x.

In (ii) above, if $U_\alpha \cap U_\beta \neq \emptyset$, the following sheaf isomorphism holds:

$$\begin{array}{ccc} \mathscr{O}_{A_\alpha \cap \varphi_\alpha(U_\alpha\cap U_\beta)} & \xrightarrow{\Phi_\beta\circ\Phi_\alpha^{-1}} & \mathscr{O}_{A_\beta \cap \varphi_\beta(U_\alpha\cap U_\beta)} \\ \downarrow & & \downarrow \\ A_\alpha \cap \varphi_\alpha(U_\alpha \cap U_\beta) & \xrightarrow{\varphi_\beta\circ\varphi_\alpha^{-1}} & A_\beta \cap \varphi_\beta(U_\alpha \cap U_\beta) . \end{array}$$

Here, $\Phi_\beta \circ \Phi_\alpha^{-1}$, $\varphi_\beta \circ \varphi_\alpha^{-1}$, etc. are restricted to open subsets where they are defined. In this sense,

$$\varphi_\beta \circ \varphi_\alpha^{-1} : A_\alpha \cap \varphi_\alpha(U_\alpha \cap U_\beta) \longrightarrow A_\beta \cap \varphi_\beta(U_\alpha \cap U_\beta)$$

is a holomorphic map with the holomorphic inverse, i.e., a holomorphic isomorphism.

Let $(X, \mathscr{O}_X)$ be a complex space. We sometimes simply call X a complex space. A section of $\mathscr{O}_X$ over an open set U of X is called a *holomorphic function* on U, and the set of all of them is denoted by

$$\mathscr{O}_X(U) = \Gamma(U, \mathscr{O}_X).$$

We write the restriction as $\mathscr{O}_U = \mathscr{O}_X|_U$. For a given $f \in \mathscr{O}_X(U)$ and each $x \in U$, there correspond $f(x) \in \mathscr{O}_{X,x}$, and also the "value" $f(x) \in \mathbf{C}$; that is, a function $f : U \to \mathbf{C}$ is determined. The notation $f(x)$ is used in the both senses unless confusion occurs, but we clarify which sense it is used, as it is necessary.

A sheaf $\mathscr{S} \to X$ which is a coherent sheaf of $\mathscr{O}_X$-modules is called a *coherent sheaf* over X. The following is fundamental:

Theorem 6.9.2 (Oka) *The structure sheaf $\mathscr{O}_X$ of a complex space $(X, \mathscr{O}_X)$ is coherent.*

This is immediate from Oka's coherence Theorem 6.5.17. As Proposition 3.3.5 we have:

Proposition 6.9.3 *For a sheaf $\mathscr{S}$ of $\mathscr{O}_X$-modules over X, the following are equivalent.*

(i) *$\mathscr{S}$ is coherent.*
(ii) *For every $x \in X$ there are a neighborhood $U \subset X$ and an exact sequence over it:*

$$\mathscr{O}_U^q \xrightarrow{\phi} \mathscr{O}_U^p \xrightarrow{\psi} \mathscr{S}|_U \to 0.$$

Let $x \in X$ be a point and let $(U_\alpha, \varphi_\alpha, A_\alpha)$ be a local chart neighborhood of x. We call x a *regular* (resp. *singular*) *point* of X if $\varphi_\alpha(x) \in A_\alpha$ is a regular (resp. singular) point of A_α; this is independent from the choice of the local chart neighborhood. We denote the set of all singular points of X by $\Sigma(X)$. When $\Sigma(X) = \emptyset$, X is said to be *non-singular*; in this case, X is a complex manifold, which may be disconnected in general.

A subset $Y \subset X$ is defined to be an *analytic subset* of X if for every chart $(U_\alpha, \varphi_\alpha, A_\alpha)$, $\varphi_\alpha(Y \cap U_\alpha)$ is an analytic subset of A_α. We define naturally the *geometric ideal sheaf* $\mathscr{I}\langle Y\rangle \subset \mathscr{O}_X$ of germs of holomorphic functions taking values 0 on Y. With the quotient sheaf $\mathscr{O}_Y = \mathscr{O}_X/\mathscr{I}\langle Y\rangle$, the ringed space $(Y, \mathscr{O}_Y)$ is a complex space, provided that Y is an analytic subset of X: We call $(Y, \mathscr{O}_Y)$ a *complex subspace* of $(X, \mathscr{O}_X)$.

By Theorem 6.5.10 we immediately have:

Theorem 6.9.4 *The singularity set $\Sigma(X)$ of a complex space $(X, \mathscr{O}_X)$ is a thin analytic subset of X.*

Let $(X, \mathscr{O}_X)$ and $(Y, \mathscr{O}_Y)$ be two complex spaces. A continuous map $f : X \to Y$ is called a *holomorphic map* if for every point $x \in X$ there are local chart neighborhoods, $(U_\alpha, \varphi_\alpha, A_\alpha)$ of x and $(V, \lambda, \psi_\lambda, B_\lambda)$ of $f(x)$ with $f(U_\alpha) \subset V_\lambda$ such that

$$\psi_\lambda \circ f \circ \varphi_\alpha^{-1} : A_\alpha \to B_\lambda$$

is a holomorphic map between analytic sets.

If a holomorphic map $f : X \to Y$ is bijective, and the inverse $f^{-1} : Y \to X$ is also holomorphic, X is said to be *holomorphically isomorphic* to Y, and f is called a *holomorphic isomorphism* or a *biholomorphism*.

For a holomorphic map $f : X \to Y$ between general complex spaces, the properness is different to the finiteness (cf. Proposition 6.7.1). For instance, the projection $p : \mathbf{C}^n \times \mathbf{P}^m(\mathbf{C}) \to \mathbf{C}^n$ is proper, but not finite.

Example 6.9.5 The so-called blow-up is a procedure which is frequently used in analytic function theory of several variables and complex geometry. We introduce it in the simplest case of a blow-up at one point. Let $n \geq 2$ and let $z = (z_1, \ldots, z_n) \in \mathbf{C}^n$ be the standard coordinate system, and let $w = [w_1, \ldots, w_n] \in \mathbf{P}^{n-1}(\mathbf{C})$ be the homogeneous coordinate system. Let $X \subset \mathbf{C}^n \times \mathbf{P}^{n-1}(\mathbf{C})$ be an analytic subset defined by the following equations:

$$(6.9.6) \qquad z_j w_k - z_k w_j = 0, \quad 1 \leq j, k \leq n.$$

Let $\pi : (z, w) \in X \to z \in \mathbf{C}^n$ be the first projection. If $z \neq 0$, e.g., $z_1 \neq 0$,

$$w_k = \frac{w_1}{z_1} z_k, \quad 1 \leq k \leq n.$$

Since $(w_k) \neq 0$, $w_1 \neq 0$ follows in this case, and $\pi^{-1} z = (z, w) \in X$ is determined as one point. For $z = 0$, clearly, $\pi^{-1} 0 = \{0\} \times \mathbf{P}^{n-1}(\mathbf{C})$. It follows that $\pi : X \to \mathbf{C}^n$ is proper but not finite. We write $X = \widehat{\mathbf{C}^n}$, and call it a one-point blow-up of $\mathbf{C}^n$. It is easily checked that $\widehat{\mathbf{C}^n}$ is non-singular, and the restriction

$$\pi|_{\widehat{\mathbf{C}^n} \setminus \pi^{-1} 0} : \widehat{\mathbf{C}^n} \setminus \pi^{-1} 0 \to \mathbf{C}^n \setminus \{0\}$$

is biholomorphic.

The local properties for analytic sets and holomorphic maps between them which we proved in the previous section remain to be valid for complex spaces and holomorphic maps between complex spaces. We present without proof those theorems which remain to hold for proper holomorphic maps, not only for finite ones (cf. Grauert–Remmert [28] for the proofs).

We define a *higher direct image sheaf* denoted by $\mathrm{dir}^q f_* \mathscr{S}$ or $f_* \mathscr{H}^q(X, \mathscr{S})$ for $q \geq 1$. In general, let $f : X \to Y$ be a continuous map between topological spaces, let $\mathscr{S} \to X$ be a sheaf, and let $q \geq 0$ in general. For an open set $V \subset Y$, we associate $H^q(f^{-1} V, \mathscr{S})$, and produce a presheaf, which induces a sheaf over Y, called the q-th *direct image sheaf* of $\mathscr{S}$, and denoted by $\mathrm{dir}^q f_* \mathscr{S}$ or $f_* \mathscr{H}^q(X, \mathscr{S})$; thus for $q = 0$, $\mathrm{dir}^0 f_* \mathscr{S} = f_* \mathscr{H}^0(X, \mathscr{S}) = f_* \mathscr{S}$.

In what follows, X and Y denote complex spaces.

Theorem 6.9.7 (Proper map theorem) *Let $f : X \to Y$ be a proper holomorphic map. Then the image $f(X)$ is an analytic subset of Y. Moreover, if X is irreducible, so is $f(X)$.*

Theorem 6.9.8 (Grauert's Direct Image Theorem) *Let $f : X \to Y$ be a proper holomorphic map, and let $\mathscr{S} \to X$ be a coherent sheaf. Then, the direct image sheaf $f_*\mathscr{S} \to Y$ is coherent. Furthermore generally,* $\mathrm{dir}^q f_*\mathscr{S}$ *is coherent for $q \geq 0$.*

Definition 6.9.9 Let $\mathscr{S} \to X$ be a sheaf of modules. The *support* Supp $\mathscr{S}$ of $\mathscr{S}$ is defined by

$$\mathrm{Supp}\ \mathscr{S} = \overline{\{x \in X;\ \mathscr{S}_x \neq 0\}}.$$

By definition, Supp $\mathscr{S}$ is closed.

Theorem 6.9.10 (Support theorem) *If $\mathscr{S} \to X$ is a coherent sheaf, then* Supp $\mathscr{S}$ *is an analytic set.*

Proof It follows from Proposition 6.9.3 that for every point $a \in X$ there are a neighborhood $U \ni a$ and an exact sequence over U:

$$\mathscr{O}_U^q \xrightarrow{\phi} \mathscr{O}_U^p \xrightarrow{\psi} \mathscr{S}|_U \to 0.$$

Then, $\mathscr{S}_x \neq 0$ $(x \in U)$ is equivalent to $\phi(\mathscr{O}_{U,x}^q) \neq \mathscr{O}_{U,x}^p$. The homomorphism ϕ is expressed by a (q, p)-matrix $A(x) = (a_{ij}(x))$ with $a_{ij} \in \mathscr{O}(U)$. Therefore we have

$$\{x \in U; \mathscr{S}_x \neq 0\} = \{x \in U; \mathrm{rank}\, A(x) < p\}.$$

Since rank $A(x) < p$ is equivalent to the vanishing of all p-minor determinants of $A(x)$, $\{x \in U; \mathscr{S}_x \neq 0\}$ is closed, and Supp $\mathscr{S}$ is analytic. □

The notion of a (un)ramified covering domain (Definition 4.5.6) is extended to the case of complex spaces. A connected open subset of X is called a *domain* of X.

Definition 6.9.11 Let $\pi : X \to Y$ be a holomorphic map between complex spaces, X and Y. We call $\pi : X \to Y$ an *unramified covering domain* over Y if for every point $x \in X$ there are neighborhoods, $U \ni x$ and $V \ni \pi(x)$ such that the restriction $\pi|_U : U \to V$ is biholomorphic; we say for this property that π is locally biholomorphic.

If $\pi|_U : U \to V$ is not necessarily biholomorphic in general, but a finite map, then $\pi : X \to Y$ is called a *ramified covering domain* over Y. In case π is injective, we call it a *univalent domain*; in this case, X may be regarded as a domain of Y.

6.10 Normal Complex Spaces and Oka's Third Coherence Theorem

6.10.1 Normal Complex Space

Let $(X, \mathscr{O}_X)$ be a complex space. We denote by $\mathscr{M}_X$ the quotient sheaf of rings of $\mathscr{O}_{X,x}$ $(x \in X)$ divided by the non-zero divisors. The set of all of elements $f \in \mathscr{M}_{X,x}$, integral over $\mathscr{O}_{X,x}$, i.e., for which there are elements $a_1, \ldots, a_d \in \mathscr{O}_{X,x}$ with

(6.10.1)
$$f^d + \sum_{j=1}^{d} a_j f^{d-j} = 0,$$

is denoted by $\widehat{\mathscr{O}}_{X,x}$, which form a sheaf $\widehat{\mathscr{O}}_X$ of rings over X. It naturally follows that $\mathscr{O}_X \subset \widehat{\mathscr{O}}_X$. We say that $\mathscr{O}_{X,x}$ is *integrally closed* if $\mathscr{O}_{X,x} = \widehat{\mathscr{O}}_{X,x}$.

Definition 6.10.2 If $\mathscr{O}_{X,x}$ is integrally closed for a point $x \in X$, we call the point x a *normal point* and say that X is *normal* at x. We denote by $\check{\mathscr{N}}(X)$ the set of all *non*-normal points of X. If $\check{\mathscr{N}}(X) = \emptyset$, $(X, \mathscr{O}_X)$ or simply X is called a *normal complex space*.

Proposition 6.10.3 *A regular point $x \in X \setminus \Sigma(X)$ is a normal point; that is, $\check{\mathscr{N}}(X) \subset \Sigma(X)$.*

Proof By the assumption, $\mathscr{O}_{X,x} \cong \mathscr{O}_{n.0}$ $(\dim_x X = n)$, and $\mathscr{O}_{n,0}$ is a unique factorization domain (Theorem 2.2.12). In general, a unique factorization domain is integrally closed. For we take an element $f = g/h \in \widehat{\mathscr{O}}_{n,0} \setminus \mathscr{O}_{n,o}$, where g and h have no common divisor. By (6.10.1),

$$-g^d = (a_1 g^{d-1} + \cdots + a_{d-1} g h^{d-1} + a_d h^{d-1}) h.$$

Then g must contain an irreducible factor of h; this contradicts the choice of g and h. □

Proposition 6.10.4 *If X is reducible at a point $x \in X$, then there is an element $h \in \mathscr{M}_{X,x}$ such that*

$$h \notin \mathscr{O}_{X,x}, \quad h^2 - h \in \mathscr{O}_{X,x}.$$

In particular, x is not normal, and X is irreducible at a normal point.

Proof Since the problem is local, we may assume that X is an analytic subset of a neighborhood of the origin 0 of $\mathbf{C}^n$. Let $\underline{X}_0 = \bigcup_{j=1}^{l} \underline{X_{j}}_0$ $(l \geq 2)$ be the irreducible decomposition. We take

$$f \in \mathscr{I}\langle \underline{X_1}_0 \rangle \setminus \mathscr{I}\left\langle \bigcup_{j=2}^{l} \underline{X_j}_0 \right\rangle, \quad g \in \mathscr{I}\left\langle \bigcup_{j=2}^{l} \underline{X_j}_0 \right\rangle \setminus \mathscr{I}\langle \underline{X_1}_0 \rangle.$$

Since $f+g \notin \mathscr{I}\langle \underline{X_{j_0}} \rangle$ $(1 \le {}^\forall j \le l)$, $f+g$ is a non-zero divisor of $\mathscr{O}_{X,0}$. With setting $h = \frac{f}{f+g} \in \mathscr{M}_{X,0}$, we obtain

$$h^2 - h = \frac{f^2 - f(f+g)}{(f+g)^2} = \frac{-fg}{(f+g)^2} = 0 \in \mathscr{O}_{X,0}. \qquad \square$$

Definition 6.10.5 (*Weakly holomorphic function*) (i) Let $U \subset X$ be an open set. A *weakly holomorphic function* f on U is a function $f \in \mathscr{O}_X(U \setminus \Sigma(X))$ which is locally bounded at every point $x \in U \cap \Sigma(X)$; i.e., there is a neighborhood $V \subset U$of x such that the restriction $f|_{V\setminus\Sigma(X)}$ is bounded. We write $\widetilde{\mathscr{O}}_X(U)$ for the set of all weakly holomorphic functions on U.

(ii) The presheaf $\{\widetilde{\mathscr{O}}_X(U)\}_U$ induces a sheaf of rings over X denoted by $\widetilde{\mathscr{O}}_X$, which is called the sheaf of germs of weakly holomorphic functions over X.

Proposition 6.10.6 *We have*

$$\widehat{\mathscr{O}}_{X,x} \subset \widetilde{\mathscr{O}}_{X,x}, \qquad {}^\forall x \in X.$$

Proof Take an arbitrary element $\underline{f}_x = \underline{g}_x/\underline{h}_x \in \widehat{\mathscr{O}}_{X,x}$. There is a neighborhood U with the representatives,

$$f, g, h \in \mathscr{O}_X(U).$$

Because of (6.10.1) we may assume that f is bounded. Since $\underline{h}_x$ is a non-zero divisor, $S = \{h = 0\}$ is a thin analytic subset of U. Therefore, f is analytic in $(U \setminus \Sigma(X)) \setminus S$. By Riemann's Extension Theorem 2.3.4, f is a bounded analytic function in $U \setminus \Sigma(X)$. Thus, $\underline{f}_x \in \widetilde{\mathscr{O}}_{X,x}$. $\square$

In fact, we will prove $\widehat{\mathscr{O}}_X = \widetilde{\mathscr{O}}_X$ in the next subsection (Theorem 6.10.20).

Remark 6.10.7 A holomorphic function f on a complex space X is, by definition, such a function that it is continuous on X and at every point $a \in X$ there is a local chart neighborhood $(U_\alpha, \varphi_\alpha, A_\alpha)$ of a with $A_\alpha \subset \Omega_\alpha \subset \mathbf{C}^{n_\alpha}$, and then the restriction $f|_{U_\alpha}$ regarded as a function on A_α, must extend holomorphically over a neighborhood of a in Ω_α. This property cannot be checked only on the space X. On the other hand, the property of being weakly holomorphic can be checked only on X (cf. Remark 6.11.11), but in fact we confirm below by an example that these are different.

Example 6.10.8 Set $X = \{(z, w) \in \mathbf{C}^2;\ w^2 = z^3\}$. Then, $\Sigma(X) = \{0\}$. The function $f(z, w) = w/z$ is continuous on X, and $f(0, 0) = 0$; in fact, for $(z, w) \in X \setminus \{(0, 0)\}$ we have

$$(6.10.9) \qquad |f(z, w)| = \left|\frac{w}{z}\right| = \sqrt{|z|} \to 0 \quad ((z, w) \to 0)).$$

Since $f \in \mathscr{O}(X \setminus \Sigma(X))$, f is a weakly holomorphic function on X, but it cannot be expressed as a restriction of a holomorphic function in a neighborhood of $0 \in \mathbf{C}^2$. For, if it is so, there is a holomorphic function F in a polydisk neighborhood $\mathrm{P}\Delta \subset (\mathbf{C}^2)$ about 0, such that $F|_{X \cap \mathrm{P}\Delta} = f|_{X \cap \mathrm{P}\Delta}$. Since $F(0,0) = 0$, with $\mathrm{P}\Delta$ taken smaller if necessary, there is a constant $M > 0$ such that

$$|F(z,w)| \leq M \max\{|z|, |w|\}, \quad (z,w) \in \mathrm{P}\Delta. \tag{6.10.10}$$

Let $(z,w) \in X \cap \mathrm{P}\Delta$ be a point with $|z| < 1$, $|w| < 1$. Since $|w|^2 = |z|^3$, $|w| \leq |z|$, and from (6.10.9) and (6.10.10) it follows that

$$\sqrt{|z|} = |f(z,w)| = |F(z,w)| \leq M|z|.$$

This is impossible, but f satisfies the following monic polynomial equation:

$$f^2 - z = 0.$$

Therefore, $\underline{f}_0 \in \widehat{\mathscr{O}}_{X,0}$, and the origin 0 is not a normal point of X.

Example 6.10.11 We consider an analytic set X defined by the following equations in $(u,v,w,t) \in \mathbf{C}^4$:

$$ut = vw, \quad w^3 = t(t-w), \quad u^2 w = v(v-u).$$

It is easy to see by computations that the holomorphic map

$$\Phi : (x,y) \in \mathbf{C}^2 \to (x, xy, y(y-1), y^2(y-1)) = (u,v,w,t) \in \mathbf{C}^4$$

is proper, $\Phi(\mathbf{C}^2) \subset X$, and

$$\Phi^{-1}0 = \{(0,0), (0,1)\}.$$

We consider Φ^{-1} on $X \setminus \{0\}$. If $u \neq 0$, then $x = u$ and $y = v/u$; if $v \neq 0$, then $x \neq 0$ and $y \neq 0$. thus, the functions $x = u$ and $y = v/u$ are holomorphic. If $w \neq 0$, $x = u$ and $y = t/w$ are holomorphic; if $t \neq 0$, then $w \neq 0$, and $x = u$ and $y = t/w$ are holomorphic. Thus we see that

$$\Phi|_{\mathbf{C}^2 \setminus \{(0,0),(0,1)\}} : \mathbf{C}^2 \setminus \{(0,0),(0,1)\} \to X \setminus \{0\}$$

is biholomorphic. We consider a function $f = v/u$ restricted to X. For $x = y \neq 0$,

$$\lim_{x \to 0} \Phi(x,x) = 0, \quad \lim_{x \to 0} f(\Phi(x,x)) = 0.$$

For $y = 1 + x$ $(x \neq 0)$, $\Phi(x, 1+x) \in X \setminus \{0\}$ and $\lim_{x \to 0} \Phi(x, 1+x) = 0$. By computation we get

$$\lim_{x \to 0} f(\Phi(x, 1+x)) = 1.$$

Thus, f is not even continuous on X, but f satisfies the following monic polynomial equation:

$$f^2 - f - w = 0.$$

Therefore, f is a weakly holomorphic function on X, and $\underline{f}_0 \in \widehat{\mathscr{O}}_{X,0}$. The origin 0 is not a normal point of X.

6.10.2 Universal Denominators

We begin with a local lemma.

Lemma 6.10.12 *Let $U \subset \mathbf{C}^n$ be a neighborhood of 0, let $X \subset U$ be an analytic subset containing 0. Assume that $\underline{X}_0$ is irreducible.*

(i) *Every element $\underline{f}_0 \in \widetilde{\mathscr{O}}_{X,0}$ is integral over $\mathscr{O}_{X,0}$.*
(ii) *There are a standard polydisk neighborhood $\mathrm{P}\Delta_n$ of X at 0 and a holomorphic function,*

$$u \in \mathscr{O}(\mathrm{P}\Delta_n), \quad \underline{u|_{X_0}} \neq 0 \in \mathscr{O}_{X,0},$$

satisfying the following:

$$\underline{u|_{X_a}} \cdot \widetilde{\mathscr{O}}_{X,a} \subset \mathscr{O}_{X,a}, \quad {}^\forall a \in X \cap \mathrm{P}\Delta_n,$$

$$\underline{g}_a \in \widetilde{\mathscr{O}}_{X,a} \hookrightarrow \underline{u|_{X_a}} \cdot \underline{g}_a \in \mathscr{O}_{X,a},$$

$$\Sigma(X) \cap \mathrm{P}\Delta_n \subset \{u = 0\}.$$

Proof (i) (a) By the assumption, the ideal $\mathscr{I}\langle \underline{X}_0 \rangle \subset \mathscr{O}_{\mathbf{C}^n,0} = \mathscr{O}_{n,0}$ is prime. Let $\mathrm{P}\Delta_n = \mathrm{P}\Delta_p \times \mathrm{P}\Delta_{n-p} \subset U$ be the standard polydisk neighborhood with coordinate system $z = (z', z_{p+1}, z'') \in \mathrm{P}\Delta_p \times \mathrm{P}\Delta_{n-p}$ ($z' \in \mathbf{C}^p$, $z'' \in \mathbf{C}^{n-p-1}$) taken in Theorems 6.4.14 and 6.4.16. Moreover, we consider the Weierstrass polynomial $P_{p+1}(z', z_{p+1})$ in z_{p+1} and its discriminant $\delta(z')$. Henceforth, we deal with the problem in $\mathrm{P}\Delta_n$, so that we simply write X for $X \cap \mathrm{P}\Delta_n$.

The projection $\pi : (z', z_{p+1}, z'') \in X \to z' \in \mathrm{P}\Delta_p$ is finite surjective. We set

$$Z = \{\delta = 0\} \subset \mathrm{P}\Delta_p, \quad X' = X \setminus \pi^{-1}Z, \ \mathrm{P}\Delta'_p = \mathrm{P}\Delta_p \setminus Z.$$

Then Z and $\pi^{-1}Z$ are both thin analytic subsets, and the projection

$$\pi|_{X'} : X' \to \mathrm{P}\Delta'_p$$

is a d ($\in \mathbf{N}$)-sheeted unramified connected covering.

(b) We take any element $\underline{f}_0 \in \widetilde{\mathscr{O}}_{X,0}$. Taking the above $\mathrm{P}\Delta_n = \mathrm{P}\Delta_p \times \mathrm{P}\Delta_{n-p}$ smaller if necessary, and taking the restriction of f there, we may assume that f is a bounded holomorphic function in X'. For $z' \in \mathrm{P}\Delta'_p$ we define a polynomial in T by

$$\prod_{z \in \pi^{-1}z'} (T - f(z)) = T^d + c_1(z')T^{d-1} + \cdots + c_d(z'), \tag{6.10.13}$$

$$c_1(z') = -\sum_{z \in \pi^{-1}z'} f(z),$$

$$\vdots$$

$$c_d(z') = (-1)^d \prod_{z \in \pi^{-1}z'} f(z).$$

Since $c_\nu(z')$ is bounded holomorphic in $\mathrm{P}\Delta_p \backslash Z$, it is holomorphic in $\mathrm{P}\Delta_p$ (Riemann's Extension Theorem 2.3.4). Substituting $T = f(z)$ in (6.10.13) with $z = (z', z_{p+1}, z'') \in X$, we obtain

$$(f(z))^d + c_1(z')(f(z))^{d-1} + \cdots + c_d(z') = 0.$$

Therefore, $\underline{f}_0$ is integral over $\mathscr{O}_{p,0}$. Since $\mathscr{O}_{p,0} \overset{\pi^*}{\hookrightarrow} \mathscr{O}_{X,0}$, $\underline{f}_0$ is integral over $\mathscr{O}_{X,0}$.

(ii) We use (a) of (i) above. We mimic (6.3.7). Take arbitrarily a point $a = (a', a_{p+1}, a''') \in X$ and an element $\underline{g}_a \in \widetilde{\mathscr{O}}_{X,a}$. Set $\pi^{-1}a' = \{(a', b''_j)\}_{j=1}^{d'} \subset X$, where $a = (a', b''_1)$. Since $\pi : X \to \mathrm{P}\Delta_p$ is finite, by Proposition 6.1.8 with a polydisk neighborhood $V' \ni a'$ smaller if necessary, there are mutually disjoint polydisk neighborhoods, $b''_j \in V''_j \subset \mathrm{P}\Delta_{n-p}$ such that every restriction

$$\pi|_{X \cap (V' \times V''_j)} : X \cap (V' \times V''_j) \to V', \quad 1 \le j \le d',$$

is finite; with restricting this to X', we obtain a finitely sheeted unramified covering,

$$\pi|_{X' \cap (V' \times V''_j)} : X' \cap (V' \times V''_j) \to \mathrm{P}\Delta_p(a') \backslash Z, \quad 1 \le j \le d'.$$

(This covering may be disconnected, but it does not matter.) Put $U_1 = X \cap (V' \times V''_1)$. The function g may be assumed to be bounded on $X' \cap U_1$. Let $P_{p+1}(z', z_{p+1})$be the one taken in (a) of (i) above. With a new variable T and $z' \in V' \backslash Z$, we consider the following polynomial:

$$\sum_{z \in U_1 \cap \pi^{-1}z'} \frac{P_{p+1}(T)}{T - z_{p+1}} g(z) = \sum_{\nu=0}^{d-1} b_\nu(z') T^\nu. \tag{6.10.14}$$

Since $b_\nu(z')$ are bounded holomorphic in $V' \setminus Z$, they are holomorphic in V'. Let $z = (z', z_{p+1}, z'') \in U_1 \cap X'$ and put $T = z_{p+1}$ into (6.10.14). Then we have

$$P'_{p+1}(z_{p+1}) g(z) = \sum_{\nu=0}^{d-1} b_\nu(z') z_{p+1}^\nu \in \Gamma(U_1, \mathscr{O}_X). \tag{6.10.15}$$

Therefore, $u(z) = P'_{p+1}(z_{p+1})$ satisfies the required conditions. □

N.B. Because of (6.4.10) one may take $u(z) = \delta(z')$ as well in the last part of the proof above. The function u taken in (ii) is called a *universal denominator* of $\widetilde{\mathscr{O}}_X$ at 0 or in $\mathrm{P}\Delta_n \cap X$. For the later use in the proof of Oka's Normalization Theorem 6.10.35, the content of this lemma will be sufficient, since it is applied for each irreducible component of $\underline{X}_0$ after the irreducible decomposition.

Let X be a complex space.

Theorem 6.10.16 (Universal denominator) *For every point $a \in X$ there are a neighborhood U of a and a holomorphic function $u \in \mathscr{O}(U)$ such that:*

(i) *The analytic subset $\{u = 0\}$ is thin in U, and*

$$\Sigma(X) \cap U \subset \{u = 0\}. \tag{6.10.17}$$

(ii) *At any point $x \in U$,*

$$\underline{f}_0 \in \widetilde{\mathscr{O}}_{X,x} \overset{\underline{u}_x \cdot}{\hookrightarrow} \underline{u}_x \cdot \underline{f}_0 \in \underline{u}_x \cdot \widetilde{\mathscr{O}}_{X,x} \subset \mathscr{O}_{X,x}. \tag{6.10.18}$$

Proof Since the problem is local, we may assume that X is an analytic subset of an open set of $\mathbf{C}^n$, and $a = 0$. Let $\underline{X}_0 = \bigcup_{\alpha=1}^{l} \underline{X_{\alpha}}_0$ be the irreducible decomposition at 0. Let $\mathfrak{a}_\alpha = \mathscr{I}\langle \underline{X_{\alpha}}_0 \rangle$ be the geometric ideal of $\underline{X_{\alpha}}_0$. Take an element $\underline{v_{\alpha}}_0 \in \bigcap_{\beta \neq \alpha} \mathfrak{a}_\beta \setminus \mathfrak{a}_\alpha$ for each α. By Lemma 6.10.12 each X_α has a universal denominator u_α in a neighborhood U of 0. Set

$$u = \sum_{\alpha=1}^{l} v_\alpha u_\alpha. \tag{6.10.19}$$

For each X_α, $u|_{X_\alpha} = (v_\alpha u_\alpha)|_{X_\alpha} \not\equiv 0$. Therefore, $\{u = 0\}$ is a thin set. If $x \in X_\alpha \cap X_\beta$ for distinct α, β, then $u(x) = 0$. We obtain (6.10.17) and (6.10.18) from the construction and Lemma 6.10.12 (ii). □

Theorem 6.10.20 $\widetilde{\mathscr{O}}_X = \widehat{\mathscr{O}}_X$.

Proof We show that $\widetilde{\mathscr{O}}_{X,x} = \widehat{\mathscr{O}}_{X,x}$ for every $x \in X$. From Theorem 6.10.16 follows $\widetilde{\mathscr{O}}_{X,x} \subset \mathscr{M}_{X,x}$. By Proposition 6.10.6, it remains to show that every $f \in \widetilde{\mathscr{O}}_{X,x}$ is integral over $\mathscr{O}_{X,x}$. Let $\underline{X}_x = \bigcup_{\alpha=1}^{l} \underline{X_{\alpha}}_x$ be the irreducible decomposition. It follows from Lemma 6.10.12 (i) that for each X_α there is a monic polynomial $P_\alpha(f)$ with coefficients in $\mathscr{O}_{X,x}$ satisfying

$$P_\alpha(f) = 0 \in \mathscr{O}_{X_\alpha, x}.$$

Since $P(f) = \prod_\alpha P_\alpha(f)$ is monic and $P(f) = 0$ ($\in \mathscr{O}_{X,x}$), f is integral over $\mathscr{O}_{X.x}$. □

Lemma 6.10.21 *Let X be a complex space and let $a \in X$ be a point such that X is irreducible at a. Let f be a weakly holomorphic function in a neighborhood U of a. Then the limit*

$$\lim_{\substack{x \to a \\ x \in U \setminus \Sigma(X)}} f(x)$$

exists.

Proof Since the problem is local, we may assume that X is an analytic subset of a polydisk $\mathrm{P}\Delta$ of $\mathbf{C}^n$ with center 0, $a = 0$, and $U = \mathrm{P}\Delta$. By Theorem 6.4.16 there is a fundamental neighborhood system of $0 \in X$,

$$V_\nu \subset V, \nu = 1, 2, \ldots, \quad V_\nu \Supset V_{\nu+1}, \quad \bigcap_\nu V_\nu = \{0\},$$

such that $V_\nu \setminus \Sigma(X)$ is connected. By Theorem 6.10.20 $\underline{f}_0$ is integral over $\mathscr{O}_{X,0}$. Therefore the following monic equation holds:

$$f(x)^d + \sum_{j=1}^{d} a_j(x) f(x)^{d-j} = 0, \tag{6.10.22}$$
$$x \in X \setminus \Sigma(X), \quad a_j \in \mathscr{O}(X).$$

Assume that the assertion does not hold. Then there are sequences of points, $x_\nu, y_\nu \in V_\nu \setminus \Sigma(X)$, $\nu = 1, 2, \ldots$ such that

$$\lim_{\nu \to \infty} f(x_\nu) = \alpha, \quad \lim_{\nu \to \infty} f(y_\nu) = \beta, \quad \alpha \neq \beta.$$

Since $V_\nu \setminus \Sigma(X)$ is connected, there is a curve $\Gamma_\nu \subset V_\nu \setminus \Sigma(X)$ connecting x_ν and y_ν. Take an arbitrary closed disk $\overline{\Delta(\alpha; \rho)} \not\ni \beta$, and denote the boundary circle by C_ρ. For $\nu \gg 1$,

$$f(x_\nu) \in \Delta(\alpha; \rho), \quad f(y_\nu) \notin \overline{\Delta(\alpha; \rho)}.$$

Therefore there is a point $w_\nu \in \Gamma_\nu$ with $f(w_\nu) \in C_\rho$. After taking a subsequence $\{w_{\nu_\mu}\}_\mu$, the limit $\gamma_\rho = \lim_{\mu \to \infty} f(w_{\nu_\mu}) \in C_\rho$ exists. Since $\Gamma_{\nu_\mu} \to 0$, it follows from (6.10.22) that

$$\gamma_\rho^d + \sum_{j=1}^{d} a_j(0) \gamma_\rho^{d-j} = 0. \tag{6.10.23}$$

Varying $\rho > 0$, we obtain infinitely many roots of (6.10.23); this is absurd. □

We immediately have the following.

Proposition 6.10.24 *A weakly holomorphic function f on a complex space X has a unique continuous extension over $(X \setminus \Sigma(X)) \cup \{a \in \Sigma(X);\ \underline{X}_a$ is irreducible$\}$.*

6.10.3 Analyticity of Non-normal Points

Let X be a complex space. In the proof of Oka's Normalization Theorem 6.10.35, proved in the next subsection, it is an important point that $X \setminus \check{\mathscr{N}}(X)$ is an open set, i.e., $\check{\mathscr{N}}(X)$ is a closed set.

Theorem 6.10.25 *The non-normal point set $\check{\mathscr{N}}(X)$ of a complex space X is an analytic subset; in particular, $\check{\mathscr{N}}(X)$ is closed.*

Proof Take an arbitrary point $a \in X$. There are a neighborhood $U \ni a$ and a universal denominator $u \in \Gamma(U, \mathscr{O}_X)$. Since the problem is local, we let $X = U \subset \Omega$, where $\Omega \subset \mathbf{C}^n$ is an open subset.

Set $S = \{u = 0\}$. Then S is a thin set containing $\Sigma(X)$. Let $\mathscr{I} = \mathscr{I}\langle S \rangle$ be the geometric ideal of S in $\mathscr{O}_X$. By applying Rückert's Nullstellen Theorem 6.4.20, for S in Ω, and then by taking the restriction to X (i.e., modulo $\mathscr{I}\langle X \rangle$) we have

(6.10.26) $$\mathscr{I} = \sqrt{u \cdot \mathscr{O}_X}\,.$$

We denote the sheaf of endomorphisms of $\mathscr{I}$ over $\mathscr{O}_X$ by

(6.10.27) $$\mathscr{F} = \mathscr{H}om_{\mathscr{O}_X}(\mathscr{I}, \mathscr{I}).$$

It follows from Oka's Second Coherence Theorem 6.5.1 that $\mathscr{I}$ is coherent, and so is $\mathscr{F}$. For an arbitrary $\alpha \in \mathscr{F}_x$ ($x \in X$) we set

(6.10.28) $$g_\alpha = \frac{\alpha\left(\underline{u}_x\right)}{\underline{u}_x} \in \mathscr{M}_{X,x}.$$

For an element $h \in \mathscr{I}_x$,

$$g_\alpha \cdot h = \frac{\alpha\left(\underline{u}_x\right)}{\underline{u}_x} h = \frac{\alpha\left(\underline{u}_x h\right)}{\underline{u}_x} = \frac{\underline{u}_x}{\underline{u}_x} \alpha(h) = \alpha(h) \in \mathscr{I}_x.$$

Thus, we have an expression "$\alpha = g_\alpha$".

Now we show that g_α is uniquely determined. Suppose that "$\alpha = g$". For every $h \in \mathscr{I}_x$,

$$(g_\alpha - g) \cdot h = 0.$$

In particular, taking a non-zero divisor h of $\mathscr{O}_{X,x}$, we get $g_\alpha = g$. Thus, the homomorphism

$$\alpha \in \mathscr{F}_x \to g_\alpha \in \mathscr{M}_{X,x}$$

is injective.

Since $g_\alpha \cdot \mathscr{I}_x \subset \mathscr{I}_x$ and $\mathscr{I}_x$ is finitely generated, by Theorem 6.3.4, there is a monic polynomial $D(g_\alpha)$ in g_α with coefficients in $\mathscr{O}_{X,x}$ such that

$$D(g_\alpha) \cdot \mathscr{I}_x = 0.$$

Since $\mathscr{I}_x$ contains a non-zero divisor (e.g., $\underline{u}_x$), $D(g_\alpha) = 0$. Thus we see that $g_\alpha \in \widehat{\mathscr{O}}_{X,x}$. Hence we obtain an injective homomorphism

$$\alpha \in \mathscr{F} \hookrightarrow g_\alpha \in \widehat{\mathscr{O}}_X,$$

by which we regard $\mathscr{F} \subset \widehat{\mathscr{O}}_X$.

For an element $f \in \mathscr{O}_{X,x}$, we assign an element of $\mathscr{F}_x$ by multiplication endomorphism, $h \in \mathscr{I}_x \to f \cdot h \in \mathscr{I}_x$; thus we regard $\mathscr{O}_{X,x} \subset \mathscr{F}_x$. The inclusion relations,

$$\mathscr{O}_X \subset \mathscr{F} \subset \widehat{\mathscr{O}}_X, \tag{6.10.29}$$

hold. The following, together with the support Theorem 6.9.10, finishes the proof.

Claim $\check{\mathscr{N}}(X) = \{x \in X; \mathscr{O}_{X,x} \neq \mathscr{F}_x\} = \operatorname{Supp} \mathscr{F}/\mathscr{O}_X$.

$\because$) Let $x \in \operatorname{Supp} \mathscr{F}/\mathscr{O}_X$. We infer from (6.10.29) that $\mathscr{O}_{X,x} \neq \widehat{\mathscr{O}}_{X,x}$, so that $x \in \check{\mathscr{N}}(X)$.

Conversely, let $x \in \check{\mathscr{N}}(X)$. Let $\{v_1, \ldots, v_N\}$ be a finite generator system of $\mathscr{I}_x$. By Rückert Nullstellen Theorem 6.4.20, there is a number $d \in \mathbf{N}$ such that $v_j^d \in \underline{u}_x \cdot \mathscr{O}_{X,x}$, $1 \leq j \leq N$. Since every $f \in \mathscr{I}_x$ is written as $f = \sum_{j=1}^N c_j v_j$ ($c_j \in \mathscr{O}_{X,x}$), with sufficiently large $k \in \mathbf{N}$ we have

$$\mathscr{I}_x^k \subset \underline{u}_x \cdot \mathscr{O}_{X,x}.$$

Therefore we obtain the following inclusion relations:

$$\mathscr{I}_x^k \cdot \widehat{\mathscr{O}}_{X,x} \subset \underline{u}_x \cdot \widehat{\mathscr{O}}_{X,x} \subset \mathscr{O}_{X,x}.$$

Because of $\widehat{\mathscr{O}}_{X,x} \supsetneqq \mathscr{O}_{X,x}$, $k \in \mathbf{N}$ may be chosen so that

$$\mathscr{I}_x^k \cdot \widehat{\mathscr{O}}_{X,x} \subset \mathscr{O}_{X,x}, \quad \mathscr{I}_x^{k-1} \cdot \widehat{\mathscr{O}}_{X,x} \not\subset \mathscr{O}_{X,x}. \tag{6.10.30}$$

We take an element $\underline{w}_x \in \mathscr{I}_x^{k-1} \cdot \widehat{\mathscr{O}}_{X,x} \setminus \mathscr{O}_{X,x}$. From (6.10.30) it follows that

$$\underline{w}_x \cdot \mathscr{I}_x \subset \mathscr{O}_{X,x}.$$

Suppose that $\underline{f}_x \in \mathscr{I}_x$. Then, $\underline{w}_x \cdot \underline{f}_x \in \mathscr{O}_{X,x}$. We consider the holomorphic function $w(z)f(z)$ in a small neighborhood V of x. For any point $y \in V \cap S$, we take $z \in V \setminus S$ ($\subset V \setminus \Sigma(X)$), and let $z \to y$: Then, $f(z) \to 0$. On the other hand, since $w(z)$ is bounded, $w(y)f(y) = 0$. Therefore, we infer that $\underline{wf}_x \in \mathscr{I}_x$. Setting

$$\alpha : \underline{f}_x \in \mathscr{I}_x \to \underline{w}_x \cdot \underline{f}_x \in \mathscr{I}_x,$$

we have that $\alpha \in \mathscr{F}_x$. Since $\underline{w}_x \notin \mathscr{O}_{X,x}$, $\mathscr{F}_x \neq \mathscr{O}_{X,x}$; that is, $x \in \operatorname{Supp} \mathscr{F}/\mathscr{O}_X$. □

6.10.4 Oka's Normalization and Third Coherence Theorem

We begin with the definition of the normalization of a complex space X.

Definition 6.10.31 Let $\widehat{X}$ be a normal complex space, and let $\pi : \widehat{X} \to X$ be a finite holomorphic map such that the restriction

$$\pi|_{\widehat{X} \setminus \pi^{-1}\Sigma(X)} : \widehat{X} \setminus \pi^{-1}\Sigma(X) \to X \setminus \Sigma(X) \tag{6.10.32}$$

is biholomorphic. Then, the triple $(\widehat{X}, \pi, X)$, or $\pi : \widehat{X} \to X$, or more simply, $\widehat{X}$ is called the *normalization* of X.

If $\pi : \widehat{X} \to X$ is the normalization of X, it follows from (6.10.32) that $\pi_* \widetilde{\mathscr{O}}_{\widehat{X}} = \widetilde{\mathscr{O}}_X$, so that from Theorem 6.10.20 we obtain the following sheaf isomorphism:

$$\pi_* \mathscr{O}_{\widehat{X}} = \pi_* \widetilde{\mathscr{O}}_{\widehat{X}} \cong \widetilde{\mathscr{O}}_X = \widehat{\mathscr{O}}_X. \tag{6.10.33}$$

Theorem 6.10.34 (Uniqueness of normalization) *The normalization of a complex space, if it exists, is unique up to biholomorphisms.*

Proof Let X be a complex space. Suppose that (Y, π, X) and (Z, η, X) are the normalizations of X. By the definition, $Y \setminus \pi^{-1}\Sigma(X)$, $X \setminus \Sigma(X)$ and $Z \setminus \eta^{-1}\Sigma(X)$ are mutually biholomorphic. Therefore there is a biholomorphic map

$$\varphi : Y \setminus \pi^{-1}\Sigma(X) \to Z \setminus \eta^{-1}\Sigma(X),$$
$$\pi|_{Y \setminus \pi^{-1}\Sigma(X)} = \eta \circ \varphi|_{Y \setminus \pi^{-1}\Sigma(X)}.$$

Taking an arbitrary point $x \in X$, we set $\pi^{-1}x = \{y_j\}_{j=1}^M$ and $\eta^{-1}x = \{z_k\}_{k=1}^N$. By Proposition 6.1.8, there are a neighborhood $U \ni x$, and mutually disjoint neighborhoods $V_j \ni y_j$, $W_k \ni z_k$ such that $\pi^{-1}U = \bigcup_j V_j$, $\eta^{-1}U = \bigcup_k W_k$ and

$$\pi|_{V_j} : V_j \to U, \qquad \eta|_{W_k} : W_k \to U$$

is a finite surjection. Since $\underline{Y}_{y_j}$, $\underline{Z}_{z_k}$ is irreducible, $V_j \setminus \pi^{-1}\Sigma(X)$ and $W_k \setminus \eta^{-1}\Sigma(X)$ may be assumed to be connected. It follows that

$$\varphi|_{\pi^{-1}(U\setminus\Sigma(X))} : \pi^{-1}(U\setminus\Sigma(X)) \to \eta^{-1}(U\setminus\Sigma(X))$$

is biholomorphic. Therefore, $M = N$ and after changing indices,

$$\varphi|_{V_j\setminus\pi^{-1}\Sigma(X)} : V_j\setminus\pi^{-1}\Sigma(X) \to W_j\setminus\eta^{-1}\Sigma(X), \quad 1 \le j \le M$$

are biholomorphic. We may assume that V_j and W_j are contained relatively compact in some chart neighborhoods, $\varphi|_{V_j\setminus\pi^{-1}\Sigma(X)}$, $\varphi^{-1}|_{W_j\setminus\eta^{-1}\Sigma X}$ are holomorphic maps represented by bounded holomorphic functions. Since V_j and W_j are normal, it follows from Theorem 6.10.20 that $\varphi|_{V_j\setminus\pi^{-1}\Sigma(X)}$ extends holomorphically over V_j, and $\varphi^{-1}|_{W_j\setminus\eta^{-1}\Sigma(X)}$ over W_j.

Since $x \in X$ is arbitrary, φ extends biholomorphically to $\varphi : Y \to Z$. □

Theorem 6.10.35 (Oka's Normalization and Third Coherence Theorem, Oka [62] VIII)

(i) *Every complex space has the normalization.*
(ii) *The sheaf* $\widehat{\mathscr{O}}_X$ $(= \widetilde{\mathscr{O}}_X)$ *is coherent over* $\mathscr{O}_X$.

Proof Suppose that (i) is shown, Since the normalization $\pi : \widehat{X} \to X$ is a finite holomorphic map, the direct image sheaf $\pi_*\mathscr{O}_{\widehat{X}} = \widehat{\mathscr{O}}_X$ is coherent over $\mathscr{O}_X$ (Theorem 6.7.16). From (6.10.33), follows the coherence of $\widehat{\mathscr{O}}_X$ over $\mathscr{O}_X$.

Below, we present the proof of (i). If there exists a neighborhood of each point $a \in X$ carrying the normalization, by Theorem 6.10.34 they patch together biholomorphically, and the normalization of X is obtained. Therefore, we may assume that X is an analytic subset of an open set Ω of $\mathbf{C}^n$, and $a = 0 \in X$. Let $\underline{X}_0 = \bigcup_\alpha \underline{X_{\alpha 0}}$ be the irreducible decomposition. Taking Ω smaller if necessary, analytic subsets $X_\alpha \subset \Omega$ represent the germs $\underline{X_{\alpha 0}}$. Each $X_\alpha\setminus\Sigma(X)$ is a connected complex manifold. If each X_α carries the normalization $\widehat{X}_\alpha$, then the disjoint union $\widehat{X} = \bigsqcup_\alpha \widehat{X}_\alpha$ gives rise to the normalization of X. Therefore we may assume that $\underline{X}_0$ is irreducible.

By Lemma 6.10.12 there are a standard polydisk neighborhood $\mathrm{P}\Delta_n$ of X at 0 such that $X \cap \mathrm{P}\Delta\setminus\Sigma(X)$ is connected, and a universal denominator u of $\mathrm{P}\Delta_n \cap X$, so that the following holds:

$$\widehat{\mathscr{O}}_{X,0} \hookrightarrow \underline{u}_0 \cdot \widehat{\mathscr{O}}_{X,0} \subset \mathscr{O}_{X,0}. \tag{6.10.36}$$

Because of the Noetherian property of $\mathscr{O}_{X,0}$, the ideal $\underline{u}_0 \cdot \widehat{\mathscr{O}}_{X,0}$ of $\mathscr{O}_{X,0}$ carries a finite generator system $\{\underline{u}_1 \cdot \underline{f_{\nu 0}}\}_{\nu=1}^l$ with $\underline{f_{\nu 0}} \in \widehat{\mathscr{O}}_{X,0}$. With $\mathrm{P}\Delta_n$ taken smaller if necessary, f_ν are bounded holomorphic functions on $X \cap \mathrm{P}\Delta_n\setminus\Sigma(X)$, and there are holomorphic functions g_ν $(1 \le \nu \le l)$ on $X \cap \mathrm{P}\Delta_n$ such that

$$u(x) f_\nu(x) = g_\nu(x), \quad x \in X \cap \mathrm{P}\Delta_n\setminus\Sigma(X). \tag{6.10.37}$$

Henceforth, since we consider X restricted to $X \cap \mathrm{P}\Delta_n$, we write X for $X \cap \mathrm{P}\Delta_n$. We set $S = \{u = 0\} \subsetneqq X$ and a holomorphic map

$$F : x \in X\backslash\Sigma(X) \to (x, f_1(x), \ldots, f_l(x)) = (x, w_1, \ldots, w_l) \in X \times \mathbf{C}^l.$$

It follows from Lemma 6.10.12 that $\Sigma(X) \subset S$. We consider an analytic subset

$$Z = \{(x, w_1 \ldots, w_l) \in X \times \mathbf{C}^l;\ u(x)w_\nu = g_\nu(x), 1 \le \nu \le l\} \subset X \times \mathbf{C}^l,$$
$$Z' = S \times \mathbf{C}^l.$$

Note that $g_\nu = 0$ on S by (6.10.37). Therefore,

$$F : X\backslash S \to Z\backslash Z'$$

is biholomorphic.

Since $X\backslash S$ is connected, so is $Z\backslash Z'$, which is non-singular. There is a connected component Y of $Z\backslash\Sigma(Z)$ with $Y \supset Z\backslash Z'$. The closure $\bar{Y}$ is an irreducible component of Z, and $Z\backslash Z' = Y \setminus S$ is dense in Y. Thus we infer that $\overline{F(X\backslash S)} = \bar{Y}$. Let

$$\eta : \bar{Y}(\subset X \times \mathbf{C}^l) \longrightarrow X$$

denote the projection.

Claim *$\bar{Y}$ is normal at* 0 $(\in X \times \mathrm{P}\Delta_l)$.

$\because$) This is an assertion at one point 0. Take an arbitrary element $\underline{g}_0 \in \widehat{O}_{\bar{Y},0}$. Taking a sufficiently small neighborhood $V = V_1 \times V_2 \subset X \times \mathbf{C}^l$ of 0, $g|_{Y\cap V}$ is a bounded holomorphic function. Since $g \circ F$ is bounded in $X\backslash\pi^{-1}S$, $\underline{g \circ F}_0 \in \widehat{\mathscr{O}}_{X,0}$, and then by (6.10.36)

$$\underline{g \circ F}_0 = \sum_{\nu=1}^{l} a_\nu \cdot \underline{f_\nu}_0, \quad a_\nu \in \mathscr{O}_{X,0}.$$

This induces the following on $\bar{Y}$:

$$\underline{g}_0 = \sum_{\nu=1}^{l} \eta^* a_\nu \cdot \underline{w_\nu}_0.$$

Therefore, $\underline{g}_0 \in \mathscr{O}_{\bar{Y},0}$. △

By Theorem 6.10.25, the set of normal points is an open set; thus $\bar{Y}$ is normal in a neighborhood W of 0. Since $\underline{X}_0$ is irreducible and $\eta : \bar{Y} \to X$ is a finite surjection, Theorem 6.7.13 implies that $\eta(W)$ contains a neighborhood ω of $0 \in X$. With ω taken smaller if necessary, we have $\eta^{-1}\omega \subset W$. Thus the normalization

$$\eta|_{\eta^{-1}\omega} : \eta^{-1}\omega \to \omega$$

is locally constructed. □

Proposition 6.10.38 *Let $\pi : \widehat{X} \to X$ be the normalization. If X is irreducible at every point $x \in \Sigma(X)$, then π is a homeomorphism.*

Proof The restriction of π, $\pi_0 = \pi|_{\widehat{X}\setminus\pi^{-1}\Sigma(X)} : \widehat{X}\setminus\pi^{-1}\Sigma(X) \to X\setminus\Sigma(X)$ is biholomorphic. The inverse π_0^{-1} is continuously extended at all $x \in \Sigma(X)$ by Proposition 6.10.24. Thus, π^{-1} gives rise to a continuous map, and hence π is a homeomorphism. □

Remark 6.10.39 Theorem 6.10.35 was first proved by Oka VIII. In Oka's proof, the order of (i) and (ii) above was opposite: If Theorem 6.10.35 (ii) is shown, the support Theorem 6.9.10 implies that $\check{\mathscr{N}}(X)$ is an analytic subset and in particular, a closed subset. As seen above, the proof of the existence of the normalization is essentially due to the closedness of $\check{\mathscr{N}}(X)$; thus, in Theorem 6.10.35, (ii) implies (i).

6.11 Singularities of Normal Complex Spaces

The aim of this section is to prove that the singularity set of a normal complex space is always of codimension ≥ 2.

6.11.1 Rank of Maximal Ideals

Let X be a complex space, and let $x \in X$. We call the smallest number of generators of a module $\mathscr{F}_x$ over $\mathscr{O}_{X,x}$ the *rank* of $\mathscr{F}_x$, denoted by

$$\mathrm{rk}_{\mathscr{O}_{X,x}} \mathscr{F}_x = \mathrm{rk}\, \mathscr{F}_x.$$

Proposition 6.11.1 *Let $\mathfrak{m}_{X,x} (\subset \mathscr{O}_{X,x})$ be the maximal ideal at $x \in X$. Then we have*

$$\mathrm{rk}\, \mathfrak{m}_{X,x} \geq \dim_x X.$$

Here, the equality holds only if $x \notin \Sigma(X)$.

Proof Since the problem is local, X may be assumed to be an analytic subset of a neighborhood U of $0 \in \mathbf{C}^n$ with $x = 0 \in X$ and $\dim_0 X = p$.

If $0 \notin \Sigma(X)$, then there is a suitable local coordinate system $(z_1, \ldots, z_n)$ about 0 such that

$$X = \{z_{p+1} = \cdots = z_n = 0\},$$

$$\mathfrak{m}_{X,0} = \sum_{j=1}^{p} \mathscr{O}_{X,0} z_j.$$

Therefore, $\operatorname{rk} \mathfrak{m}_{X,x} = p$.

Conversely, we assume that $\operatorname{rk} \mathfrak{m}_{X,0} = d$. Let $g_1, \ldots, g_d \in \mathfrak{m}_{X,0}$ be the generators. We consider the geometric ideal sheaf $\mathscr{I}\langle X \rangle$ of X. Since $\mathfrak{m}_{\mathbf{C}^n,0} = \mathfrak{m}_{X,0} + \mathscr{I}\langle X \rangle_0$, there are elements $f_j \in \mathscr{I}\langle X \rangle_0$ $(1 \le j \le n)$ satisfying

$$z_j = \sum_{k=1}^{d} c_{jk} g_k + f_j, \quad c_{jk} \in \mathscr{O}_{n,0}. \tag{6.11.2}$$

We denote by

$$\operatorname{rk} \langle df_1(0), \ldots, df_n(0) \rangle$$

the rank as vectors of differential 1-forms at 0. It follows from (6.11.2) that

$$dz_j = \sum_{k=1}^{d} c_{jk}(0) dg_k(0) + df_j(0), \quad 1 \le j \le n. \tag{6.11.3}$$

Hence, $\operatorname{rk} \langle df_1(0), \ldots, df_n(0) \rangle \ge n - d$. After reordering, we have that $\operatorname{rk} \langle df_1(0), \ldots, df_{n-d}(0) \rangle = n - d$. Then,

$$Y = \{f_1 = \cdots = f_{n-d} = 0\}$$

is a d-dimensional complex submanifold. Since $\underline{X}_0 \subset \underline{Y}_0$,

$$d = \operatorname{rk} \mathfrak{m}_{Y,0} = \dim_0 Y \ge \dim_0 X = p.$$

If the equality $d = p$ holds here, then $\underline{X}_0 = \underline{Y}_0$, and so $0 \notin \Sigma(X)$. □

Corollary 6.11.4 *Let $Y \subset X$ be an analytic subset. Suppose that X is pure p-dimensional and* $\operatorname{codim}_X Y \ge q$. *Then the following hold:*

$$\begin{aligned} &\operatorname{rk}_{\mathscr{O}_{X,x}} \mathscr{I}\langle Y \rangle_x \ge q, \quad {}^{\forall} x \in Y, \\ &\operatorname{rk}_{\mathscr{O}_{X,x}} \mathscr{I}\langle Y \rangle_x \ge q + 1, \quad {}^{\forall} x \in \Sigma(X) \cap Y \setminus \Sigma(Y). \end{aligned}$$

Proof Since the problem is local about $x \in Y$, X and Y may be assumed to be analytic subsets of a neighborhood of $x = 0 \in \mathbf{C}^n$. With $s = \operatorname{rk}_{\mathscr{O}_{X,0}} \mathscr{I}\langle Y \rangle_0$, we take the generators $\underline{g_1}_0, \ldots, \underline{g_s}_0 \in \mathscr{I}\langle Y \rangle_0$ over $\mathscr{O}_{X,0}$. Since $\mathscr{I}\langle Y \rangle$ is coherent by Oka's Second Coherence Theorem 6.5.1, there is a neighborhood $U \ni 0$ such that at every $a \in U$, $\underline{g_1}_a, \ldots, \underline{g_s}_a$ generate $\mathscr{I}\langle Y \rangle_a$. We take a point $a \in U \cap Y \setminus \Sigma(Y)$ and put $\dim_a Y = r$. Then there is a local coordinate system $(z_1, \ldots, z_n)$ about a such that with $a = (a_j)_{1 \le j \le n}$,

$$Y = \{z_{r+1} - a_{r+1} = \cdots = z_n - a_n = 0\}$$

in a neighborhood of a. Note that $\mathfrak{m}_{Y,a} = \mathfrak{m}_{X,a}/\mathscr{I}\langle Y\rangle_0$ is generated by $\underline{z_1 - a_1}_a, \ldots, \underline{z_r - a_r}_a$. Therefore, $\mathfrak{m}_{X,a}$ is generated by

$$\underline{z_1 - a_1}_a, \ldots, \underline{z_r - a_r}_a, \underline{g_1}_a, \ldots, \underline{g_s}_a.$$

By Proposition 6.11.1, $r + s \geq p$. Hence,

$$s \geq p - r = \operatorname{codim}_X Y \geq q.$$

Here, if $a \in \Sigma(X)$, the equality cannot hold, so that $s \geq q + 1$. □

6.11.2 *Higher Codimension of the Singularity Sets of Normal Complex Spaces*

Theorem 6.11.5 *If X is a normal complex space, then* codim $\Sigma(X) \geq 2$.

Proof We set $S = \Sigma(X)$ and assume that X is normal and $\operatorname{codim}_X S = 1$. We are going to deduce a contradiction.

There is a point $x \in S$ with

$$\dim_x X = p, \quad \dim_x S = p - 1.$$

Since X is normal, X is irreducible at x, and hence pure dimensional in a neighborhood U of x. After moving x in $U \cap S$, we may assume that $S \cap U$ is also pure dimensional. By the coherence of $\mathscr{I}\langle S\rangle$ (Oka's Second Coherence Theorem 6.5.1), we may assume that there is a finite generator system $\{f_j\}_{j=1}^{l} \subset \Gamma(U, \mathscr{I}\langle S\rangle)$ of $\mathscr{I}\langle S\rangle$ on U with $\underline{f_j}_x \neq 0$ such that

$$S = \bigcap_{j=1}^{l} \{f_j = 0\} \subset \bigcup_{j=1}^{l} \{f_j = 0\}.$$

Both the sides above are of $p - 1$ dimension, and hence there are a point $a \in S \setminus \Sigma(S)$ and its neighborhood V with $V \cap \Sigma(S) = \emptyset$ such that

$$S \cap V = \{f_j = 0\} \cap V, \quad 1 \leq j \leq l.$$

In the sequel, we consider inside V, so that taking restrictions already to V we abbreviate the notation "$\cap V$".

Since $\operatorname{codim}_X S = 1$ and $a \in S \setminus \Sigma(S)$, we infer from Corollary 6.11.4 that any generator system of $\mathscr{I}\langle S\rangle_a$ must consist of at least two elements. Let l be the least of such numbers of elements in the generator systems; Since $l \geq 2$, with f_1 and f_2 we set

$$h = \frac{f_2}{f_1} \in \Gamma(X, \mathscr{M}_{X,a}).$$

By the construction we have

$$\underline{h}_a \notin \mathscr{O}_{X,a}.$$

For, if $\underline{h}_a \in \mathscr{O}_{X,a}$, then $\underline{f_2}_a \in \underline{f_1}_a \cdot \mathscr{O}_{X,a}$, so that the number l of elements in the generator system can be made strictly smaller; this contradicts the minimality of l.

The function h however is holomorphic in $X \backslash S$. We may take a sequence $a_\nu \in X \backslash S$, $\nu = 1, 2, \ldots$, of points with $a_\nu \to a$ such that $\{h(a_\nu)\}_{\nu=1}^{\infty}$ is bounded. For, if this is unbounded, after taking a subsequence we exchange f_1 and f_2; then we have the required property.

Since $S = \{f_1 = 0\}$, by Rückert Nullstellen Theorem 6.4.20 there is a number $m \in \mathbf{N}$ such that

$$(\mathscr{I}\langle S \rangle_a)^m \subset \underline{f_1}_a \cdot \mathscr{O}_{X,a}.$$

Therefore,

$$\underline{h}_a \cdot (\mathscr{I}\langle S \rangle_a)^m \subset \underline{f_2}_a \cdot \mathscr{O}_{X,a} \subset \mathscr{O}_{X,a}.$$

We take the number m as the minimum one for which

$$\underline{h}_a \cdot (\mathscr{I}\langle S \rangle_a)^m \subset \mathscr{O}_{X,a}$$

holds. Then, there is a multi-index $\alpha = (\alpha_1, \ldots, \alpha_l) \in (\mathbf{Z}^+)^l$ satisfying the following:

(6.11.6)
$$\begin{aligned} |\alpha| &= m - 1, \\ \underline{u}_a &:= \underline{h}_a \cdot \underline{f^\alpha}_a \notin \mathscr{O}_{X,a}, \\ \underline{u}_a \cdot \underline{f_j}_a &\in \mathscr{O}_{X,a}, \quad 1 \leq {}^\forall j \leq l. \end{aligned}$$

Here, we put $f^\alpha = f_1^{\alpha_1} \cdots f_l^{\alpha_l}$. By the choice of $\{a_\nu\}$ we have

$$u(a_\nu) \cdot f_j(a_\nu) \to 0 \ (\nu \to \infty),$$

so that $u \cdot f_j$ vanishes at a. It follows from the choice of u and f_j that

$$a \in \{u \cdot f_j = 0\} \subset \bigcup_{j=1}^{l} \{f_j = 0\} = S.$$

Because of $\dim \{u \cdot f_j = 0\} = \dim S = p - 1$, we see that

$$\underline{u}_a \cdot \underline{f_j}_a \in \mathscr{I}\langle S \rangle_a.$$

Since $\{\underline{f_j}_a\}_{j=1}^l$ is a generator system of $\mathscr{I}\langle S\rangle_a$,

$$\underline{u}_a \cdot \mathscr{I}\langle S\rangle_a \subset \mathscr{I}\langle S\rangle_a.$$

Since $\mathscr{I}\langle S\rangle_a$ is finitely generated over $\mathscr{O}_{X,a}$, it follows from Theorem 6.3.4 that $\underline{u}_a$ is integral over $\mathscr{O}_{X,a}$. That is, $\underline{u}_a \in \widehat{\mathscr{O}}_{X,a}$. Now, $\mathscr{O}_{X,a} = \widehat{\mathscr{O}}_{X,a}$ implies $\underline{u}_a \in \mathscr{O}_{X,a}$; this contradicts (6.11.6). □

Corollary 6.11.7 *Let X be a normal complex space. If* $\dim X = 1$*, then X is non-singular. If* $\dim X = 2$*, any singular point of X is isolated.*

Remark 6.11.8 Let X be a complex space. If there is a non-singular complex space $\tilde{X}$ with a proper holomorphic map $\pi : \tilde{X} \to X$ satisfying that the restriction

$$\pi|_{\tilde{X}\setminus\pi^{-1}\Sigma(X)} : \tilde{X}\setminus\pi^{-1}\Sigma(X) \to X\setminus\Sigma(X)$$

is biholomorphic, the triple $(\tilde{X}, \pi, X)$ is called a *resolution of singularities* or *desingularization* of X. If $\dim X = 1$, the normalization $(\tilde{X}, \pi, X)$ of X (Theorem 6.10.35) gives rise to a desingularization of X by Corollary 6.11.7. Hironaka's resolution of singularities [30] guarantees the existence of a resolution of singularities of any complex space, but the details exceed the level of this book.

Theorem 6.11.9 *On a normal complex space X, a function, which is holomorphic in the regular point set $X\setminus\Sigma(X)$, is holomorphic on X.*

Proof Let f be a holomorphic function on $X\setminus\Sigma(X)$. It is sufficient to show that f extends holomorphically on a neighborhood $U(\subset X)$ of every point $x \in \Sigma(X)$. By the assumption for X to be normal, it suffices to show that f is locally bounded on $U\setminus\Sigma(X)$.

Taking a local chart about x, we may assume that X is an analytic subset of an open set of $\mathbf{C}^n$, and $x = 0$ by translation. It is noted that $\underline{X}_0$ is irreducible, and of pure dimension. Take a standard polydisk neighborhood $\mathrm{P}\Delta_n = \mathrm{P}\Delta_p \times \mathrm{P}\Delta_{n-p}$ of X about 0 ($p = \dim_0 X$). Then the projection

$$\pi : (z', z'') \in X \cap (\mathrm{P}\Delta_p \times \mathrm{P}\Delta_{n-p}) \to z' \in \mathrm{P}\Delta_p$$

is finite and surjective. From now on we consider only the restriction $X \cap \mathrm{P}\Delta_n$, and so we write X for it. There is an analytic subset $R \subsetneqq \mathrm{P}\Delta_p$ such that

$$\pi|_{X\setminus\pi^{-1}R} : (z', z'') \in X\setminus\pi^{-1}R \to z' \in \mathrm{P}\Delta_p\setminus R$$

is a d-sheeted connected unramified covering (Theorem 6.4.16). For $z' \in \mathrm{P}\Delta_p\setminus R$, we set $\pi^{-1}z' = \{z^{(1)}, \ldots, z^{(d)}\}$, and denote the elementary symmetric polynomials of degree ν $(1 \le \nu \le d)$ in $f(z^{(1)}), \ldots, f(z^{(d)})$ by

$$c_\nu(z') = (-1)^\nu \sum_{1\le i_1<\cdots<i_\nu\le d} f\left(z^{(i_1)}\right)\cdots f\left(z^{(i_\nu)}\right).$$

Then the following holds:

$$(6.11.10)\qquad \begin{aligned}&(f(z))^d + \sum_{\nu=1}^{d} c_\nu(z')(f(z))^{d-\nu} = 0,\\ &z \in X\backslash\pi^{-1}R,\ z' = \pi(z) \in \mathrm{P}\Delta_p\backslash R.\end{aligned}$$

The image $\pi(\Sigma(X))$ is an analytic subset of $\mathrm{P}\Delta_p$, the dimension of which is the same as $\Sigma(X)$ (Theorem 6.7.2). Since $f(z)$ is holomorphic in $z \in X\backslash\Sigma(X)$, each $c_\nu(z')$ extends holomorphically in $z' \in \mathrm{P}\Delta_p\backslash\pi(\Sigma(X))$. By Theorem 6.11.5,

$$\mathrm{codim}_{\mathrm{P}\Delta_p}\pi(\Sigma(X)) = \mathrm{codim}_X \Sigma(X) \geq 2,$$

so that Hartogs' Extension Theorem 6.5.15 implies $c_\nu \in \mathscr{O}(\mathrm{P}\Delta_p)$. With $\mathrm{P}\Delta_n$ taken smaller if necessary, f is bounded by (6.11.10). □

Remark 6.11.11 Let $(X, \mathscr{O}_X)$ be a complex space in general. As mentioned in Remark 6.10.7, it requires information outside the space X to check if a function on X is holomorphic. But, if X is normal, Theorem 6.11.9 implies that it is sufficient to check if f is holomorphic in $X\backslash\Sigma(X)$. Since $X\backslash\Sigma(X)$ is a complex manifold by itself (possibly disconnected), f can be determined to be holomorphic or not only on $X\backslash\Sigma(X)$ without any information outside X. Oka's Normalization Theorem 6.10.35 is significant in the sense that it guarantees the unique existence of such complex space for every complex space without any condition.

6.12 Stein Spaces and Oka–Cartan Fundamental Theorem

It is easy to extend the notion of Stein manifolds to the case of complex spaces.

Definition 6.12.1 A connected complex space X is called a *Stein space*, if X satisfies the second countability axiom and the following, so-called Stein conditions (i)–(iii):

(i) (Holomorphic separation) For two distinct points $x, y \in X$, there is a holomorphic function $f \in \Gamma(X, \mathscr{O}_X)$ with $f(x) \neq f(y)$.
(ii) (Local chart) For every point $x \in X$ there is a local chart neighborhood (U, φ, A) about x such that $\varphi = (\varphi_j)_{1\leq j\leq n}$ is given by $\varphi_j \in \Gamma(X, \mathscr{O}_X)$.
(iii) (Holomorphic convexity) For every compact subset $K \Subset X$, the holomorphically convex hull of K,

$$\hat{K}_X = \{x \in X;\, |f(x)| \leq \max_K |f|,\ {}^\forall f \in \Gamma(X, \mathscr{O}_X)\}$$

is compact, too.

The theory developed in Chap. 4 remains valid for complex spaces; in particular, as Theorem 4.5.8, we have:

Theorem 6.12.2 (Oka–Cartan Fundamental Theorem) *Let X be a Stein space. For a coherent sheaf $\mathscr{F}$ over X,*

$$H^q(X, \mathscr{F}) = 0, \qquad q \geq 1.$$

Remark 6.12.3 In the proof of the above theorem for $q = 1$, one needs an approximation theorem of Runge–Oka type. See Sect. 8.1.3 for the details of the convergence of holomorphic functions on a general complex space.

As for Stein's Definition 6.12.1, the holomorphic convexity (iii) might be inevitable, but it is natural to consider if condition (i) or (ii) could be simplified, and modified to one more easily checked. The following *weak separation* condition is due to H. Grauert, who called it "*K-complete*".

Definition 6.12.4 (*Weak separation*) Let X be a complex space in general (without the second countability condition). We say that X satisfies the *weak separation condition* if for every point $x \in X$ there are finitely many holomorphic functions $f_j \in \Gamma(X, \mathscr{O}_X)$ with $f_j(x) = 0$, $1 \leq j \leq N$, such that x is isolated in the analytic subset $\bigcap_{j=1}^{N} \{f_j = 0\}$.

It is clear that a Stein space satisfies the weak separation condition. For the inverse, the following is known; the proof exceeds the level of this book and so is omitted.

Theorem 6.12.5 (Grauert [21]) *If a complex space X satisfies the conditions of holomorphic convexity and weak separation, then X is a Stein space, satisfying the second countability axiom.*

In Grauert's Theorem, it is noticed that even the second countability axiom is not assumed. By making use of this theorem, we easily see that if there is a finite holomorphic map $Y \to X$ between complex spaces and if X is Stein, then Y is Stein. In particular, applying this fact to the normalization, we have the following fundamental theorem.

Theorem 6.12.6 *The normalization of a Stein space is again Stein.*

Similarly to Theorem 4.5.11 we have:

Theorem 6.12.7 (General interpolation) *Let X be a Stein space and let Y be an analytic subset of X. Then, for any holomorphic function g on Y there exists a holomorphic function f on X with $f|_Y = g$.*

Historical Supplements

We refer the readers to Chap. 9 for terming Oka's Second and Third Coherence Theorems. In writing here, the author feels still the difficulty of the proof of Oka's Third Coherence Theorem, more difficult than that of Levi's Problem (Hartogs' Inverse Problem) dealt with in the next chapter.

In the comment to Oka VIII of Oka's Collected Works [67], H. Cartan begins with writing

> Ce Mémoire VIII, de lecture très difficile, est ...

The proof presented here is due to Grauert–Remmert [28], which is considerably simplified, but still requires a lot to catch the proof. This proof is already introduced in R. Narasimhan [48] p. 121 (1966), so it should have been known rather a long time before.

K. Oka invented the notion and the theory of "ideals or modules with undetermined domains", i.e., of "Coherent Sheaf", in order to solve Levi's Problem (Hartogs' Inverse Problem) even for ramified covering domains over $\mathbf{C}^n$. But it did not go well, and he restricted himself to the case of unramified domains (i.e., Riemann domains) and wrote up Oka IX, giving a complete solution of Levi's Problem (Hartogs' Inverse Problem), which was then a big problem unsolved for a long time (cf. the next chapter). Therefore, after all, Oka's Second and Third Coherence Theorems that are essential in the singular (ramified) case, were not used there.

On the other hand, taking a look into H. Cartan's paper [9] (1944), one may find that he seems to have been concerned with problems of Cousin type, and the name "Cousin" appears quite frequently, but no mention of Levi's Problem (Hartogs' Inverse Problem). Therefore there was a difference in what they were looking for there.

The above facts may be interesting in view of mathematical developments.

Exercises

1. Let $U \subset \mathbf{C}^n$ be an open set, and let $f \in \mathscr{O}(U)$. Let $X = \{f = 0\}$ and let $a \in X$. Assume that $\mathscr{I}\langle X\rangle_a = \mathscr{O}_{\mathbf{C}^n,a} \cdot \underline{f}_a$. Show that there is a neighborhood V of a such that $\Sigma(X) \cap V = \{f = df = 0\} \cap V$.
2. Prove Proposition 6.5.19.
3. (Analytic Sard's Theorem). Let $\Omega \subset \mathbf{C}^n$ be a domain and let $f \in \mathscr{O}(\Omega)$. Then, show the existence of an at most countable subset $Z \subset \mathbf{C}$ such that for all $w \in \mathbf{C} \setminus Z$,
$$df(z) \neq 0, \quad {}^\forall z \in f^{-1}w.$$
4. Let $X = \{(z, w) \in \mathbf{C}^2;\ w^2 = z^3\}$. Then, $\Sigma(X) = \{0\}$ and X is not normal at 0, as shown in Example 6.10.8. Find the normalization of X.
5. Let $X = \{(u, v, w) \in \mathbf{C}^3;\ w^2 = uv\}$. Show that $\Sigma(X) = \{0\}$ and that X is normal.

(Hint: Consider a holomorphic map $\pi : (u, v) \in \mathbf{C}^2 \to (u^2, v^2, uv) \in X$ and the pull-back $\pi^* f$ for f with $\underline{f}_0 \in \tilde{\mathscr{O}}_{X,0}$.)

6. Let $U \subset \mathbf{C}^n$ be an open set, and let $Y \subset X \subset U$ be analytic subsets. Let $f : X \to \mathbf{C}$ be a continuous function which is weakly holomorphic on X. Show that the restriction $f|_Y$ is weakly holomorphic on Y. (This is not trivial when $Y \subset \Sigma(X)$.)
7. Let X be a normal complex space, and let Y be a thin analytic subset of X. Let $f : X \backslash Y \to \mathbf{C}$ be a holomorphic function such that f is locally bounded near every point of Y. Show that f extends holomorphically on X.
8. In $\mathbf{C}^3$ we define a subset A given by points (z_1, z_2, z_3) as follows:

$$z_1 = u, \quad z_2 = uv, \quad z_3 = uve^v$$

for $(u, v) \in \mathrm{P}\Delta_2 \subset \mathbf{C}^2$, where $\mathrm{P}\Delta_2$ is a polydisk neighborhood about 0. Show that $A \cap \{z_1 \neq 0\}$ is an analytic subset of $\mathrm{P}\Delta_2 \cap \{z_1 \neq 0\}$, but that A is not an analytic subset in any neighborhood of $0 \in \mathrm{P}\Delta_2$.
(Hint: Show that if $\underline{f}_0 \in \mathscr{O}_{2,0}$ satisfies $\underline{f}_0|_{\underline{A}_0} = 0$, then $\underline{f}_0 = 0$.)

Chapter 7
Pseudoconvex Domains and Oka's Theorem

In Chap. 4 we saw that the Oka–Cartan Fundamental Theorem 4.4.2 holds on *holomorphically convex domains*, and in Chap. 5 that a holomorphically convex domain is equivalent to a *domain of holomorphy*. In the present chapter we introduce a notion of *pseudoconvex domains* which includes those domains as a subclass; it is characterized by a real function without making use of holomorphic functions. The aim of this chapter is to show Oka's Theorem, claiming that pseudoconvex domains are domains of holomorphy, due to Oka IX published in 1953; however, it had been solved in a Japanese research report written in 1943. In an example of Chap. 5 we saw that a domain in $\mathbf{C}^n$ may have the envelope of holomorphy, not univalent but infinitely sheeted multivalent over $\mathbf{C}^n$. This example implies that it is insufficient to deal with this problem only for univalent domains.

In this chapter we introduce plurisubharmonic functions defined by Oka VI to solve Levi's Problem (Hartogs' Inverse Problem) and to solve the problem first for univalent domains. Afterwards we solve it in the general case of Riemann domains. As a consequence we give the proofs of the case of univalent domains twice. The reason to do so is because there is a difference in the contents used on coherent sheaves: The method employed here is not Oka's original, but one by the finite dimensionality of the cohomologies of coherent sheaves by H. Grauert, which has a broad application and will be used also in the next chapter.

7.1 Plurisubharmonic Functions

7.1.1 Subharmonic Functions

We begin with Jensen's formula in one variable. This elementary formula is extended for several variables and will be used in a number of places. We use the notation as

J. Noguchi, *Analytic Function Theory of Several Variables*,
DOI 10.1007/978-981-10-0291-5_7

in (1.2.1). Let $z = x + iy$ ($x, y \in \mathbf{R}$) be the coordinate of the complex plane $\mathbf{C}$. Let φ be a real-valued C^2 function in z. By a computation one gets

$$\begin{aligned} dd^c\varphi &= \frac{i}{2\pi}\partial\bar{\partial}\varphi = \frac{\partial^2\varphi}{\partial z\partial\bar{z}}\frac{i}{2\pi}dz \wedge d\bar{z} \\ &= \frac{1}{4\pi}\left(\frac{\partial^2\varphi}{\partial x^2} + \frac{\partial^2\varphi}{\partial y^2}\right) dx \wedge dy. \end{aligned}$$

The notation, $dd^c\varphi \geq 0$ (resp. > 0) means that

$$\frac{\partial^2\varphi}{\partial z\partial\bar{z}} = \frac{1}{4}\left(\frac{\partial^2\varphi}{\partial x^2} + \frac{\partial^2\varphi}{\partial y^2}\right) \geq 0 \text{ (resp. } > 0).$$

In terms of the polar coordinate $z = re^{i\theta}$ ($\log z = \log r + i\theta$),

$$d^c\varphi = \frac{1}{4\pi}\left\{\frac{\partial\varphi}{\partial(\log r)}d\theta - \frac{\partial\varphi}{\partial\theta}d(\log r)\right\} = \frac{1}{4\pi}\left(r\frac{\partial\varphi}{\partial r}d\theta - \frac{1}{r}\frac{\partial\varphi}{\partial\theta}dr\right). \tag{7.1.1}$$

In particular, with $\varphi = \log|z| = \log r$,

$$d^c \log|z| = \frac{1}{4\pi}d\theta. \tag{7.1.2}$$

Let $D \Subset \mathbf{C}$ be a bounded domain with (positively oriented) ∂D of C^1-class. For a 1-form $\eta = Pdz + Qd\bar{z}$ of C^1-class in a neighborhood of the closure $\bar{D}$, Stokes' Theorem is stated as follows:

$$\int_{\partial D}\eta = \int_D d\eta = \int_D\left(-\frac{\partial P}{\partial\bar{z}} + \frac{\partial Q}{\partial z}\right) dz \wedge d\bar{z}. \tag{7.1.3}$$

Let $\varphi(z)$ be a function in an open set $U \subset \mathbf{C}$, and let $\Delta(a; r) \Subset U$. We write

$$\frac{1}{2\pi}\int_{|\zeta|=r}\varphi(a+\zeta)d\theta = \frac{1}{2\pi}\int_0^{2\pi}\varphi(a+re^{i\theta})d\theta.$$

We regard $\frac{1}{2\pi}\int_{|\zeta|=0}\varphi(a+\zeta)d\theta = \varphi(a)$.

Lemma 7.1.4 (Jensen's formula) *Let $\varphi(z)$ be a C^2 function in a neighborhood of $\overline{\Delta(a; r)}$. For $0 \leq s < r$ we have*

$$\frac{1}{2\pi}\int_{|\zeta|=r}\varphi(a+\zeta)d\theta - \frac{1}{2\pi}\int_{|\zeta|=s}\varphi(a+\zeta)d\theta = 2\int_s^r\frac{dt}{t}\int_{\Delta(a;t)}\frac{i}{2\pi}\partial\bar{\partial}\varphi.$$

Proof By a translation, one may put $a = 0$. By (7.1.2) and by using Stokes' Theorem several times, one obtains

$$\begin{aligned}
&\frac{1}{2\pi}\int_{|z|=r}\varphi(z)d\theta - \frac{1}{2\pi}\int_{|z|=s}\varphi(z)d\theta \\
&= 2\int_{|z|=r}\varphi(z)d^c\log|z| - 2\int_{|z|=s}\varphi(z)d^c\log|z| \\
&= 2\int_{\Delta(r)\setminus\Delta(s)} d\varphi\wedge d^c\log|z| = 2\int_{\Delta(r)\setminus\Delta(s)} d\log|z|\wedge d^c\varphi \\
&= 2\int_s^r\frac{dt}{t}\int_{|z|=t}d^c\varphi = 2\int_s^r\frac{dt}{t}\int_{\Delta(t)}dd^c\varphi.
\end{aligned}$$

□

Let U be an open set of $\mathbf{C}$, and consider a function $\varphi : U \to [-\infty, \infty)$ which is allowed to have value "$-\infty$".

Definition 7.1.5 A function $\varphi : U \to [-\infty, \infty)$ is called a *subharmonic function* if φ is upper semi-continuous and has a sub-mean property; that is,

(i) (Upper semi-continuity) $\overline{\lim_{z\to a}}\,\varphi(z) \le \varphi(a), \qquad \forall a \in U.$

(ii) (Sub-mean property) For every disk $\Delta(a; r) \Subset U$,

$$\varphi(a) \le \frac{1}{2\pi}\int_0^{2\pi}\varphi(a + re^{i\theta})d\theta.$$

If both $\pm\varphi$ are subharmonic, φ is called a *harmonic function*; i.e., $\varphi : U \to \mathbf{R}$ is continuous and satisfies the *mean property*,

$$\text{(7.1.6)}\qquad \varphi(a) = \frac{1}{2\pi}\int_0^{2\pi}\varphi(a + re^{i\theta})d\theta, \qquad \Delta(a; r) \Subset U.$$

Remark 7.1.7 (i) If $\varphi : U \to [-\infty, \infty)$ is upper semi-continuous, φ is bounded from above on every compact subset $K \Subset U$.

(ii) For $\varphi : U \to [-\infty, \infty)$ to be upper semi-continuous it is necessary and sufficient that for every $c \in \mathbf{R}$, $\{z \in U; \varphi(z) < c\}$ is open.

(iii) $\varphi : U \to [-\infty, \infty)$ being upper semi-continuous is equivalent to the existence of a monotone decreasing sequence of continuous functions, $\psi_\nu : U \to \mathbf{R}$, $\nu = 1, 2, \ldots$ such that $\lim_{\nu\to\infty}\psi_\nu(z) = \varphi(z)$ at every $z \in U$.

(iv) It is immediate from Definition 7.1.5 (ii) above that

$$\text{(7.1.8)}\qquad \begin{aligned}
\varphi(a) &\le \frac{1}{\pi r^2}\int_0^r t\,dt\int_0^{2\pi}\varphi(a + te^{i\theta})d\theta \\
&= \frac{1}{r^2}\int_{|\zeta|<r}\varphi(a+\zeta)\frac{i}{2\pi}d\zeta\wedge d\bar{\zeta} < \infty.
\end{aligned}$$

Theorem 7.1.9 (i) *Let φ be a subharmonic function in U. If $\varphi(a) > -\infty$ at a point $a \in U$, then φ is locally integrable in a connected component U' of U containing a.*

(ii) (Maximum principle) *Let φ be a subharmonic function in U. If φ takes the maximum value at a point $a \in U$, then φ is constant in a connected component U' of U containing a.*

(iii) *Assume that $\varphi \in C^2(U)$. Then, φ is subharmonic if and only if $dd^c\varphi = (i/2\pi)\partial\bar{\partial}\varphi \geq 0$.*

(iv) *Let $\varphi : U \to [-\infty, \infty)$ be subharmonic, and let λ be a monotone increasing convex function defined on $[\inf \varphi, \sup \varphi)$. Then, $\lambda \circ \varphi$ is subharmonic. Here, $\lambda(-\infty) = \lim_{t \to -\infty} \lambda(t)$.*

(v) *Let $\varphi_\nu : U \to [-\infty, \infty)$, $\nu = 1, 2, \ldots$, be a monotone decreasing sequence of subharmonic functions. Then the limit function $\varphi(z) = \lim_{\nu\to\infty} \varphi_\nu(z)$ is subharmonic.*

(vi) *Let $\{\varphi_\lambda\}_{\lambda\in\Lambda}$ be a family of subharmonic functions in U. If*

$$\varphi(z) := \sup_{\lambda\in\Lambda} \varphi_\lambda(z)$$

is upper semi-continuous, $\varphi(z)$ is subharmonic. In particular, if Λ is a finite set, $\varphi(z)$ is upper semi-continuous, and so subharmonic.

Proof (i) Without loss of generality we may assume that U is connected. If $\varphi(a) > -\infty$, it follows from (7.1.8) that φ is integrable on every $\Delta(a; r) \Subset U$. Suppose the existence of a point $a \in U$ with $\varphi(a) > -\infty$. Let U_0 be the set of all points $z \in U$ satisfying the following property: There is a neighborhood W of z, on which the restriction $\varphi|_W$ is integrable. Clearly, U_0 is non-empty and open. We show that U_0 is closed in U. Let $a \in U$ be an accumulation point of U_0. Take a sequence of points $z_\nu \in U_0$, $\nu = 1, 2, \ldots$, converging to a. We may assume that $\varphi(z_\nu) > -\infty$, $\nu = 1, 2, \ldots$. Then, there are some $r > 0$ and a large ν such that $a \in \Delta(z_\nu; r) \Subset U$. As remarked at the beginning, $\varphi|_{\Delta(z_\nu;r)}$ is integrable. Hence, $a \in U_0$ and so U_0 is open and closed in U. Since U is connected, $U_0 = U$.

(ii) Assume that U is connected and $\varphi(a)$ is the maximum. It follows from (7.1.8) that

$$\int_{\Delta(a;r)} \{\varphi(\zeta) - \varphi(a)\} \frac{i}{2\pi} d\zeta \wedge d\bar{\zeta} = 0 \tag{7.1.10}$$

for every $\Delta(a; r) \Subset U$. Note that $\varphi(\zeta) - \varphi(a) \leq 0$. If there is a point $b \in \Delta(a; r)$ with $\varphi(b) - \varphi(a) = \delta_0 < 0$, the upper semi-continuity of φ implies the existence of a neighborhood of b, $\Delta(b; \varepsilon)(\subset \Delta(a; r))$, where $\varphi(\zeta) - \varphi(a) < \frac{\delta_0}{2}$. It follows that

$$\begin{aligned}&\int_{\Delta(a;r)} \{\varphi(\zeta) - \varphi(a)\} \frac{i}{2\pi} d\zeta \wedge d\bar{\zeta} \\ &\leq \int_{\Delta(b;\varepsilon)} \{\varphi(\zeta) - \varphi(a)\} \frac{i}{2\pi} d\zeta \wedge d\bar{\zeta} \leq \frac{\delta_0 \pi \varepsilon^2}{2} < 0 :\end{aligned}$$

This contradicts (7.1.10). Therefore, $\varphi|_{\Delta(a;r)} \equiv \varphi(a)$. Let U_1 denote the set of all points $z \in U$ for which there is a neighborhood W satisfying $\varphi|_W \equiv \varphi(a)$. Clearly, U_1 is not empty. By a similar argument as in (i) above, U_1 is open and closed in U. Therefore, $U_1 = U$.

(iii) In a neighborhood of each point $a \in U$ we expand φ to the second order:

$$\begin{aligned}\varphi(a+\varepsilon e^{i\theta}) &= \varphi(a) + \frac{\partial\varphi}{\partial z}(a)\varepsilon e^{i\theta} + \frac{\partial\varphi}{\partial\bar z}(a)\varepsilon e^{-i\theta}\\ &\quad + \varepsilon^2\left(\frac{\partial^2\varphi}{\partial z^2}(a)e^{2i\theta} + 2\frac{\partial^2\varphi}{\partial z\partial\bar z}(a) + \frac{\partial^2\varphi}{\partial \bar z^2}(a)e^{-2i\theta}\right)(1+o(1)).\end{aligned}$$

Taking the mean of the integration with respect to θ, we obtain

$$\frac{1}{2\pi}\int_0^{2\pi}\varphi(a+\varepsilon e^{i\theta})d\theta = \varphi(a) + \varepsilon^2(1+o(1))2\frac{\partial^2\varphi}{\partial z\partial\bar z}(a).$$

The submean property implies that $\frac{\partial^2\varphi}{\partial z\partial\bar z}(a) \ge 0$.

Conversely, we assume that $\frac{\partial^2\varphi}{\partial z\partial\bar z} \ge 0$. Set

$$d(a, \partial U) = \inf\{|a-w|; w \in \partial U\}.$$

It follows from Lemma 7.1.4 that about every point $a \in U$,

$$(7.1.11)\qquad \frac{1}{2\pi}\int_{|\zeta|=s}\varphi(a+\zeta)d\theta \le \frac{1}{2\pi}\int_{|\zeta|=r}\varphi(a+\zeta)d\theta$$

for $0 \le s < r < d(a, \partial U)$. With $s = 0$ we get

$$\varphi(a) \le \frac{1}{2\pi}\int_{|\zeta|=r}\varphi(a+\zeta)d\theta.$$

(iv) The boundedness implies the continuity of λ, and so $\lambda\circ\varphi$ is upper semi-continuous. Let $\Delta(a;r) \Subset U$. Then, the convexity of λ yields that

$$\int_0^{2\pi}\lambda(\varphi(a+re^{i\theta}))\frac{d\theta}{2\pi} \ge \lambda\left(\int_0^{2\pi}\varphi(a+re^{i\theta})\frac{d\theta}{2\pi}\right).$$

From the submean property of φ and the monotone increasingness of λ it follows that

$$\lambda\left(\int_0^{2\pi}\varphi(a+re^{i\theta})\frac{d\theta}{2\pi}\right) \ge \lambda(\varphi(a)).$$

Therefore we see the submean property of $\lambda\circ\varphi$, and that it is subharmonic.

(v) It is immediate by the assumption that φ is upper semi-continuous. An upper semi-continuous function is bounded from above on every compact subset. Therefore, φ_ν is bounded from above on any relatively compact subset of U. Taking a disk $\Delta(a; r) \Subset U$, we have by Fatou's Lemma that

$$\varphi(a) = \lim_{\nu\to\infty} \varphi_\nu(a) \le \overline{\lim_{\nu\to\infty}} \frac{1}{2\pi}\int_0^{2\pi} \varphi_\nu(a + re^{i\theta})d\theta$$
$$\le \frac{1}{2\pi}\int_0^{2\pi} \overline{\lim_{\nu\to\infty}} \varphi_\nu(a + re^{i\theta})d\theta = \frac{1}{2\pi}\int_0^{2\pi} \varphi(a + re^{i\theta})d\theta.$$

(vi) If Λ is finite, $\varphi(z)$ is upper semi-continuous. We show the submean property. For any $\Delta(a; r) \Subset U$ and for any $\lambda \in \Lambda$ we get

$$\varphi_\lambda(a) \le \int_0^{2\pi} \varphi_\lambda(a + re^{i\theta})\frac{d\theta}{2\pi} \le \int_0^{2\pi} \varphi(a + re^{i\theta})\frac{d\theta}{2\pi}.$$

Taking the supremum of the left-hand side, we obtain the submean property:

$$\varphi(a) \le \int_0^{2\pi} \varphi(a + re^{i\theta})\frac{d\theta}{2\pi}.$$

□

Example 7.1.12 (i) For a holomorphic function $f : U \to \mathbf{C}$, $\log|f|$ and $|f|^c$ with $c > 0$ are both subharmonic. An easy direct computation implies the subharmonicity of $\log(|f|^2 + C)$ $(C > 0)$. Letting $C = 1/\nu$, $\nu = 1, 2, \ldots$, and taking the limit, we infer from Theorem 7.1.9 (v) that $\log|f|^2 = 2\log|f|$ is subharmonic, and so is $\log|f|$. Since the exponential function e^{ct} in $t \in \mathbf{R}$ is monotone increasing and convex, Theorem 7.1.9 (iv) implies that $|f|^c$ is subharmonic.

(ii) (Discontinuous example) Let $\alpha_k \in \Delta(0; 1)\setminus\{0\}$, $k = 1, 2, \ldots$, be any sequence converging to 0. Taking c_k such that

$$0 < c_k \le \frac{1}{2^k}, \qquad c_k \left|\log\left|\frac{\alpha_k}{2}\right|\right| < \frac{1}{2^k}, \qquad k = 1, 2, \ldots,$$

we set

$$\varphi_N(z) = \sum_{k=1}^{N} c_k \log\frac{|z - \alpha_k|}{2}, \quad z \in \mathbf{C} \quad N = 1, 2, \ldots.$$

By (i) above, $\varphi_N(z)$ is subharmonic in $\mathbf{C}$. For $z \in \Delta(0; 1)$, $\frac{|z-\alpha_k|}{2} < 1$, so that $\varphi_N(z)$ converges decreasingly point-wisely to $\varphi(z) = \lim_{N\to\infty}\varphi_N(z)$. By Theorem 7.1.9 (v), $\varphi(z)$ is subharmonic in $\Delta(0; 1)$. On $\mathbf{C}\setminus\Delta(0; 1)$ the series $\varphi(z)$ normally converges. Thus, $\varphi(z)$ is subharmonic in $\mathbf{C}$. Note that

$$\varphi(0) = \sum_{k=1}^{\infty} c_k \log\frac{|\alpha_k|}{2} \in \mathbf{R},$$

but $\varphi(\alpha_k) = -\infty$, $k = 1, 2, \ldots$.

To avoid $-\infty$ for a value, we set $\psi(z) = \exp \varphi(z)$. Since $\exp(\cdot)$ is a monotone increasing convex function, it follows from Theorem 7.1.9 (iv) that $\psi(z)$ is subharmonic; now, $\psi(0) > 0$, $\psi(\alpha_k) = 0$, $k = 1, 2, \ldots$.

We take $\chi \in C_0^\infty(\mathbf{C})$ such that $\operatorname{Supp} \chi \subset \Delta(0; 1)$, $\chi(z) = \chi(|z|) \geq 0$ and

$$\int \chi(z) \frac{i}{2} dz \wedge d\bar{z} = 1.$$

Setting $\chi_\varepsilon(z) = \chi(\varepsilon^{-1} z)\varepsilon^{-2}$ with $\varepsilon > 0$, we get

$$\int \chi_\varepsilon(z) \frac{i}{2} dz \wedge d\bar{z} = 1.$$

Let φ be a subharmonic function in U such that $\varphi \not\equiv -\infty$ on any connected component of U. By Theorem 7.1.9 (i), φ is locally integrable in U. Set

$$U_\varepsilon = \{z \in U;\, d(z, \partial U) > \varepsilon\}.$$

Then, the *smoothing* $\varphi_\varepsilon(z)$ of $\varphi(z)$ for $z \in U_\varepsilon$ is defined by

$$\begin{aligned}
\varphi_\varepsilon(z) = \varphi * \chi_\varepsilon(z) &= \int_{\mathbf{C}} \varphi(w)\chi_\varepsilon(w - z)\frac{i}{2} dw \wedge d\bar{w} \\
&= \int_{\mathbf{C}} \varphi(z + w)\chi_\varepsilon(w)\frac{i}{2} dw \wedge d\bar{w} \\
&= \int_0^1 \chi(t) t\, dt \int_0^{2\pi} \varphi(z + \varepsilon t e^{i\theta}) d\theta \\
&\geq \varphi(z) \int_0^1 2\pi \chi(t) t\, dt = \varphi(z).
\end{aligned} \tag{7.1.13}$$

We see that $\varphi_\varepsilon(z)$ is C^∞ in U_ε, and has a submean property because of the following computation: For $\Delta(z; r) \Subset U_\varepsilon$,

$$\begin{aligned}
&\frac{1}{2\pi} \int_{|\zeta|=r} \varphi_\varepsilon(z + \zeta) d\theta \\
&= \int_{|\zeta|=r} \frac{d\theta}{2\pi} \int_0^1 \chi(t) t\, dt \int_0^{2\pi} \varphi(z + \zeta + \varepsilon t e^{i\vartheta}) d\vartheta \\
&= \int_0^1 \chi(t) t\, dt \int_0^{2\pi} d\vartheta \int_{|\zeta|=r} \varphi(z + \zeta + \varepsilon t e^{i\vartheta}) \frac{d\theta}{2\pi} \\
&\geq \int_0^1 \chi(t) t\, dt \int_0^{2\pi} \varphi(z + \varepsilon t e^{i\vartheta}) d\vartheta = \varphi_\varepsilon(z).
\end{aligned} \tag{7.1.14}$$

Therefore, $\varphi_\varepsilon(z)$ is subharmonic. It follows from Theorem 7.1.9 (iii) that

$$\frac{\partial^2}{\partial z \partial \bar{z}} \varphi_\varepsilon(z) \geq 0.$$

Next, we let $\varepsilon_1 > \varepsilon_2 > 0$ and $\delta > 0$, and take the double smoothing $(\varphi_\delta)_{\varepsilon_i} = (\varphi_{\varepsilon_i})_\delta, i = 1, 2$. Since φ_δ is C^∞ and subharmonic, one may apply (7.1.11) to obtain

$$\int_0^{2\pi} \varphi_\delta(z + \zeta + \varepsilon_1 t e^{i\vartheta}) d\vartheta \geq \int_0^{2\pi} \varphi_\delta(z + \zeta + \varepsilon_2 t e^{i\vartheta}) d\vartheta.$$

This together with computations in (7.1.14) yields that

$$(\varphi_\delta)_{\varepsilon_1} \geq (\varphi_\delta)_{\varepsilon_2}.$$

Changing the order of integrations, we get $(\varphi_{\varepsilon_1})_\delta \geq (\varphi_{\varepsilon_2})_\delta$. As $\delta \to 0$, $\varphi_{\varepsilon_1} \geq \varphi_{\varepsilon_2}$. Thus, as $\varepsilon \searrow 0$, $\varphi_\varepsilon(z)$ monotonously decreases. It follows from (7.1.13) that

$$\varphi(z) \leq \lim_{\varepsilon \to 0} \varphi_\varepsilon(z).$$

Now, we show the equality by making use of the upper semi-continuity:

If $\varphi(z) = -\infty$, for every $K < 0$ there is a neighborhood disk $\Delta(z; r) \subset U$ such that $\varphi|_{\Delta(z;r)} < K$. By Definition 7.1.13, $\varphi_\varepsilon(z) \leq K$ for $\varepsilon < r$. Therefore, $\lim_{\varepsilon \to 0} \varphi_\varepsilon(z) \leq K$. Since $K < 0$ is arbitrary, $\lim_{\varepsilon \to 0} \varphi_\varepsilon(z) = -\infty$.

If $\varphi(z) > -\infty$, for every $\varepsilon' > 0$ there is a neighborhood disk $\Delta(z; r) \subset U$ such that $\varphi|_{\Delta(z;r)} < \varphi(z) + \varepsilon'$. Because of the same reason as above, $\varphi_\varepsilon(z) \leq \varphi(z) + \varepsilon'$ for $\varepsilon < r$. Therefore, $\lim_{\varepsilon \to 0} \varphi_\varepsilon(z) = \varphi(z)$. Thus we see the monotone convergence, $\varphi_\varepsilon(z) \searrow \varphi(z)$ $(\varepsilon \searrow 0)$.

We apply (7.1.11) for the C^∞ subharmonic function φ_ε: For $\Delta(a; r) \Subset U$ with $0 < s < r$ and for a sufficiently small $\varepsilon > 0$,

$$\frac{1}{2\pi} \int_{|\zeta|=s} \varphi_\varepsilon(a + \zeta) d\theta \leq \frac{1}{2\pi} \int_{|\zeta|=r} \varphi_\varepsilon(a + \zeta) d\theta.$$

Letting $\varepsilon \searrow 0$, by Lebesgue's monotone convergence theorem we have

$$\frac{1}{2\pi} \int_{|\zeta|=s} \varphi(a + \zeta) d\theta \leq \frac{1}{2\pi} \int_{|\zeta|=r} \varphi(a + \zeta) d\theta. \tag{7.1.15}$$

By Theorem 7.1.9 (i), φ is locally integrable in U. By Fubini's theorem, for almost all $s \in (0, r)$ with respect to the Lebesgue measure we have

$$\frac{1}{2\pi} \int_{|\zeta|=s} \varphi(a + \zeta) d\theta > -\infty. \tag{7.1.16}$$

We infer from this and (7.1.15) that (7.1.16) holds for all $s \in (0, r]$.

Summarizing the above, we obtain:

Theorem 7.1.17 *Let $\varphi : U \to [-\infty, \infty)$ be a subharmonic function in U such that $\varphi \not\equiv -\infty$ in every connected component of U.*

(i) *The smoothing $\varphi_\varepsilon(z)$ is subharmonic. As $\varepsilon \searrow 0$, $\varphi_\varepsilon(z)$ monotonously decreases and converges to $\varphi(z)$.*
(ii) *For every $\Delta(a; r) \Subset U$ and $s \in (0, r)$,*

$$-\infty < \frac{1}{2\pi} \int_{|\zeta|=s} \varphi(a+\zeta) d\theta \le \frac{1}{2\pi} \int_{|\zeta|=r} \varphi(a+\zeta) d\theta < \infty.$$

Corollary 7.1.18 *A function $\varphi : U \to \mathbf{R}$ is harmonic if and only if φ is of C^∞-class and satisfies the Laplace equation*

$$\Delta\varphi = \left(\frac{\partial^2}{\partial x^2} + \frac{\partial^2}{\partial y^2}\right) \varphi = 0 \quad \text{in } U.$$

Proof Assume that φ is harmonic. We consider the above smoothing $\varphi_\varepsilon(z)$. The functions $\pm\varphi_\varepsilon(z)$ are of C^∞ class, and as $\varepsilon \searrow 0$ they both monotonously decrease, and converge to $\varphi(z)$. Therefore, $\varphi_\varepsilon(z) = \varphi(z)$. It follows also that $\frac{\partial^2}{\partial z \partial \bar{z}} \varphi = 0$.

The converse follows from the definition and Theorem 7.1.9 (iii). □

Example 7.1.19 Let $f(z)$ be a holomorphic function in U. Then the real part $\Re f(z)$ satisfies the mean property, and also $\Delta \Re f = 0$, so that it is harmonic. In particular, the real part $\Re P(z)$ of a polynomial $P(z)$ is harmonic in $\mathbf{C}$.

Theorem 7.1.20 (i) *The subharmonicity is a local property: i.e., if a function $\varphi : U \to [-\infty, \infty)$ is subharmonic in a neighborhood of every point $a \in U$, then φ is subharmonic in U.*
(ii) *Let U, V be open subsets of $\mathbf{C}$, and let $f : V \to U$ be a holomorphic map. Then the pull-back $f^*\varphi = \varphi \circ f$ of a subharmonic function φ in U is subharmonic in V. If f is biholomorphic, the converse is also true.*

Proof (i) We consider the smoothing $\varphi_\varepsilon(z)$. If φ is subharmonic in $\Delta(a; r) \subset U$, then φ_ε with $0 < \varepsilon < r/2$ is subharmonic in $\Delta(a; r/2)$. Therefore, $dd^c \varphi_\varepsilon(z) \ge 0$. It follows from Theorem 7.1.9 (iii) that $\varphi_\varepsilon(z)$ is subharmonic in U_ε.

To show Definition 7.1.5 (ii), it suffices to prove that in every U_δ ($\delta > 0$) fixed, φ is subharmonic. Let δ ($> \varepsilon \searrow 0$) be any fixed positive number. Since $\varphi_\varepsilon(z) \searrow \varphi(z)$, monotonously in U_δ, by Theorem 7.1.9 (v), φ is subharmonic in U_δ.

(ii) If φ is of C^2-class,

$$\frac{\partial^2 f^*\varphi}{\partial\zeta \partial\bar{\zeta}}(\zeta) = \frac{\partial^2 \varphi}{\partial z \partial \bar{z}}(f(\zeta)) \cdot \left|\frac{df}{d\zeta}\right|^2 \ge 0.$$

Hence, $f^*\varphi$ is subharmonic. In general, by (i) it suffices to show the subharmonicity in a neighborhood of each point $\zeta \in V \mapsto f(\zeta) \in U$. We may assume that there is a connected neighborhood V' of ζ such that $f^*\varphi \not\equiv -\infty$. It follows that $\varphi \not\equiv -\infty$ in a connected neighborhood $U'(\subset f(V'))$ of $f(\zeta)$. Taking smaller V' and U' if necessary, we deduce from Theorem 7.1.17 (i) that $\varphi|_{U'}$ is a monotone decreasing limit of a sequence $\{\phi_\nu\}$ of C^∞ subharmonic functions. Therefore, $f^*\varphi|_{V'}$ is the monotone decreasing limit of $\{f^*\phi_\nu\}$. It follows from Theorem 7.1.9 (v) that $f^*\varphi$ is subharmonic in V'. □

Theorem 7.1.21 *For a upper semi-continuous function $\varphi : U \to [-\infty, \infty)$ the following conditions are equivalent:*

(i) *φ is subharmonic in U.*
(ii) *For every $\Delta(a; r) \subset U$,*

$$\varphi(a) \le \frac{1}{r^2} \int_{\Delta(a;r)} \varphi(z) \frac{i}{2\pi} dz \wedge d\bar{z}. \tag{7.1.22}$$

(iii) *Let $K \Subset U$ be an arbitrary compact subset, and let $h : K \to \mathbf{R}$ be a continuous function which is harmonic in the interior point set K° of K. If $\varphi(z) \le h(z)$ on the boundary $\partial K = K \setminus K^\circ$ of K, then $\varphi(z) \le h(z)$ on K.*

Proof (i)⇒(ii): This was shown already by (7.1.8).

(ii)⇒(iii): Assume that the inequality $\varphi(z) \le h(z)$ $(z \in K)$ claimed by (iii) does not hold. Then,

$$\eta := \sup\{\varphi(z) - h(z); z \in K\} > 0.$$

Since K is compact, there is a sequence $\{z_\nu\}_{\nu=1}^\infty$ in K such that

$$\eta = \lim_{\nu\to\infty} \{\varphi(z_\nu) - h(z_\nu)\},$$
$$a = \lim_{\nu\to\infty} z_\nu \in K.$$

The upper semi-continuity implies that $\eta \le \varphi(a) - h(a)$. By the definition of η, $\eta = \varphi(a) - h(a)$ (> 0). Therefore, $a \in K^\circ$. It follows from the assumption of (ii) and the choice of η that for $\Delta(a; r) \subset K$,

$$\eta = \varphi(a) - h(a) \le \frac{1}{r^2} \int_{\Delta(a;r)} \{\varphi(z) - h(z)\} \frac{i}{2\pi} dz \wedge d\bar{z} \le \eta.$$

Thus,

$$\frac{1}{r^2} \int_{\Delta(a;r)} \{\varphi(z) - h(z) - \eta\} \frac{i}{2\pi} dz \wedge d\bar{z} = 0. \tag{7.1.23}$$

Note that $\varphi(z) - h(z) - \eta \le 0$, $z \in \Delta(a; r)$. Suppose the existence of $b \in \Delta(a; r)$ with $\gamma := \varphi(b) - h(b) - \eta < 0$. Then, $\{z \in \Delta(a; r); \varphi(z) - h(z) - \eta < \gamma/2\}$

contains a neighborhood of b, and then (7.1.23) is impossible. (cf. the argument after (7.1.10)). Therefore,

$$(\varphi - h)|_{\Delta(a;r)} \equiv \eta.$$

Then, by the same argument of the proof of Theorem 7.1.9 (ii) we see that $(\varphi - h)|_V \equiv \eta$ in the connected component V of K° containing a. Hence, for $z \in \bar{V} \cap \partial K \neq \emptyset$, $\varphi(z) - h(z) \geq \eta > 0$; this is a contradiction.

(iii) $\Rightarrow$(i): Taking an arbitrary disk $\Delta(a;r) \Subset U$, we apply (iii) for $K = \overline{\Delta(a;r)}$. For the sake of simplicity, by making use of Theorem 7.1.20 (ii) we may assume that $a = 0$ and $r = 1$.

Take a monotone decreasing sequence $\{h_\nu(e^{i\theta})\}_{\nu=1}^{\infty}$ of continuous functions on $\partial\Delta(0;1)$, converging to $\varphi(e^{i\theta})$. We consider the Poisson integral of h_ν:

$$\hat{h}_\nu(z) = \int_0^{2\pi} \frac{h_\nu(e^{i\theta})(1 - |z|^2)}{|e^{i\theta} - z|^2} \frac{d\theta}{2\pi} \tag{7.1.24}$$

(cf., e.g., [52] Chap. 3 Sect. 6). The function $\hat{h}_\nu(z)$ is continuous on $\overline{\Delta(0;1)}$, harmonic in the interior $\Delta(0;1)$ and $\hat{h}_\nu(e^{i\theta}) = h_\nu(e^{i\theta}) \geq \varphi(e^{i\theta})$ on the boundary. By the assumption, in particular at the center,

$$\varphi(0) \leq \hat{h}_\nu(0) = \int_{|\zeta|=1} h_\nu(\zeta)\frac{d\theta}{2\pi}.$$

It follows from the monotone convergence theorem of Lebesgue that

$$\varphi(0) \leq \int_{|\zeta|=r} \varphi(\zeta)\frac{d\theta}{2\pi};$$

the subharmonicity is shown. □

Theorem 7.1.25 *A upper semi-continuous function $\varphi : U \to [-\infty, \infty)$ is subharmonic if and only if the following property is satisfied:*

Property 7.1.26 If for every disk $\Delta(a;r) \Subset U$ and for every polynomial $P(z)$

$$\varphi(z) \leq \Re P(z), \qquad |z - a| = r,$$

then, $\varphi(a) \leq \Re P(a)$.

Proof If φ is subharmonic, $\varphi(z) - \Re P(z)$ is subharmonic, too (cf. Example 7.1.19). It follows from the maximum principle (Theorem 7.1.9 (ii)) that the validity of $\varphi(z) - \Re P(z) \leq 0$ on the boundary $|z - a| = r$ implies that of $\varphi(z) - \Re P(z) \leq 0$ also in the interior $|z - a| < r$. In particular, $\varphi(a) - \Re P(a) \leq 0$.

Next, we assume Property 7.1.26. What we have to show is the submean property:

$$\varphi(a) \leq \frac{1}{2\pi} \int_{|\zeta - a| = r} \varphi(a + \zeta) d\theta, \quad \Delta(a; r) \Subset U.$$

For simplicity, by Theorem 7.1.20 (ii) we may assume that $a = 0$ and $r = 1$.

We take a monotone decreasing sequence $\{h_\nu(e^{i\theta})\}_{\nu=1}^{\infty}$ of continuous functions on the boundary $\partial\Delta(0; 1)$ which converges to $\varphi(e^{i\theta})$ (cf. Remark 7.1.7 (iii)). By (7.1.24) we take the Poisson integral $\hat{h}_\nu(z)$ of each $h_\nu(e^{i\theta})$. The function $\hat{h}_\nu(z)$ is continuous in $|z| \leq 1$, harmonic in the interior $|z| < 1$, and on the boundary

$$\hat{h}_\nu(z) = h_\nu(z), \qquad |z| = 1.$$

Taking the associated harmonic function $\hat{h}_\nu^*(z)$ of $\hat{h}_\nu(z)$, we put

$$g_\nu(z) = \hat{h}_\nu(z) + i\hat{h}_\nu^*(z), \qquad |z| < 1.$$

Then it is holomorphic and satisfies

$$\Re\, g_\nu(z) = \hat{h}_\nu(z).$$

For a moment, we fix ν arbitrarily. For every $\varepsilon > 0$, taking $0 < r < 1$ sufficiently close to 1, we have

$$\hat{h}_\nu(re^{i\theta}) - \varepsilon < h_\nu(e^{i\theta}) < \hat{h}_\nu(re^{i\theta}) + \varepsilon = \Re\, g_\nu(re^{i\theta}) + \varepsilon. \tag{7.1.27}$$

Since $g_\nu(z)$ can be uniformly approximated by the finite sums of the power series expansion $g_\nu(z) = \sum c_\mu z^\mu$ on $|z| \leq r$, there is a polynomial $P_\nu(z)$, $P_\nu(0) = g_\nu(0)$ such that

$$|g_\nu(z) - P_\nu(z)| < \varepsilon, \qquad |z| \leq r. \tag{7.1.28}$$

It follows that

$$h_\nu(e^{i\theta}) < \Re P_\nu(re^{i\theta}) + 2\varepsilon.$$

Therefore,

$$\varphi(z) < \Re P_\nu(rz) + 2\varepsilon, \qquad |z| = 1.$$

By the assumption we have

$$\varphi(0) \leq \Re P_\nu(0) + 2\varepsilon.$$

Since $\Re P_\nu(rz)$ is harmonic, the mean property holds:

$$\Re P_\nu(0) = \int_0^{2\pi} \Re P_\nu(re^{i\theta}) \frac{d\theta}{2\pi}.$$

Therefore it follows from (7.1.28) and (7.1.27) that

$$\begin{aligned}\varphi(0) &\le \int_0^{2\pi} \Re P_\nu(re^{i\theta})\frac{d\theta}{2\pi} + 2\varepsilon \le \int_0^{2\pi} \Re g_\nu(re^{i\theta})\frac{d\theta}{2\pi} + 3\varepsilon \\ &= \int_0^{2\pi} \hat{h}_\nu(re^{i\theta})\frac{d\theta}{2\pi} + 3\varepsilon \le \int_0^{2\pi} h_\nu(e^{i\theta})\frac{d\theta}{2\pi} + 4\varepsilon.\end{aligned}$$

Since $\varepsilon > 0$ is arbitrary, with $\varepsilon \searrow 0$ we get

$$\varphi(0) \le \int_0^{2\pi} h_\nu(e^{i\theta})\frac{d\theta}{2\pi}.$$

Now, by the monotone convergence theorem of Lebesgue, we see, as $\nu \to \infty$, that

$$\varphi(0) \le \int_0^{2\pi} \varphi(e^{i\theta})\frac{d\theta}{2\pi}.$$

□

7.1.2 Plurisubharmonic Functions

We consider the case of several variables. Let $U \subset \mathbf{C}^n$ be an open subset, and let $z = (z_1, \ldots, z_n)$ denote the standard coordinate system of $\mathbf{C}^n$, and set

$$\begin{aligned} d(z, \partial U) &= \inf\{\|z - w\|; w \in \partial U\}, \qquad z \in U, \\ U_\varepsilon &= \{z \in U; d(z, \partial U) > \varepsilon\}, \qquad \varepsilon > 0. \end{aligned}$$

We write $z_j = x_j + iy_j$ $(1 \le j \le n)$ with $x_j, y_j \in \mathbf{R}$ and use the symbols in (1.2.1). Moreover, we introduce the following:

(7.1.29)
$$\begin{aligned} &B(r) = B(0; r), \\ &\alpha = dd^c\|z\|^2, \quad \beta = dd^c \log \|z\|^2, \\ &\gamma = d^c \log \|z\|^2 \wedge \beta^{m-1}. \end{aligned}$$

The pull-backed form to the hypersphere $\{\|z\| = r\}(\subset \mathbf{C}^n)$ of a differential form on $\mathbf{C}^n$ by the inclusion map $\iota : \{\|z\| = r\} \hookrightarrow \mathbf{C}^n$ is called an *induced differential form* on $\{\|z\| = r\}$. Since $\iota^*(d\|z\|^2) = 0$, as induced differential forms on $\{\|z\| = r\}$, $d\|z\|^2 = \partial\|z\|^2 + \bar{\partial}\|z\|^2 = 0$. It follows from this that

$$\partial\|z\|^2 \wedge \bar{\partial}\|z\|^2 = 0,$$

as induced differential forms on $\{\|z\| = r\}$. Therefore, the induced differential forms of α and β on the hypersphere $\{\|z\| = t\}$ satisfy

$$\beta = \frac{1}{t^2}\alpha. \tag{7.1.30}$$

With the notation above we have

$$\int_{B(r)} \alpha^n = r^{2n}, \qquad \int_{\|z\|=r} \gamma = 1.$$

Definition 7.1.31 (*Oka VI 1942*) A function $\varphi : U \to [-\infty, \infty)$ is called a *plurisubharmonic function* or a *pseudoconvex function*[1] if the following conditions are satisfied:

(i) φ is upper semi-continuous.
(ii) For every $z \in U$ and every $v \in \mathbf{C}^n$, the function

$$\zeta \in \mathbf{C} \to \varphi(z + \zeta v) \in [-\infty, \infty)$$

is subharmonic in ζ where it is defined.

From Example 7.1.12 (i) we obtain the following.

Example 7.1.32 For a holomorphic function $f : U \to \mathbf{C}$, $\log |f|$ and $|f|^c$ with $c > 0$ are both plurisubharmonic.

Let φ be a plurisubharmonic function in U, and let $B(a; r) \Subset U$. By making use of the invariance of α with respect to the rotation $z \to e^{i\theta} z$ ($\theta \in [0, 2\pi]$), we infer from Definition 7.1.5 (ii) that

$$\begin{aligned}
\int_{B(r)} \varphi(a + z)\alpha^n &= \int_{B(r)} \varphi(a + e^{i\theta} z)\alpha^n \\
&= \frac{1}{2\pi}\int_0^{2\pi} d\theta \int_{B(r)} \varphi(a + e^{i\theta} z)\alpha^n \\
&= \int_{z\in B(r)} \left(\frac{1}{2\pi}\int_0^{2\pi} \varphi(a + e^{i\theta} z) d\theta\right) \alpha^n \\
&\geq \int_{B(r)} \varphi(a)\alpha^n = r^{2n}\varphi(a).
\end{aligned}$$

[1] Cf. "Historical supplements" at the end of this section.

Therefore we obtain the following inequality, similar to (7.1.8):

$$\begin{aligned}(7.1.33)\qquad \varphi(a) &\le \frac{1}{r^{2n}}\int_0^r 2nt^{2n-1}dt \int_{\|z\|=t} \varphi(a+z)\gamma(z) \\ &= \frac{1}{r^{2n}}\int_{B(a;r)} \varphi(z)\alpha^m \qquad (B(a;r)\Subset U).\end{aligned}$$

This means that φ is a *subharmonic function*[2] in an open subset U of $\mathbf{R}^{2n}$ with identification $\mathbf{C}^n \cong \mathbf{R}^{2n}$.

Let φ be of C^2-class. Then,

$$dd^c\varphi = \sum_{1\le j,k\le m} \frac{\partial^2\varphi}{\partial z_j \partial \bar{z}_k}\frac{i}{2\pi} dz_j \wedge d\bar{z}_k.$$

We write $dd^c\varphi \ge 0$ if the hermitian matrix $\left(\frac{\partial^2\varphi}{\partial z_j\partial\bar{z}_k}\right)$ is positive semidefinite; i.e., for every vector $(\xi_j) \in \mathbf{C}^n$,

$$\sum_{j,k} \frac{\partial^2\varphi}{\partial z_j \partial \bar{z}_k}\xi_j\bar{\xi}_k \ge 0.$$

In this case, considering $\varphi(z+\zeta v)$ $(\zeta \in \mathbf{C})$ with $v = (v_1,\ldots,v_n) \in \mathbf{C}^n \setminus \{0\}$, one gets

$$\left.\frac{\partial^2}{\partial\zeta\partial\bar{\zeta}}\right|_{\zeta=0} \varphi(z+\zeta v) = \sum_{j,k}\frac{\partial^2\varphi}{\partial z_j\partial\bar{z}_k}(z)v_j\bar{v}_k \ge 0.$$

From the above and a similar argument of the proof of Theorem 7.1.9 the following is obtained.

Theorem 7.1.34 (i) *A plurisubharmonic function is subharmonic with* $\mathbf{C}^n \cong \mathbf{R}^{2n}$.

(ii) *If φ is plurisubharmonic in U, and $\varphi(a) > -\infty$ for a point $a \in U$, then φ is locally integrable in the connected component U' of U containing a.*

(iii) *Let φ be a plurisubharmonic function in U. If φ takes the maximum value at a point $a \in U$, φ is constant in the connected component U' of U containing a.*

(iv) *For φ of C^2-class, the plurisubharmonicity is equivalent to $dd^c\varphi \ge 0$.*

(v) *Let $\varphi : U \to [-\infty,\infty)$ be plurisubharmonic, and let λ be a monotone increasing convex function defined on* $[\inf\varphi, \sup\varphi)$.[3] *Then, $\lambda\circ\varphi$ is plurisubharmonic, where $\lambda(-\infty) = \lim_{t\to-\infty}\lambda(t)$.*

(vi) *Let $\varphi_\nu : U \to [-\infty,\infty)$, $\nu = 1,2,\ldots$, be a monotone decreasing sequence of plurisubharmonic functions. Then, the limit function $\varphi(z) = \lim_{\nu\to\infty}\varphi_\nu(z)$ is plurisubharmonic.*

---

[2]In general, a function $\psi : W \to [-\infty,\infty)$ in an open subset W of $\mathbf{R}^n$ is said to be subharmonic if ψ is upper semi-continuous, and satisfies a submean property in the sense of (7.1.33).

[3]*Remark*: If $\lambda(t)$ is of C^2-class in an interval $I \subset \mathbf{R}$, $\lambda(t)$ is convex if and only if $\lambda''(t) \ge 0$.

(vii) *Let* $\{\varphi_\lambda\}_{\lambda\in\Lambda}$ *be a family of plurisubharmonic functions in* U. *If* $\varphi(z) := \sup_{\lambda\in\Lambda}\varphi_\lambda(z)$ *is upper semi-continuous,* $\varphi(z)$ *is plurisubharmonic; in particular, if* Λ *is finite,* $\varphi(z)$ *is upper semi-continuous, and so plurisubharmonic.*

Definition 7.1.35 A function $\varphi : U \to \mathbf{R}$ is said to be *strongly pseudoconvex* or *strongly plurisubharmonic* if it is of C^2-class and satisfies

$$dd^c\varphi(z) > 0, \qquad z \in U.$$

That is, the hermitian matrix $\left(\frac{\partial^2\varphi}{\partial z_j\partial\bar{z}_k}\right)$ is positive definite:

$$\sum_{j,k}\frac{\partial^2\varphi}{\partial z_j\partial\bar{z}_k}\xi_j\bar{\xi}_k > 0, \quad {}^\forall(\xi_j) \in \mathbf{C}^n\setminus\{0\}.$$

By Theorem 7.1.34 (iv), strongly plurisubharmonic functions are plurisubharmonic.

Example 7.1.36 (i) The function $\varphi(z) = \|z\|^2 = \sum_{j=1}^n |z_j|^2$ is strongly plurisubharmonic in $z \in \mathbf{C}^n$. By computations, the matrix $\left(\frac{\partial^2\varphi}{\partial z_j\partial\bar{z}_k}\right)$ is a unit one, and so positive definite.

(ii) For $z = (z_j) = (x_j + iy_j)$, $(1 \le j \le n)$ we take the sum of the squares of the imaginary parts,

$$\phi(z) = \sum_{j=1}^n |y_j|^2 = \sum_{j=1}^n \frac{|z_j - \bar{z}_j|^2}{4}.$$

A direct computation yields that

$$dd^c\phi(z) = \sum_{j=1}^n \frac{i}{4\pi}dz_j \wedge d\bar{z}_j > 0.$$

Therefore, $\phi(z)$ is strongly plurisubharmonic. Similarly for the real parts, $\sum_{j=1}^n |x_j|^2$ is strongly plurisubharmonic.

Lemma 7.1.37 *Let* $\varphi(z)$ *be a strongly plurisubharmonic function in* U. *Let* $\lambda(t)$ *be a* C^2 *function defined in an open interval containing the range* $\varphi(U)$ *such that* $\lambda'(t) > 0$ *and* $\lambda''(t) \ge 0$. *Then the composite* $\lambda\circ\varphi(z)$ *is strongly plurisubharmonic.*

Proof In fact, by computations we see that

$$\begin{aligned} dd^c\lambda\circ\varphi(z) &= \frac{i}{2\pi}\partial\bar{\partial}\lambda\circ\varphi(z) \\ &= \sum_{j,k}\left(\lambda''(\varphi(z))\frac{\partial\varphi}{\partial z_j}\cdot\overline{\frac{\partial\varphi}{\partial z_k}} + \lambda'(\varphi(z))\frac{\partial^2\varphi}{\partial z_j\partial\bar{z}_k}\right)\frac{i}{2\pi}dz_j\wedge d\bar{z}_k. \end{aligned}$$

For a vector $(\xi_j) \in \mathbf{C}^n \setminus \{0\}$,

$$\sum_{j,k} \left(\lambda''(\varphi(z)) \frac{\partial \varphi}{\partial z_j} \cdot \overline{\frac{\partial \varphi}{\partial z_k}} + \lambda'(\varphi(z)) \frac{\partial^2 \varphi}{\partial z_j \partial \bar{z}_k} \right) \xi_j \bar{\xi}_k$$
$$= \lambda''(\varphi(z)) \sum_j \left| \frac{\partial \varphi}{\partial z_j} \xi_j \right|^2 + \lambda'(\varphi(z)) \sum_{j,k} \frac{\partial^2 \varphi}{\partial z_j \partial \bar{z}_k} \xi_j \bar{\xi}_k$$
$$\geq \lambda'(\varphi(z)) \sum_{j,k} \frac{\partial^2 \varphi}{\partial z_j \partial \bar{z}_k} \xi_j \bar{\xi}_k > 0.$$

Therefore, $dd^c \lambda \circ \varphi > 0$. □

We take a function $\chi(z) = \chi(|z_1|, \ldots, |z_n|) \in C_0^\infty(\mathbf{C}^n)$ such that

$$\chi(z) \geq 0, \quad \operatorname{Supp} \chi \subset B(1), \quad \int \chi(z) \alpha^m = 1,$$

and set

$$\chi_\varepsilon(z) = \chi(\varepsilon^{-1} z) \varepsilon^{-2n}, \quad \varepsilon > 0.$$

Definition 7.1.38 We set the smoothing of a locally integrable function φ in U by

$$\varphi_\varepsilon(z) = \varphi * \chi_\varepsilon(z) = \int_{\mathbf{C}^n} \varphi(w) \chi_\varepsilon(w - z) \alpha^n(w)$$
$$= \int_{\mathbf{C}^n} \varphi(z + w) \chi_\varepsilon(w) \alpha^n(w), \quad z \in U_\varepsilon.$$

Then, $\varphi_\varepsilon(z)$ is of C^∞-class in U_ε.

Let φ be plurisubharmonic in U and $\varphi \not\equiv -\infty$ in any connected component of U. The rotation invariance $\chi(w) = \chi(e^{i\theta} w)$ for $0 \leq \theta \leq 2\pi$ implies that

$$\varphi_\varepsilon(z) = \int_{\mathbf{C}^n} \varphi(z + \varepsilon w) \chi(w) \alpha^n(w)$$
$$= \int_{\mathbf{C}^n} \alpha^n(w) \frac{1}{2\pi} \int_0^{2\pi} d\theta \, \varphi(z + \varepsilon e^{i\theta} w) \chi(w)$$
$$\geq \varphi(z) \int_{\mathbf{C}^n} \chi(w) \alpha^n = \varphi(z).$$

Thus, it follows from Theorem 2.1.9 (iii) that as $\varepsilon \searrow 0$, φ_ε monotonously decreases. Since φ is upper semi-continuous, by a similar argument to the proof of Theorem 2.1.9 (ii), $\varphi_\varepsilon(z) \searrow \varphi(z)$.

Thus we obtain the following.

Theorem 7.1.39 *Let $\varphi : U \to [-\infty, \infty)$ be a plurisubharmonic function in U such that $\varphi \not\equiv -\infty$ in every connected component of U.*

(i) *The smoothing $\varphi_\varepsilon(z)$ is C^∞ plurisubharmonic in U_ε, and converges monotone decreasingly to $\varphi(z)$.*
(ii) *For every $B(a; R) \subset U$ and $0 < s < r < R$,*

$$-\infty < \int_{\|z\|=s} \varphi(a+z)\gamma(z) \le \int_{\|z\|=r} \varphi(a+z)\gamma(z) < \infty. \tag{7.1.40}$$

Proof (i) There remains to show the plurisubharmonicity. For a vector $v = (v_1, \ldots, v_n) \in \mathbf{C}^n$ we consider $\varphi_\varepsilon(z+\zeta v)$ in $\zeta \in \mathbf{C}$, where it is defined. Taking a sufficiently small $r > 0$, we have

$$\varphi(z+\zeta v) \le \int_0^{2\pi} \varphi\left(z + \left(\zeta + re^{i\theta}\right) v\right) \frac{d\theta}{2\pi}.$$

We take the smoothing of each side of the above equation with respect to z, and apply Fubini's theorem for the right-hand side to change the order of integrations; then, we obtain

$$\varphi_\varepsilon(z+\zeta v) \le \int_0^{2\pi} \varphi_\varepsilon\left(z + \left(\zeta + re^{i\theta}\right) v\right) \frac{d\theta}{2\pi}.$$

This means that $\varphi_\varepsilon(z)$ is plurisubharmonic.

(ii) Firstly, by Theorem 7.1.34 (ii), it is noted that φ is locally integrable. It follows from (7.1.33) and Fubini's theorem that there is a subset $E \subset (0, R)$ of Lebesgue's measure zero such that for all $t \in (0, R)\backslash E$, $\int_{\|z\|=t} \varphi(a+z)\gamma(z)$ is finite.

On the other hand, by the $\mathbf{C}^*$-invariance of $\gamma(te^{i\theta}z) = \gamma(z)$ with $t \in (0, R)$ and $\theta \in [0, 2\pi]$, we have

$$\begin{aligned}\int_{\|z\|=t} \varphi(a+z)\gamma(z) &= \int_{\|z\|=1} \varphi\left(a + te^{i\theta}z\right)\gamma(z) \\ &= \int_{\|z\|=1}\int_0^{2\pi} \varphi\left(a + te^{i\theta}z\right)\frac{d\theta}{2\pi}\gamma(z).\end{aligned}$$

This right-hand side together with Theorem 2.1.9 (iii) implies that for every $0 < s < r < R$,

$$\int_{\|z\|=s} \varphi(a+z)\gamma(z) \le \int_{\|z\|=r} \varphi(a+z)\gamma(z) < \infty.$$

We apply this for $0 < t < s$, $t \notin E$ to get

$$-\infty < \int_{\|z\|=t} \varphi(a+z)\gamma(z) \le \int_{\|z\|=s} \varphi(a+z)\gamma(z).$$

Therefore, (7.1.40) is valid. □

Similarly to Theorem 7.1.20, the following holds.

Theorem 7.1.41 (i) *The plurisubharmonicity is a local property.*

(ii) *Let U, V be open subsets of $\mathbf{C}^n$, and let $f : V \to U$ be a holomorphic map. Then the pull-back $f^*\varphi = \varphi \circ f$ of a plurisubharmonic function φ in U is plurisubharmonic in V. If f is biholomorphic, the converse is also true.*

Supplements: We would like to state three theorems on plurisubharmonic functions. Each of them is very interesting, but independent of the theoretical development of the present book and not necessary there, so that we omit the proof. Interested readers should consult the referred books.

(a) There is an interesting property of the separate analyticity for functions in several complex variables, which is due to Hartogs and is stated as follows.

Theorem 7.1.42 (Hartogs' separate analyticity) *Let $f(z_1, \ldots, z_n)$ be a function in a domain Ω of $\mathbf{C}^n$. If at every point of Ω, $f(z_1, \ldots, z_j, \ldots, z_n)$ is holomorphic in each variable z_j with other variables fixed, then $f(z_1, \ldots, z_n)$ is holomorphic in n variables.*

It should be noted that even the continuity of f is not assumed. This does not hold for real analytic functions as shown by the following example:

$$f(x, y) = \begin{cases} \dfrac{x^2y^2}{x^2+y^2}, & (x, y) \ne (0, 0), \\ 0, & (x, y) = (0, 0). \end{cases}$$

For the proof, cf. Nishino [49] Chap. 1 Sect. 1.4, or Hörmander [33] Chap. 2.

(b) For plurisubharmonic functions, Riemann's extension theorem and Hartogs' extension theorem hold.

Theorem 7.1.43 (Riemann's extension) *Let $U \subset \mathbf{C}^n$ be an open subset and let $A \subset U$ be a thin set. If a plurisubharmonic function φ in $U \setminus A$ is locally bounded from above about every point a of A, i.e., there is a neighborhood V of a in U and a constant M such that*

$$\varphi(z) \le M, \quad {}^{\forall}z \in V \setminus A,$$

then φ is extended to a unique plurisubharmonic function in U.

Theorem 7.1.44 (Hartogs' extension; Grauert–Remmert) *Let $U \subset \mathbf{C}^n$ be an open subset, let $A \subset U$ be an analytic subset, and let φ be a plurisubharmonic function* *in* $U \backslash A$. *If* $\operatorname{codim}_U A \geq 2$, *then* φ *is extended to a unique plurisubharmonic function in* U.

For the proof, cf., e.g., Noguchi–Ochiai [56] Chap. 3.

Historical Supplements

The notion of a plurisubharmonic function was first defined by K. Oka VI 1942 (received Oct. 1941). Oka called this "fonction pseudoconvexe" and discussed its properties by setting one section titled "Nouvelle classe de fonctions réelles". The purpose was to solve Levi's Problem (Hartogs' Inverse Problem); in fact, the new notion is more flexible than that due to Levi, and played an essential role in the solution of the problem.

The differential form $dd^c\varphi$ of type $(1, 1)$ in Definition 7.1.35 is called, in general, the *Levi form*, and is denoted by

$$L[\varphi] = dd^c\varphi = \frac{i}{2\pi}\partial\bar{\partial}\varphi,$$

but this form is due to Oka VI, and is different from the original definition of E.E. Levi. E.E. Levi dealt with the case of two complex variables, and wrote the form in real four variables; by making use of two complex variables (z, w), it is written for a real C^2 function $\varphi(z, w)$ as

$$\operatorname{Levi}[\varphi] = \begin{vmatrix} 0 & \varphi_z & \varphi_w \\ \varphi_{\bar{z}} & \varphi_{z\bar{z}} & \varphi_{w\bar{z}} \\ \varphi_{\bar{w}} & \varphi_{z\bar{w}} & \varphi_{w\bar{w}} \end{vmatrix}.$$

And Levi defined a domain $\{\varphi < 0\}$ to be pseudoconvex if at every point of the boundary $\{\varphi = 0\}$, $\operatorname{Levi}[\varphi] \geq 0$. By this definition, the condition "$\operatorname{Levi}[\varphi_j] > 0$, $(j = 1, 2)$" does not imply the same property "$\operatorname{Levi}[\varphi_1 + \varphi_2] > 0$" for the addition $\varphi_1 + \varphi_2$. The form $L[\varphi]$ preserves the sign for the addition. K. Oka had looked for a notion of such functions that was easier to deal with and that would play the same role as Levi's functions (see Oka VI), and had moved backward to the logarithmic subharmonicity of Hartogs' functions; then the notion he obtained was that of *plurisubharmonic (pseudoconvex)* functions which was introduced here, and played a crucial role in the solution of Levi's (Hartogs' Inverse) Problem.

According to H. Cartan's record[4] the news of the solution of the Problem was brought with a preprint from K. Oka, via H. Behnke's hand, to H. Cartan in Paris 1941. In Paris, P. Lelong published two announcement papers of about two pages in *Comptes Rendus* 1942 (received November of the same year) which described only the definitions of this kind of function, calling it "fonction plurisousharmonique"; there, no mention on the relation with "pseudoconvexity" was given. But now, probably because of the goodness of the wording or the location of the activities, the naming "plurisubharmonic function" is more popular than "pseudoconvex function".

[4]H. Cartan, "Quelques Souvenirs" presented to H. Behnke's 80th birthday in October 1978 at Münster (Springer-Verlag).

7.2 Pseudoconvex Domains

Let $\Omega \subset \mathbf{C}^n$ be a domain. Fixing an arbitrary polydisk $\mathrm{P}\Delta = \mathrm{P}\Delta(0; (r_1, \ldots, r_n))$, we consider the boundary distance function $\delta_{\mathrm{P}\Delta}(z, \partial\Omega)$ of Ω with respect to $\mathrm{P}\Delta$ (cf. (5.3.2)).

The key Lemma 5.3.7 of the proof of the Cartan–Thullen Theorem 5.3.1, showing that domains of holomorphy are holomorphically convex, has been used so far only for constant f. Here, applying it for non-constant holomorphic functions, we show that domains of holomorphy (equivalently, holomorphically convex domains) are pseudoconvex domains, which we are going to define.

Theorem 7.2.1 (Oka) *For a domain of holomorphy Ω, the function "$-\log \delta_{\mathrm{P}\Delta}(z, \partial\Omega)$" is continuous and plurisubharmonic in Ω.*

Proof The continuity of $-\log \delta_{\mathrm{P}\Delta}(z, \partial\Omega)$ follows from (5.3.4).

Let $L \subset \mathbf{C}^n$ be any complex line. We show the restriction $-\log \delta_{\mathrm{P}\Delta}(z, \partial\Omega)|_L$ is subharmonic where it is defined in $L \cong \mathbf{C}$. We use the criterion of Theorem 7.1.25. We take an arbitrary closed disk

$$E = \{a + \zeta \mathrm{v};\ |\zeta| \le R\} \subset L \cap \Omega,$$

where $\mathrm{v} \in \mathbf{C}^n \setminus \{0\}$ is a directional vector of L. We set

$$K = \{a + \zeta \mathrm{v};\ |\zeta| = R\},$$

and take a polynomial $P(\zeta)$ with

(7.2.2) $$-\log \delta_{\mathrm{P}\Delta}(a + \zeta \mathrm{v}, \partial\Omega) \le \Re P(\zeta), \quad |\zeta| = R.$$

It suffices to show the following.

Claim 7.2.3 $-\log \delta_{\mathrm{P}\Delta}(a, \partial\Omega) \le \Re P(0)$.

Since L is an affine linear subspace of $\mathbf{C}^n$, there is a polynomial $\hat{P}(z)$ on $\mathbf{C}^n$ with $\hat{P}|_L = P$. It follows from (7.2.2) that

$$\delta_{\mathrm{P}\Delta}(z, \partial\Omega) \ge \left| e^{-\hat{P}(z)} \right|, \quad z \in K.$$

By the maximum principle, $\hat{K}_\Omega \supset E$, and then by Lemma 5.3.7,

$$\delta_{\mathrm{P}\Delta}(z, \partial\Omega) \ge \left| e^{-\hat{P}(z)} \right|, \quad z \in E.$$

At $z = a$, $\delta_{\mathrm{P}\Delta}(a, \partial\Omega) \ge \left| e^{-\hat{P}(a)} \right| = \left| e^{-P(0)} \right|$; therefore, Claim 7.2.3 is proved. □

Remark 7.2.4 Oka's Theorem 7.2.1 in the case of $n = 1$ is almost trivial. For, with $\mathrm{P}\Delta$ a unit disk with center at the origin, we write $\delta(z, \partial\Omega) = \delta_{\mathrm{P}\Delta}(z, \partial\Omega)$. Then,

$$-\log\delta(z, \partial\Omega) = \sup_{w\in\partial\Omega} -\log|z - w|.$$

For $w \in \partial\Omega$, $-\log|z - w|$ is subharmonic in $z \in \Omega$ (harmonic in this case). It is already known that $-\log\delta(z, \partial\Omega)$ is continuous. It follows from Theorem 7.1.9 (vi) that $-\log\delta(z, \partial\Omega)$ is subharmonic in Ω.

Definition 7.2.5 In general, a continuous function $\varphi : \Omega \to \mathbf{R}$ is called an *exhaustion function* if

$$\{\varphi < c\} = \{z \in \Omega; \varphi(z) < c\} \Subset \Omega, \quad {}^{\forall}c \in \mathbf{R}.$$

Definition 7.2.6 A domain Ω is said to be *pseudoconvex* if there is a plurisubharmonic exhaustion function $\varphi : \Omega \to \mathbf{R}$.

Definition 7.2.7 Let $\Omega \Subset \mathbf{C}^n$ be a bounded domain. If there is a strongly plurisubharmonic function ψ defined in a neighborhood of the boundary $\partial\Omega$ of Ω such that

$$\Omega \cap U = \{\psi < 0\},$$

then $\partial\Omega$ is called a *strongly pseudoconvex boundary*, and ψ is called a defining function of the boundary $\partial\Omega$. In this case, Ω is called a *strongly pseudoconvex domain*.

Remark 7.2.8 In the definition above, taking a sufficiently small $\delta > 0$, we have that $\{\psi = -\delta\} \Subset U$. Setting $\psi(z) = -\infty$, $z \notin U$ for a moment, we define

$$\tilde{\psi}(z) = \max\{-\delta, \psi(z)\}$$

for $z \in \Omega \cup U$. Then, $\tilde{\psi}(z)$ is continuous and plurisubharmonic in $\Omega \cup U$, and

$$\Omega = \{\tilde{\psi} < 0\}.$$

Note that $\tilde{\psi}(z)$ is strongly plurisubharmonic in a neighborhood of $\partial\Omega$.

Proposition 7.2.9 *A strongly pseudoconvex domain is pseudoconvex.*

Proof Let $\tilde{\psi}(z)$ be the function defined in Remark 7.2.8, which is continuous and plurisubharmonic, and satisfies

$$\{\tilde{\psi} < 0\} = \Omega.$$

Since the function $\lambda(t) = -1/t$ is increasing and convex in $t < 0$, it follows from Theorem 7.1.34 (v) that $\varphi(z) = -1/\tilde{\psi}(z)$ is a plurisubharmonic exhaustion function of Ω. □

Example 7.2.10 The ball $B(0; R)$ is strongly pseudoconvex; in fact, it suffices to take $\psi = \|z\|^2 - R^2$.

Theorem 7.2.11 (Hartogs, Levi, Oka) *Let Ω be a domain of $\mathbf{C}^n$. If $-\log \delta_{\mathrm{P}\Delta}(z, \partial\Omega)$ is plurisubharmonic, then Ω is pseudoconvex. In particular, a domain of holomorphy Ω $(\subset \mathbf{C}^n)$ (or equivalently, a holomorphically convex domain) is pseudoconvex.*

Proof First, note Theorem 7.2.1 for the last statement, and that the square of the norm $\|z\|^2$ is a strongly plurisubharmonic function. We set

$$\varphi(z) = \max\{\|z\|^2, -\log \delta_{\mathrm{P}\Delta}(z, \partial\Omega)\}.$$

By Theorem 7.1.34 (vii) $\varphi(z)$ is continuous and plurisubharmonic. For every $c \in \mathbf{R}$, the sub-level set $\{\varphi < c\}$ is bounded and does not accumulate to a point of $\partial\Omega$. Therefore, $\{\varphi < c\} \Subset \Omega$ is satisfied, and so φ is an exhaustion function. □

In Theorem 7.2.11 the plurisubharmonicity assumption for $-\log \delta_{\mathrm{P}\Delta}(z, \partial\Omega)$ can be localized at the boundary $\partial\Omega$:

Theorem 7.2.12 *Let Ω be a domain of $\mathbf{C}^n$. Assume the following: For every boundary point $b \in \partial\Omega$ there is a neighborhood U of a in $\mathbf{C}^n$ such that $-\log \delta_{\mathrm{P}\Delta}(z, \partial\Omega)$ is plurisubharmonic in $z \in U \cap \Omega$. Then Ω is pseudoconvex.*

Proof The assumption immediately implies the existence of a neighborhood V of $\partial\Omega$ such that $-\log \delta_{\mathrm{P}\Delta}(z, \partial\Omega)$ is plurisubharmonic in $z \in V \cap \Omega$. We set

$$F = \Omega \setminus V,$$
$$F_j = \{z \in F;\ j \le \|z\| \le j+1\}, \quad j = 0, 1, 2, \ldots.$$

Then, all F_j are compact, and $F = \bigcup_{j=0}^{\infty} F_j$. Let $c_0 > 0$ be a constant such that with $\phi_0(t) = c_0(t+1)$ for $t \le 1$

$$\phi_0(\|z\|^2) > -\log \delta_{\mathrm{P}\Delta}(z, \partial\Omega), \quad {}^{\forall} z \in F_0.$$

Inductively, for $j \ge 1$ we set

$$\phi_j(t) = c_j(t - j^2) + \phi_{j-1}(j^2), \quad j^2 \le t \le (j+1)^2,$$

where $c_j > c_{j-1}$ and

$$\phi_j(\|z\|^2) > -\log \delta_{\mathrm{P}\Delta}(z, \partial\Omega), \quad {}^{\forall} z \in F_j.$$

Then the function $\phi(t)$ in $t \ge 0$ defined by

$$\phi(t) = \phi_j(t), \quad j^2 \le t \le (j+1)^2$$

is a monotone increasing convex function. By Theorem 7.1.34 (v), $\phi(\|z\|^2)$ is a continuous plurisubharmonic function on $\mathbf{C}^n$ with $\lim_{\|z\|\to\infty} \phi(\|z\|^2) = \infty$. Set

$$\psi(z) = \max\{-\log \delta_{\mathrm{P}\Delta}(z, \partial\Omega), \phi(\|z\|^2)\}, \quad z \in \Omega.$$

Then, $\psi(z)$ is an exhaustion function of Ω. It is immediate that $\psi(z)$ is plurisubharmonic in $\Omega \setminus F$, and that $\psi(z) = \phi(\|z\|^2) > -\log \delta_{\mathrm{P}\Delta}(z, \partial\Omega)$ for $z \in F$. Therefore, $\psi(z)$ is plurisubharmonic in the whole Ω, and so Ω is pseudoconvex. □

The problem to ask for the validity of the converse of Theorem 7.2.11 is called *Levi's Problem* (*Hartogs' Inverse Problem*).

Definition 7.2.13 A boundary point $b \in \partial\Omega$ of a domain Ω is called a *holomorphically convex (boundary) point* (of Ω) if there is an element $f \in \mathscr{O}(\Omega)$ with $\lim_{z\to b} |f(z)| = \infty$.

Example 7.2.14 Note that $\varphi(z) = -r^2 + \|z - a\|^2$ is strongly plurisubharmonic and $B(a; r) = \{\varphi < 0\}$. Let $b \in \partial B(a; r)$ and set

$$L(z) = -r^2 + \sum_{j=1}^{n} (z_j - a_j)\overline{(b_j - a_j)}.$$

Then, this is holomorphic in z and $L(b) = 0$; geometrically, $\{L = 0\}$ is the complex hyperplane passing through b, tangent to $\partial B(a; r)$. It follows from the Cauchy–Schwarz inequality that for $z \in B(a; r)$,

$$|L(z)| \geq r^2 - \|z - a\| \cdot \|b - a\| > 0.$$

With $f(z) = 1/L(z)$, $f \in \mathscr{O}(B(a; r))$ and $\lim_{z\to b} |f(z)| = \infty$. Therefore, b is a holomorphically convex boundary point of $B(a; r)$.

Lemma 7.2.15 *If every boundary point of a domain Ω is a holomorphically convex point, then, Ω is holomorphically convex.*

Proof Let $K \Subset \Omega$ be a compact subset. It is immediate that $\hat{K}_\Omega$ is bounded. If $\hat{K}_\Omega$ were not relatively compact in Ω, there is a sequence of points $z_\nu \in \hat{K}_\Omega$, $\nu = 1, 2, \ldots$, which converges to a boundary point $b \in \partial\Omega$. By the assumption, there is an $f \in \mathscr{O}(\Omega)$ with $\lim_{z\to b} |f(z)| = \infty$. Thus,

$$\infty = \lim_{\nu\to\infty} |f(z_\nu)| \leq \max_K |f| < \infty.$$

This is absurd. □

We are going to prove that every boundary point of a strongly pseudoconvex domain is a holomorphically convex point by an argument from the local to the global by making use of cohomology theory (cf. Theorem 7.4.12): Here we prepare the local.

Lemma 7.2.16 *Let Ω be a strongly pseudoconvex domain. For every point $b \in \partial\Omega$, there is a number $\delta > 0$ such that all points of $\partial(B(b;\delta) \cap \Omega)$ are holomorphically convex points of $B(b;\delta) \cap \Omega$.*

Proof Let φ be a strongly plurisubharmonic function defining $\partial\Omega$. We expand $\varphi(z)$ about b up to the second order. For simplicity, after a translation we may set $b = 0$ ($\varphi(0) = 0$):

$$\varphi(z) = \Re\left\{2\sum_{j=1}^{n}\frac{\partial\varphi}{\partial z_j}(0)z_j + \sum_{j,k}\frac{\partial^2\varphi}{\partial z_j\partial z_k}(0)z_j z_k\right\} + \sum_{j,k}\frac{\partial^2\varphi}{\partial z_j\partial\bar{z}_k}(0)z_j\bar{z}_k + o(\|z\|^2).$$

Since $\left(\frac{\partial^2\varphi}{\partial z_j\partial\bar{z}_k}(0)\right)$ is positive definite, there are $\varepsilon, \delta > 0$ such that for $\|z\| \le \delta$,

$$\sum_{j,k}\frac{\partial^2\varphi}{\partial z_j\partial\bar{z}_k}(0)z_j\bar{z}_k + o(\|z\|^2) \ge \varepsilon\|z\|^2.$$

Therefore,

$$\varphi(z) \ge \Re\left\{2\sum_{j=1}^{n}\frac{\partial\varphi}{\partial z_j}(0)z_j + \sum_{j,k}\frac{\partial^2\varphi}{\partial z_j\partial z_k}(0)z_j z_k\right\} + \varepsilon\|z\|^2 \tag{7.2.17}$$

for $\|z\| \le \delta$. Since φ is of C^2-class, taking ε and δ smaller if necessary, we can keep (7.2.17) valid when the center 0 of the expansion is moved locally in a neighborhood of 0. That is, for every $c \in B(0;\delta) \cap \partial\Omega$ and $z \in \overline{B(c;\delta)}$,

$$\varphi(z) \ge \Re\left\{2\sum_{j=1}^{n}\frac{\partial\varphi}{\partial z_j}(c)(z_j - c_j) + \sum_{j,k}\frac{\partial^2\varphi}{\partial z_j\partial z_k}(c)(z_j - c_j)(z_k - c_k)\right\} + \varepsilon\|z - c\|^2. \tag{7.2.18}$$

We set a polynomial of degree 2 in z by

$$Q_c(z) = 2\sum_{j=1}^{n}\frac{\partial\varphi}{\partial z_j}(c)(z_j - c_j) + \sum_{j,k}\frac{\partial^2\varphi}{\partial z_j\partial z_k}(c)(z_j - c_j)(z_k - c_k).$$

It follows from (7.2.18) that for $c \in \overline{B(0;\delta/2)} \cap \partial\Omega$ (cf. Fig. 7.1),

$$\bar{\Omega} \cap B(0;\delta/2) \cap \{Q_c(z) = 0\} = \{c\} \tag{7.2.19}$$

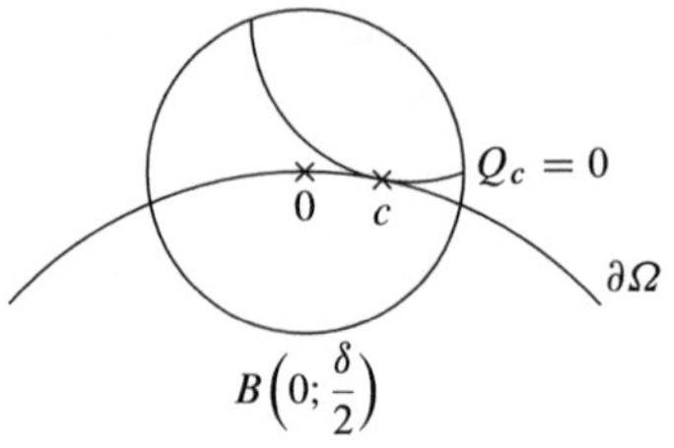

Fig. 7.1 Local holomorphically convex point

Set

$$U = \Omega \cap B(0; \delta/2).$$

For a boundary point $c \in \partial U$ of U, either $c \in \partial\Omega$ or $c \in \partial B(0; \delta/2)$: In the first case, $f(z) = 1/Q_c(z)$ satisfies that $f \in \mathscr{O}(U)$ and $\lim_{z \to c} |f(z)| = \infty$, so that c is a holomorphically convex point of U; in the second case, by Example 7.2.10, c is a holomorphically convex point of $B(0; \delta/2)$, and hence of U. □

7.3 L. Schwartz's Finiteness Theorem

In this section all vector spaces are defined over **C**. The aim of the present section is to prove a finite-dimensional theorem due to L. Schwartz (cf. Theorem 7.3.23). It is indispensable for the proofs of Grauert's Theorems 7.4.1 and 8.4.1, and the Cartan–Serre Theorem 8.2.1. For a general reference of topological vector spaces, cf., e.g., Yamanaka [79], Maeda [40], and Trèves [76].

7.3.1 Topological Vector Spaces

A function $\|x\|$ defined on a vector space E $(\ni x)$ is called a *semi-norm* if the following are satisfied:

(i) $\|x\| \geq 0, \quad x \in E.$
(ii) $\|\lambda x\| = |\lambda| \cdot \|x\|, \quad \lambda \in \mathbf{C},\ x \in E.$
(iii) $\|x + y\| \leq \|x\| + \|y\|, \quad x, y \in E.$

If $\|x\| = 0$ implies $x = 0$, then $\|x\|$ is called a norm; a norm naturally induces a metric in E.

If $\|x\|$ is a complete norm, i.e., $d(x, y) = \|x - y\|$ for $x, y \in E$ is a complete metric, E is called a Banach space.

Let E be given a family $\{\|x\|_\alpha\}_{\alpha \in \Gamma}$ of semi-norms. Then, the subsets

$$U(\varepsilon; \alpha_j, 1 \leq j \leq N) = \{x \in E;\ \|x\|_{\alpha_j} < \varepsilon, 1 \leq j \leq N\}$$

with $\varepsilon > 0$ and finitely many $\alpha_j \in \Gamma, 1 \le j \le N \in \mathbf{N}$ satisfy the axiom of a fundamental system of neighborhoods of 0.

Definition 7.3.1 We endow E with a topology such that a neighborhood system of each point $a \in E$ is given by $\{a + U(\varepsilon; \alpha_j, 1 \le j \le N)\}$. Then, E is called a *topological vector space*. Because the neighborhood $U(\varepsilon; \alpha_j, 1 \le j \le N)$ is convex, E is called also a locally convex topological vector space.

Assumption 7.3.2 In what follows in this section, Γ is *at most countable*, and the topology of E defined as above satisfies the Hausdorff separation axiom.

Let E be a topological vector space with a semi-norm family $\{\|\cdot\|_j\}_{j\in\mathbf{N}}$. Then the fundamental neighborhood system of $0 \in E$ is given by

$$U(\varepsilon, N) = \{x \in E;\ \|x\|_j < \varepsilon, 1 \le j \le N\}, \quad 0 < \varepsilon < 1. \tag{7.3.3}$$

Lemma 7.3.4 *Let $M \subset E$ be a closed vector subspace, and let $v \in E \setminus M$. Then there are numbers $N \in \mathbf{N}$ and $C > 0$ such that for every $x = y + \alpha v$ with $y \in M, \alpha \in \mathbf{C}$,*

$$|\alpha| \le C \max_{1\le j\le N} \|x\|_j.$$

Proof Since M is closed, there are $N \in \mathbf{N}$ and $\varepsilon > 0$ such that

$$\big(v + \{w \in E;\ \|w\|_j < \varepsilon, 1 \le j \le N\}\big) \cap M = \emptyset. \tag{7.3.5}$$

If $\alpha = 0$, it is trivial; we assume $\alpha \ne 0$. It follows from the assumption that

$$v - \frac{1}{\alpha}x = -\frac{1}{\alpha}y \in M.$$

It follows from (7.3.5) that

$$\max_{1\le j\le N} \left\| \frac{1}{\alpha}x \right\|_j \ge \varepsilon.$$

Therefore,

$$|\alpha| \le C \max_{1\le j\le N} \|x\|_j, \qquad C := \frac{1}{\varepsilon}.$$ □

Theorem 7.3.6 *Let M be a closed vector subspace of E, and let M_0 be a finite-dimensional vector subspace of E. Then $M + M_0$ is a closed vector subspace.*

In particular, any finite-dimensional subspace is closed.

Proof With $m = \dim M_0$ we use induction on m.

(a) $m = 1$: With a vector $v \in M_0 \setminus \{0\}$, $M_0 = \langle v \rangle$ (the vector subspace spanned by v). Take a point $x \in \overline{M + \langle v \rangle}$. Then there is a sequence of points converging to x:

$$x_\nu = y_\nu + \alpha_\nu v, \quad y_\nu \in M, \ \alpha_\nu \in \mathbf{C}, \ \nu = 1, 2, \ldots.$$

It follows from Lemma 7.3.4 that there are numbers $N \in \mathbf{N}$ and $C > 0$ such that

$$|\alpha_\nu - \alpha_\mu| \le C \max_{1 \le j \le N} \|x_\nu - x_\mu\|_j \to 0, \quad \nu, \mu \to \infty.$$

Thus, $\{\alpha_\nu\}_\nu$ is a Cauchy sequence, and has a limit $\alpha = \lim_{\nu \to \infty} \alpha_\nu$. Since

$$M \ni y_\nu = x_\nu - \alpha_\nu v \to x - \alpha v, \quad \nu \to \infty,$$

the closedness of M implies $y := x - \alpha v \in M$. It follows that $x = y + \alpha v \in M + \langle v \rangle$, and hence $M + \langle v \rangle$ is closed.

(b) $m \ge 2$: Assume that it holds for $\dim M_0 = m - 1$. We take an $(m-1)$-dimensional vector subspace $M_1 \subset M_0$ and a vector $v \in M_0 \setminus M_1$. Naturally,

$$M_0 = M_1 + \langle v \rangle.$$

By the induction hypothesis, $M + M_1$ is closed. Since

$$M + M_0 = (M + M_1) + \langle v \rangle,$$

by (a) above, $M + M_0$ is closed.

The last assertion follows with $M = \{0\}$. □

We say that a subset Z of E is *relatively compact*, written as $Z \Subset E$, if the closure $\bar{Z}$ is a compact subset, in the sense that any open covering $\bigcup_{\gamma \in \Gamma} V_\gamma \supset \bar{Z}$ has a finite covering $\bigcup_{\gamma \in \Gamma'} V_\gamma \supset \bar{Z}$ with a finite subset $\Gamma' \subset \Gamma$.

Remark 7.3.7 It is known that if the space E is metrizable, the notion of this compactness is equivalent to the so-called point sequential compactness; i.e., every sequence of points of $\bar{Z}$ has a subsequence converging to a point of $\bar{Z}$.

Definition 7.3.8 (*Completely continuous*) Let E, F be two topological vector spaces. A continuous linear map $h : E \to F$ is said to be *completely continuous* if there is a neighborhood U of $0 \in E$ with $h(U) \Subset F$.

Definition 7.3.9 A *Baire space* is a topological space such that any countable union of closed subsets without interior points contains no interior point. If a topological vector space E is Baire, then E is called a *Baire vector space*.

N.B. In general, a complete metric space is a Baire space.

7.3.2 *Fréchet Spaces*

Let E be a topological vector space endowed with a countable system of semi-norms $\|x\|_j, j = 1, 2, \ldots$. In the case where the number of semi-norms is finite, we consider it as a case of $\|x\|_{n_0} = \|x\|_{n_0+1} = \cdots = 0$ with some number n_0. For two points $x, y \in E$, we set

$$d_E(x, y) = \sum_{j=1}^{\infty} \frac{1}{2^j} \cdot \frac{\|x - y\|_j}{1 + \|x - y\|_j}. \tag{7.3.10}$$

The function $t/(1 + t)$ is monotone increasing in $t \geq 0$. Since for $t, s \geq 0$,

$$\frac{t+s}{1+t+s} - \frac{t}{1+t} - \frac{s}{1+s} = \frac{-2ts - ts^2 - t^2 s}{(1+t+s)(1+t)(1+s)} \leq 0,$$

$d_E(x, y)$ satisfies the axioms of a metric (distance). The distance $d_E(x, y)$ satisfies the following invariant properties (with $w \in E$):

$$\begin{aligned} & d_E(x + w, y + w) = d_E(x, y), \quad d_E(-x, 0) = d_E(x, 0), \\ & d_E(x + y, 0) \leq d_E(x + y, y) + d_E(y, 0) = d_E(x, 0) + d_E(y, 0). \end{aligned} \tag{7.3.11}$$

Lemma 7.3.12 *The topology of E is homeomorphic to the metric topology defined by* $d_E(x, y)$.

Proof It suffices to consider the neighborhood system of the origin $0 \in E$. Take a neighborhood $U(\varepsilon, N)$ of $0 \in E$ given by (7.3.3). Note that if $d_E(x, 0) < \delta$ ($\delta > 0$), then by definition,

$$\frac{1}{2^j} \cdot \frac{\|x\|_j}{1 + \|x\|_j} < d_E(x, 0) < \delta.$$

We take $\delta(> 0)$ such that $\delta 2^N < \varepsilon$. Then,

$$\|x\|_j < \frac{\delta 2^j}{1 - \delta 2^j} \leq \frac{\delta 2^N}{1 - \delta 2^N}, \quad 1 \leq j \leq N.$$

Moreover, we let $\delta > 0$ satisfy $\frac{\delta 2^N}{1 - \delta 2^N} < \varepsilon$. Then,

$$\{x \in E; d_E(x, 0) < \delta\} \subset U(\varepsilon, N).$$

Conversely, for an arbitrary $\varepsilon > 0$ we consider a neighborhood $V_\varepsilon := \{x \in E; d_E(x, 0) < \varepsilon\}$ of 0 with respect to the metric topology by $d_E(x, y)$. Since

$$\begin{aligned} d_E(x,0) &= \sum_{j=1}^{N} \frac{1}{2^j} \cdot \frac{\|x\|_j}{1+\|x\|_j} + \sum_{j=N+1}^{\infty} \frac{1}{2^j} \cdot \frac{\|x\|_j}{1+\|x\|_j} \\ &\le \sum_{j=1}^{N} \frac{1}{2^j}\|x\|_j + \sum_{j=N+1}^{\infty} \frac{1}{2^j} \\ &\le \max_{1\le j\le N} \|x\|_j + \frac{1}{2^N}, \end{aligned}$$

with N such that $\frac{1}{2^N} < \frac{\varepsilon}{2}$, a neighborhood $U(\frac{\varepsilon}{2}, N)$ of 0 satisfies

$$U\left(\frac{\varepsilon}{2}, N\right) \subset V_\varepsilon. \qquad \square$$

Definition 7.3.13 If the metric $d_E(x, y)$ above is complete, E is called a *Fréchet space*.

Remark 7.3.14 A Fréchet space is a Baire vector space.

Example 7.3.15 There are a number of examples of Fréchet spaces, but the important one in complex analysis is the following.

(i) Let $\Omega \Subset \mathbf{C}^n$ be a bounded domain, let $\mathscr{C}(\bar{\Omega})$ be the set of all continuous functions on the closure of Ω, and set a norm

$$\|f\|_{\bar{\Omega}} = \max_{z\in\bar{\Omega}} |f(z)|.$$

Then, $\mathscr{C}(\bar{\Omega})$ is a separable Banach space. We set

$$E(\bar{\Omega}) = \mathscr{C}(\bar{\Omega}) \cap \mathscr{O}(\Omega);$$

that is, an element of $E(\bar{\Omega})$ is a function, continuous on $\bar{\Omega}$ and holomorphic in the interior Ω. Since a uniform limit of holomorphic functions is holomorphic, $E(\bar{\Omega})$ with the same norm $\|f\|_{\bar{\Omega}}$ forms a separable Banach space.

(ii) Let $\Omega \subset \mathbf{C}^n$ be domain, and let $\{\Omega_j\}_{j=1}^{\infty}$ be an increasing covering by subdomains such that

$$\Omega_j \Subset \Omega_{j+1}, \qquad \Omega = \bigcup_{j=1}^{\infty} \Omega_j.$$

We take a system of semi-norms of $\mathscr{O}(\Omega)$ defined by

$$\|f\|_{\bar{\Omega}_j} = \max_{\bar{\Omega}_j} |f|, \quad j = 1, 2, \ldots, \quad f \in \mathscr{O}(\Omega). \tag{7.3.16}$$

With this system of semi-norms $\|f\|_{\bar{\Omega}_j}$, $j \in \mathbf{N}$, $\mathscr{O}(\Omega)$ is a Fréchet space.

The topology of $\mathscr{O}(\Omega)$ thus defined is equivalent to that of the uniform convergence on compact subsets.

Theorem 7.3.17 *Let $\Omega_1 \Subset \Omega_2 \subset \mathbf{C}^n$ be a pair of domains. Then the restriction map*

$$\rho : f \in \mathscr{O}(\Omega_2) \to f|_{\Omega_1} \in \mathscr{O}(\Omega_1)$$

is completely continuous.

Proof We take a domain Ω' with $\Omega_1 \Subset \Omega' \Subset \Omega_2$. The set $U = \{f \in \mathscr{O}(\Omega_2); \|f\|_{\bar{\Omega}'} < 1\}$ is a neighborhood of $0 \in \mathscr{O}(\Omega_2)$. An arbitrary sequence of elements of U contains a subsequence converging uniformly on $\bar{\Omega}_1$ by Montel's Theorem 1.2.19. The limit function is holomorphic in the interior Ω_1. Therefore, $\rho(U)$ is relatively compact in $\mathscr{O}(\Omega_1)$. □

7.3.3 Banach's Open Mapping Theorem

We begin with a preparatory lemma.

Lemma 7.3.18 *Let $A : E \to F$ be a continuous linear surjection between topological vector spaces. Assume that F is a Baire vector space. Then for every neighborhood U of $0 \in E$, the closure $\overline{A(U)}$ contains $0 \in F$ as an interior point.*

Proof By the continuity of algebraic operation $(x, y) \in E \times E \to x - y \in E$ there is a neighborhood W of $0 \in E$ such that

$$W - W \subset U.$$

Since $E = \bigcup_{\nu=1}^{\infty} \nu W$, $F = \bigcup_{\nu=1}^{\infty} \nu \overline{A(W)}$. Since $\nu \overline{A(W)}$ are closed, the assumption implies that there is some ν_0 such that $\nu_0 \overline{A(W)}$ contains an interior point. Therefore, $\overline{A(W)}$ also contains an interior point x_0. Noting that $0 = x_0 - x_0 \in \overline{A(W)} - x_0$, we see that 0 is an interior point of $\overline{A(W)} - x_0$. Now, since

$$\begin{aligned} 0 \in \overline{A(W)} - x_0 \subset \overline{A(W)} - \overline{A(W)} \subset \overline{A(W) - A(W)} \\ = \overline{A(W - W)} \subset \overline{A(U)}, \end{aligned}$$

0 is an interior point of $\overline{A(U)}$. □

Theorem 7.3.19 (Banach's Open Mapping Theorem) *Let E be a Fréchet space and let F be a Baire vector space. If $A : E \to F$ is a continuous linear surjection, then A is an open mapping (i.e., the image of any open subset by A is open).*

Proof Let $d_E(x, x')$ be the complete metric of the Fréchet space E defined by (7.3.10). We set

(7.3.20) $$U(\varepsilon) = \{x \in E; d_E(x, 0) < \varepsilon\}, \quad \varepsilon > 0.$$

It suffices to prove that for every $\varepsilon > 0$, $A(U(\varepsilon))$ contains $0 \in F$ as an interior point. By Lemma 7.3.18 there is a neighborhood V of $0 \in F$ such that $V \subset \overline{A(U(\varepsilon))}$. Set

$$U_\nu = U\left(\frac{\varepsilon}{2^{\nu+1}}\right), \qquad \nu = 1, 2, \ldots.$$

For each $\overline{A(U_\nu)}$ we can take a neighborhood V_ν of $0 \in F$ with $V_\nu \subset \overline{A(U_\nu)}$. We may assume that $V_\nu \supset V_{\nu+1}$ and $\bigcap_{\nu=1}^{\infty} V_\nu = \{0\}$. We show:

Claim 7.3.21 $A(U(\varepsilon)) \supset V_1$.

$\because$) Take an arbitrary point $y = y_1 \in V_1$. Since $y_1 \in \overline{A(U_1)}$,

$$(y_1 - V_2) \cap A(U_1) \neq \emptyset.$$

There is a point $y_2 \in V_2$, $x_1 \in U_1$ with $y_1 - y_2 = A(x_1)$. Since $y_2 \in \overline{A(U_2)}$,

$$(y_2 - V_3) \cap A(U_2) \neq \emptyset.$$

Thus, there is a point $y_3 \in V_3$, $x_2 \in U_2$ such that

$$y_2 - y_3 = A(x_2).$$

In this way we inductively define $x_\nu \in U_\nu$ and $y_\nu \in V_\nu$ so that

$$y_\nu - y_{\nu+1} = A(x_\nu), \quad \nu = 1, 2, \ldots.$$

By the choice of $\{V_\nu\}_\nu$, $\lim_{\nu\to\infty} y_\nu = 0$, and we have

$$\begin{aligned} y = y_1 &= A(x_1) + y_2 = A(x_1) + A(x_2) + y_3 \\ &= \cdots = \sum_{j=1}^{\nu} A(x_j) + y_{\nu+1} = A\left(\sum_{j=1}^{\nu} x_j\right) + y_{\nu+1}. \end{aligned} \tag{7.3.22}$$

We check the convergence of $\sum_{\nu=1}^{\infty} x_\nu$. Since $x_\nu \in U_\nu = U(\frac{\varepsilon}{2^{\nu+1}})$, it follows from (7.3.11) that for every $\nu, \mu \in \mathbf{N}$

$$\begin{aligned} d_E\left(\sum_{j=1}^{\nu} x_j, \sum_{j=1}^{\nu+\mu} x_j\right) &= d_E\left(0, \sum_{j=\nu+1}^{\nu+\mu} x_j\right) \leq \sum_{j=\nu+1}^{\nu+\mu} d_E(0, x_j) \\ &< \sum_{j=\nu+1}^{\nu+\mu} \frac{\varepsilon}{2^{j+1}} < \frac{\varepsilon}{2^{\nu+1}} \to 0 \quad (\nu \to \infty). \end{aligned}$$

Therefore $\sum_{\nu=1}^{\infty} x_\nu$ forms a Cauchy series, and hence it is convergent. We set the limit $w = \sum_{\nu=1}^{\infty} x_\nu$. Then by (7.3.22), $y = A(w)$, and

$$d_E(0, w) \leq \sum_{\nu=1}^{\infty} d_E(0, x_\nu) \leq \sum_{\nu=1}^{\infty} \frac{\varepsilon}{2^{\nu+1}} = \frac{1}{2}\varepsilon < \varepsilon.$$

Therefore the proof is completed. □

7.3.4 L. Schwartz's Finiteness Theorem

Theorem 7.3.23 (L. Schwartz's Finiteness Theorem) *Let E be a Fréchet space and let F be a Baire vector space. Let $A : E \to F$ be a continuous linear surjection, and let $B : E \to F$ be a completely continuous linear map. Then, $(A + B)(E)$ is closed and the cokernel* $\mathrm{Coker}(A + B)$ *is finite dimensional.*

Proof (a) Set $A_0 = A + B : E \to F$. By the assumption there is a convex neighborhood U_0 of $0 \in E$ such that $-U_0 = U_0$ and $K := \overline{B(U_0)}$ is compact. Since A is surjective, $V_0 := A(U_0)$ is open by Theorem 7.3.19. We consider an open covering

$$K \subset \bigcup_{b \in K} \left(b + \frac{1}{2} V_0 \right).$$

Since K is compact, there are finitely many points $b_j \in K$, $1 \leq j \leq l$, such that

$$K \subset \bigcup_{j=1}^{l} \left(b_j + \frac{1}{2} V_0 \right).$$

Let $S = \langle b_1, \ldots, b_l \rangle$ denote the vector subspace of finite dimension spanned by b_j, $1 \leq j \leq l$.

We are going to show:

Claim 7.3.24 $A_0(E) + S = F$ *(algebraically).*

Assuming this claim, we finish the proof of the present theorem: We decompose $S = S' \oplus (S \cap A_0(E))$ as a direct sum. Then $F = A_0(E) \oplus S'$, algebraically. Since $\mathrm{Ker}\, A_0$ is closed, the quotient space $E/\mathrm{Ker}\, A_0$ satisfies Assumption 7.3.2, and so does $\widehat{E} := (E/\mathrm{Ker}\, A_0) \oplus S'$. We consider the following continuous linear surjection and bijection:

$$\widetilde{A}_0 : x \oplus y \in E \oplus S' \to A_0(x) + y \in F,$$
$$\widehat{A}_0 : [x] \oplus y \in \widehat{E} = (E/\mathrm{Ker}\, A_0) \oplus S' \to A_0(x) + y \in F.$$

Note that $E \oplus S'$ is Fréchet. By Theorem 7.3.19, $\widetilde{A}_0$ is open, and so is $\widehat{A}_0$. Thus, $\widehat{A}_0$ is a homeomorphism, so that $A_0(E) = \widehat{A}_0((E/\mathrm{Ker}\, A_0) \oplus \{0\})$ is a closed linear subspace. Since $F/A_0(E) \cong S'$, the dimension of $\mathrm{Coker}\, A_0$ is finite.

(b) Proof of Claim 7.3.24: Since S is closed by Theorem 7.3.6, the quotient space F/S satisfies Assumption 7.3.2 and is Baire. Let $\pi : F \to F/S$ be the quotient map. Set $\tilde{V}_0 = \pi(V_0)$. Note that $\tilde{K} = \pi(K)$ is compact, and $\tilde{K} \subset \frac{1}{2}\tilde{V}_0$. Replacing F by F/S, we may assume from the beginning that

$$K \subset \frac{1}{2}V_0,$$

and then prove that $A_0(E) = F$. Since $A_0(E)$ is a vector subspace of F, $A_0(E) = F$ is deduced from the following:

Claim 7.3.25 $A_0(E) \supset V_0$.

∵) Take an arbitrary point $y_0 \in V_0$. There is a point $x_0 \in U_0$ with $A(x_0) = y_0$. Since $y_1 = y_0 - A_0(x_0) = -B(x_0) \in K \subset \frac{1}{2}V_0 = A(\frac{1}{2}U_0)$, there is a point $x_1 \in \frac{1}{2}U_0$ with $A(x_1) = y_1$, and then

$$\begin{aligned} y_2 &:= y_1 - A_0(x_1) = -B(x_1) \in B\left(\frac{1}{2}U_0\right) = \frac{1}{2}B(U_0) \\ &\subset \frac{1}{2}K \subset \frac{1}{2^2}V_0 = A\left(\frac{1}{2^2}U_0\right). \end{aligned}$$

Hence there is a point $x_2 \in \frac{1}{2^2}U_0$ with $y_2 = A(x_2)$. Inductively we choose $x_\nu \in \frac{1}{2^\nu}U_0$, $y_\nu = A(x_\nu)$, $\nu = 1, 2, \ldots$, so that

$$y_{\nu+1} = y_\nu - A_0(x_\nu) \in \frac{1}{2^\nu}K \subset A\left(\frac{1}{2^{\nu+1}}U_0\right).$$

Thus, $\lim_{\nu\to\infty} y_\nu = 0$ and

(7.3.26)
$$\begin{aligned} y_{\nu+1} &= y_\nu - A_0(x_\nu) = y_{\nu-1} - A_0(x_{\nu-1}) - A_0(x_\nu) \\ &= \cdots = y_0 - A_0\left(\sum_{j=0}^{\nu} x_j\right). \end{aligned}$$

We would like to re-choose them so that $\sum_{j=0}^{\infty} x_j$ converges.

Let d_E (resp. $U(\varepsilon)$) be as in (7.3.10) (resp. (7.3.20)). We take a fundamental neighborhood system $\{U_p\}_{p=0}^{\infty}$ of $0 \in E$ as follows:

(i) U_0 above may be assumed to satisfy $U_0 \subset U(1)$; moreover, $U_p \subset U(2^{-p})$, $p = 1, 2, \ldots$.
(ii) Every U_p is convex and symmetric; i.e., $-U_p = U_p$.
(iii) $U_{p+1} \subset \frac{1}{2}U_p$, $p = 0, 1, \ldots$.

We consider an open covering of K,

$$K \subset A\left(\left(\bigcup_{\mu=1}^{\infty} 2^{\mu} U_p\right) \cap \frac{1}{2} U_0\right) = \bigcup_{\mu=1}^{\infty} A\left((2^{\mu} U_p) \cap \frac{1}{2} U_0\right).$$

Then there is a number $N(p)(\geq 1)$ such that

$$K \subset A\left((2^{N(p)} U_p) \cap \frac{1}{2} U_0\right). \tag{7.3.27}$$

We may assume that $N(p) < N(p+1)$ $(p = 1, 2, \ldots)$. For $0 \leq \nu \leq N(1)$ we take x_ν chosen above, and set

$$\tilde{x}_0 = x_0 + \cdots + x_{N(1)}.$$

For $N(p) < \nu \leq N(p+1)$ $(p = 1, 2, \ldots)$ we have by (7.3.27)

$$\frac{1}{2^{\nu-1}} K \subset A\left((2^{N(p)-\nu+1} U_p) \cap \frac{1}{2^\nu} U_0\right).$$

Since $y_\nu \in \frac{1}{2^{\nu-1}} K$, we take $x_\nu \in \left(2^{N(p)-\nu+1} U_p\right) \cap \frac{1}{2^\nu} U_0$ with $A(x_\nu) = y_\nu$. Then,

$$\begin{aligned} \tilde{x}_p &:= x_{N(p)+1} + \cdots + x_{N(p+1)} \in \left(1 + \frac{1}{2} + \cdots + \frac{1}{2^{N(p+1)-N(p)-1}}\right) U_p \\ &\subset 2U_p \subset U_{p-1} \subset U\left(\frac{1}{2^{p-1}}\right); \end{aligned} \tag{7.3.28}$$

$$d_E(\tilde{x}_p, 0) < \frac{1}{2^{p-1}}. \tag{7.3.29}$$

For every $p > q > q_0$ we have by (7.3.11) and (7.3.29)

$$\begin{aligned} d_E\left(\sum_{\nu=0}^{p} \tilde{x}_\nu, \sum_{\nu=0}^{q} \tilde{x}_\nu\right) &\leq \sum_{\nu=q+1}^{p} d_E(\tilde{x}_\nu, 0) < \sum_{\nu=q+1}^{p} \frac{1}{2^{\nu-1}} \\ &< \frac{1}{2^{q-1}} \leq \frac{1}{2^{q_0}} \to 0 \quad (q_0 \to \infty). \end{aligned}$$

Therefore $\sum_{\nu=0}^{\infty} \tilde{x}_\nu$ is a Cauchy series. Since d_E is complete, there is a limit $w = \sum_{\nu=0}^{\infty} \tilde{x}_\nu \in E$. By (7.3.26)

$$y_{N(p+1)+1} = y_0 - A_0\left(\sum_{\nu=0}^{p} \tilde{x}_\nu\right),$$

and letting $p \to \infty$, we get $y_0 = A_0(w)$. Hence, $A_0(E) \supset V_0$. □

Remark 7.3.30 (i) The above statement of L. Schwartz's Theorem 7.3.23 is slightly generalized from the original one, where F was assumed also to be Fréchet (cf. L. Schwartz [71], S. Hitotsumatsu [31], L. Bers [6], Gunning–Rossi [29], Grauert–Remmert [27], J.-P. Demailly [13], etc.).

(ii) The above proof of Theorem 7.3.23 is due to Demailly's notes [13], Chap. IX Theorem 1.8 (b). It is considerably simplified, compared with those in the references listed above (other than [13]).

7.4 Oka's Theorem

The most essential part of Oka's original proof of Levi's Problem (Hartogs' Inverse Problem) was a Connecting Lemma for two domains of holomorphy (it was called Oka's Heftungslemma). Afterwards, A. Andreotti and R. Narasimhan [2] extended this over complex spaces with applications. T. Nishino [49] Chap. 4 dealt with this problem over complex spaces.

The proof that we are going to give here is another one due to H. Grauert [22]. It relies on L. Schwartz's Finitenss Theorem 7.3.23 proved in the previous section. The proof of Grauert reveals the essence of Oka's Coherence Theorem, and his theorem has broad applications (see the next chapter).

Theorem 7.4.1 (Grauert) *Let Ω be a strongly pseudoconvex domain of $\mathbf{C}^n$. Then, as a complex vector space,*

$$\dim_{\mathbf{C}} H^1(\Omega, \mathscr{O}_\Omega) < \infty.$$

N.B. The theorem itself holds more generally for coherent sheaves over complex spaces (for the case of complex manifolds, see Theorems 7.5.26 and 8.4.1). For *univalent* domains, the above form is sufficient for our purpose.

Proof Let φ be a strongly plurisubharmonic function defining $\partial\Omega$ (cf. Definition 7.2.7). The proof is divided into several steps:

Step 1. For each point $a \in \partial\Omega$ we take double ball neighborhoods $U = B(a; \delta)$ of Lemma 7.2.16 and $V = B(a; \delta/2) \Subset U$. Since $\partial\Omega$ is compact, we can cover it with finitely many such $V_i \Subset U_i$ with center $a_i \in \partial\Omega$:

$$\partial\Omega \subset \bigcup_{i=1}^{l} V_i \Subset \bigcup_{i=1}^{l} U_i. \tag{7.4.2}$$

Since $\Omega \setminus \bigcup_{i=1}^{l} V_i$ is compact, it is covered by finitely many double ball neighborhoods:

$$V_i = B(a_i; \delta_i/2) \Subset U_i = B(a_i; \delta_i) \Subset \Omega, \qquad i = l+1, \ldots, L. \tag{7.4.3}$$

Since all $\Omega \cap V_i$ and $\Omega \cap U_i$, $1 \leq i \leq L$, are holomorphically convex, $\mathscr{V} = \{\Omega \cap V_i\}$ and $\mathscr{U} = \{\Omega \cap U_i\}$ are Leray coverings with respect to the sheaf $\mathscr{O}_\Omega$. Therefore, in particular,

$$H^1(\Omega, \mathscr{O}_\Omega) \cong H^1(\mathscr{V}, \mathscr{O}_\Omega) \cong H^1(\mathscr{U}, \mathscr{O}_\Omega). \tag{7.4.4}$$

Step 2. Take a C^∞ function $c_1(z) \geq 0$ such that

$$\operatorname{Supp} c_1 \subset U_1, \qquad c_1|_{V_1} = 1.$$

With a sufficiently small $\varepsilon > 0$, $\varphi_\varepsilon(z) := \varphi(z) - \varepsilon c_1(z)$ is strongly plurisubharmonic in a neighborhood of $\bigcup_{i=1}^{l} \bar{U}_i$. We set

$$W_1 = U_1 \cap \{\varphi_\varepsilon < 0\} \tag{7.4.5}$$

(cf. Fig. 7.2). Letting $\varepsilon > 0$ smaller if necessary, we see that

$$\overline{V_1 \cap \Omega} \Subset W_1$$

and every point b of $\partial W_1 \setminus \Omega$ is a holomorphically convex point of W_1 (Lemma 7.2.16), and moreover that b is a holomorphically convex point of $(U_j \cap W_1) \cup (U_j \cap \Omega)$ for other U_j having non-empty intersection with U_1 (cf. Fig. 7.3). We set

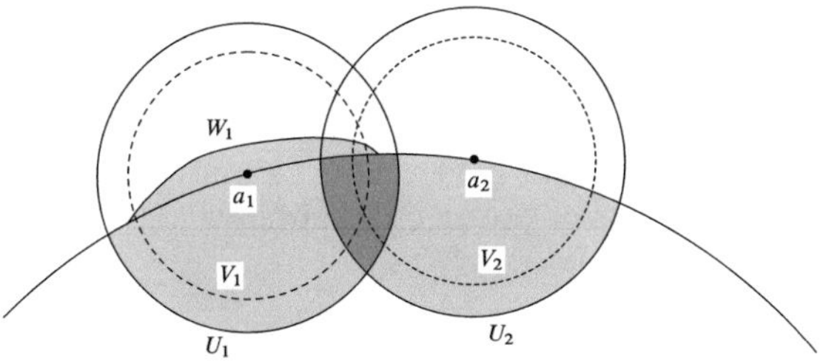

Fig. 7.2 Boundary bumping method 1

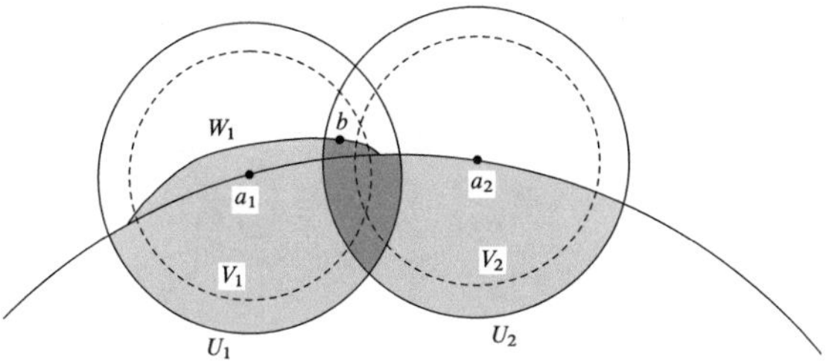

Fig. 7.3 Boundary bumping method 2

$$\begin{aligned} U_1^{(1)} &= W_1, \qquad U_j^{(1)} = U_j \cap \Omega, \quad j \geq 2, \\ \mathscr{U}^{(1)} &= \{U_j^{(1)}\}_{j=1}^L, \\ \Omega^{(1)} &= \bigcup_{j=1}^L U_j^{(1)}. \end{aligned}$$

It follows that $\mathscr{U}^{(1)}$ is a Leray covering of $\Omega^{(1)}$ with respect to $\mathscr{O}_{\Omega^{(1)}}$, and for 1-simplexes $\sigma = (U_{j_0}, U_{j_1}) \in N_1(\mathscr{U})$ and $\tau = \left(U_{j_0}^{(1)}, U_{j_1}^{(1)}\right) \in N_1(\mathscr{U}^{(1)})$ with the same pair of indices (j_0, j_1), where $j_0 \neq j_1$, we have that $|\sigma| = |\tau|$. Therefore we obtain the following equality and an exact sequence:

$$(7.4.6) \qquad \begin{aligned} &Z^1(\mathscr{U}, \mathscr{O}_\Omega) = Z^1(\mathscr{U}^{(1)}, \mathscr{O}_{\Omega^{(1)}}), \\ &H^1(\Omega^{(1)}, \mathscr{O}_{\Omega^{(1)}}) \cong H^1(\mathscr{U}^{(1)}, \mathscr{O}_{\Omega^{(1)}}) \to H^1(\mathscr{U}, \mathscr{O}_\Omega) \cong H^1(\Omega, \mathscr{O}_\Omega) \to 0. \end{aligned}$$

Step 3. We change the covering of $\Omega^{(1)}$ as follows. Now, W_1 is already taken in (7.4.4). Set

$$\begin{aligned} W_j &= \Omega^{(1)} \cap U_j, \qquad j \geq 2, \\ \mathscr{W} &= \{W_j\}_{j=1}^L. \end{aligned}$$

Since all W_j are holomorphically convex, $\{W_j\}_{j=1}^L$ is a Leray covering with respect to $\mathscr{O}_{\Omega^{(1)}}$. Therefore,

$$(7.4.7) \qquad H^1(\Omega^{(1)}, \mathscr{O}_{\Omega^{(1)}}) \cong H^1(\mathscr{W}, \mathscr{O}_{\Omega^{(1)}}).$$

Step 4. For $\Omega^{(1)} = \bigcup_j W_j$ and W_2 we practice the procedures of Step 2 and Step 3. Repeating this procedure l-times, we enlarge outward all $U_i \cap \partial\Omega$, $i = 1, 2, \ldots, l$, and denote the resulting covering of $\partial\Omega$ by

$$\tilde{U}_1, \tilde{U}_2, , \ldots, \tilde{U}_l$$

(cf. Fig. 7.4), and after the $(l + 1)$-th we put without change,

$$\tilde{U}_i = U_i, \qquad l + 1 \leq i \leq L.$$

Now we set

$$\tilde{\mathscr{U}} = \{\tilde{U}_i\}_{i=1}^L, \qquad \tilde{\Omega} = \bigcup_{i=1}^L \tilde{U}_i.$$

From the construction and (7.4.6) we infer that

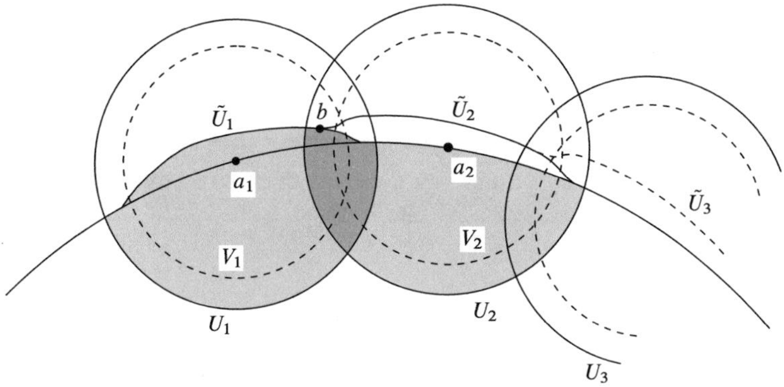

Fig. 7.4 Boundary bumping method 3

$$(7.4.8) \qquad V_i \Subset \tilde{U}_i, \qquad 1 \le i \le L,$$
$$\tilde{\rho} : H^1(\tilde{\mathscr{U}}, \mathscr{O}_{\tilde{\Omega}}) \to H^1(\mathscr{V}, \mathscr{O}_{\Omega}) \to 0.$$

Here, $\tilde{\rho}$ is the homomorphism naturally induced from the restriction morphisms. Therefore we obtain the following surjective homomorphism:

$$(7.4.9) \ \Psi : \xi \oplus \eta \in Z^1(\tilde{\mathscr{U}}, \mathscr{O}_{\tilde{\Omega}}) \oplus C^0(\mathscr{V}, \mathscr{O}_{\Omega}) \to \rho(\xi) + \delta\eta \in Z^1(\mathscr{V}, \mathscr{O}_{\Omega}) \to 0,$$

where ρ denotes the restriction morphism from $\tilde{U}_\alpha \cap \tilde{U}_\beta$ to $V_\alpha \cap V_\beta$, and by definition

$$H^1(\mathscr{V}, \mathscr{O}_{\tilde{\Omega}}) = Z^1(\mathscr{V}, \mathscr{O}_{\tilde{\Omega}}) / \delta C^0(\mathscr{V}, \mathscr{O}_{\tilde{\Omega}}).$$

Since $V_\alpha \cap V_\beta \Subset \tilde{U}_\alpha \cap \tilde{U}_\beta$, ρ is completely continuous by Theorem 7.3.17. It follows from Schwartz's Theorem 7.3.23 that

$$\mathrm{Coker}(\Psi - \rho) = Z^1(\mathscr{V}, \mathscr{O}_{\Omega}) / \delta C^0(\mathscr{V}, \mathscr{O}_{\Omega}) = H^1(\mathscr{V}, \Omega_{\Omega})$$

is finite dimensional. We see by (7.4.4) that $\dim H^1(\Omega, \mathscr{O}_\Omega) < \infty$. □

With the preparation above we prove Oka's Theorem for univalent domains:

Theorem 7.4.10 *Every pseudoconvex domain Ω ($\subset \mathbf{C}^n$) is holomorphically convex (equivalently, a domain of holomorphy).*

The steps of the proof are as follows: We construct an open covering of Ω by increasing strongly pseudoconvex domains, $\Omega_\nu \Subset \Omega_{\nu+1}$, $\nu = 1, 2, \ldots$. We next show that a strongly pseudoconvex domain is a holomorphically convex domain; together with Behnke–Stein's Theorem 5.4.10, we conclude that Ω is a holomorphically convex domain.

Lemma 7.4.11 *A pseudoconvex domain Ω carries an open covering by increasing strongly pseudoconvex domains Ω_ν of Ω $(\nu = 1, 2, \ldots)$ such that $\Omega_\nu \Subset \Omega_{\nu+1}$ and $\bigcup_{\nu=1}^{\infty} \Omega_\nu = \Omega$.*

Proof Let $\varphi : \Omega \to [-\infty, \infty)$ be a plurisubharmonic exhaustion function. We set $\Omega'_\nu = \{\varphi < \nu\}$, $\nu = 1, 2, \ldots$. Fix any $\nu \in \mathbf{N}$. Let φ_ε be a smoothing of φ. With sufficiently small $\varepsilon > 0$, φ_ε is of C^∞-class in $\Omega'_{\nu+1}$, and

$$\Omega'_\nu \Subset \left\{\varphi_\varepsilon < \nu + \frac{1}{2}\right\} \Subset \Omega'_{\nu+1}.$$

With another sufficiently small $\varepsilon' > 0$, $\psi_\nu(z) = \varphi_\varepsilon(z) + \varepsilon'\|z\|^2$ satisfies that

$$\Omega'_\nu \Subset \left\{\psi_\nu < \nu + \frac{1}{2}\right\} \Subset \Omega'_{\nu+1}.$$

We set

$$\Omega''_\nu = \left\{\psi_\nu < \nu + \frac{1}{2}\right\}.$$

Since ψ_ν is strongly plurisubharmonic in $\Omega'_{\nu+1}$, Ω''_ν is a relatively compact open subset consisting of finitely many strongly pseudoconvex domains. Defining Ω''_ν, $\nu = 1, 2, \ldots$, in this way, we obtain

$$\Omega''_\nu \Subset \Omega_{\nu+1}, \quad \bigcup_{\nu=1}^{\infty} \Omega''_\nu = \Omega.$$

To choose domains, we take a point $a_0 \in \Omega''_1$, and denote by Ω_ν the connected component of Ω''_ν containing a_0. Then, $\Omega_\nu \Subset \Omega_{\nu+1}$ for $\nu \geq 1$. For every $a \in \Omega$, there is a curve C in Ω connecting a_0 and a. Since C is compact, there is some ν such that $\Omega''_\nu \supset C$, and so $\Omega_\nu \supset C$. Therefore we see that $a \in \Omega_\nu$ and $\{\Omega_\nu\}_{\nu=1}^{\infty}$ is an open covering of Ω. □

Theorem 7.4.12 *Every boundary point of a strongly pseudoconvex domain Ω is a holomorphically convex point of Ω; hence, Ω is holomorphically convex.*

Proof Let φ be a plurisubharmonic function defining Ω such that φ is strongly plurisubharmonic in a neighborhood of $\partial\Omega$ and $\Omega = \{\varphi < 0\}$. Take a point $b \in \partial\Omega$. By a translation, we may put $b = 0$. Set

$$Q(z) = 2\sum_{j=1}^{n} \frac{\partial \varphi}{\partial z_j}(0) z_j + \sum_{j,k} \frac{\partial^2 \varphi}{\partial z_j \partial z_k}(0) z_j z_k.$$

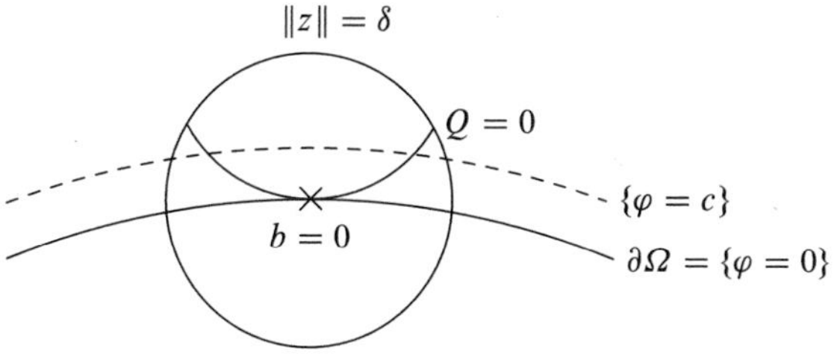

Fig. 7.5 Strongly pseudoconvex boundary point

It follows from (7.2.17) that there are positive numbers $\varepsilon, \delta > 0$ satisfying

$$\varphi(z) \geq \Re Q(z) + \varepsilon\|z\|^2, \quad \|z\| \leq \delta,$$
$$\inf\{\varphi(z); Q(z) = 0, \|z\| = \delta\} \geq \varepsilon\delta^2 > 0.$$

Taking c with $0 < c < \varepsilon\delta^2$, we set $\Omega' = \{\varphi < c\}$ (cf. Fig. 7.5). We set, moreover, $U_0 = B(0; \delta) \cap \Omega'$ and $Y = \{Q = 0\} \cap U_0$. By the choice, Y is an analytic hypersurface of Ω'; in particular, Y is closed and $U_1 = \Omega' \setminus Y$ is open. We set

$$f_{01}(z) = \frac{1}{Q(z)}, \quad z \in U_0 \cap U_1,$$
$$f_{10}(z) = -f_{01}(z), \quad z \in U_1 \cap U_0.$$

Then, $\mathscr{U} = \{U_0, U_1\}$ is an open covering of Ω', and a 1-cocyle $f = (f_{01}(z), f_{10}(z)) \in Z^1(\mathscr{U}, \mathscr{O}_{\Omega'})$ is obtained. For $k \in \mathbf{N}$ we define

$$f_{01}^{[k]}(z) = (f_{01}(z))^k, \quad z \in U_0 \cap U_1,$$
$$f_{10}^{[k]}(z) = -f_{01}^{[k]}(z), \quad z \in U_1 \cap U_0.$$

Then $(f^{[k]}) \in Z^1(\mathscr{U}, \mathscr{O}_{\Omega'})$. Thus we obtain cohomology classes,

$$[f^{[k]}] \in H^1(\mathscr{U}, \mathscr{O}_{\Omega'}) \hookrightarrow H^1(\Omega', \mathscr{O}_{\Omega'}), \quad k \in \mathbf{N}.$$

(see Proposition 3.4.11 for the injection "$\hookrightarrow$"). Since Ω' is strongly pseudoconvex, Theorem 7.4.1 implies the finite dimensionality of $H^1(\Omega', \mathscr{O}_{\Omega'})$. Therefore, for N large, there is a non-trivial linear relation,

$$\sum_{k=1}^{N} c_k[f^{[k]}] = 0 \in H^1(\mathscr{U}, \mathscr{O}_{\Omega'}) \quad (c_k \in \mathbf{C}).$$

We may suppose that $c_N \neq 0$. Then there exists elements $g_i \in \mathscr{O}(U_i)$, $i = 0, 1$, such that

$$\sum_{k=1}^{N} \frac{c_k}{Q^k(z)} = g_1(z) - g_0(z), \quad z \in U_0 \cap U_1.$$

Therefore,

$$g_0(z) + \sum_{k=1}^{N} \frac{c_k}{Q^k(z)} = g_1(z), \quad z \in U_0 \cap U_1, \quad c_N \neq 0.$$

It is inferred that there is a meromorphic function F in Ω' with poles of order N on Y. Since $Y \cap \Omega = \emptyset$, the restriction $F|_\Omega$ of F to Ω is holomorphic, and

$$\lim_{z \to 0} |F(z)| = \infty.$$

Therefore, $b = 0 \in \partial\Omega$ is a holomorphically convex point of Ω.

It follows from this and Lemma 7.2.15 that Ω is holomorphically convex. □

Proof of Theorem 7.4.10. Let Ω be a pseudoconvex domain. Let Ω_ν, $\nu \in \mathbf{N}$, be an increasing sequence of strongly pseudoconvex subdomains of Ω obtained in Lemma 7.4.11. By Theorem 7.4.12, all Ω_ν are domains of holomorphy. It follows from the Behnke–Stein Theorem 5.4.10 that Ω is a domain of holomorphy. □

The following is immediate from Theorem 7.4.10 and the Oka–Cartan Fundamental Theorem 4.4.2.

Corollary 7.4.13 *Let Ω be a pseudoconvex domain, and let $\mathscr{F} \to \Omega$ be a coherent sheaf. Then,*

$$H^q(\Omega, \mathscr{F}) = 0, \quad q \geq 1.$$

In particular, we have the following (cf. Corollary 4.4.20).

Corollary 7.4.14 ($\bar\partial$-equation) *Let Ω be a pseudoconvex domain. Then, for every $f \in \Gamma(\Omega, \mathscr{E}_\Omega^{(p,q)})$ with $q \geq 1$ and $\bar\partial f = 0$ there is an element $g \in \Gamma(\Omega, \mathscr{E}_\Omega^{(p,q-1)})$ such that $\bar\partial g = f$.*

Remark 7.4.15 (i) In the above proof of Theorem 7.4.10 it is a nice idea of H. Grauert to claim only the finite dimensionality of $H^1(\Omega, \mathscr{O}_\Omega)$ (Theorem 7.4.1) for strongly pseudoconvex Ω ($\Subset \mathbf{C}^n$), which is weaker than the *a posteriori* statement, $H^1(\Omega, \mathscr{O}_\Omega) = 0$, just as above. That weakened claim made the proof considerably easier than the original ones of Oka VI and IX.

(ii) L. Hörmander [33] solves first Corollary 7.4.14 by the so-called $\bar\partial$–L^2 method, and then apply it to solve Levi's Problem and the Cousin I, II Problems. The approach is very opposite to the one of this book.

7.5 Oka's Theorem on Riemann Domains

It is known now that there are a number of proofs of Oka's Theorem (Oka IX) to solve Levi's Problem (Hartogs' Inverse Problem) for Riemann domains. In T. Nishino [49] the proof is based on Oka's original one. Other proofs with including extensions are those in Docquier–Grauert [14], R. Narasimhan [48], Gunning–Rossi [29], L. Hörmander [33] (in this book the holomorphic separation condition is already assumed in the definition of a Riemann domain, which is stronger than in the present text). In the present section we give a proof as easily as possible by making use of a specialty of Riemann domains, based on [53].

7.5.1 *Riemann Domains*

We begin with the definition. Let X be a complex manifold, and let $\pi : X \to \mathbf{C}^n$ be a holomorphic map.

Definition 7.5.1 An unramified covering domain $\pi : X \to \mathbf{C}^n$ (Definition 4.5.6) over $\mathbf{C}^n$ is called a *Riemann domain.*

It is noted that X is connected, and π is locally biholomorphic.

A Riemann domain X can be endowed with a Riemannian metric, the pull-backed Euclidean metric on $\mathbf{C}^n$ through π, which we call the *Euclidean metric on X*. Therefore X is metrizable, and hence satisfies the second countability axiom.

In the present section, X always denotes a Riemann domain. Take a polydisk $\mathrm{P}\Delta = \mathrm{P}\Delta(0; r_0)$ $(r_0 = (r_{0j}))$ with center at the origin $0 \in \mathbf{C}^n$. By definition, for any point $x \in X$ there are a number $\rho > 0$ and a neighborhood $U_\rho(x) \ni x$ such that

$$\pi|_{U_\rho(x)} : U_\rho(x) \to \pi(x) + \rho\mathrm{P}\Delta$$

is biholomorphic. We denote the supremum of such $\rho > 0$ by

$$\delta_{\mathrm{P}\Delta}(x, \partial X) = \sup\{\rho > 0; {}^{\exists}U_\rho(x)\} \leq \infty,$$

and call it the *boundary distance function* of X.[5]

If $\delta_{\mathrm{P}\Delta}(x, \partial X) = \infty$, then π is a biholomorphism, and there is nothing to discuss further. In what follows, we assume that $\delta_{\mathrm{P}\Delta}(x, \partial X) < \infty$.

The function $\delta_{\mathrm{P}\Delta}(x, \partial X)$ is a continuous function satisfying Lipschitz' condition (the proof is similar to the case of univalent domains; cf. (5.3.4)). For a subset $A \subset X$ we set

---

[5]Here we use "∂X" just symbolically without defining "∂X", but we may define ∂X as the "ideal boundary" or the "accessible boundary" of X relative to the mapping $\pi : X \to \mathbf{C}^n$; cf., e.g., Fritzsche–Grauert [19] Chap. 2.

$$\delta_{\mathrm{P}\Delta}(A, \partial X) = \inf_{x \in A} \delta_{\mathrm{P}\Delta}(x, \partial X).$$

For an open subset $\Omega \subset X$ we use $\delta_{\mathrm{P}\Delta}(x, \partial\Omega)$ and $\delta_{\mathrm{P}\Delta}(A, \partial\Omega)$ $(A \subset \Omega)$ defined similarly as above.

Definition 7.5.2 (*cf. Definitions* 1.2.30 *and* 5.1.1) Let $\pi : X \to \mathbf{C}^n$ be a Riemann domain satisfying the following condition:

7.5.3 For any distinct points $a, b \in X$ with $\pi(a) = \pi(b)$ there is an element $f \in \mathscr{O}(X)$ with $f(a) \neq f(b)$ (cf. Exc. 5 at the end of this chapter).

Then, we define as follows:

(i) A Riemann domain $\tilde{\pi} : \tilde{X} \to \mathbf{C}^n$ is called an *extension of holomorphy* of $\pi : X \to \mathbf{C}^n$, if there is a holomorphic open embedding $\varphi : X \hookrightarrow \tilde{X}$ such that $\pi = \varphi \circ \tilde{\pi}$ and with regarding $X \subset \tilde{X}$ through φ, every $f \in \mathscr{O}(X)$ is analytically extended over $\tilde{X}$ $(\varphi^* \mathscr{O}(\tilde{X}) = \mathscr{O}(X))$.
(ii) The maximal one among the extensions of holomorphy of X is called the *envelope of holomorphy* of X.
(iii) X is called a *domain of holomorphy* if X itself is the envelope of holomorphy of X.

Example 7.5.4 We give an example which explains the necessity of condition 7.5.3. Let $\mathrm{P}\Delta = \Delta(0; 1)^2 \subset \mathbf{C}^2$ be the unit polydisk, and set

$$\begin{aligned} K &= \left\{(z_1, z_2) \in \mathrm{P}\Delta;\ \tfrac{1}{4} \le |z_j| \le \tfrac{3}{4}\right\}, \\ \Omega &= \mathrm{P}\Delta \setminus K, \\ \Omega_{\mathrm{H}} &= \left(\Delta(0; 1) \times \Delta(0; \tfrac{1}{4})\right) \cup \left(\{\tfrac{3}{4} < |z_1| < 1\} \times \Delta(0; 1)\right), \end{aligned}$$

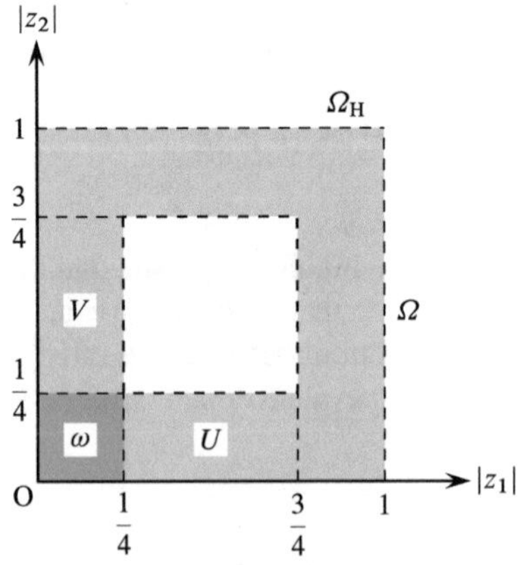

Fig. 7.6 Riemann domain X

$$\omega = \Delta(0; \tfrac{1}{4})^2,$$
$$U = \Delta(0; 1) \times \Delta(0; \tfrac{1}{4}),$$
$$V = \Delta(0; \tfrac{1}{4}) \times \Delta(0; 1).$$

Then, U and V are subdomains of Ω, and $U \cap V = \omega$. We form a Riemann domain X by distinguishing ω of U and ω of V with the natural projection $\pi : X \to \Omega \subset \mathbf{C}^2$, which is 2-sheeted over ω (cf. Fig. 7.6). Since Ω_{H} $(\subset X)$ is a Hartogs domain with the envelope of holomorphy $\mathrm{P}\Delta$, we canonically have

$$\mathscr{O}(X) \cong \mathscr{O}(\mathrm{P}\Delta).$$

Therefore, X does not satisfy Condition 7.5.3, and there is no domain of holomorphy that contains X as a subdomain.

Lemma 7.5.5 *Let $\pi : X \to \mathbf{C}^n$ be a domain of holomorphy. Let $K \Subset X$ be a compact subset, and let $\hat{K}_X$ be the holomorphically convex hull of K. If for a holomorphic function $f \in \mathscr{O}(X)$,*

$$\delta_{\mathrm{P}\Delta}(x, \partial X) \geq |f(x)|, \quad x \in K,$$

then

$$\delta_{\mathrm{P}\Delta}(x, \partial X) \geq |f(x)|, \quad x \in \hat{K}_X.$$

In particular, with constant f,

$$\delta_{\mathrm{P}\Delta}(K, \partial X) = \delta_{\mathrm{P}\Delta}(\hat{K}_X, \partial X). \tag{7.5.6}$$

The proof depends only on a property of convergent domains of power series expansions of holomorphic functions, and hence the proof in the case of univalent domains is applied (cf. Lemma 5.3.7). By making this in the same way as in the proof of Theorem 5.3.1 we see:

Theorem 7.5.7 *The following three conditions are equivalent for a Riemann domain X.*

(i) *X is a domain of holomorphy.*
(ii) *There is an element $f \in \mathscr{O}(X)$ whose domain of existence is X.*
(iii) *X is holomorphically convex.*

It is also the same as in the case of univalent domains that Lemma 7.5.5 implies:

Theorem 7.5.8 (Oka) *If X is a domain of holomorphy, $-\log \delta_{\mathrm{P}\Delta}(x, \partial X)$ is plurisubharmonic.*

7.5.2 Pseudoconvexity

We define the pseudoconvexity for a general complex manifold.

Definition 7.5.9 A complex manifold M is said to be *pseudoconvex* if there is a plurisubharmonic exhaustion function $\phi : M \to \mathbf{R}$.

Lemma 7.5.10 (Oka IX) *Let X be a Riemann domain. If $-\log \delta_{\mathrm{P}\Delta}(x, \partial X)$ is plurisubharmonic, then X is pseudoconvex.*

Proof Although the proof is elementary, it is a little bit long; it is divided into several steps.

Step 1. Take a point $x_0 \in X$ and a number, $0 < \rho < \delta_{\mathrm{P}\Delta}(x_0, \partial X)$. We denote by X_ρ the connected component of $\{x \in X; \delta_{\mathrm{P}\Delta}(x, \partial X) > \rho\}$, containing x_0. It follows that $X_\rho \subset X_{\rho'}$ for $0 < \rho' < \rho$ and $\bigcup_{\rho>0} X_\rho = X$.

For a point $x \in X_\rho$ we denote by $L(C(x))$ the length of a piece-wise C^1 curve $C(x)$ joining x and x_0 in X_ρ with respect to the Euclidean metric on X. Define a function d_ρ for $x \in X_\rho$:

$$d_\rho(x) = \inf_{C(x) \subset X_\rho} L(C(x)).$$

The Lipschitz continuity

$$\begin{aligned} (7.5.11) \qquad |d_\rho(x') - d_\rho(x'')| &\le \|\pi(x') - \pi(x'')\| \\ &= \|x' - x''\|, \quad x', x'' \in U_\rho(x) \end{aligned}$$

holds. Here we identified points x', x'' contained in a univalent subdomain $U_\rho(x)$ and $\pi(x'), \pi(x'')$ in $\mathbf{C}^n$. For simplicity of notation we use this identification to deal with points in a univalent subdomain unless confusion occurs.

Since X_ρ is not necessarily relatively compact in X in general, we would like to exhaust it by relatively compact subdomains by making use of d_ρ.

Lemma 7.5.12 *For every $b > 0$, $\{x \in X_\rho;\ d_\rho(x) < b\} \Subset X$.*

Proof Let $b = \rho$. It follows from the choice that

$$\{x \in X_\rho;\ d_\rho(x) \le \rho\} \subset \bar{U}_\rho(x_0) \Subset U_{\delta_{\mathrm{P}\Delta}(x_0)}(x_0).$$

Therefore, $\{x \in X_\rho; d_\rho(x) \le \rho\} \Subset X$, so that the assertion holds for $b = \rho$, and hence for $0 < b \le \rho$.

Suppose that the assertion holds for a number $b \ge \rho$. That is, $K := \{x \in \bar{X}_\rho;\ d_\rho(x) \le b\}$ is compact. For any point $x \in K$, $\bar{U}_{\rho/2}(x) \Subset X$. We infer that

$$K' = \bigcup_{x \in K} \bar{U}_{\rho/2}(x)$$

is compact. For, taking a sequence of points $y_\nu \in K'$, $\nu \in \mathbf{N}$, we find $x_\nu \in K$, $w_\nu \in \mathbf{C}^n$ with $\|w_\nu\| \le \rho/2$ such that

$$y_\nu = x_\nu + w_\nu, \quad \nu \in \mathbf{N}.$$

Since K is compact, after taking a subsequence, we have that $\lim_{\nu\to\infty} x_\nu = x_0 \in K$ and $\lim_{\nu\to\infty} w_\nu = w_0$ with $\|w_0\| \le \rho/2$. Thus,

$$\lim_{\nu\to\infty} y_\nu = x_0 + w_0 \in K'.$$

Since $\{x \in X_\rho;\ d_\rho(x) < b+\rho/2\} \subset K'$, the assertion holds for $b + \rho/2$. Inductively, for $\rho + \nu\rho/2$, $\nu = 1, 2, \ldots$, the assertion holds. Therefore, it holds for all $b > 0$. □

Step 2. As in Definition 7.1.38, we take a C^∞ function $\chi(z) \ge 0$ on $\mathbf{C}^n$ as follows:

$$\text{(7.5.13)} \qquad \operatorname{Supp} \chi \subset \mathrm{P}\Delta,$$
$$\int_{w\in\mathbf{C}^n} \chi(w)\alpha^n(w) = 1, \quad \alpha = \frac{i}{2\pi}\partial\bar\partial\,\|w\|^2.$$

For $\varepsilon > 0$ we set

$$\chi_\varepsilon(w) = \chi\left(\frac{w}{\varepsilon}\right)\frac{1}{\varepsilon^{2n}}.$$

One gets

$$\operatorname{Supp} \chi_\varepsilon \subset \varepsilon\mathrm{P}\Delta, \quad \int_{\mathbf{C}^n} \chi_\varepsilon(w)\alpha^n(w) = 1.$$

For $0 < \varepsilon \le \rho$ the smoothing of d_ρ is defined by

$$(d_\rho)_\varepsilon(x) = (d_\rho) * \chi_\varepsilon(x) = \int_{w\in\varepsilon\mathrm{P}\Delta} d_\rho(x+w)\chi_\varepsilon(w)\alpha^n(w), \quad x \in X_\rho$$

(cf. Definition 7.1.38), which is of C^∞-class on X_ρ.

We write $\mathrm{P}\Delta = \mathrm{P}\Delta(0; (r_{0j}))$ and put $C_0 = \sqrt{\sum_j r_{0j}^2}$. It follows from (7.5.11) that

$$|(d_\rho)_\varepsilon(x) - d_\rho(x)| \le \varepsilon C_0, \quad x \in X_\rho.$$

Therefore we see that

$$\text{(7.5.14)} \qquad \{x \in X_\rho;\ (d_\rho)_\varepsilon(x) < b\} \Subset X, \quad {}^\forall b > 0.$$

About a point $x \in X$, we have a local coordinate system $\pi(x) = (z_j) = (x_j + iy_j)$, and write $\frac{\partial}{\partial\xi}$ for the directional differential of a direction ξ in x_j, y_j, $1 \le j \le n$, with $\|\xi\| = 1$:

$$\lim_{h\to 0}\frac{(d_\rho)_\varepsilon(x+h\xi)-(d_\rho)_\varepsilon(x)}{h}=\frac{\partial (d_\rho)_\varepsilon}{\partial \xi}(x).$$

On the other hand, we deduce the following estimate from (7.5.11) that

$$\begin{aligned}
&\left|\frac{(d_\rho)_\varepsilon(x+h\xi)-(d_\rho)_\varepsilon(x)}{h}\right|\\
&=\left|\frac{1}{h}\int_W \{(d_\rho)(x+h\xi+w)-(d_\rho)(x+w)\}\chi_\varepsilon(w)\alpha^n(w)\right|\\
&\le \frac{1}{|h|}\int_W |(d_\rho)(x+h\xi+w)-(d_\rho)(x+w)|\,\chi_\varepsilon(w)\alpha^n(w)\\
&\le \frac{1}{|h|}C_0|h|\cdot\|\xi\|=C_0.
\end{aligned}$$

Therefore we obtain

$$\left|\frac{\partial (d_\rho)_\varepsilon}{\partial \xi}(x)\right|\le C_0,\quad x\in X_\rho,\ 0<\varepsilon\le\rho. \tag{7.5.15}$$

Taking $0<2\varepsilon\le\rho$, we consider

$$\tilde{d}_{\rho,\varepsilon}(x)=\big((d_\rho)_\varepsilon\big)_\varepsilon(x),\quad x\in X_\rho.$$

It follows that

$$\begin{aligned}
\frac{\partial \tilde{d}_{\rho,\varepsilon}}{\partial \xi}(x)&=\int_W \frac{\partial (d_\rho)_\varepsilon}{\partial \xi}(x+w)\chi_\varepsilon(w)\alpha^n(w)\\
&=\int_W \frac{\partial (d_\rho)_\varepsilon}{\partial \xi}(w)\chi\left(\frac{w-x}{\varepsilon}\right)\frac{1}{\varepsilon^{2n}}\alpha^n(w).
\end{aligned}$$

Letting $\frac{\partial}{\partial\eta}$ be another directional differential, similar to $\frac{\partial}{\partial\xi}$, we obtain

$$\frac{\partial^2 \tilde{d}_{\rho,\varepsilon}}{\partial\eta\partial \xi}(x)=\int_W \frac{\partial (d_\rho)_\varepsilon}{\partial \xi}(w)\frac{\partial\chi}{\partial\eta}\left(\frac{w-x}{\varepsilon}\right)\frac{-1}{\varepsilon^{2n+1}}\alpha^n(w).$$

This with (7.5.15) implies that

$$\begin{aligned}
\left|\frac{\partial^2 \tilde{d}_{\rho,\varepsilon}}{\partial\eta\partial \xi}(x)\right|&\le\int_W \left|\frac{\partial (d_\rho)_\varepsilon}{\partial \xi}(w)\right|\cdot\left|\frac{\partial\chi}{\partial\eta}\left(\frac{w-x}{\varepsilon}\right)\right|\frac{1}{\varepsilon^{2n+1}}\alpha^n(w)\\
&\le\frac{C_0}{\varepsilon}\int_W\left|\frac{\partial\chi}{\partial\eta}(w)\right|\alpha^n(w)=\frac{C_1}{\varepsilon}.
\end{aligned} \tag{7.5.16}$$

Note that C_1 is a positive constant independent from ε, ρ. Set

$$\hat{d}_\rho(x) = \tilde{d}_{\rho, \frac{\rho}{2}}(x), \quad x \in X_\rho.$$

By (7.5.14) we see also for $\hat{d}_\rho(x)$ that

$$\{x \in X_\rho; \hat{d}_\rho(x) < b\} \Subset X, \quad {}^\forall b > 0. \tag{7.5.17}$$

Setting

$$\varphi_\rho(x) = \hat{d}_\rho(x) + C_2 \|\pi(x)\|^2$$

with $C_2 \gg 2\frac{C_1}{\rho}$, we infer from (7.5.16) that

$$\sum_{j,k} \frac{\partial^2 \varphi_\rho}{\partial z_j \partial \bar{z}_k} \xi_j \bar{\xi}_k \geq \|(\xi_j)\|^2.$$

Thus, summarizing the above, we have:

Lemma 7.5.18 *There is a strongly plurisubharmonic C^∞ function $\varphi_\rho(x) > 0$ on X_ρ such that*

$$\{x \in X_\rho;\ \varphi_\rho(x) < b\} \Subset X, \quad {}^\forall b > 0.$$

Step 3. Now, we assume $-\log \delta_{\mathrm{P}\Delta}(x, \partial X)$ to be plurisubharmonic. Notice that $-\log \delta_{\mathrm{P}\Delta}(x, \partial X)$ is continuous.

Take an increasing sequence of relatively compact subdomains $\{\Omega_j\}_{j=1}^\infty$ of X such that $x_0 \in \Omega_1 \Subset \Omega_2 \Subset \cdots$ and

$$X = \bigcup_j \Omega_j. \tag{7.5.19}$$

Then we take a monotone increasing divergent sequence, $a_1 < a_2 < \cdots < a_j \nearrow \infty$ such that

$$\bar{\Omega}_j \Subset X_j := X_{e^{-a_j}}, \quad j = 1, 2, \ldots.$$

It follows that $X_j \subset X_{j+1}$ and $X = \bigcup_{j=1}^\infty X_j$. Lemma 7.5.18 applied to X_j yields a strongly plurisubharmonic C^∞ function $\varphi_j(x) > 0$ on X_j; here, however, only $\varphi_j(x)$ being continuous plurisubharmonic will be used. It is noticed that for every $b > 0$

$$\{x \in X_j;\ \varphi_j(x) < b\} \Subset X_{j+1}.$$

We choose a monotone increasing sequence, $b_1 < b_2 < \cdots \nearrow \infty$ as follows: Let $b_1 > 0$ be a number such that

$$\Delta_1 := \{x \in X_1;\ \varphi_4(x) < b_1\} \Supset \bar{\Omega}_1.$$

One sees that

$$\partial \Delta_1 \subset \{-\log \delta_{P\Delta}(x) = a_1\} \cup \{\varphi_4(x) = b_1\}. \tag{7.5.20}$$

Since $\Delta_1 \cup \Omega_2 \Subset X_2$, there is a number $b_2 > \max\{2, b_1\}$ such that

$$\Delta_2 = \{x \in X_2;\ \varphi_5(x) < b_2\} \Supset \Delta_1 \cup \Omega_2.$$

Inductively, one choose $b_j > \max\{j, b_{j-1}\}$ so that

$$\Delta_j = \{x \in X_j;\ \varphi_{j+3}(x) < b_j\} \Supset \Delta_{j-1} \cup \Omega_j$$

holds. It follows from (7.5.19) that

$$X = \bigcup_{j=1}^{\infty} \Delta_j.$$

Set $\Phi_1(x) = \varphi_4(x) + 1\ (> 1), x \in \Delta_4$. Assume that for $j \geq 1$, $\Phi_h(x)$, $1 \leq h \leq j$, are defined so that the following are satisfied:

7.5.21 (i) $\Phi_h(x)$ is a continuous plurisubharmonic function in Δ_{h+3}.
(ii) $\Phi_h(x) > h, {}^{\forall}x \in \Delta_{h+2} \backslash \Delta_{h+1}$, $1 \leq h \leq j$.
(iii) $\Phi_h(x) = \Phi_{h-1}(x), {}^{\forall}x \in \Delta_h$, $2 \leq h \leq j$.

We define a continuous plurisubharmonic function in Δ_{j+4} by

$$\psi_{j+1}(x) = \max\{-\log \delta_{P\Delta}(x, \partial X) - a_{j+1}, \varphi_{j+4}(x) - b_{j+1}\}, \quad x \in \Delta_{j+4}.$$

Then,

$$\begin{aligned} &\psi_{j+1}(x) < 0, \quad x \in \Delta_{j+1}, \\ &\min_{\bar{\Delta}_{j+3} \backslash \Delta_{j+2}} \psi_{j+1}(x) > 0. \end{aligned} \tag{7.5.22}$$

It follows from this that for a sufficiently large $k_{j+1} > 0$

$$\min_{\bar{\Delta}_{j+3} \backslash \Delta_{j+2}} k_{j+1}\psi_{j+1}(x) > \max\{j+1, \max_{\bar{\Delta}_{j+2}} \Phi_j(x)\}. \tag{7.5.23}$$

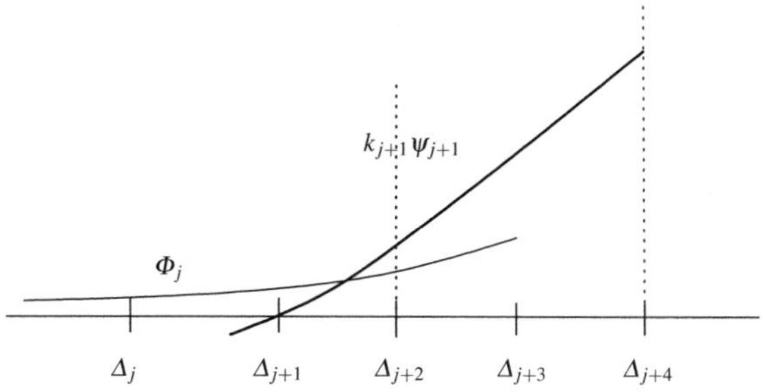

Fig. 7.7 Graph of exhaustion function

Set

$$\Phi_{j+1}(x) = \begin{cases} \max\{\Phi_j(x), k_{j+1}\psi_{j+1}(x)\}, & x \in \Delta_{j+2}, \\ k_{j+1}\psi_{j+1}(x), & x \in \Delta_{j+4}\setminus\Delta_{j+2}, \end{cases}$$

(cf. Fig. 7.7). By (7.5.23), $\Phi_{j+1}(x) = k_{j+1}\psi_{j+1}(x)$ in a neighborhood of $\partial\Delta_{j+2}$, and hence $\Phi_{j+1}(x)$ is a continuous plurisubharmonic function in Δ_{j+4}. It follows from (7.5.22) and (7.5.23) that

$$\Phi_{j+1}(x) = \Phi_j(x), \quad x \in \Delta_{j+1},$$
$$\Phi_{j+1}(x) > j+1, \quad x \in \Delta_{j+4}\setminus\Delta_{j+2}.$$

Thus inductively, $\Phi_j(x)$, $j = 1, 2, \ldots$, are obtained, so that they satisfy 7.5.21.

With setting

$$\Phi(x) = \lim_{j\to\infty} \Phi_j(x), \quad x \in X,$$

$\Phi(x)$ is a continuous plurisubharmonic function in X. It follows from 7.5.21 (ii) that

$$\Phi(x) > j, \quad x \in X\setminus\Delta_{j+1}, \quad j = 1, 2, \ldots.$$

and hence that $\Phi(x)$ is an exhaustion function. This completes the proof of Lemma 7.5.10. □

Remark 7.5.24 As in Theorem 7.2.12 it is possible to localize the plurisubharmonicity assumption of $-\log\delta_{P\Delta}(z, \partial\Omega)$ in Lemma 7.5.10 for a Riemann domain X at the "ideal boundary ∂X": For the details, cf. Oka [62] IX, Nishino [49], Chap. 4, Docquier–Grauert [14].

7.5.3 Strongly Pseudoconvex Domains

In general, let M be a complex manifold, and let $\Omega \Subset M$ be a relatively compact domain.

Definition 7.5.25 (*cf. Definition* 7.2.7) We say that Ω is *strongly pseudoconvex* if there is a strongly plurisubharmonic function $\phi : U \to \mathbf{R}$ in a neighborhood U of the boundary $\partial\Omega$ satisfying

$$\{x \in U;\ \phi(x) < 0\} = \Omega \cap U.$$

We extend Grauert's Theorem 7.4.1 a little bit as follows: In fact, this theorem holds for any coherent sheaf and for the cohomologies of all positive degrees, but the present form is sufficient for our purpose to solve Levi's Problem (Hartogs' Inverse Problem).

Theorem 7.5.26 (Grauert) *Let $\Omega \Subset M$ be a strongly pseudoconvex domain.*

(i) $\dim_{\mathbf{C}} H^1(\Omega, \mathscr{O}_\Omega) < \infty$.
(ii) *For the geometric ideal sheaf $\mathscr{I}\langle N\rangle$ of a complex submanifold $N \subset M$,*

$$\dim_{\mathbf{C}} H^1(\Omega, \mathscr{I}\langle N\rangle) < \infty.$$

Proof We look first at the proof of Theorem 7.4.1. It is observed that the proof relies only on the local informations and the local operations in a neighborhood $\partial\Omega$. Therefore, with the bumping method inside holomorphic local coordinate neighborhoods of $U \subset M$ that proof remains valid on M; this finishes (i).

In the proof of Theorem 7.4.1 the bumping method and L. Schwartz's Theorem 7.3.23 are used. In the proof of L. Schwartz's Theorem 7.3.23 the uniform convergence on compact subsets is used for the topology of the vector space of holomorphic functions. In (ii), $\mathscr{I}\langle N\rangle \subset \mathscr{O}_M$. For every open subset $U \subset M$, $\Gamma(U, \mathscr{O}_M)$ is endowed with the topology of uniform convergence on compact subsets, which naturally induces a topology in $\Gamma(U, \mathscr{I}\langle N\rangle) \subset \Gamma(U, \mathscr{O}_M)$. Note that $\Gamma(U, \mathscr{I}\langle N\rangle)$ is closed in $\Gamma(U, \mathscr{O}_M)$; i.e., if a sequence of $f_\nu \in \Gamma(U, \mathscr{I}\langle N\rangle)$ ($\nu = 1, 2, \ldots$) converges uniformly on compact subsets to $f \in \Gamma(U, \mathscr{O}_M)$, then necessarily $f \in \Gamma(U, \mathscr{I}\langle N\rangle)$. Thus (ii) is proved. □

Theorem 7.5.27 *Assume that $\Omega \Subset M$ is strongly pseudoconvex. Then every boundary point $x \in \partial\Omega$ is a holomorphically convex point. In particular, Ω is holomorphically convex.*

Proof The proof is the same as that of Theorem 7.4.12; there were used only two facts, the finite dimensionality of Theorem 7.5.26 (i) and that every boundary point $a \in \partial\Omega$ is locally a holomorphically convex point. □

The following lemma is a key:

Lemma 7.5.28 *Let X be a Riemann domain, and let $\Omega \Subset X$ be a strongly pseudoconvex domain. Then Ω is Stein.*

Proof Stein condition (iii) (holomorphic convexity) was finished by Theorem 7.5.27. Stein condition (ii) (holomorphic local coordinates) is contained in the definition of Riemann domain. There remains only Stein condition (i) (holomorphic separation).

We proceed by induction on the dimension $n \geq 1$.

(a) The case of $n = 1$: We take two distinct points $a, b \in \Omega$. If $\pi(a) \neq \pi(b)$, it is finished. Let $\pi(a) = \pi(b)$. By a translation of $\mathbf{C}^n$, we may assume that $\pi(a) = \pi(b) = 0 \in \mathbf{C}^n$. Take a neighborhood $U_0 \subset \Omega$ of a such that $U_0 \not\ni b$ and $\pi|_{U_0} : U_0 \to \Delta(0; \delta)$ with $\delta > 0$ is biholomorphic. Putting $U_1 = \Omega \setminus \{a\}$, we get an open covering $\mathscr{U} = \{U_0, U_1\}$ of Ω. For $k \in \mathbf{N}$ we set

$$\gamma_k(x) = \frac{1}{\pi(x)^k}, \quad x \in U_0 \cap U_1.$$

Then γ_k defines an element $[\gamma_k]$ of $H^1(\mathscr{U}, \mathscr{O}_\Omega)$. Note that $H^1(\mathscr{U}, \mathscr{O}_\Omega) \hookrightarrow H^1(\Omega, \mathscr{O}_\Omega)$ (injective). By the finite dimensionality of Theorem 7.5.26 (i) there is a non-trivial linear relation:

$$\sum_{k=1}^{h} c_k[\gamma_k] = 0, \quad c_k \in \mathbf{C}, \ c_h \neq 0.$$

Therefore, there are elements $f_j \in \mathscr{O}(U_j)$, $j = 0, 1$, satisfying

$$f_1(x) - f_0(x) = \sum_{k=1}^{h} c_k \frac{1}{\pi(x)^k}, \quad x \in U_0 \cap U_1.$$

Thus, we get a meromorphic function F on Ω which has a pole only at a of order precisely h:

$$F(x) = \begin{cases} f_0(x) + \displaystyle\sum_{k=1}^{h} c_k \frac{1}{\pi(x)^k}, & x \in U_0, \\ f_1(x), \quad x \in U_1. \end{cases}$$

It follows from the construction that

$$\pi(x)^h F(x) \in \mathscr{O}(\Omega),$$
$$\pi(a)^h F(a) = c_h \neq 0,$$
$$\pi(b)^h F(b) = 0.$$

Therefore, a, b are separated by an element of $\mathscr{O}(\Omega)$.

(b) The case of $\dim X = n \geq 2$: Suppose that the assertion holds in $\dim X = n-1$.

(1) Let $a, b \in \Omega$, $a \neq b$, be arbitrary points. If $\pi(a) \neq \pi(b)$, the proof is over. Suppose that $\pi(a) = \pi(b)$. By a translation we may assume $\pi(a) = \pi(b) = 0$. Taking a hyperplane $L = \{z_n = 0\}$, we consider a restriction:

$$\pi_{X'} : X' = \pi^{-1}L \longrightarrow L.$$

Since $L \cong \mathbf{C}^{n-1}$ (biholomorphic), each connected component X'' of X' is a Riemann domain of dimension $(n-1)$. By the induction hypothesis every connected component of $X' \cap \Omega$ is Stein.

(2) Let $\mathfrak{m}_a \subset \mathscr{O}_{X',a}$ be the maximal ideal of the local ring $\mathscr{O}_{X',a}$ and let $\mathfrak{m}_a^k$ be its k-th power. Set

$$\mathfrak{m}^k\langle a, b\rangle = \mathfrak{m}_a^k \otimes \mathfrak{m}_b^k \subset \mathscr{O}_{X'},$$

which is a coherent ideal sheaf of $\mathscr{O}_{X'}$.

Since every connected component of $X' \cap \Omega$ is Stein, by Theorem 4.5.8 there exists an element $g_k \in \mathscr{O}(X' \cap \Omega)$ for every $k \in \mathbf{N}$ satisfying the following properties:

$$\begin{aligned} \underline{g_k}_a &\equiv 0 \pmod{\mathfrak{m}^{k-1}\langle a, b\rangle_a}, \\ \underline{g_k}_a &\not\equiv 0 \pmod{\mathfrak{m}^k\langle a, b\rangle_a}, \\ \underline{g_k}_b &\equiv 0 \pmod{\mathfrak{m}^k\langle a, b\rangle_b}. \end{aligned} \tag{7.5.29}$$

Here, $\underline{g_k}_a$ denotes the germ of g_k at a.

(3) We set $\Omega' = \Omega \cap X'$. We denote by $\mathscr{I}$ the geometric ideal sheaf of the complex submanifold $X' \subset X$. Since X' is a complex manifold, $\mathscr{O}_{X'}$ is coherent (Oka's First Coherence Theorem 2.5.1). By Theorem 4.4.6 (i), $\mathscr{I}$ is also coherent. Restricting this to Ω, we have a short exact sequence:

$$0 \to \mathscr{I} \to \mathscr{O}_\Omega \to \mathscr{O}_{\Omega'} \to 0.$$

This yields the following exact sequence:

$$\mathscr{O}(\Omega) \to \mathscr{O}(\Omega') \overset{\delta}{\to} H^1(\Omega, \mathscr{I}). \tag{7.5.30}$$

From $\{g_k\}_{k\in\mathbf{N}}$ we get $\{\delta(g_k)\}_{k\in\mathbf{N}} \subset H^1(\Omega, \mathscr{I})$. By Theorem 7.5.26 (ii), $H^1(\Omega, \mathscr{I})$ is of finite dimension, and there is a non-trivial linear relation,

$$\sum_{k=k_0}^{N} c_k \delta(g_k) = 0, \quad c_k \in \mathbf{C},\ N < \infty.$$

We may assume that $c_{k_0} \neq 0$. It follows from (7.5.30) that there exists an element $f \in \mathscr{O}(\Omega)$ with

$$f|_{\Omega'} = \sum_{k=k_0}^{N} c_k g_k.$$

In a small neighborhood of $a \in \Omega$, $\pi = (z_1, \ldots, z_n)$ serves a holomorphic local coordinate system, and with $z' = (z_1, \ldots, z_{n-1})$ there is an expression

(7.5.31) $$f(z) = \sum_{k=k_0}^{N} c_k g_k(z') + h(z) \cdot z_n.$$

We infer from (7.5.29) the existence of a differential operator of order k_0 in z',

$$\mathrm{D} = \frac{\partial^{k_0}}{\partial z_1^{\alpha_1} \cdots \partial z_{n-1}^{\alpha_{n-1}}}, \quad \sum_{j=1}^{n-1} \alpha_j = k_0,$$

such that

(7.5.32) $$\begin{aligned} \mathrm{D}g_{k_0}(a) &\neq 0, \\ \mathrm{D}g_k(a) &= 0, \quad k > k_0, \\ \mathrm{D}g_k(b) &= 0, \quad k \geq k_0. \end{aligned}$$

By the definition of D and (7.5.31),

$$\mathrm{D}f(z) = \sum_{k=k_0}^{N} c_k \mathrm{D}g_k(z') + (\mathrm{D}h(z)) \cdot z_n.$$

At a and b, $z_n = 0$, and so by (7.5.32)

$$\mathrm{D}f(a) \neq 0, \quad \mathrm{D}f(b) = 0.$$

Since $\mathrm{D}f \in \mathscr{O}(\Omega)$, the holomorphic separability of Ω is shown. □

Lemma 7.5.33 (i) *Let $\Omega_1 \Subset \Omega_2 \Subset \Omega_3 \Subset X$ be a sequence of subdomains. Assume that Ω_3 is Stein. If*

$$\delta_{\mathrm{P}\Delta}(\partial\Omega_1, \partial\Omega_3) > \max_{x \in \partial\Omega_2} \delta_{\mathrm{P}\Delta}(x, \partial\Omega_3),$$

then there is an $\mathscr{O}(\Omega_3)$-analytic polyhedron P such that

$$\Omega_1 \Subset P \Subset \Omega_2.$$

(ii) *Every $f \in \mathscr{O}(P)$ can be approximated uniformly on compact subsets in P by elements of $\mathscr{O}(\Omega_3)$; that is, (P, Ω_3) is a Runge pair.*

Proof (i) The proof is the same as in the case of univalent domains (Lemma 5.4.8).

(ii) Since Ω_3 is assumed to be Stein, Oka's Fundamental Lemma 4.3.15 and Oka's Jôku-Ikô reduce the domain to a polydisk and the proof is done. □

By this we may extend the Behnke–Stein Theorem 5.4.10 to the case of Riemann domains.

Theorem 7.5.34 *Let $\pi : X \to \mathbf{C}^n$ be a Riemann domain, and let $X_\nu \subset X_{\nu+1}$, $\nu = 1, 2, \ldots$, be a sequence of Riemann subdomains with $X = \bigcup_\nu X_\nu$. If all X_ν are Stein, then so is X.*

The proof of this theorem is left to the readers. It is possible to use this theorem in the proof below, but we prefer another path. Conversely, by making use of our goal, Oka's Theorem 7.5.43, we may prove Theorem 7.5.34: Note first that

$$-\log \delta_{\mathrm{P}\Delta}(x, \partial X_\nu) \searrow -\log \delta_{\mathrm{P}\Delta}(x, \partial X), \quad \nu \to \infty.$$

"X_ν being Stein" implies the plurisubharmonicity of $-\log \delta_{\mathrm{P}\Delta}(x, \partial X_\nu)$, and then the limit, $-\log \delta_{\mathrm{P}\Delta}(x, \partial X)$, is plurisubharmonic. By Oka's Theorem 7.5.43 X is Stein.

Theorem 7.5.35 (Oka's Theorem, IX (1953)) *A pseudoconvex Riemann domain is Stein.*

Proof By assumption, there is an exhaustion plurisubharmonic function $\phi : X \to \mathbf{R}$. Fix a point $x_0 \in X$, and for $c > \phi(x_0) = c_0$ set

$$\Omega_c = \text{ the connected component of } \{x \in X; \phi(x) < c\} \text{ containing } x_0.$$

It follows that

$$\Omega_c \Subset \Omega_b \Subset X, \quad c_0 < c < b,$$
$$\bigcup_{c > c_0} \Omega_c = X.$$

For the Steinness of X, with the help of Theorem 5.4.13 it suffices to prove:

Lemma 7.5.36 (i) *For every $c > c_0$, Ω_c is Stein.*
(ii) *For every pair $(c_0 \leq) c < b$, (Ω_c, Ω_b) is a Runge pair.*

Proof (i) Let $K \Subset \Omega_c$ be a compact subset. Set

$$\eta = \delta_{\mathrm{P}\Delta}(K, \partial \Omega_c) \, (> 0).$$

We take $b > c$ so that

$$\max_{x \in \partial \Omega_c} \delta_{\mathrm{P}\Delta}(x, \partial \Omega_b) < \eta. \tag{7.5.37}$$

We fix a number, $0 < \rho < \delta_{\mathrm{P}\Delta}(\bar{\Omega}_b, \partial X)$. Take a C^∞ function $\chi(z)$ in (7.5.13) satisfying $\chi(z) = \chi(|z_1|, \ldots, |z_n|)$, and take the smoothing, $\phi_\varepsilon(x) = \phi * \chi_\varepsilon(x)$ with $0 < \varepsilon < \rho$. The function $\phi_\varepsilon(x)$ is C^∞ plurisubharmonic in X_ρ ($\Supset \Omega_b$) (Theorem 7.1.39). Then

$$\psi_\varepsilon(x) := \phi_\varepsilon(x) + \varepsilon \|\pi(x)\|^2$$

is strongly plurisubharmonic in $X_\rho(\Supset \bar{\Omega}_b)$, and as $\varepsilon \to 0$, it converges uniformly to ϕ on $\bar{\Omega}_b$. Taking a sufficiently small $\varepsilon > 0$, we denote by Ω the connected component of $\{x \in \Omega_b;\ \psi_\varepsilon(x) < \frac{b+c}{2}\}$ containing Ω_c. Then, Ω is strongly pseudoconvex, and

$$\Omega_c \Subset \Omega \Subset \Omega_b.$$

By Lemma 7.5.28, Ω is Stein. Therefore, Ω_c satisfies Stein conditions (i) and (ii). There remains Stein condition (iii) (holomorphic convexity):

Claim 7.5.38 $\hat{K}_{\Omega_c} \Subset \Omega_c$.

$\because$) Applying (7.5.6) to $K \Subset \Omega$, we get

$$\delta_{\mathrm{P}\Delta}(\hat{K}_\Omega, \partial\Omega) = \delta_{\mathrm{P}\Delta}(K, \partial\Omega) > \eta.$$

On the other hand, by (7.5.37),

$$\max_{x \in \partial\Omega_c} \delta_{\mathrm{P}\Delta}(x, \partial\Omega) < \eta.$$

It follows from the two equations above that

(7.5.39) $$\hat{K}_{\Omega_c} \subset \hat{K}_\Omega \Subset \Omega_c.$$ $\triangle$

(ii) We keep the notation as above.

(1) Now it has been shown that each Ω_c $(c > c_0)$ is Stein; in the arguments of (i) the role of Ω can be played by Ω_b. Therefore,

(7.5.40) $$\hat{K}_{\Omega_c} \subset \hat{K}_{\Omega_b} \Subset \Omega_c \Subset \Omega_b.$$

Claim 7.5.41 $\hat{K}_{\Omega_c} = \hat{K}_{\Omega_b}$.

$\because$) It follows from (7.5.40) that there is an $\mathscr{O}(\Omega_b)$-analytic polyhedron P satisfying

$$\hat{K}_{\Omega_c} \subset \hat{K}_{\Omega_b} \Subset P \Subset \Omega_c \Subset \Omega_b.$$

If there is a point $\zeta \in \hat{K}_{\Omega_b} \setminus \hat{K}_{\Omega_c}$, then there is an element $g \in \mathscr{O}(\Omega_c)$ with

$$\max_K |g| < |g(\zeta)|.$$

By Lemma 7.5.33 (ii), g can be uniformly approximated on $\hat{K}_{\Omega_b}$ by elements of $\mathscr{O}(\Omega_b)$. Thus, there is an $f \in \mathscr{O}(\Omega_b)$ such that

$$\max_K |f| < |f(\zeta)|;$$

this is absurd. △

(2) By Claim 7.5.41,

$$\hat{K}_{\Omega_c} = \hat{K}_{\Omega_t}, \quad c \leq {}^{\forall}t \leq b. \tag{7.5.42}$$

Set

$$E = \{t \geq c\,;\, \hat{K}_{\Omega_t} = \hat{K}_{\Omega_c}\} \subset [c, \infty).$$

By definition, if $t \in E$, then $[c, t] \subset E$. By the result of (1), E is open in $[c, \infty)$.

(3) We put $a = \sup E$.

Claim $a = \infty$; *i.e.*, $E = [c, \infty)$.

∵) Suppose $a < \infty$. By definition,

$$K_1 = \hat{K}_{\Omega_c} = \hat{K}_{\Omega_t}, \quad c \leq {}^{\forall}t < a.$$

Taking $t < a$ sufficiently close to a, we have

$$\delta_{\mathrm{P}\Delta}(K_1, \partial\Omega_a) > \max_{x \in \partial\Omega_t} \delta_{\mathrm{P}\Delta}(x, \partial\Omega_a).$$

Since Ω_a is Stein,

$$\delta_{\mathrm{P}\Delta}(\hat{K}_{1\Omega_a}, \partial\Omega_a) = \delta_{\mathrm{P}\Delta}(K_1, \partial\Omega_a) > \max_{x \in \partial\Omega_t} \delta_{\mathrm{P}\Delta}(x, \partial\Omega_a).$$

Thus, $\hat{K}_{1\Omega_a} \Subset \Omega_t$ follows. Therefore,

$$\hat{K}_{\Omega_t} \subset \hat{K}_{\Omega_a} \subset \hat{K}_{1\Omega_a} \Subset \Omega_t \Subset \Omega_a.$$

Similarly to the arguments of (1), we deduce that $\hat{K}_{\Omega_t} = \hat{K}_{\Omega_a}$. Therefore, $a \in E$. Since E is open, there is a number $a' > a$ with $a' \in E$; this contradicts the choice of a. △

(4) As a consequence of (2) above, for every pair $c < b$ and for every compact subset $K \Subset \Omega_c$,

$$\hat{K}_{\Omega_c} = \hat{K}_{\Omega_b}.$$

Oka's Fundamental Lemma 4.3.15 together with Oka's Jôku-Ikô implies that the pair (Ω_c, Ω_b) is Runge. □

The following is the main goal of Oka [62] IX (1953):

Theorem 7.5.43 (Oka's Theorem) *Let X be a Riemann domain. Then X is Stein if and only if $-\log \delta_{P\Delta}(x, \partial X)$ is a plurisubharmonic function.*

Proof If X is Stein, then it is holomorphically convex; Theorems 7.5.7 and 7.5.8 imply the plurisubharmonicity of $-\log \delta_{P\Delta}(x, \partial X)$.

Conversely, if $-\log \delta_{P\Delta}(x, \partial X)$ is plurisubharmonic, Oka's Lemma 7.5.10 implies that X is pseudoconvex. It follows from Theorem 7.5.35 that X is Stein. □

Remark 7.5.44 Because of Theorem 7.5.35, Corollaries 4.4.20, 4.4.21 and 4.4.29 hold for a pseudoconvex Riemann domain X: In particular, $\bar{\partial}$-equation

$$\bar{\partial} g = f \in \Gamma(X, \mathscr{E}_X^{(p,q)}), \quad \bar{\partial} f = 0, \ p \geq 0, \ q \geq 1$$

carries always a solution $g \in \Gamma(X, \mathscr{E}_X^{(p,q-1)})$. L. Hörmander [32], [33] proved X being a domain of holomorphy (Stein) by solving $\bar{\partial}$-equation with functional analysis method.

Remark 7.5.45 As for the generalization of Oka's Theorem for Riemann domains and counter-examples in the case of ramified domains, the following results are known:

(i) The generalization for Riemann domains over $\mathbf{P}^n(\mathbf{C})$ was obtained by R. Fujita [20] and A. Takeuchi [74].

(ii) In the case of a ramified domains over $\mathbf{P}^n(\mathbf{C})$, a counter-example was given by H. Grauert [47], [25].

(iii) A counter-example of a ramified domain over $\mathbf{C}^n$ was given by J.E. Fornæss [16]. In a little bit more precise, we let X be an n-dimensional complex manifold and a holomorphic map $\pi : X \to \mathbf{C}^n$, which is a ramified covering domain. It is said that X is locally Stein, if for every point $z \in \mathbf{C}^n$ there is a neighborhood $U \ni z$ such that $\pi^{-1}U$ is Stein. In [16] Fornæss constructed an example of a 2-sheeted ramified covering domain X over $\mathbf{C}^2$ such that it is locally Stein, but X itself is not Stein. Cf. [17] for more counter-examples.

(iv) There is recently a positive result in the ramified case: See [55].

Historical Supplements

As mentioned at the end of Sect. 7.1, the condition which E.E. Levi [39] himself used to formulate a problem that bears his name is different to that described by plurisubharmonic functions. K. Oka called this problem "Hartogs' Inverse Problem"; it is reduced to asking that if $-\log \delta_{P\Delta}(x, \partial X)$ defined by the boundary distance function $\delta_{P\Delta}(x, \partial X)$ of a domain X, univalent or multivalent, is plurisubharmonic, then X is a domain of holomorphy. Here a function defining the boundary is unnecessary. In this sense, Hartogs' Inverse Problem has a nuance more general than Levi's Problem.

Levi's Problem (Hartogs' Inverse Problem) was then regarded as the most difficult problem. K. Oka announced the solution of this problem in dimension 2 in 1941 ([63]), and published the full paper in 1942 ([62] VI); then after some interval he proved it for Riemann domains of arbitrary dimension in 1953 ([62] IX); however, he had solved it in a Japanese research report written in 1943.[6] This fact was written twice at the beginning of the introductions of Oka VIII (1951) and IX (1953), but it has been disregarded historically. In 1954, H.J. Bremermann [7] and F. Norguet [58] gave independently the proofs of Levi's Problem for univalent domains of general dimension by generalizing Oka's method (Heftungslemma, Oka VI) to the general dimensional case. As mentioned already, in this problem, even if a domain is given univalently in $\mathbf{C}^n$, the envelope of holomorphy may be necessarily multivalent over $\mathbf{C}^n$ (see Example 5.1.5). Therefore, the solution of this problem for univalent domains is not complete. As shown in the proof of this chapter, the difficulty essentially increases from dealing with univalent domains to dealing with multivalent domains (Riemann domains).

It is now made clear by records that the essential part of the solution of Levi's Problem (Hartogs' Inverse Problem) in Oka IX is that of his research report written in Japanese, sent to Teiji Takagi (Professor, The Imperial University of Tokyo, well-known as the founder of class field theory) in 1943. During the time Oka wrote and published two papers VII (1950) and VIII (1951), proving Three Coherence Theorems. That purpose was to solve Levi's Problem (Hartogs' Inverse Problem) including ramified domains over $\mathbf{C}^n$ (cf. Introductions of Oka [62] VII, VIII). In the end, Oka restricted himself to the case of unramified domains to write up a paper, and published it (IX); therefore, the Second and Third Coherence Theorems were not used.

As mentioned at the beginning of Sect. 7.4, the proof here given is due to Grauert's Theorem 7.5.26, which is different to Oka's original one. As for this proof, H. Grauert wrote in his Collected Volume [26], Vol. I, pp. 155-156 a comment introducing an observation of C.L. Siegel:

> Oka's methods are very complicated. At first he proved (rather simply) that in any unbranched pseudoconvex domain X there is a continuous strictly plurisubharmonic function $p(x)$ which converges to $+\infty$ as x goes to the (ideal) boundary of X. Then he got the existence of holomorphic functions f from this property. In [19] (this is [22] at the end of the present book) the existence of the f comes from a theorem of L. Schwartz in functional analysis (topological vector spaces, see: H. Cartan, Séminaire E.N.S. 1953/54, Exposés XVI and XVII). The approach is much simpler, but my predecessor in Göttingen C.L. Siegel nevertheless did not like it: Oka's method is constructive and this one is not!

For ramified domains the counter-examples were found later on; in this sense the choice of Oka was right. But it remains in mind that H. Grauert put an emphasis on Levi's Problem (Hartogs' Inverse Problem) yet unsolved for ramified domains in his talk at the Memorial Conference of Kiyoshi Oka's Centennial Birthday on Complex Analysis in Several Variables, Kyoto/Nara 2001.

---

[6]Now, one can find the manuscript in [66].

Exercises

1. Show Remark 7.1.7.
2. Show Theorem 7.1.42, assuming the continuity of f.
3. Show that a complete metric space is Baire.
4. Let $\pi : X \to \mathbf{C}^n$ be a Riemann domain. Show that there are a Riemannn domain $\pi_0 : X_0 \to \mathbf{C}^n$ satisfying 7.5.3, and an unramified cover $\lambda : X \to X_0$ such that $\pi = \pi_0 \circ \lambda$ and $\lambda^* \mathscr{O}(X_0) = \mathscr{O}(X)$.
5. Let $\pi : X \to \mathbf{C}^n$ be a Riemann domain. Show that X satisfies 7.5.3 if and only if for any distinct points $a, b \in X$ with $\pi(a) = \pi(b)$ there is an element $f \in \mathscr{O}(X)$ with $\underline{f \circ \pi_a^{-1}}_z \neq \underline{f \circ \pi_b^{-1}}_z$, where π_a (resp. π_b) is the local biholomorphism defined by π from a neighborhood of a (resp. b) to a neighborhood of z.
6. Let $B \subset \mathbf{C}^n$ be an open ball with center at the origin, and let $X \to \mathbf{C}^n$ be a Riemann domain. Define $\delta_B(x, \partial X)$ in the same way as $\delta_{\mathrm{P}\Delta}(x, \partial X)$ with replacing $\mathrm{P}\Delta$ by B.

   Show Lemma 7.5.5 with $\delta_B(z, \partial X)$ in place of $\delta_{\mathrm{P}\Delta}(x, \partial X)$.
7. Show Theorem 7.5.8 with $\delta_B(z, \partial X)$ in place of $\delta_{\mathrm{P}\Delta}(x, \partial X)$.
8. Prove Theorem 7.5.34 (cf. the proof of Theorem 5.4.10).
9. Let $\pi : X \to \mathbf{C}^n$ be a Riemann domain, and let $\mathrm{P}\Delta$ be a polydisk with center at the origin. Let $\tilde{\pi} : \tilde{X} \to \mathbf{C}^n$ be a pseudoconvex Riemann domain such that $\tilde{X} \supset X$ and $\tilde{\pi}|_X = \pi$. Assume that for every point $b \in \partial X$ (in $\tilde{X}$) there is a neighborhood U of b in $\tilde{X}$, satisfying that $-\log \delta_{\mathrm{P}\Delta}(x, \partial X)$ is plurisubharmonic in $x \in U \cap X$. Then, show that X is pseudoconvex. (Note that for a given Riemann domain X there exists always such an $\tilde{X}$ by taking the envelope of holomorphy of X.)
10. Let X be a Riemann surface (1-dimensional complex manifold), and let $\Omega \Subset X$ be a subdomain. Referring to the proofs of Grauert's Theorem 7.4.1 and Lemma 7.5.28 with $n = 1$, show the following:
    a. $\dim_{\mathbf{C}} H^1(\Omega, \mathscr{O}_\Omega) < \infty$.
    b. Let $\tilde{\Omega}$ be a domain such that $\Omega \Subset \tilde{\Omega} \Subset X$. Infer from 10a for $\tilde{\Omega}$ that for a point $p \in \partial\Omega$ there is a meromorphic function on $\tilde{\Omega}$ with a pole only at p; show that Ω is holomorphically convex.
    c. Show that for a point $q \in \Omega$ there is a meromorphic function f on $\tilde{\Omega}$ with a pole only at p_0 such that $df(q) \neq 0$.
    d. Show that for distinct points $q, q' \in \Omega$ there is a meromorphic function g on $\tilde{\Omega}$ with a pole only at p_0 such that $g(q) \neq g(q')$; therefore, Ω is Stein. (Thus, we have that $H^1(\Omega, \mathscr{O}_\Omega) = 0$.)

    (Note that every domain of $\mathbf{C}$ is holomorphically convex.)
11. Show that for every domain $\Omega \subset \mathbf{C}^n$ there exists a Stein Riemann domain $X \xrightarrow{\pi} \mathbf{C}^n$ such that $\Omega \subset X$ and $\pi|_\Omega$ is the inclusion map of $\Omega \subset \mathbf{C}^n$.

Chapter 8
Cohomology of Coherent Sheaves and Kodaira's Embedding Theorem

Up to the present we have been dealt with open domains and open complex manifolds. In this chapter we also deal with compact ones. We will introduce a topology in the space of sections of a coherent sheaf. As a consequence we will see that all cohomologies of a coherent sheaf over a compact complex space are finite dimensional (Cartan–Serre Theorem). Furthermore, we will extend Grauert's Theorem 7.5.26 for a general coherent sheaf. Then, as an application, we prove Kodaira's Embedding Theorem to embed a Hodge manifold into a complex projective space.

Kodaira's Embedding Theorem provides a bridge between the theory of compact Kähler manifolds and that of complex projective algebraic varieties; it is nice to see such a theorem being naturally proved on the extended line of the theory of coherent sheaves.

8.1 Topology of the Space of Sections of a Coherent Sheaf

It was a key point in the proof of Grauert's Theorems 7.4.1 and 7.5.26 to apply L. Schwartz's Theorem 7.3.23 to the topological vector space of holomorphic functions induced from the convergence uniform on compact subsets. To deal with general coherent sheaves, it is necessary to introduce a suitable topology on the space of sections.

8.1.1 Domains of $\mathbf{C}^n$

To begin with a preparation we deal locally with a domain of $\mathbf{C}^n$. We consider the following setting.

8.1.1 (Setting) As in Chap. 2, we write $\mathscr{O}_n = \mathscr{O}_{\mathbf{C}^n}$, $\mathscr{O}_{n,a} = \mathscr{O}_{\mathbf{C}^n,a}$ ($a \in \mathbf{C}^n$). Let $\mathscr{O}_n^p$ be the p-th direct product of $\mathscr{O}_n$, let $M_0 \subset \mathscr{O}_{n,0}^p$ be an $\mathscr{O}_{n,0}$-submodule, and let $\Omega \subset \mathbf{C}^n$ be a domain containing 0.

J. Noguchi, *Analytic Function Theory of Several Variables*,
DOI 10.1007/978-981-10-0291-5_8

By Theorem 2.2.20 (Noetherian property) M_0 is finitely generated.

Lemma 8.1.2 *Let $p \geq 1$ and let $\{\underline{U_j}_0\}_{j=1}^l$ be a finite generator system of M_0. Then there are a polydisk $\mathrm{P}\Delta \subset \Omega$ with center at 0 and a positive constant $C > 0$ such that for $f \in \mathscr{O}(\Omega)^p$ with $\underline{f}_0 \in M_0$ there exist $F_j \in \mathscr{O}(\mathrm{P}\Delta)$, $1 \leq j \leq l$, satisfying*

$$\underline{f}_0 = \sum_{j=1}^{l} \underline{F_j}_0 \underline{U_j}_0,$$
$$\|F_j\|_{\mathrm{P}\Delta} \leq C\|f\|_{\mathrm{P}\Delta}.$$

Here, $\|\cdot\|_{\mathrm{P}\Delta}$ stands for the supremum norm on $\mathrm{P}\Delta$.

Proof The proof follows after the induction part of the proof of Oka's First Coherence Theorem 2.5.1 with double induction on $n \geq 0$ and p. The case of $n = 0$ and $p \geq 1$ is clear.

(a) Suppose that $p > 1$, and that the assertion holds for $\mathscr{O}_{n,0}^q$ with all $q < p$. Let

$$\pi : \mathscr{O}_{n,0}^p \to \mathscr{O}_{n,0},$$
$$\pi_* : \mathscr{O}(\Omega)^p \to \mathscr{O}(\Omega)$$

be the projections to the first elements. We apply the induction hypothesis to $\pi(M_0)$ and $\pi_* f$. Since $\pi(M_0)$ is generated by $\{\pi(U_j)\}$, there are a polydisk $\mathrm{P}\Delta$ with center at 0, a constant $C' > 0$, and $g_j \in \mathscr{O}(\mathrm{P}\Delta)$ such that

$$\underline{\pi_* f}_0 = \sum_{j=1}^{l} \underline{g_j}_0 \pi(\underline{U_j}_0),$$
$$\|g_j\|_{\mathrm{P}\Delta} \leq C'\|\pi_* f\|_{\mathrm{P}\Delta} \leq C'\|f\|_{\mathrm{P}\Delta}.$$

Take $\mathrm{P}\Delta$ small enough so that all $\underline{U_j}_0$ have representatives U_j there. It follows that

$$\underline{f - \sum_{j=1}^{l} g_j U_j}_0 \in M_0 \cap \operatorname{Ker} \pi \subset \operatorname{Ker} \pi \cong \mathscr{O}_{n,0}^{p-1}.$$

By Theorem 2.2.20, $M_0 \cap \operatorname{Ker} \pi$ has a finite generator system $\{\underline{V_k}_0\}_k$. Since $f - \sum_{j=1}^{l} g_j U_j \in \mathscr{O}(\mathrm{P}\Delta)^{p-1}$, again by the induction hypothesis, after $\mathrm{P}\Delta$ taken smaller if necessary, with the above C' taken larger if necessary, there are constants $C'' > 0$, $C''' > 0$ and functions $h_k \in \mathscr{O}(\mathrm{P}\Delta)$ such that

$$f - \sum_{j=1}^{l} g_j U_j = \sum_k h_k V_k,$$

$$\|h_k\|_{\mathrm{P}\Delta} \leq C'' \left\| f - \sum_{j=1}^{l} g_j U_j \right\|_{\mathrm{P}\Delta} \leq C''' \|f\|_{\mathrm{P}\Delta}.$$

Thus we have

$$f = \sum_j g_j U_j + \sum_k h_k V_k,$$

$$\underline{V_k}_0 = \sum_j \underline{a_{kj}}_0 \underline{U_j}_0 \quad (\underline{a_{kj}}_0 \in \mathscr{O}_{n,0}).$$

We take $\mathrm{P}\Delta$ smaller if necessary, so that $\underline{a_{kj}}_0$ have representatives $a_{kj} \in \mathscr{O}(\overline{\mathrm{P}\Delta})$, and set

$$F_j = g_j + \sum_k h_k a_{kj}.$$

Then there is a constant $C > 0$ such that

$$\|F_j\|_{\mathrm{P}\Delta} \leq C \|f\|_{\mathrm{P}\Delta},$$

$$f = \sum_j F_j U_j.$$

(b) As the induction hypothesis we assume that the assertion holds for $\mathscr{O}^p_{n-1}$ with any $p \geq 1$. We show the case of $p = 1$ and $n \geq 1$. Take an element $P \in M_0$, $P \neq 0$. By Weierstrass' Preparation Theorem 2.1.3, there is a standard polydisk $\mathrm{P}\Delta = \mathrm{P}\Delta' \times \Delta_n \Subset \Omega$ of P; we may assume that $P(z', z_n)$ $((z', z_n) \in \mathrm{P}\Delta = \mathrm{P}\Delta' \times \Delta_n)$ is a Weierstrass polynomial in z_n of degree d. Every $f \in \mathscr{O}(\Omega)$ is written as

(8.1.3)
$$f(z', z_n) = a(z)P(z', z_n) + \sum_{\lambda=1}^{d} b_\lambda(z') z_n^{d-\lambda},$$

$$a(z) \in \mathscr{O}(\mathrm{P}\Delta), \quad \|a\|_{\mathrm{P}\Delta} \leq C' \|f\|_{\mathrm{P}\Delta},$$

$$b_\lambda(z') \in \mathscr{O}(\mathrm{P}\Delta'), \quad \|b_\lambda\|_{\mathrm{P}\Delta'} \leq C' \|f\|_{\mathrm{P}\Delta},$$

where $C' > 0$ is a constant independent of f (here and below constants are different to those in (a)). If $\underline{f}_0 \in M_0$,

$$\underline{\sum_{\lambda=1}^{d} b_\lambda(z') z_n^{d-\lambda}}_0 \in M_0.$$

Set

$$M'_0 = \left\{ (B_1, \ldots, B_d) \in \mathscr{O}^d_{n-1,0};\ \sum_{\lambda=1}^{d} B_\lambda \left(\underline{z_n}_0\right)^{d-\lambda} \in M_0 \right\}.$$

Let $\{\underline{W_k}_0\}_{k=1}^m$ be a generator system of M'_0 as $\mathscr{O}_{n-1,0}$-module. For a given $f \in \mathscr{O}(\Omega)$ with $\underline{f}_0 \in M_0$ we define a, b_λ by (8.1.3). By the induction hypothesis we have a polydisk $\mathrm{P}\Delta'$, which may be chosen smaller, and a constant $C'' > 0$ such that $\underline{W_k}_0$ have representatives $W_k(z') \in \mathscr{O}(\mathrm{P}\Delta')$, and

$$(b_\lambda(z'))_\lambda = \sum_{k=1}^{m} e_k(z')(W_{k\lambda}(z'))_\lambda,$$
$$e_k(z') \in \mathscr{O}(\mathrm{P}\Delta'), \qquad \|e_k\|_{\mathrm{P}\Delta'} \le C''\|(b_\lambda)_\lambda\|_{\mathrm{P}\Delta'}.$$

Since

$$f = aP + \sum_{k=1}^{l} e_k \sum_{\lambda=1}^{d} W_{k\lambda} z_n^{d-\lambda},$$

and $\underline{P}_0$ and $\underline{\sum_{\lambda=1}^{d} W_{k\lambda} z_n^{d-\lambda}}_0$ are written as a linear sum of $\underline{U_j}_0$ with $\mathscr{O}_{n,0}$ coefficients, we obtain the required coefficient functions $F_j \in \mathscr{O}(\mathrm{P}\Delta)$ with $\mathrm{P}\Delta$, which is chosen smaller if necessary. □

Lemma 8.1.4 *If a sequence of $f_\nu \in \mathscr{O}(\Omega)^p$, $\nu = 1, 2, \ldots$, converges uniformly on compact subsets to $f \in \mathscr{O}(\Omega)^p$, and all $\underline{f_\nu}_0 \in M_0$, then $\underline{f}_0 \in M_0$.*

Proof The difference sequence $\{f_\nu - f_{\nu'}\}_{\nu,\nu'}$ converges uniformly on compact subsets to 0, and $\underline{f_\nu - f_{\nu'}}_0 \in M_0$. Take and fix a finite generator system $\{\underline{U_j}_0\}_{j=1}^l$ of M_0. Let $\mathrm{P}\Delta \Subset \Omega$ and $C > 0$ be those in Lemma 8.1.2. The representatives U_j may be assume to be holomorphic in a neighborhood of $\overline{\mathrm{P}\Delta}$, and

$$\lim_{N\to\infty} \sup_{\nu,\nu'\ge N} \|f_\nu - f_{\nu'}\|_{\mathrm{P}\Delta} = 0$$

holds. Therefore one can choose a sequence of $\mathbf{N} \ni N_\lambda < N_{\lambda+1} < \cdots$ such that

$$\|f_{N_{\lambda+1}} - f_{N_\lambda}\|_{\mathrm{P}\Delta} \le \frac{1}{2^\lambda}, \quad \lambda = 1, 2, \ldots.$$

By Lemma 8.1.2, there are $a_{\lambda j} \in \mathscr{O}(\mathrm{P}\Delta)$ satisfying

$$f_{N_{\lambda+1}} - f_{N_\lambda} = \sum_{j=1}^{l} a_{\lambda j} U_j, \tag{8.1.5}$$
$$\|a_{\lambda j}\|_{\mathrm{P}\Delta} \le C\|f_{N_{\lambda+1}} - f_{N_\lambda}\|_{\mathrm{P}\Delta} \le \frac{C}{2^\lambda}.$$

Therefore the series $\sum_{\lambda=1}^{\infty} a_{\lambda j}$ is of majorant convergence, and determines an element of $\mathscr{O}(\mathrm{P}\Delta)$. It follows from the construction that

$$f = \lim_{\lambda\to\infty} f_{N_\lambda} = f_{N_1} + \sum_{j=1}^{l} \left(\sum_{\lambda=1}^{\infty} a_{\lambda j} \right) U_j.$$

Thus we see that $\underline{f}_0 \in M_0$. □

Theorem 8.1.6 *Let $\mathscr{S} \subset \mathscr{O}_\Omega^p$ be an arbitrary subsheaf of $\mathscr{O}_\Omega$-modules. Then $\Gamma(\Omega, \mathscr{S})$ is closed in $\mathscr{O}(\Omega)^p$ with respect to the topology of uniform convergence on compact subsets.*

Proof Let a sequence of $f_\nu \in \Gamma(\Omega, \mathscr{S})$, $\nu = 1, 2, \ldots$, converge uniformly on compact subsets to $f \in \mathscr{O}(\Omega)^p$. By Lemma 8.1.4, $\underline{f}_x \in \mathscr{S}_x$ at every point $x \in \Omega$. Therefore, $f \in \Gamma(\Omega, \mathscr{S})$. □

N.B. In this theorem it is unnecessary to assume the coherence for $\mathscr{S}$.

Let $\mathscr{F} \to \Omega$ be a coherent sheaf over Ω. We take polydisks

(8.1.7) $$\mathrm{P}\Delta \Subset \mathrm{P}\Delta_0 \Subset \Omega.$$

By Lemma 4.3.9 (Oka's Syzygies) there is a finite generator system $\{\sigma_j\}_{j=1}^{l}$ of $\mathscr{F}$ on $\mathrm{P}\Delta_0$, which we fix. We infer from Oka's Fundamental Lemma 4.3.15 that for every $f \in \Gamma(\mathrm{P}\Delta, \mathscr{F})$, there are elements $f_j \in \mathscr{O}(\mathrm{P}\Delta)$, $1 \leq j \leq l$, with

(8.1.8) $$f = \sum_{j=1}^{l} f_j \sigma_j.$$

Here we define a semi-norm by

(8.1.9) $$\|f\|_{\mathrm{P}\Delta} = \inf_{(f_j)} \sup\{|f_j(x)|; x \in \mathrm{P}\Delta\},$$

where $\{f_j\}$ runs over all such coefficient functions. It may happen that $\|f\|_{\mathrm{P}\Delta} = \infty$. If f is defined on a neighborhood of the closure $\overline{\mathrm{P}\Delta}$, then f_j are taken to be holomorphic in a neighborhood of $\overline{\mathrm{P}\Delta}$, so that $\|f\|_{\mathrm{P}\Delta} < \infty$.

Let $\{\tau_k\}_{k=1}^{m}$ be another finite generator system of $\mathscr{F}$ on $\mathrm{P}\Delta_0$. Then, with $f = \sum_k g_k \tau_k$, a semi-norm

(8.1.10) $$\|f\|'_{\mathrm{P}\Delta} = \inf_{(g_k)} \sup\{|g_k(x)|; x \in \mathrm{P}\Delta\}$$

is defined. Since every σ_j (resp. τ_k) is written as a linear sum of τ_k (resp. σ_j) with coefficients in $\mathscr{O}(\mathrm{P}\Delta_0)$ on $\mathrm{P}\Delta_0$, there is a constant $C > 0$ independent from f such that

$$C^{-1}\|f\|'_{\mathrm{P}\Delta} \le \|f\|_{\mathrm{P}\Delta} \le C\,\|f\|'_{\mathrm{P}\Delta}, \quad {}^{\forall} f \in \Gamma(\mathrm{P}\Delta, \mathscr{F}). \tag{8.1.11}$$

Therefore, $\|f\|_{\mathrm{P}\Delta}$ and $\|f\|'_{\mathrm{P}\Delta}$ are mutually equivalent semi-norms, and define the same topology on $\Gamma(\mathrm{P}\Delta, \mathscr{F})$.

Lemma 8.1.12 *Let the notation be as above. If $f \in \Gamma(\mathrm{P}\Delta, \mathscr{F})$ satisfies $\|f\|_{\mathrm{P}\Delta} = 0$, then $f(x) = 0$ at all $x \in \mathrm{P}\Delta$.*

Proof Let $f_j \in \mathscr{O}(\mathrm{P}\Delta)$ be as in (8.1.8). On the other hand, by the assumption, we can write $f = \sum_j a_{\nu j}\sigma_j$ with $a_{\nu j} \in \mathscr{O}(\mathrm{P}\Delta)$, $\nu = 1, 2, \ldots$, such that $\{a_{\nu j}\}_\nu$ converges uniformly to 0 on $\mathrm{P}\Delta$. Let $\mathscr{R} = \mathscr{R}(\sigma_1, \ldots, \sigma_l) \subset \mathscr{O}^l_{\mathrm{P}\Delta}$ be the relation sheaf. Then,

$$(f_1 - a_{\nu 1}, \ldots, f_l - a_{\nu l}) \in \Gamma(\mathrm{P}\Delta, \mathscr{R}), \quad \nu = 1, 2, \ldots.$$

We deduce from Theorem 8.1.6 that $(f_j) \in \Gamma(\mathrm{P}\Delta, \mathscr{R})$; thus, $f \equiv 0$ follows. □

The complex vector space $\{f \in \mathscr{O}(\mathrm{P}\Delta);\ \|f\|_{\mathrm{P}\Delta} < \infty\}$ is a Banach space with norm $\|\cdot\|_{\mathrm{P}\Delta}$. Therefore we see by Lemma 8.1.12:

Proposition 8.1.13 *Let the notation be as above. Then $\{f \in \Gamma(\mathrm{P}\Delta, \mathscr{F});\ \|f\|_{\mathrm{P}\Delta} < \infty\}$ endowed with norm $\|\cdot\|_{\mathrm{P}\Delta}$ is a Banach space.*

8.1.2 Complex Manifolds

In this chapter we assume that a complex manifold satisfies the *second countability axiom*. Let M be a complex manifold. In a holomorphic local coordinate neighborhood Ω we take double polydisks $P \Subset Q \Subset \Omega$ as in (8.1.7), and collect countably many such doubles

$$P_\alpha \Subset Q_\alpha \Subset \Omega_\alpha, \quad \alpha = 1, 2, \ldots, \tag{8.1.14}$$

so that $\{P_\alpha\}_\alpha$, as well $\{\Omega_\alpha\}_\alpha$ forms a base of open subsets of M.

On an open subset $U \subset M$ we introduce a countable system of semi-norms for $f \in \Gamma(U, \mathscr{F})$ by

$$\|f\|_{P_\alpha}, \quad P_\alpha \Subset \Omega_\alpha \subset U. \tag{8.1.15}$$

We call the topology of $\Gamma(U, \mathscr{F})$ defined by this semi-norm system the *topology of locally uniform convergence*, or the *topology of convergence uniform on compact subsets*. It follows from Proposition 8.1.13 that:

Proposition 8.1.16 *The complex vector space $\Gamma(U, \mathscr{F})$ endowed with the topology of locally uniform convergence is a Fréchet space.*

Proposition 8.1.17 *For a relatively compact open subset $V \Subset U$ the restriction map*

$$\rho : f \in \Gamma(U, \mathscr{F}) \to f|_V \in \Gamma(V, \mathscr{F})$$

is completely continuous.

Proof This is immediate from Lemma 8.1.2 and Montel's Theorem 1.2.19 (cf. Theorem 7.3.17, too). □

8.1.3 Complex Spaces

We assume below that a complex space satisfies the second countability axiom. Let $(X, \mathscr{O}_X)$ be a complex space. We deal with the locally uniform convergence of holomorphic functions in an open subset U of X. Let $\{f_\mu\}_{\mu=1}^{\infty}$ be a sequence of $\mathscr{O}(U)(=\Gamma(U, \mathscr{O}_X))$ converging locally uniformly to a function f in U. It is immediate to conclude that f is weakly holomorphic in U; more precisely, it is continuous in U and holomorphic in $U \setminus \Sigma(X)$. If X is normal, then $f \in \mathscr{O}(U)$ (Theorem 6.10.20). We show that this remains valid without the condition "normal" on X.

Theorem 8.1.18 *Let X be a complex space and let U be an open subset of X.*

(i) *Let $\{f_\mu\}_{\mu=1}^{\infty}$ a sequence of holomorphic functions $f_\mu \in \mathscr{O}(U)$, converging locally uniformly in U. Then the limit function $f = \lim f_\mu \in \mathscr{O}(U)$.*

In particular, $\mathscr{O}(X)$ endowed with topology of locally uniform convergence is Fréchet.

(ii) *Let X be a Stein space and let $Y \subset X$ be a complex subspace. Assume that the number of the irreducible components of Y is finite. Then for every compact subset $K \Subset X$, there exist a compact subset $L \Subset Y$ and a positive constant C such that for any $g \in \mathscr{O}(Y)$ there is an element $f \in \mathscr{O}(X)$ satisfying*

$$f|_Y = g, \quad \|f\|_K \leqq C\|g\|_L. \tag{8.1.19}$$

Proof (i) Since the problem is local, we consider in a neighborhood of a point $a \in U$. Let $\underline{X}_a = \bigcup_{\alpha=1}^{l} \underline{X_{\alpha}}_a$ be the irreducible decomposition of X at a. We may assume that U is an analytic subset of an open subset $\Omega \subset \mathbf{C}^n$ with $a = 0$ and X_α are irreducible analytic subsets of Ω. We take a polydisk $\mathrm{P}\Delta$ about 0 such that $\mathrm{P}\Delta \Subset \Omega$ and for every X_α there is a decomposition of the polydisk by the coordinate indices such as $\mathrm{P}\Delta = \mathrm{P}\Delta' \times \mathrm{P}\Delta''$ giving rise to a standard polydisk neighborhood of X_α.

Let $g \in \tilde{\mathscr{O}}(X \cap \mathrm{P}\Delta)$ be weakly holomorphic. We use (6.10.14), (6.10.15) and (6.10.19) for g. Let $\delta_\alpha = v_\alpha u_\alpha$ be as in (6.10.19), which is a universal denominator for X_α and is vanishing on all the other X_β. By (6.10.14) and (6.10.15) we have

$$\delta_\alpha g = B_\alpha|_{X\cap \mathrm{P}\Delta}, \quad B_\alpha \in \mathscr{O}(\mathrm{P}\Delta), \tag{8.1.20}$$

$$\|B_\alpha\|_{\mathrm{P}\Delta} \leq C\|g\|_{X\cap \mathrm{P}\Delta}, \tag{8.1.21}$$

where C is a positive constant independent from g.

Now, let $f = \lim f_\mu$ be as assumed. It suffices to show:

Claim 8.1.22 $\underline{f}_0 \in \mathscr{O}_{X,0}$.

After choosing a subsequence we may assume that

$$\|f_\mu - f_{\mu-1}\|_{X\cap \mathrm{P}\Delta} < \frac{1}{2^\mu}, \quad \mu = 2, 3, \ldots.$$

Then,

$$f = f_1 + \sum_{\mu=2}^{\infty} \left(f_\mu - f_{\mu-1}\right).$$

It follows from (8.1.20) that there are holomorphic functions $B_{\alpha\mu} \in \mathscr{O}(\mathrm{P}\Delta)$ such that for $\mu \geq 2$

$$\delta_\alpha \left(f_\mu - f_{\mu-1}\right) = B_{\alpha\mu}|_{X\cap \mathrm{P}\Delta}, \qquad \|B_{\alpha\mu}\|_{\mathrm{P}\Delta} \leq C\|f_\mu - f_{\mu-1}\|_{X\cap \mathrm{P}\Delta} \leq \frac{C}{2^\mu}.$$

Take $B_{\alpha 1} \in \mathscr{O}(\mathrm{P}\Delta)$ with $\delta_\alpha f_1 = B_{\alpha 1}|_{X\cap \mathrm{P}\Delta}$ and set

$$g_{\alpha N} = B_{\alpha 1} + \sum_{\mu=2}^{N} B_{\alpha\mu} \to g_\alpha \in \mathscr{O}(\mathrm{P}\Delta) \quad (N \to \infty).$$

Note that the above convergence is uniform, and that

$$\delta_\alpha f_N = g_{\alpha N}|_{X\cap \mathrm{P}\Delta}.$$

Let $\mathscr{I}\langle X\rangle (\subset \mathscr{O}(\mathrm{P}\Delta))$ be the geometric ideal sheaf of $X \cap \mathrm{P}\Delta$. We define a homomorphism

$$\varphi : \underline{h}_0 \in \mathscr{O}_{n,0} \to (\underline{\delta_1}_0 \underline{h}_0, \underline{\delta_2}_0 \underline{h}_0, \ldots, \underline{\delta_l}_0 \underline{h}_0) \in (\mathscr{O}_{n,0})^l,$$
$$M_0 = \varphi(\mathscr{O}_{n,0}) + (\mathscr{I}\langle X\rangle_0)^l.$$

Then M_0 is a submodule of $(\mathscr{O}_{n,0})^l$. Set

$$g_{N,0} = (\underline{g_{1N}}_0, \ldots, \underline{g_{lN}}_0) \in M_0,$$
$$g_0 = (\underline{g_1}_0, \ldots, \underline{g_l}_0).$$

Since the convergence $g_{\alpha N} \to g_\alpha$ $(N \to \infty)$ is uniform, it follows from Lemma 8.1.4 that $g_0 \in M_0$; that is, there exists an element $\underline{h}_0 \in \mathscr{O}_{n,0}$ such that

$$\varphi(\underline{h}_0) - g_0 \in (\mathscr{I}\langle X\rangle_0)^l.$$

Therefore, $\delta_\alpha(x) f(x) = \delta_\alpha(x) h(x)$ $(1 \leq \alpha \leq l)$ in a neighborhood $0 \in X$ in X. It follows that $f = h$ on every $X_\alpha \backslash \{\delta_\alpha = 0\}$, which is dense in X_α. Thus, $f = h$ in a neighborhood of 0 in X: This implies that $f \in \mathscr{O}_{n,0}$.

(ii) Let $\mathscr{I}\langle Y\rangle$ be the geometric ideal sheaf of Y in $\mathscr{O}_X$. Then we have the following short exact sequence of coherent sheaves:

$$0 \to \mathscr{I}\langle Y\rangle \to \mathscr{O}_X \to \mathscr{O}_X/\mathscr{I}\langle Y\rangle = \mathscr{O}_Y \to 0.$$

Then it leads to a long exact sequence:

$$0 \to \Gamma(X, \mathscr{I}\langle Y\rangle) \to \mathscr{O}(X) \to \mathscr{O}(Y) \to H^1(X, \mathscr{I}\langle Y\rangle) \to \cdots.$$

Since X is Stein, $H^1(X, \mathscr{I}\langle Y\rangle) = 0$ by the Oka–Cartan Fundamental Theorem 6.12.2. Therefore, the restriction map is surjective:

$$\rho : f \in \mathscr{O}(X) \to f|_Y \in \mathscr{O}(Y) \to 0.$$

As a consequence of (i) just above, $\mathscr{O}(X)$ and $\mathscr{O}(Y)$ are Fréchet. It follows from Banach's Open Mapping Theorem 7.3.19 that with a compact subset $K \Subset X$, $\rho(\{f \in \mathscr{O}(X);\ \|f\|_K < 1\})$ contains a neighborhood of 0 in $\mathscr{O}(Y)$. Thus, there are a compact subset $L \Subset Y$ and $\varepsilon > 0$ such that

$$\rho(\{f \in \mathscr{O}(X);\ \|f\|_K < 1\}) \supset V := \{g \in \mathscr{O}(Y);\ \|g\|_L < \varepsilon\}.$$

Since the number of the irreducible components of Y is finite, L may be taken so that L contains a non-empty open subset of every irreducible component of Y. It follows that for $g \in \mathscr{O}(Y)$, $g = 0$ if and only if $\|g\|_L = 0$.

Now, take any $g \in \mathscr{O}(Y)$. If $\|g\|_L = 0$, $g = 0$. Then one may take $f = 0$, so that (8.1.19) holds with any $C > 0$. Suppose that $\|g\|_L \neq 0$. Then $\frac{\varepsilon}{2\|g\|_L} g \in V$. There is an $h \in \mathscr{O}(X)$ with $\|h\|_K < 1$ such that $\rho(h) = \frac{\varepsilon}{2\|g\|_L} g$. With $f = \frac{2\|g\|_L}{\varepsilon} h$ and $C = \frac{2}{\varepsilon}$ we get

$$\rho(f) = g,$$
$$\|f\|_K = \frac{2}{\varepsilon}\|g\|_L\|h\|_K < C\|g\|_L. \qquad \square$$

Remark 8.1.23 The finiteness of the number of irreducible components of Y in (ii) above is necessary. For example, let X be irreducible and let K contain a non-empty open subset of X. If the number of irreducible components of Y is infinite, then for any compact subset $L \Subset Y$ there is a $g \in \mathscr{O}(Y)$ such that $g \neq 0$ but $\|g\|_L = 0$. Then any $f \in \mathscr{O}(X)$ with $\rho(f) = g$ cannot be 0, so that $\|f\|_K > 0$. Thus, there is no $C > 0$ such that $\|f\|_K \leq C\|g\|_L$.

Theorem 8.1.24 (Montel's Theorem on complex space) *Let X be a complex space and let $\{f_\nu\}_{\nu=1}^\infty$ be a sequence of holomorphic functions $f_\nu \in \mathscr{O}(X)$. Assume that $\{f_\nu\}_{\nu=1}^\infty$ is locally uniformly bounded in X. Then, $\{f_\nu\}_{\nu=1}^\infty$ contains a subsequence which converges locally uniformly to a holomorphic function $f \in \mathscr{O}(X)$.*

Proof Take any point $a \in X$ and a local chart neighborhood $U(\Subset X)$ of a such that $\{f_\nu\}_{\nu=1}^\infty$ is uniformly bounded by $M > 0$ in U. It suffices to show that there are a neighborhood V $(\subset U)$ of a and a subsequence of $\{f_\nu\}_{\nu=1}^\infty$, converging locally uniformly to a function $f \in \mathscr{O}(V)$ in V.

Now, we identify U with an analytic subset of a polydisk $\mathrm{P}\Delta$ with center $a = 0$ in some $\mathbf{C}^n$. Let $K = \overline{\mathrm{P}\Delta'} \Subset \mathrm{P}\Delta$ be a closed polydisk with center 0 in $\mathrm{P}\Delta$. Then there are only a finite number of the irreducible components of $U = X \cap \mathrm{P}\Delta$ $(\Subset X)$. By applying Theorem 8.1.18 (ii), there are a compact subset $L \Subset U \cap \mathrm{P}\Delta$, a constant C, and $F_\nu \in \mathscr{O}(\mathrm{P}\Delta)$, $\nu = 1, 2, \ldots$ such that

$$F_\nu|_U = f_\nu, \quad \|F_\nu\|_K \leq C\|f_\nu\|_L \leq CM.$$

By Montel's Theorem 1.2.19 there is a subsequence $\{F_{\nu_\mu}\}$ of $\{F_\nu\}$ which converges locally uniformly to a holomorphic function $F \in \mathscr{O}(\mathrm{P}\Delta')$ in $\mathrm{P}\Delta'$. Then the subsequence $\{f_{\nu_\mu}\}$ with $f_{\nu_\mu}|_{U\cap\mathrm{P}\Delta'} = F_{\nu_\mu}|_{U\cap\mathrm{P}\Delta'}$ converges locally uniformly to $f := F|_{U\cap\mathrm{P}\Delta'} \in \mathscr{O}(U \cap \mathrm{P}\Delta')$. With $V = \mathrm{P}\Delta'$ this finishes the proof. $\square$

Let $\mathscr{F} \to X$ be a coherent sheaf over X. One may define two kinds of topologies on the space of sections of $\mathscr{F}$. Let U be an open subset of X and $f \in \Gamma(U, \mathscr{F})$. Take any point $a \in U$ and local chart neighborhoods $V' \Subset V(\Subset U)$ of a such that:

(i) $\mathscr{F}|_V$ is generated by finitely many $\sigma_j \in \Gamma(V, \mathscr{F})$, $1 \leq j \leq l$, over V,
(ii) V is biholomorphic to an analytic subset W of a polydisk $\mathrm{P}\Delta$ in some $\mathbf{C}^n$, and $V' = V \cap \mathrm{P}\Delta'$, where V is identified with W and $\mathrm{P}\Delta'$ is a polydisk with $\mathrm{P}\Delta' \Subset \mathrm{P}\Delta$.

We denote the simple extension of $\mathscr{F}|_W$ over $\mathrm{P}\Delta$ by $\widehat{\mathscr{F}}_{\mathrm{P}\Delta}$. Then $\widehat{\mathscr{F}}_{\mathrm{P}\Delta}$ is coherent over $\mathrm{P}\Delta$ (Proposition 6.5.19) and $\Gamma(W, \mathscr{F})$ is canonically identified with $\Gamma(\mathrm{P}\Delta, \widehat{\mathscr{F}}_{\mathrm{P}\Delta})$. Then we write

$$f|_{\mathrm{P}\Delta} = \sum_{j=1}^{l} f_j\sigma_j, \quad f_j \in \mathscr{O}(\mathrm{P}\Delta),$$

and then set, as in (8.1.9) and (8.1.15),

$$\|f\|_{\mathrm{P}\Delta'} = \inf_{(f_j)} \sup\{|f_j(x)|;\ x \in \mathrm{P}\Delta'\}. \tag{8.1.25}$$

In this way, $\Gamma(U, \mathscr{F})$ is endowed with countably many semi-norms, where the collections $\{V\}$ and $\{V'\}$ form open bases of X.

On the other hand, writing

$$f|_V = \sum_{j=1}^{l} f'_j \sigma_j, \quad f'_j \in \mathscr{O}(V),$$

we set

$$\|f\|_{V'} = \inf_{(f'_j)} \sup\{|f'_j(x)|;\ x \in V'\}. \tag{8.1.26}$$

Endowed with countably many semi-norms of this type, where $\{V\}$ and $\{V'\}$ form open bases of X, $\Gamma(U, \mathscr{F})$ gives rise to a topological vector space.

By definition,

$$\|f\|_{V'} \le \|f\|_{\mathrm{P}\Delta'},$$

and by Theorem 8.1.18 (ii) there are a compact subset L with $L \Subset V$ and a positive constant C such that

$$\|f\|_{\mathrm{P}\Delta'} \le C\|f\|_L.$$

Therefore we have:

Lemma 8.1.27 *The topology on $\Gamma(U, \mathscr{F})$ defined by countably many semi-norms given by* (8.1.25) *is equivalent to the one defined by* (8.1.26).

Now, we consider $\Gamma(U, \mathscr{F})$ as a topological vector space defined by semi-norms as in Lemma 8.1.27 just above. As in Sect. 8.1.2 above, together with Theorems 8.1.18 (i) and 8.1.24, we have the following:

Theorem 8.1.28 *Let $\mathscr{F} \to X$ be a coherent sheaf over a complex space X.*

(i) *Let U be an open subset of X. Then $\Gamma(U, \mathscr{F})$ is a Fréchet space.*
(ii) *For a relatively compact open subset $V \Subset U$ the restriction map*

$$\rho : f \in \Gamma(U, \mathscr{F}) \to f|_V \in \Gamma(V, \mathscr{F})$$

is completely continuous.

8.2 Cartan–Serre Theorem

Theorem 8.2.1 (Cartan–Serre [12]) *Let X be a compact complex space, and let $\mathscr{F}$ be a coherent sheaf over X. Then, as complex vector spaces,*

$$\dim_{\mathbf{C}} H^q(X, \mathscr{F}) < \infty, \qquad {}^\forall q \geq 0.$$

Proof By making use of relatively compact Stein open subsets,

$$V_\alpha \Subset U_\alpha \Subset X,$$

we take finite coverings $X = \bigcup_\alpha V_\alpha = \bigcup_\alpha U_\alpha$. Set $\mathscr{V} = \{V_\alpha\}$ and $\mathscr{U} = \{U_\alpha\}$. It follows from Leray's Theorem 3.4.40 that

(8.2.2) $$H^q(X, \mathscr{F}) \cong H^q(\mathscr{U}, \mathscr{F}) \cong H^q(\mathscr{V}, \mathscr{F}), \quad q \geq 0.$$

We consider applying the proof of Grauert's Theorem 7.4.1[1]: Since there is no boundary of X, the bumping of the boundary is not necessary, and it is just sufficient to apply it for $\mathscr{V}$ and $\mathscr{U}$. In the proof of Grauert's Theorem 7.4.1 we dealt with only $q = 1$, but the argument is completely the same for $q \geq 0$, and we obtain the following (cf. (7.4.9)):

(8.2.3) $$\Psi : \xi \oplus \eta \in Z^q(\mathscr{U}, \mathscr{F}) \oplus C^{q-1}(\mathscr{V}, \mathscr{F}) \to \rho(\xi) + \delta\eta \in Z^q(\mathscr{V}, \mathscr{F}) \to 0.$$

Here, when $q = 0$, we set $C^{-1}(*) = 0$. The map ρ consists of restrictions from simplexes $U_{\alpha_0} \cap \cdots \cap U_{\alpha_q}$ belonging to $N_q(\mathscr{U})$ to $V_{\alpha_0} \cap \cdots \cap V_{\alpha_q}$ $(\Subset U_{\alpha_0} \cap \cdots \cap U_{\alpha_q})$ belonging to $N_q(\mathscr{V})$, and it is completely continuous by Theorem 8.1.28 (ii). By definition,

$$H^q(\mathscr{V}, \mathscr{F}) = Z^q(\mathscr{V}, \mathscr{F}) / \delta C^{q-1}(\mathscr{V}, \mathscr{F}).$$

Therefore it follows from L. Schwartz's Theorem 7.3.23 that

$$\mathrm{Coker}(\Psi - \rho) = Z^q(\mathscr{V}, \mathscr{F}) / \delta C^{q-1}(\mathscr{V}, \mathscr{F}) = H^q(\mathscr{V}, \mathscr{F})$$

is finite dimensional. By (8.2.2), $\dim_{\mathbf{C}} H^q(X, \mathscr{F}) < \infty$. □

8.3 Positive Line Bundles and Hodge Manifolds

Let $L \to M$ be a (holomorphic) line bundle over a complex manifold M. We write $\mathscr{O}(L) \to M$ for the sheaf of germs of holomorphic sections of $L \to M$. Let $M = \bigcup_\alpha U_\alpha$ be an open covering of local trivialization of L, let $\{\xi_{\alpha\beta}\}$ be the system of

[1] Historically, Grauert's proof followed after that of the Cartan–Serre Theorem.

transition functions of L, and let $h = \{h_\alpha\}$ be a hermitian metric in L. The transition relations

$$h_\alpha(x) = |\xi_{\alpha\beta}(x)|^2 h_\beta(x), \quad x \in U_\alpha \cap U_\beta$$

are satisfied. Then, as in (5.5.37), the curvature form of h

$$\omega_h = \frac{i}{2\pi} \partial\bar{\partial} \log h_\alpha$$

is well defined on M. Let $(z_\alpha^1, \dots, z_\alpha^n)$ be a holomorphic local coordinate system in U_α. Then,

$$\omega_h = \frac{i}{2\pi} \sum_{j,k} \frac{\partial^2 \log h_\alpha}{\partial z_\alpha^j \partial \bar{z}_\alpha^k} dz_\alpha^j \wedge d\bar{z}_\alpha^k . \tag{8.3.1}$$

When the hermitian form $\left(\frac{\partial^2 \log h_\alpha}{\partial z_\alpha^j \partial \bar{z}_\alpha^k}\right)_{jk}$ is positive definite, we write $\omega_h > 0$, and say that (L, h) is *positive*. If a line bundle L carries a hermitian metric with positive curvature form, we call L a *positive line bundle*, and write $L > 0$; if L^{-1} is positive, L is said to be *negative* and is written $L < 0$.

In general, let

$$g = \sum_{j,k} g_{\alpha j\bar{k}} dz_\alpha^j \otimes d\bar{z}_\alpha^k,$$

$$\left(g_{\alpha j\bar{k}}\right) > 0 \quad \text{(positive definite)}$$

be a hermitian metric in the holomorphic tangent bundle $\mathbf{T}(M)$ of a complex manifold M. If the associated hermitian form

$$\omega = \frac{i}{2} \sum_{j,k} g_{\alpha j\bar{k}} dz_\alpha^j \wedge d\bar{z}_\alpha^k \tag{8.3.2}$$

is closed, i.e., $d\omega = 0$, then g is called a *Kähler metric* and ω of (8.3.2) is called the *Kähler form* of g or of M. If M carries a Kähler metric, it is called a *Kähler manifold*.

If a hermitian line bundle (L, h) is positive, the form ω_h of (8.3.1) gives a Kähler form, which is a special Kähler metric on M. A complex manifold endowed with this kind of special Kähler metric is called a *Hodge manifold*.

Example 8.3.3 Let $[z^0, \dots, z^n]$ be a homogeneous coordinate system of $\mathbf{P}^n(\mathbf{C})$. Let

$$U_j = \{[z^0, \dots, z^n];\, z^j \neq 0\} \cong \left\{\left(\frac{z^0}{z^j}, \dots, \check{1}, \dots, \frac{z^n}{z^j}\right)\right\} \cong \mathbf{C}^n, \; 0 \leq j \leq n$$

be holomorphic local coordinate neighborhoods. A line bundle $H \to \mathbf{P}^n(\mathbf{C})$ determined by the following transition functions

$$\xi_{jk} = \frac{z^j}{z^k} \quad \text{on } U_j \cap U_k$$

is called the *hyperplane bundle* over $\mathbf{P}^n(\mathbf{C})$. On each U_j we set

$$h_j = 1 + \sum_{k \neq j} \left| \frac{z^k}{z^j} \right|^2 > 0.$$

Then, $h = (h_j)$ gives rise to a hermitian metric in H. Now we compute the curvature form ω_h of H; e.g., with $j = 0$ and $z_0 = 1$ we get

$$h_0 = 1 + \sum_{k=1}^{n} |z^k|^2 = 1 + \|(z^k)\|^2.$$

From this it follows that

$$\begin{aligned}
&\partial\bar{\partial} \log h_0 \\
&= \partial\bar{\partial} \log \left(1 + \|(z^k)\|^2\right) = \partial \frac{\sum_k z^k d\bar{z}^k}{1 + \|(z^k)\|^2} \\
&= \frac{\sum_k dz^k \wedge d\bar{z}^k + \|(z^k)\|^2 \sum_k dz^k \wedge d\bar{z}^k - \left(\sum_k \bar{z}^k dz^k\right) \wedge \left(\sum_k z^k d\bar{z}^k\right)}{(1 + \|(z^k)\|^2)^2}.
\end{aligned}$$

Taking a holomorphic tangent vector $X = \sum_k \xi^k \frac{\partial}{\partial z^k}$, we compute the quadratic form $\partial\bar{\partial} \log h_0(X, \bar{X})$:

$$\begin{aligned}
\partial\bar{\partial} \log h_0(X, \bar{X}) &= \frac{\|(\xi^k)\|^2 + \|(z^k)\|^2 \|(\xi^k)\|^2 - |\sum_k \bar{z}^k \xi^k|^2}{(1 + \|(z^k)\|^2)^2} \\
&\geq \frac{\|(\xi^k)\|^2 + \|(z^k)\|^2 \|(\xi^k)\|^2 - \|(\bar{z}^k)\|^2 \|(\xi^k)\|^2}{(1 + \|(z^k)\|^2)^2} \\
&= \frac{\|(\xi^k)\|^2}{(1 + \|(z^k)\|^2)^2} > 0, \quad (\xi^k) \neq 0.
\end{aligned}$$

Therefore we see that

$$\omega_h > 0,$$

and the hyperplane bundle $H \to \mathbf{P}^n(\mathbf{C})$ is positive. We often write simply $\mathcal{O}(H) = \mathcal{O}(1)$ and $\mathcal{O}(H^k) = \mathcal{O}(k)$, and $\{h_j\}_{j=0}^n$ (resp. ω_h) is called the *Fubini–Study metric* (resp. *form*), with which $\mathbf{P}^n(\mathbf{C})$ is a Hodge manifold.

Proposition 8.3.4 *Any complex submanifold of* $\mathbf{P}^n(\mathbf{C})$ *is Hodge.*

Proof Let $M \subset \mathbf{P}^n(\mathbf{C})$ be a complex submanifold. Let $L = H|_M \to M$ be the restriction of the hyperplane bundle $H \to \mathbf{P}^n(\mathbf{C})$ to M. Then, $L > 0$, and M is Hodge. □

The converse of this fact is Kodaira's Embedding Theorem.

Now in general, let $L \to M$ be a line bundle. We take a covering $\{U_\alpha\}$ of local trivializations of L, and the transition function system $\{\xi_{\alpha\beta}\}$. We take holomorphic sections of L:

$$\sigma_0 = (\sigma_{0\alpha}), \ldots, \sigma_N = (\sigma_{N\alpha}) \in \Gamma(M, L).$$

The zero set $\{\sigma_j = 0\}$ of σ_j is well defined by setting $\sigma_{j\alpha} = 0$ in each U_α. We obtain an analytic subset

$$B = \bigcap_{j=0}^{N} \{\sigma_j = 0\},$$

which is called the *base locus* of $\{\sigma_j\}$.

For a point $x \in U_\alpha \backslash B$ we associate $(\sigma_{0\alpha}(x), \ldots, \sigma_{N\alpha}(x)) \in \mathbf{C}^{N+1} \backslash \{0\}$. If $x \in U_\beta \backslash B$,

$$(\sigma_{0\alpha}(x), \ldots, \sigma_{N\alpha}(x)) = \xi_{\alpha\beta}(x)(\sigma_{0\beta}(x), \ldots, \sigma_{N\beta}(x)).$$

Therefore, a point of $\mathbf{P}^N(\mathbf{C})$ is defined independently from the choice of U_α. We denote this map by

(8.3.5) $$\Phi : x \in M \backslash B \to [\sigma_0(x), \ldots, \sigma_N(x)] \in \mathbf{P}^N(\mathbf{C}).$$

If $B = \emptyset$, then naturally Φ is holomorphic in the whole M.

Here we assume that M is compact. By Theorem 8.2.1,

$$\dim \Gamma(M, L) = \dim H^0(M, \mathscr{O}(L)) < \infty.$$

Let $\{\sigma_j\}_{j=0}^N$ be a basis of the finite-dimensional vector space $\Gamma(M, L)$, where $N = \dim \Gamma(M, L) - 1$. The analytic subset of M

$$B(L) = \bigcap_j \{\sigma_j = 0\}$$

is called the *base locus* of L; this is independent from the choice of the basis $\{\sigma_j\}$, and determined only by L. We write

(8.3.6) $$\Phi_L : x \in M \backslash B(L) \to [\sigma_0(x), \ldots, \sigma_N(x)] \in \mathbf{P}^N(\mathbf{C})$$

for the holomorphic map given by (8.3.5).

8.4 Grauert's Theorem

8.4.1 Strongly Pseudoconvex Domains

In this section we let M be a complex manifold in general. The following is due to [22]:

Theorem 8.4.1 (Grauert) *Let $\Omega \Subset M$ be a strongly pseudoconvex domain. Let $\mathscr{F}$ be a coherent sheaf defined in a neighborhood of the closure $\bar{\Omega}$. Then*

$$\dim_{\mathbf{C}} H^q(\Omega, \mathscr{F}) < \infty, \quad q \geq 1.$$

Proof We consider in a neighborhood of $\bar{\Omega}$, and so may assume that M satisfies the second countability axiom, and that $\mathscr{F}$ is defined on M.

We apply the same proof of Grauert's Theorem 7.4.1 by making use of the bumping method for the strongly pseudoconvex boundary $\partial\Omega$.

We endow the space of sections of $\mathscr{F}$ with the topology of locally uniform convergence to have a Fréchet space. Then the idea to apply L. Schwartz's Theorem is the same as in the proof of the Cartan–Serre Theorem 8.2.1.

Take Leray coverings $\mathscr{V}$ and $\tilde{\mathscr{U}}$ as in the proof of Grauert's Theorem 7.4.1. In Grauert's Theorem 7.4.1 we dealt with only $q = 1$, but it is completely similar for $q \geq 1$. We have

$$H^q(\mathscr{V}, \mathscr{F}) \cong H^q(\Omega, \mathscr{F}), \tag{8.4.2}$$

furthermore (cf. (7.4.9))

$$\Psi : \xi \oplus \eta \in Z^q(\tilde{\mathscr{U}}, \mathscr{F}) \oplus C^{q-1}(\mathscr{V}, \mathscr{F}) \to \rho(\xi) + \delta\eta \in Z^q(\mathscr{V}, \mathscr{F}) \to 0. \tag{8.4.3}$$

Here, the map ρ consists of restrictions from the simplexes of $N_q(\mathscr{U})$ to the relatively compact simplexes of $N_q(\mathscr{V})$, and it is completely continuous by Proposition 8.1.17. It is the definition that

$$H^q(\mathscr{V}, \mathscr{F}) = Z^q(\mathscr{V}, \mathscr{F}) / \delta C^{q-1}(\mathscr{V}, \mathscr{F}).$$

By L. Schwartz's Theorem 7.3.23,

$$\mathrm{Coker}(\Psi - \rho) = Z^q(\mathscr{V}, \mathscr{F}) / \delta C^{q-1}(\mathscr{V}, \mathscr{F}) = H^q(\mathscr{V}, \mathscr{F})$$

is finite dimensional, and so $\dim_{\mathbf{C}} H^q(\Omega, \mathscr{F}) < \infty$ by (8.4.2). □

8.4.2 Positive Line Bundles

In this subsection, M denotes a compact complex manifold. We first show a local lemma.

Lemma 8.4.4 *Let $U \subset \mathbf{C}^n$ be an open subset, and let $h > 0$ be a function of C^2-class in U such that* $\log h$ *is strongly plurisubharmonic in U. Then the function $|\zeta|^2 h$ is strongly plurisubharmonic in $(\zeta, z) \in \mathbf{C}^* \times U$.*

Proof By computation we get

$$\partial\bar{\partial} \log h = \frac{1}{h^2}(h\partial\bar{\partial}h - \partial h \wedge \bar{\partial}h) > 0. \tag{8.4.5}$$

Next we compute $\partial\bar{\partial}(|\zeta|^2 h)$:

$$\begin{aligned}\partial\bar{\partial}(|\zeta|^2 h) &= \partial(\zeta h d\bar{\zeta} + |\zeta|^2\bar{\partial}h) \\ &= h d\zeta \wedge d\bar{\zeta} + \zeta\partial h \wedge d\bar{\zeta} + \bar{\zeta} d\zeta \wedge \bar{\partial}h + |\zeta|^2\partial\bar{\partial}h.\end{aligned} \tag{8.4.6}$$

Applying (8.4.5) for the last term, we obtain

$$\begin{aligned}\partial\bar{\partial}(|\zeta|^2 h) &= h d\zeta \wedge d\bar{\zeta} + \zeta\partial h \wedge d\bar{\zeta} + \bar{\zeta} d\zeta \wedge \bar{\partial}h + \frac{|\zeta|^2}{h}\partial h \wedge \bar{\partial}h \\ &\quad + |\zeta|^2 h\partial\bar{\partial}\log h \\ &= \frac{1}{h}\left(h^2 d\zeta \wedge d\bar{\zeta} + \zeta h\partial h \wedge d\bar{\zeta} + \bar{\zeta} h d\zeta \wedge \bar{\partial}h + |\zeta|^2\partial h \wedge \bar{\partial}h\right) \\ &\quad + |\zeta|^2 h\partial\bar{\partial}\log h \\ &= \frac{1}{h}(h d\zeta + \zeta\partial h) \wedge \overline{(h d\zeta + \zeta\partial h)} + |\zeta|^2 h\partial\bar{\partial}\log h \geq 0.\end{aligned} \tag{8.4.7}$$

Therefore the Levi form $\partial\bar{\partial}(|\zeta|^2 h)$ is positive semidefinite. To show the positive definiteness we take a tangent vector $X = X^0\frac{\partial}{\partial\zeta} + \sum_j X^j\frac{\partial}{\partial z^j}$ with $\partial\bar{\partial}(|\zeta|^2 h)\langle X, \bar{X}\rangle = 0$. We shall prove $X = 0$. With substituting X to (8.4.7), the equality must hold. Thus,

$$(\partial\bar{\partial}\log h)\langle X, \bar{X}\rangle = 0.$$

We infer from (8.4.5) that $X^j = 0, 1 \leq j \leq n$. Hence, $X = X^0\frac{\partial}{\partial\zeta}$, and then by (8.4.7), $\partial\bar{\partial}(|\zeta|^2 h)\langle X, \bar{X}\rangle = |X^0|^2 h = 0$, which implies $X^0 = 0$. □

Theorem 8.4.8 (Grauert's Vanishing Theorem) *Let $L \to M$ be a positive line bundle. For an arbitrary coherent sheaf $\mathscr{F} \to M$, there is a number $k_0 \in \mathbf{N}$ such that*

$$H^q(M, \mathscr{O}(L^k)\otimes\mathscr{F}) = 0, \quad q \geq 1, \ k \geq k_0.$$

Proof Let $\{U_\alpha\}$ be a covering of local trivializations of L, and let $\{\xi_{\alpha\beta}\}$ be the system of transition functions. By the assumption, L carries an hermitian metric $h = \{h_\alpha\}$ such that in each U_α,

$$\partial\bar{\partial}\log h_\alpha > 0.$$

We consider the dual L^{-1} of L. With writing

$$L^{-1}|_{U_\alpha} = U_\alpha \times \mathbf{C} \ni (x, \zeta_\alpha),$$
$$\zeta_\alpha = \frac{1}{\xi_{\alpha\beta}}\zeta_\beta \qquad (\text{in } U_\alpha \cap U_\beta).$$

The function

$$\psi(x, \zeta_\alpha) = |\zeta_\alpha|^2 h_\alpha(x)$$

is a C^∞ exhaustion function on the complex manifold L^{-1}. By Lemma 8.4.4, ψ is strongly plurisubharmonic in $L^{-1}\backslash 0$, where 0 denotes the zero section. The neighborhood $\Omega = \{\psi < 1\} \subset L^{-1}$ of 0 is strongly pseudoconvex. Let

$$\pi : L^{-1} \to M, \quad \pi : \Omega \to M$$

be the projections. A holomorphic section $\sigma = (\sigma_\alpha) \in \Gamma(U, L^k)$ of L^k on an open subset $U \subset M$ defines a holomorphic function on $(\pi^{-1}U) \cap \Omega$ by

$$\pi^*\sigma : (x, \zeta_\alpha) \in (\pi^{-1}U) \cap \Omega \to \zeta_\alpha^k \cdot \sigma_\alpha(x) \in \mathbf{C}.$$

For distinct k's they are holomorphic functions linearly independent over $\mathbf{C}$. Let $\pi^*\mathscr{F} \to L^{-1}$ be the pull-back of $\mathscr{F}$ over L^{-1}. Then $\pi^*\mathscr{F}$ is a coherent sheaf of $\mathscr{O}_{L^{-1}}$-modules. Taking each U_α to be Stein, we have a Stein covering $\mathscr{U} = \{U_\alpha\}$ of M (i.e., every member of the covering is Stein), and $\pi^{-1}\mathscr{U} = \{(\pi^{-1}U_\alpha)\cap\Omega\}$ which is a Stein covering of Ω. Therefore we obtain the following exact sequence:

$$\begin{aligned} 0 \longrightarrow & \bigoplus_{k\geq 1} H^q(M, \mathscr{O}(L^k)\otimes\mathscr{F}) \cong \bigoplus_{k\geq 1} H^q(\mathscr{U}, \mathscr{O}(L^k)\otimes\mathscr{F}) \\ \overset{\pi^*}{\longrightarrow} & \; H^q(\pi^{-1}\mathscr{U}, \pi^*\mathscr{F}) \cong H^q(\Omega, \pi^*\mathscr{F}). \end{aligned}$$

By Grauert's Theorem 8.4.1, $H^q(\Omega, \pi^*\mathscr{F})$ is finite dimensional. Therefore there is a number $k_0 \in \mathbf{N}$ such that

$$H^q(M, \mathscr{O}(L^k)\otimes\mathscr{F}) = 0, \quad k \geq k_0. \qquad \square$$

Remark 8.4.9 In the proof above it was a key point that the zero section of L^{-1} has a strongly pseudoconvex neighborhood Ω. Grauert [24] called a line bundle whose zero section carries a strongly pseudoconvex neighborhood a *weakly negative* line bundle. This notion is rightly extended to vector bundles. Theorem 8.4.8 remains valid more generally for a vector bundle over M whose dual E^* is weakly negative;

in the proof we use the symmetric tensor power $S^k E$ in place of L^k (cf. Exercise 5 at the end of this chapter).

8.5 Kodaira's Embedding Theorem

In this section M denotes a compact complex manifold.

Theorem 8.5.1 (Kodaira) *A Hodge manifold M is embeddable into* $\mathbf{P}^N(\mathbf{C})$.

Proof By assumption there is a positive line bundle $L \to M$. Let $\mathfrak{m}_x$ denote the maximal ideal of $\mathscr{O}_{M,x}$ at a point $x \in M$. We consider the following exact sequence:

$$0 \to \mathfrak{m}_x^2 \to \mathscr{O}_M \to \mathscr{O}_M/\mathfrak{m}_x^2 \to 0.$$

Taking a tensor product of this with $\mathscr{O}(L^k)$ $(k \geq 1)$, we have

$$0 \to \mathfrak{m}_x^2 \otimes \mathscr{O}(L^k) \to \mathscr{O}(L^k) \to (\mathscr{O}_M/\mathfrak{m}_x^2) \otimes \mathscr{O}(L^k) \to 0.$$

This yields the following exact sequence:

$$H^0(M, \mathscr{O}(L^k)) \to H^0(M, (\mathscr{O}_M/\mathfrak{m}_x^2) \otimes \mathscr{O}(L^k)) \to H^1(M, \mathfrak{m}_x^2 \otimes \mathscr{O}(L^k)).$$

By Grauert's Vanishing Theorem 8.4.8 there is a number $k_0 \in \mathbf{N}$ such that

$$H^1(M, \mathfrak{m}_x^2 \otimes \mathscr{O}(L^k)) = 0, \quad k \geq k_0.$$

Therefore the following is exact:

$$H^0(M, \mathscr{O}(L^k)) \to H^0(M, (\mathscr{O}_M/\mathfrak{m}_x^2)\otimes\mathscr{O}(L^k)) \to 0, \quad k \geq k_0.$$

Hence, there exist an element $\sigma \in \Gamma(M, L^k)$ with

(8.5.2) $$\sigma(x) \neq 0,$$

and moreover $\sigma_1, \ldots, \sigma_n \in \Gamma(M, L^k)$ $(n = \dim M)$ satisfying

(8.5.3) $$(d\sigma_1 \wedge \cdots \wedge d\sigma_n)(x) \neq 0.$$

Then there is a neighborhood U of x where (8.5.2) and (8.5.3) hold. Since M is compact, there is an open covering of M by such U's. We see that for a sufficiently large k_0, $B(L) = \emptyset$ and

$$\Phi_{L^k} : M \to \mathbf{P}^N(\mathbf{C}), \quad N = \dim \Gamma(M, L^k) - 1, \ k \geq k_0$$

is a holomorphic immersion.

To make this a holomorphic embedding, we take arbitrarily two distinct points $x, y \in M$ and set

$$\mathfrak{m}\langle x, y\rangle = \mathfrak{m}_x \otimes \mathfrak{m}_y.$$

The following is exact:

$$0 \to \mathfrak{m}\langle x, y\rangle \otimes \mathscr{O}(L^k) \to \mathscr{O}(L^k) \to (\mathscr{O}_M/\mathfrak{m}\langle x, y\rangle) \otimes \mathscr{O}(L^k) \to 0.$$

This implies an exact sequence:

$$H^0(M, \mathscr{O}(L^k)) \to H^0(M, (\mathscr{O}_M/\mathfrak{m}\langle x, y\rangle) \otimes \mathscr{O}(L^k)) \to H^1(M, \mathfrak{m}\langle x, y\rangle \otimes \mathscr{O}(L^k)).$$

Again by Grauert's Vanishing Theorem 8.4.8, with $k_0 \in \mathbf{N}$ taken larger if necessary, we have

$$H^1(M, \mathfrak{m}\langle x, y\rangle \otimes \mathscr{O}(L^k)) = 0, \quad k \geq k_0.$$

Therefore the following is exact:

$$H^0(M, \mathscr{O}(L^k)) \to H^0(M, (\mathscr{O}_M/\mathfrak{m}\langle x, y\rangle) \otimes \mathscr{O}(L^k)) \to 0, \quad k \geq k_0.$$

It follows that there is an element $\sigma \in \Gamma(M, L^k)$ with $\sigma(x) = 0$ and $\sigma(y) \neq 0$. Thus, if $k \geq k_0$, then

$$\Phi_{L^k}(x) \neq \Phi_{L^k}(y). \tag{8.5.4}$$

We consider the product

$$\Phi_{L^k} \times \Phi_{L^k} : M \times M \to \mathbf{P}^N(\mathbf{C}) \times \mathbf{P}^N(\mathbf{C}).$$

Let $p : M \times M \to M$ be the first projection. Since Φ_{L^k} is an immersion, there is a neighborhood W of the diagonal $\{(x, y) \in M \times M;\ x = y\}$ such that the restrictions $(\Phi_{L^k} \times \Phi_{L^k})|_{W \cap p^{-1}x}$ are injective over $W \cap p^{-1}x$ for all $x \in M$. For each $(x, y) \in M \times M \setminus W$ we see by (8.5.4) that

$$\Phi_{L^k}(x) \neq \Phi_{L^k}(y)$$

for all sufficiently large k; this holds in a neighborhood of (x, y). Since $M \times M \setminus W$ is compact, with k_0 taken sufficiently large,

$$((\Phi_{L^k} \times \Phi_{L^k})(M \times M \setminus W)) \cap \Delta = \emptyset, \quad k \geq k_0, \tag{8.5.5}$$

where Δ denotes the diagonal of $\mathbf{P}^N(\mathbf{C}) \times \mathbf{P}^N(\mathbf{C})$.

Let $k \geq k_0$ and let $(x, y) \in M \times M$ be any point. Suppose that $\Phi_{L^k}(x) = \Phi_{L^k}(y)$. Then, $(x, y) \notin M \times M \setminus W$ by (8.5.5), so that $(x, y) \in W$. The choice of W implies that $x = y$. Therefore,

$$\Phi_{L^k} : M \to \mathbf{P}^N(\mathbf{C}), \quad k \geq k_0$$

is injective, and gives rise to a holomorphic embedding. □

Remark 8.5.6 The proof given above remains valid even if M is allowed to have singularities, that is, a compact complex space. In fact, H. Grauert [24] proved the extension for complex spaces. For readers who have read up to here, it will be clear what one should do for that proof. Therefore, Kodaira's Embedding Theorem 8.5.1 ([37]) is extended to singular spaces by means of coherent sheaves.

Example 8.5.7 Let $(z^1, \ldots, z^n)$ be the standard coordinate system of $\mathbf{C}^n$. Let $\gamma_j \in \mathbf{C}^n$, $1 \leq j \leq 2n$, be $2n$ linearly independent vectors over $\mathbf{R}$. Set

$$\Gamma = \sum_{j=1}^{2n} \mathbf{Z} \cdot \gamma_j \subset \mathbf{C}^n. \tag{8.5.8}$$

Then Γ is called a *lattice* of $\mathbf{C}^n$ and acts on $\mathbf{C}^n$ by translations. We set the quotient

$$M = \mathbf{C}^n / \Gamma. \tag{8.5.9}$$

Then, M is a compact complex manifold, called a *complex torus*. Since the hermitian metric form

$$\omega_0 = \sum_{j=1}^{n} \frac{i}{2} dz^j \wedge d\bar{z}^j$$

is d-closed, M is a compact Kähler manifold. It depends on the choice of the lattice Γ whether M is Hodge or not (see Exercise 7 at the end of this chapter). Cf., e.g., Weil [78] more for the theory of compact Kähler manifolds and complex tori.

Exercises

1. Let M be a paracompact complex manifold of dimension n. Show that

$$H^q(M, \mathscr{O}_M) = 0, \quad q > n.$$

2. Let X be a compact Riemann surface with universal covering $\pi : \Delta \to X$ and with deck transformation group $\Gamma \subset \mathrm{Aut}(\Delta)$, where $\Delta \subset \mathbf{C}$ is the unit disk about 0. By making use of the Poincaré metric, which is $\mathrm{Aut}(\Delta)$-invariant, show that the line bundle K_X of holomorphic 1-forms over X is positive, and hence X is projective algebraic.

   If the reader knows the "Bergman metric", show the above statement with replacing Δ (resp. 1-forms) by a bounded domain Ω of $\mathbf{C}^n$ (resp. n-forms).

3. Let X be a compact Riemann surface. Then,

$$\dim_{\mathbf{C}} H^1(X, \mathscr{O}_X) < \infty$$

(Cartan–Serre Theorem 8.2.1). By making use of this, show the following:

   a. Let $p_0 \in X$ be any given point. Then there is a non-constant meromorphic function φ on X with a pole only at p_0.
   b. Let d_0 be the order of the pole of φ in 3a above. Show that there is a meromorphic function ψ with a pole only at p_0, where the pole order of ψ is $kd_0 - 1$ with some $k \in \mathbf{N}$, so that the meromorphic function $f = \psi/\varphi^k$ is holomorphic in a neighborhood of p_0, $f(p_0) = 0$ and $df(p_0) \neq 0$.
   c. For an arbitrary point $p \in X \setminus \{p_0\}$, there is a meromorphic function φ with a pole only at p_0 such that $d\varphi(p) \neq 0$.
   d. For any distinct points $p, q \in X \setminus \{p_0\}$, there is a meromorphic function ψ with a pole only at p_0 on X such that $\psi(p) \neq \psi(q)$.
   e. X is projective algebraic.

4. Let X be a compact complex manifold (or space). Let $E \to X$ be a holomorphic vector bundle over X, and let $\mathscr{O}(E)$ denote the sheaf of germs of holomorphic sections of E over X. Show that

$$\dim_{\mathbf{C}} H^q(X, \mathscr{O}(E)) < \infty, \quad q \geq 0.$$

5. Let $E \to X$ be as above. Then the dual vector bundle E^* is naturally defined as locally $E^*|_U = U \times (\mathbf{C}^n)^*$, where $E|_U \cong U \times \mathbf{C}^n$ is a local trivialization and $(\mathbf{C}^n)^*$ stands for the dual vector space of $\mathbf{C}^n$. Assume that there is a strongly pseudoconvex neighborhood $\Omega \Subset E^*$ of the zero section of E^*. Show that for every coherent sheaf $\mathscr{F}$ over X there is a number $k_0 \in \mathbf{N}$ with

$$H^q(X, \mathscr{F} \otimes \mathscr{O}(S^k E)) = 0, \quad q \geq 1, \ k \geq k_0,$$

where $S^k E$ denotes the k-th symmetric tensor power of E.

6. Let $\alpha_j \in \mathbf{C}$, $j = 1, 2$, with $|\alpha_j| > 1$. Consider a holomorphic action on $\mathbf{C}^2 \setminus \{0\}$ defined by

$$\lambda_n : (z_1, z_2) \in \mathbf{C}^2 \setminus \{0\} \to (\alpha_1 z_1, \alpha_2 z_2) \in \mathbf{C}^2 \setminus \{0\}, \quad n \in \mathbf{Z}.$$

Then we take the quotient $X = (\mathbf{C}^2 \setminus \{0\})/\{\lambda_n; n \in \mathbf{Z}\}$ with the natural map $\pi : \mathbf{C}^2 \setminus \{0\} \to X$. Show the following:

   a. X is a compact complex manifold (called a Hopf manifold).
   b. If ϕ is a meromorphic function on X, then $\pi^*\phi$ extends meromorphically over $\mathbf{C}^2$.
   c. X is *not* projective algebraic. In fact, assuming that X is projective algebraic, show by 6b above that the closure of the graph $G(\pi)(\subset (\mathbf{C}^2 \setminus \{0\}) \times X)$ in

$\mathbf{C}^2 \times X$ is an analytic subset, and then infer a contradiction.

N.B. In fact, X is not even Kähler, since $H^2(X, \mathbf{R}) = 0$.

(Hint: Use Remmert's Extension Theorem 6.8.1 and Hartogs' Extension (Theorem 1.2.26).

7. Let Γ and $M = \mathbf{C}^n/\Gamma$ be as in (8.5.8) and (8.5.9).

a. Show that M is a compact complex manifold.

Let γ^{j*} denote the dual of γ_j over $\mathbf{R}$. Then, the real 2-forms $d\gamma^{j*} \wedge d\gamma^{k*}$, $1 \le j < k \le 2n$, constitute a generator system of $H^2(M, \mathbf{Z})$ over $\mathbf{Z}$. Let

$$\omega = \frac{i}{2} \sum_{\nu,\mu} g_{\nu\bar{\mu}} dz^\nu \wedge d\bar{z}^\mu$$

be a Kähler form on M, where $g_{\nu\bar{\mu}} \in \mathbf{C}$ and $(g_{\nu\bar{\mu}})$ is a positive definite hermitian matrix. Write

$$\omega = \sum_{1 \le j < k \le 2n} v_{jk} d\gamma^{j*} \wedge d\gamma^{k*}, \quad v_{jk} \in \mathbf{R}.$$

Then, ω is Hodge if and only if all $v_{jk} \in \mathbf{Z}$.

Set

$$\gamma_j = \begin{pmatrix} \gamma_j^1 \\ \gamma_j^2 \\ \vdots \\ \gamma_j^n \end{pmatrix} \in \mathbf{C}^n.$$

b. Show that

$$v_{jk} = -\Im \left(\sum_{\nu,\mu} g_{\nu\bar{\mu}} \gamma_j^\nu \bar{\gamma}_k^\mu \right), \quad 1 \le j < k \le n.$$

c. Show that every M is Hodge, if $n = 1$ (therefore, M is projective algebraic by Theorem 8.5.1).

d. Let $n = 2$ and set

$$(\gamma_1, \gamma_2, \gamma_3, \gamma_4) = \begin{pmatrix} 1 & 0 & \sqrt{2}i & \sqrt{3}i \\ 0 & 1 & i & \sqrt{5}i \end{pmatrix}.$$

Show that M is not Hodge; that is, there is no Hodge metric on M.

e. Let $n = 2$ and set

$$(\gamma_1, \gamma_2, \gamma_3, \gamma_4) = \begin{pmatrix} 1 & 0 & 2i & 3i \\ 0 & 1 & i & 5i \end{pmatrix}.$$

Find a Hodge metric on M.

Chapter 9
On Coherence

The purpose of the present note is to put certain fundamental works of Kiyoshi Oka in historical perspective and, in particular, to give them their deserved recognition. Let us begin with remarks from others.

In an exposition on a series of works of Oka, Lipman Bers, who is well-known, for example, in the theory of Teichmüller moduli spaces of Riemann surfaces, concluded the introduction of his lecture notes at Columbia University [6], writing:

> Every account of the theory of several complex variables is largely a report on the ideas of Oka. This one is no exception.

L. Bers was not a specialist in function theory in several complex variables, so that this can be regarded as an objective evaluation from the viewpoint of a third party.

The series of Oka's fundamental papers culminated in Oka VII where his First Coherence Theorem is proved. Every time we read the proof, we are very much impressed by its subtle beauty. In [69], p. 46, Reinhold Remmert wrote:

> It is no exaggeration to claim that Oka's theorem became a landmark in the development of function theory of several complex variables.

And he continues:

> By sheafifying one suddenly was able to obtain results one had not dared to dream of in 1950. ... In algebraic geometry, Serre proved in 1955, cf. [FAC], that the structure sheaves of algebraic varieties are coherent. This result, however, is much easier to obtain than Oka's theorem.[1]

Therefore it is no exaggeration to say that Oka's coherence theorems are essential building blocks for the theory of analytic functions of several variables, or complex analysis in several variables, after 1950. This is exemplified by Oka's Jôku-Ikô, used since Oka I, and by the fact that the Cousin Problems are naturally solved, even on singular domains. Oka wrote:

[1] It is interesting to learn the easier algebraic case was established under the influence of the difficult analytic case; [FAC] is [72].

J. Noguchi, *Analytic Function Theory of Several Variables*,
DOI 10.1007/978-981-10-0291-5_9

(From Introduction of Oka VII (Iwanami version))

> ... dans le présent Mémoire on trouvera, comme conclusion, plusieurs théorèmes et un problème bien filtré de la même nature (Voir No. 7); dont les théorèmes me sont indispensables pour traiter les problèmes depuis le Mémoire I, aux domaines contenant les points de ramification, et ils sont utiles pour les domaines moins compliqués.

In terms of contents, this means that Oka developed methods that allowed the results described from Chaps. 1 to 5 of the present book to be proved in the setting of singular complex spaces. In his view, Levi's Problem (Hartogs' Inverse Problem) was included; in the end of the introduction of Oka VII, Iwanami version, he wrote:

> Or, nous, devant le beau système de problèmes à F. Hartogs et aux successeurs, voulons léguer des nouveaux problèmes à ceux qui nous suivront; or, comme le champ de fonctions analytiques de plusieurs variables s'étend heureusement aux divers branches de mathématiques, nous serons permis de rêver divers types de nouveaux problèmes y préparant.

But, after H. Cartan's modification[2] which was not known to Oka at the time, this part was completely erased in the published version, *Bull. Soc. Math. France* (1950). This is the reason why there are now *two versions* of Oka VII, one published by *Bull. Soc. Math. France* and the other (*Oka's original*) by *Iwanami* [65]; the English translation of Oka VII published by Springer [67] was taken from the Iwanami version. In the view of Cartan, the notion of coherence would almost certainly appear to be useful for Cousin-type problems, but targeting the Levi Problem (Hartogs' Inverse Problem) by inventing the notion of coherence would seem to be beyond his realm of consideration (see [9]). In fact Oka had at that time already solved Levi's Problem (Hartogs' Inverse Problem) for unramified Riemann domains over $\mathbf{C}^n$ of general dimension $n \geq 2$, although it was unpublished (written as a research report in Japanese to Teiji Takagi in *1943*, then Professor of Imperial University of Tokyo), and was trying to extend it to the *ramified* case. With this in mind he had introduced the notion of "ideals or modules with undetermined domains", i.e., coherent sheaves. According to the records in [68], he started to think in this direction some time during *1942*.

However, restricting himself to the case of unramified domains, he wrote Oka IX to solve Levi's Problem (Hartogs' Inverse Problem) by making use of only the First Coherence Theorem. Later, counter-examples were found for ramified domains (cf. Narasimhan [47], Fornæss [16]); hence, the choice of Oka was correct.

In [9] Cartan was experimenting (word due to Grauert) with the notion of coherence which had been introduced by Oka in attempting to solve Levi's Problem (Hartogs' Inverse Problem) on ramified domains. Due to the war there was no communication between them during this period. It is interesting as a page of history that Cartan deleted that comment of Oka.

---

[2] A number of records indicate this; cf., e.g., [42], "Message from Professor Henri Cartan", or [†] p. 262, M. Audin (Ed.), *Correspondance entre Henri Cartan et André Weil, 1928-1991*, Soc. Math. France, Paris, 2011; there one may find also their frank comments on the French language of K. Oka.

The points in Oka's papers, where he wrote comments on Levi's Problem (Hartogs' Inverse Problem) are as follows:

(i) Oka IV (1941), the sentence second from last in the Introduction.
(ii) Oka VI (1942) solved the problem in univalent domains of $\mathbf{C}^2$, stating that he believes the general dimensional case to hold.
(iii) Oka VII (1948/1950), the original version (Iwanami and the English translation), in the end of Introduction.
(iv) Oka VIII (1951), the second sentence of the Introduction, he wrote that the convexity problem was solved at least for unramified domains, and sent the manuscript in Japanese to Professor T. Takagi (Tokyo) as a research report in 1943.
(v) Oka IX (1953): He solved the problem for unramified domains over $\mathbf{C}^n$ with arbitrary $n \geq 1$.

For Oka, the notion he introduced and proved to hold true in order to solve Levi's Problem (Hartogs' Inverse Problem) for ramified Riemann domains was that of "coherence". This resulted in Oka VII and VIII. Therefore, in Oka VII it was very natural for him to mention Levi's Problem (Hartogs' Inverse Problem). Here there might be a difference between Oka's eye and Cartan's.

The notion of coherence appears to be very subtle and, at first glance, perhaps mysterious. Just looking at it, one may think that it would be useful for Cousin-type problems. However, it is hard to imagine that it would provide the foundation to solve approximation problems of Runge type and Levi's Problem (Hartogs' Inverse Problem). Oka's eyes are those of a genius; for the genius' eyes, what should be seen was already seen without the availability of the theory of sheaf cohomology. R. Remmert, writing on this in the introduction of the Collected Works of K. Oka [67],[3] English translation, published in 1984, referring to Goethe's words, said:

> Okas Mathematik bedarf der Interpretation.GOETHE spricht einmal von der Dumpfheit des Genies, das Dinge schaut, ohne dem Geshauten sofort den klaren Ausdruck geben zu können. Klarheit wird erst allmählich druch später Arbeit gewonnen.

In the same volume, Cartan made the following comment:

> Mais il faut avouer que les aspects techniques de ses démonstrations et le mode de présentation de ses résultats rendent difficile la tâche du lecteur, et que ce n'est qu'au prix d'un réel effort que l'on parvient à saisir la portée de ses résultats, qui est considérable.

J.-P. Serre, who introduced the notion of coherence in algebraic geometry in 1955 [72], began the introduction of that paper as follows:

> On sait que les méthodes cohomologiques, et particulièrement la théorie des faisceaux, jouent un rôle croissant, non seulement en théorie des fonctions de plusieurs variables complexes (cf. [5][4]), mais aussi en géométrie algébrique classique (qu'il me suffise de citer les travaux récents de Kodaira-Spencer sur le théorème de

[3] It is noted that Oka VII in this volume is a translation from the Iwanami version.
[4] This means H. Cartan [11].

Riemann-Roch). Le caractère algébrique de ces méthodes laissait penser qu'il était possible de les appliquer également à la géométrie algébrique abstraite; le but du présent mémoire est de demontrer que tel est bien le cas.

The notion of coherence in algebraic geometry, which is nowadays part of the standard foundation, was therefore introduced, based on Oka's Coherence Theorems.

In later years, H. Grauert proved his Finiteness Theorem, which looks like a lotus of the coherence, and gave a proof of Levi's Problem (Hartogs' Inverse Problem), which is presented in Chap. 7 of the present book. Moreover, by this Finiteness Theorem of Grauert, Kodaira's Embedding Theorem was generalized to the setting of singular compact complex spaces (cf. Remark 8.5.6).

The significance of *Oka's First Coherence Theorem* cannot be underestimated. At first, he considered the problem of coherence as a problem of a generating system of an ideal of $\mathscr{O}_{\mathbf{C}^n}$, whose domain of existence is not determined, and he termed it as "*idéal de domaines indéterminés*". H. Grauert and R. Remmert wrote in the introduction of their basic book [28]:

> Of greatest importance in Complex Analysis is the concept of a coherent analytic sheaf.

And they continued with

> H. Cartan experimented in 1944, and Oka affirmatively solved it in *1948*.

The name "*cohérent*" is due to H. Cartan [9] (1944). The publication year of Oka's paper, [62] VII, was 1950; *1948* was the year when the paper was received.[5] As far as we know, when referring to Oka's First Coherence Theorem, Grauert and R. Remmert, always gave the year *1948*. Cartan [10] (1950) also wrote, referring Oka's First Coherence Theorem:

> Oka a écrit en *1948* un Mémoire où il étudie les mêmes questions, quoique en des termes un peu différents.

Therefore we have also given this date in the present book.

In [28] Grauert and Remmert listed the following (ii)–(v) as the *Four Fundamental Coherence Theorems in complex analysis*:

1 (Fundamental Coherence Theorems) (i) *$\mathscr{O}_{\mathbf{C}^n}$ is coherent (cf. Sect. 2.5 in this book; the same below).*

(ii) *The structure sheaf $\mathscr{O}_X$ of a complex space X is coherent (cf. Sect. 6.9).*
(iii) *A geometric ideal sheaf $\mathscr{I}\langle A\rangle$ is coherent (cf. Sect. 6.5).*
(iv) *The normalization sheaf $\hat{\mathscr{O}}_X$ of a complex space X is coherent (cf. Sect. 6.10).*
(v) *The direct image sheaves of a coherent sheaf by a proper map is coherent (cf. Theorem 6.9.8).*

[5] In this sense the received dates of Oka's papers are of importance. It is incomprehensible that the received dates of all papers are deleted in the Collected Works of Oka [67] published by Springer-Verlag.

In VII (1948/'50) and VIII (1951) Oka proved the three coherence theorems, (i), (iii) and (iv). It should be noted that Grauert and Remmert call (ii) *Oka's Theorem*, which is a consequence of (i) and (iii). The last Coherence Theorem, (v), is due to Grauert [23] (1960).

If one takes a look at Oka VII and VIII, he sees that (i), (iii) and (iv) are in one set (cf., e.g., Footnote at p. 233). In many references the coherence of geometric ideal sheaves (iii) is attributed to H. Cartan, since the first proof of (iii) was published by Cartan [10] (1950) (Oka wrote so, giving the proof in VIII). But a key part of the proof of (iii) was formulated and discussed as *Problème* (K) in Oka VII, Sect. 3, and proved already there (cf. Sects. 3, 6): It was then used by Cartan [10] and Oka VIII in the proof of (iii). After stating at the end of Sect. 2 that the coherence of geometric ideal sheaves would be proved in his next paper, Oka started Sect. 3 by considering a relation sheaf

$$f_1\tau_1 + \cdots + f_q\tau_q = 0,$$

where τ_j are given holomorphic functions and f_j are unknowns (cf. (2.5.2)). In Problème (K) he asked for the locally finite generation of the ideal generated only by the first coefficient, f_1. Up to this point, the preparation was done in Oka VII (cf. also Footnote at p. 233).

Then, in the proof of the coherence of a geometric ideal sheaf $\mathscr{I}\langle X\rangle$ of an analytic subset $X = \{F_1 = \cdots = F_q = 0\}$ with suitably chosen holomorphic functions F_j, the above preparation was applied to a relation sheaf

$$f_0\delta^N + f_1F_1 + \cdots + f_qF_q = 0,$$

where δ is the universal denominator of $\mathscr{O}_X$ and $N \in \mathbf{N}$ (cf. (6.5.5)). Then, $f_0 \in \mathscr{I}\langle X\rangle$ if and only if f_0 satisfies the above relation equation. Thus, the coherence of $\mathscr{I}\langle X\rangle$ was deduced; this part of the argumentation is the same in Cartan [10] (1950) and in Oka VIII (1951).

In Oka VII, which was received for publication in 1948, Oka was writing clearly in *two* places, in the last paragraph in Sect. 2 and at the end of the paper, in total *16 lines*, that *it is possible to prove the coherence of geometric ideal sheaves without any additional assumption, which will be dealt with in the next paper*. Although it takes a bit too long, we quote these passages, first looking at the original Iwanami version (wavy underlining was added by the present author):

Iwanami version, [Sect. 2, the last part]:

Nous verrons deux espèces d'idéaux de domaines indéterminés pour lesquelles le problème (J) est résoluble aux polycylindres fermés; dont l'une sera traitée dans le présent Mémoire.

L'autre espèce est *les idéaux géométriques de domaines indéterminés* (ce qui correspond aux idéaux géométriques aus champ de polynomes), qui deviendront indispensables à nous, lorsque nous nous occuperons des domaines admettant des points de ramifications. La démonstration pour les idéaux de cette espèce (pour que le problème (J) soit résoluble aux polycylindre fermés) demande, outre les résultats du Mémoire actuel, quelque notions sur tels domaines. Nous le traiterons donc, dans un Mémoire ultérieur.

[The last part of the paper]:

Problème (J) *Pour les idéaux de domaines indéterminés, trouver les pseudobases finies locales.*

....

Comme cas particulier de ce problème, nous venons de résoudre le *prob lème* (K). Ceci a été indispensable pour établir les théorèmes ci-dessus. Nous reviendrons encore une fois à ce problème, et montrerons, pour *les idéaux géométriques de domaines indéterminés*, qu'il est résoluble sans condition. Cela est indispensable à nous, pour traiter les problèmes depuis le Mémoire I en admettant les points de ramifications d'intervenir. Ces deux exemples parlerons de l'importance du problème.

(Juillet 1948 à Kimimura, Wakayama-Ken, Japon. Reçu le 15 octobre, 1948.)

Let us compare this to the version published in *Bull. Soc. Math. France*; the contents are similar, but there are subtle differences:

Version of *Bull. Soc. Math. France*, [Sect. 2, the last part]:

Nous étudierons deux catégories d'idéaux de domaines indéterminés pour lesquels le problème (J) peut être résolu pour les polycylindres fermés bornés. L'une d'ellesva être étudiée dansleprésent Mémoire. L'autre est celle des *idéaux géométriques de domaines indéterminés* (qui correspondent aux idéaux de polynomes attachés aux variétés algébriques), dont la considération deviendra indispensable quand nous aurons à nous occuper des domaines qui admettent des points de ramification. Pour pouvoir montrer que le problème (J) peut être résolu pour les idéaux de cette espèce (et pour les polycylindres bornés fermés), nous aurons besoin non seulement des résultats du présent Mémoire, mais de quelques notions concernant les domaines ramifiés. Nous réserverons donc cette étude pour un Mémoire ultérieur.

[The last part of the paper]:

PROBLÈME (**J**). — *Pour les idéaux de domaines indéterminés, trouver une pseudobase finie locale.*

....

C'est un cas, particulier de ce problème qui a été résolu sous la forme du problème (K). La solution du problème (K) nous était indispensable pour établir les théorèmes ci-dessus. Nous reviendrons une autre fois sur le problème (J), et montrerons qu'il est résoluble sans condition pour les *idéaux géométriques de domaines indétermi nés*. Cela nous sera indispensable si nous voulons pouvoir traiter les problèmes envisagés depuis le Mémoire I, dans le cas où des points de ramification ne sont pas exclus. Ces deux exemples mettront en évidence l'importance du problème.

(Manuscrit reçu le 15 octobre 1948).

It seems that Oka was not informed that the paper of Cartan proving the coherence of geometric ideal sheaves would be published in the same volume of *Bull. Soc. Math.*

France as Oka VII. Confirming this, we note that in a survey paper [64] (Received 19 Dec. 1949), which Oka wrote just before the publication of Oka VII, he only referred to the papers of Cartan chronologically up to [9], which is dated 1944.

Oka VII was modified by Cartan, including the above 16 lines. All of these changes would have been unthinkable from Oka's viewpoint. Later, Oka wrote a manuscript stating how he was unsatisfied with these modifications. It is titled

> *Sur les formes objectives et les contenus subjectifs — Propos postérieur — Pourquoi le présent mémoire est publié de nouveau*

(cf. [68]). He had intended to publish his original version of Oka VII together with this manuscript as an Appendix, but this was not realized.

In this regard, Cartan wrote the following commentary in the Collected Works of Oka [67] (Springer):

> Oka pose ici une série de problèmes fondamentaux. Le problème (J), en termes de faisceaux, est le suivant: "un faisceau analytique d'idéaux est-il cohérent?" Oka donne lui-même un contre-exemple. Il semble qu'à cette époque il savait que le faisceau d'idéaux défini par un sous-ensemble analytique est cohérent (cf. les 5 dernières lignes du Mémoire): mais il n'a pas publié de démonstration, ce résultat ayant éte entre temps publié par CARTAN dan son article 1950.[6]

However, as mentioned above, there is not only the last part of the paper, clearly announcing that the coherence of geometric ideal sheaves would be proved in the forthcoming paper, but also in the last part of Sect. 2 of the main text where more details were included.

Based on this historical documentation, we refer to (i), (iii) and (iv) of **1** (Fundamental Coherence Theorems) as

Oka's First Coherence Theorem, Oka's Second Coherence Theorem,
Oka's Third Coherence Theorem.

As for the Second Coherence Theorem, it would be historically correct to state that Cartan had in the meantime given his own proof based on Oka's First Coherence Theorem.

The notion of Oka's Coherence is so deep that nowdays his Coherence Theorems provide the fundamentals and guiding forces in a broad area of modern Mathematics as discussed in Sect. 4.5.4.

---

[6]The wavy underlining was added by the present author.

Erratum to: Analytic Function Theory of Several Variables

Erratum to:
J. Noguchi, ***Analytic Function Theory of Several Variables,***
DOI 10.1007/978-981-10-0291-5

The book was inadvertently published without including the following mandatory information shown in the contract, "Copyright for the Japanese edition © 2013, Asakura Publishing Company, Ltd. All Rights reserved". The erratum book has been updated with the change.

The updated original online version for this book can be found at DOI 10.1007/978-981-10-0291-5.

J. Noguchi (✉)
The University of Tokyo, Tokyo, Japan
e-mail: noguchi@ms.u-tokyo.ac.jp

J. Noguchi
Tokyo Institute of Technology, Tokyo, Japan

J. Noguchi, *Analytic Function Theory of Several Variables*,
DOI 10.1007/978-981-10-0291-5_10

Appendix
Kiyoshi Oka

We would like to introduce briefly Professor Kiyoshi Oka and his life (cf. [68]). T. Nishino has written surveys of K. Oka in [50] for Oka I–VI and in [51] for Oka VII–X. We will refer to them without specific comments.

As mentioned already several times the present textbook describes the foundation of analytic function theory of several variables, and it mainly consists of the contents of the series of papers Oka [62], although the order of the presentation is changed from the numbering of the papers.

At the time when the first two papers (Oka I, II) of the series were published, there was a postcard to K. Oka from four well-known French and German mathematicians:

> Münster i. W.
>
> 16-5-38
>
> Monsieur et cher Colligue,
>
> A l'occasion d'une petite réunion de specialistes des fonctions analytiques de plusieurs variables, nous nous sommes aperçus qu'il restait un point obscur dans vos beaux travaux sur le problème de Cousin: il s'agit de savoir si le nom de Oka doit être prononcé à l'anglaise ou à la française. C'est un problème que nous n'avons pas pu résoudre. En attendant la solution, nous vous envoyons notre meilleur souvenir.
>
> H. Cartan H. Behnke E. Peschl K. Stein

The postcard was written in Münster, a university city, where H. Behnke and K. Stein, etc., were working. It seems that a small workshop or a seminar was organized to read Oka I and II, in which H. Cartan was visiting there to participate. The card was probably inscribed by Behnke or Cartan. The sentences represent a close friendship to K. Oka, and how the first two papers of Oka were accepted with a deep influence on the research then.

Professor Kiyoshi Oka was born in 19th of April 1901 at Tajimacho Higashiku Osaka, and soon moved to the home town of his family, Kimitohge, Hashimotoshi located on the border of Naraken and Wakayamaken. For 1914–1918 he was in Kokawa Middle High School, where he read W.K. Clifford "The Common Sense of the Exact Sciences" translated into Japanese by Kikuchi Dairoku (1855–1917; Professor, the Imperial University of Tokyo). He entered into the Third High School

J. Noguchi, *Analytic Function Theory of Several Variables*,
DOI 10.1007/978-981-10-0291-5

in Kyoto 1919, and then into the Kyoto Imperial University in 1922. At the beginning he chose Physics for his major, since he did not have confidence in working on Mathematics. At a semester examination in the second year he solved a problem in Mathematics with which he was so content and impressed that he changed the course to Mathematics. Picture 1 was taken about that time. One may find this story in his essay "Shunsho Juwa (Ten talks in Spring Evenings)", which was placed in the Mainichi Daily Newspaper for ten consecutive days in February of 1963. The present author also read it, since his home was buying the newspaper. It remained something of curiosity in a boy's mind, and he read it intensively. Now, there are many records of K. Oka on a special website run by the main library of Nara Women's University [68]. One can find a considerable number of his records, not only officially published ones but also those which were not published.

After his graduation at the Kyoto Imperial University in 1925 he was employed as a lecturer there; during that time Hideki Yukawa and Shinichiro Tomonaga were in his class, both Nobel laurels later, to whom Oka left some strong impressions; they mentioned that the way of Oka's lecturing had been very different to the others'. He was interested in G. Julia's works on complex dynamics, and studied Julia's papers. Then he visited G. Julia at Paris (cf. Pic. 2 from his passport) and spent three years there, 1929–1932. During the stay at Paris he found the theme for his life's work, "Analytic Function Theory in Several Variables". After returning to Japan, he moved to Hiroshima University as associate professor in 1932, and started the research with reading Behnke and Thullen [4]. Behnke and Thullen [4] is a survey monograph and K. Oka gave the proofs for the statements there by himself, and for some which he could not, he visited the library of the Kyoto Imperial University. He lived in Hiroshima during 1932–1938, but because of a health problem, he left Hiroshima for Kimitohge (his family town) in June 1938, taking sick leave. He quit the position at Hiroshima in June 1940. (This was fortunate for him, since later in August 1945 Hiroshima was leveled by the attack of the first atomic bomb.) In October 1940 he obtained a PhD from the Kyoto Imperial University by submitting a paper summarizing the contents of Oka I–V.

Pic. 1 Student at Kyoto Imperial University, October 1923

Pic. 2 Passport Photo (1929)

In some records, he first noticed the idea of "Coherence" about 1942, and it took him for seven years to give a form. It resulted in Oka VII (1948/1950) and Oka VIII (1951); in fact, no paper was published after Oka VI (1942) for eight years. From September to December of *1943* he wrote research notes in Japanese solving Levi's Problem (Hartogs' Inverse Problem) for (unramified) Riemann domains, sent to T. Takagi as research reports (cf. the footnote (4) of Oka IX, [62]). Levi's Problem (Hartogs' Inverse Problem) in general dimension remained a big problem then. The research notes were solving it even for *Riemann domains of general dimension*. The notes in Japanese were very complete just before the translation into French: T. Nishino was writing such comments on the notes as "It is a wonder why the notes were not translated (into French)".

Oka preferred to use time not for translating the notes into French, but for thinking of an unclear new notion or concept which vaguely appeared in his mind then. During the time he just concentrated in inventing the notion and giving a form to it. At that time, no one might think that there was something essentially important but not found yet in such a fundamental object as $\mathscr{O}_{\mathbf{C}^n,a}$, the ring of convergent power series in n variables, but Oka was thinking that there should be yet something unknown, and beginning to see a shadow of what he would look for. He was not heading for the front rank of Mathematical research, but moved backward to the basis of the theme, trying to catch a truly new invention. It would require a tremendous spiritual concentration to find a really new invention, moving backward; Pic. 3 was a photo taken about that time. One may feel his spiritual tension on the night before a big historical finding. He obtained a proof of Coherence in the middle of 1946. He wrote up Oka VII in French in 1948: The paper was carried by Shizuo Kakutani on a boat crossing the Pacific Ocean and by a train on the North American continent; he traveled to Princeton, New Jersey, stopping at Chicago, where the paper was handed to André Weil (cf. [†] in Footnote 2 in p. 368, Chap. 9) and then forwarded to Henri Cartan who received it on 15th October 1948.

Pic. 3 Days thinking of "Coherence" (1943)

Around that time he had no position and lived in Kimitohge, his home town. He was receiving the support of a scholarship fund, courtesy of Teiji Takagi (then, Professor of the Imperial University of Tokyo, well-known as the originator of class field theory). In 1949 he was assigned as professor at Nara Women's University and stayed there until retirement at the age of sixty-two in 1964.

Kiyoshi Oka has received a number of awards for research achievements in Mathematics, and among them the highest should be the Order of Culture (Bunka-Kunsyō) in the fall of 1960 (Pic. 4). Rightmost is Kiyoshi Oka, and in order to left, Eiji Yoshioka (novelist), Haruo Sato (poet), and leftmost is the prime minister Hayato Ikeda.

Picture 5 was taken at the Nara Hotel together with (from left) S. Nakano, J.-P. Serre, Y. Akizuki, K. Oka, and S. Hitotsumatsu, when J.-P. Serre visited Nara to see K. Oka in 1955. Picture 6 was taken when H. Cartan visited K. Oka in Nara 1963.

In an interview published in a monthly magazine *Sugaku Seminar* (*Seminar in Mathematics*) in September 1968, when asked about mathematicians he had met, K. Oka stated that:

> During the visit to Paris from 1929 to 1932 I got a strong impressions from A. Denjoy and M. Fréchet, etc. Among those who had worked together in the field of analytic function theory of several variables and whom I met lately, it is Henri Cartan, to whom I recalled various memories. It was Henri Cartan, no one else, who had cultivated the wild field together for a long time of thirty years. I have the most remembrance thought for Cartan.[1]

[1]Cf. below "Message from Professor Henri Cartan" referred from [42].

Pic. 4 Order of Culture (1960)

Among those who I respect, I know André Weil. I had met him twice.[2] I was not personally close to him, but Akizuki liked Weil very much. Then, as listed, they are all French, but among German, I met C.L. Siegel, and H. Grauert.

Kiyoshi Oka died in Nara on 1 March 1978.

In the fall of 2001, an international conference titled "Memorial Conference of Kiyoshi Oka's Centennial Birthday on Complex Analysis in Several Variables, Kyoto/Nara 2001" for K. Oka was held at R.I.M.S. Kyoto University and at Nara Women's University (cf. [42]). The chair of the organizing committee was Toshio Nishino, a student of K. Oka. Professor Henri Cartan sent a special message to this conference [42], writing:

> Je suis heureux de m'associer à l'hommage qui sera rendu à Kiyoshi OKA à l'occasion du centième anniversaire de sa naissance. C'est en 1934 que nos relations épistolaire commencèrent. ......
>
> ......

[2]There is an interesting comment of A. Weil on K. Oka in relation to the work of Weil's integral formula at the end of Chap. 4, A. Weil "Souvenirs d'apprentissage", Birkhäuser Verlag, Basel, 1991.

Pic. 5 (With Serre), August 1955 at Nara Hotel

Pic. 6 With Cartan and Akizuki, November 1963 in Nara

> En 1963, au cours d'un long séjour au Japon, j'eus le privilège de passer une journée entière dans la ville de Nara, où enseignait alors OKA. Il me fit visiter les principaux temples de cette ville. Ce fut mon dernier contact avec cet homme exceptionnel. . . . . .

Picture 7 was taken on that occasion at a dinner on an evening during the conference. Among the overseas participants not in the picture, there were Y.-T. Siu, B. Shiffman, E. Bedford, O. Riemenschneider, among others. It was because of the name, K. Oka

Pic. 7 Names: *Top row* T. Ohsawa T. Fujimoto H. Tsuji K. Hirachi L. Lempert K. Matsumoto H. Shiga K. Miyajima T. Ueda. *Middle row* S. Matsuhara H. Oka S. Kujiraoka J. Noguchi N. Sibony H. Yamaguchi H. Kazama. *Bottom row* J.E. Fornæss P. Dolbeault J. Wermer J.J. Kohn H. Grauert M. Kuranishi S. Kobayashi T. Nishino

that so many distinguished mathematicians gathered in one place from all over the world.

Finally, the present author would like to introduce to the readers the most impressive words of K. Oka that have remained in his mind. It was when the author was an undergraduate that in a special issue of a monthly journal *Sugaku Seminar*, Aug. 1968 (*Mathematical Seminar*) a number of university professors in Mathematics were writing their recommendation books of Mathematics for undergraduate students to read for the first summer-break in university. Many were listing some well-known elementary books in mathematics to begin with. But, K. Oka wrote something completely different; he was recommending to read "Manyoshu"[3]

Read Manyoshu. For one may find for Japanese how to open the flower. .....

It has been very difficult for the author to comprehend his words, but nowadays he feels that he understands them a little, better than before.

[3]An ancient collection of Japanese poems, Waka's, by various kind of people from emperors to soldiers and farmers from the beginning of 7th C. A.C. to the middle of 8th C. A.C., which consists of about 4500 poems in 20 volumes.

Chronological List of Kiyoshi Oka

1901 April 19	: Born in Osaka
1904 Spring	: Moved to his home town, Kimitohge, Hashimoto, Wakayama
1914–1918	: Konakawa Middle High School
1919–1921	: The Third High School (Dai San Kou) in Kyoto
1922–1925	: The Kyoto Imperial University
1925–1932	: Lecturer, the Kyoto Imperial University
1929	: Married Michi Koyama
1929–1932	: Sabbatical stay in Paris
1932–1940	: Assistant Professor, Hiroshima University
1940	: PhD, the Kyoto Imperial University
1941–1942	: Research fellow, Hokkaido University
1942–1949	: Supported by Fujukai scholarship
1949–1964	: Professor, Nara Women's University
1960	: The Order of Culture (Bunka-Kunsyō)
1969–	: Professor, Kyoto Sangyo University
1978 March 1	: Died in Nara

References

1. M. Abe, S. Hamano and J. Noguchi, On Oka's extra-zero problem and examples, Math. Z. **275** (2013), 79–89.
2. A. Andreotti and R. Narasimhan, Oka's Heftungslemma and the Levi problem, Trans. Amer. Math. Soc. **111** (1964), 345–366.
3. K. Aomoto and M. Kita, Hypergeometric Function Theory, Springer, Tokyo, 1994.
4. H. Behnke and P. Thullen, Theorie der Funktionen mehrerer komplexer Veränderlichen, Ergebnisse der Mathematik und ihrer Grenzgebiete Bd. 3, Springer, Heidelberg, 1934.
5. H. Behnke and K. Stein, Entwicklung analytischer Funktionen auf Riemannschen Flächen, Math. Ann. **120** (1949), 430–461.
6. L. Bers, Introduction to Several Complex Variables, Lecture Notes, Courant Inst. Math. Sci., New York University, 1964.
7. H.J. Bremermann, Über die Äquivalenz der pseudokonvexen Gebiete und der Holomorphiegebiete im Raum von n komplexen Veränderlichen, Math. Ann. **128** (1954), 63–91.
8. H. Cartan, Sur les matrices holomorphes de n variables complexes, J. Math. pure appl. **19** (1940), 1–26.
9. H. Cartan, Idéaux de fonctions analytiques de n variables complexes, Ann. Sci. École Norm. Sup. **61** (1944), 149–197.
10. H. Cartan, Idéaux et modules de fonctions analytiques de variables complexes Bull. Soc. Math. France **78** (1950), 29–64.
11. H. Cartan, Variétés analytiques complexes et cohomologie, Colloque sur les fonctions de plusieurs variables, pp. 41–55, Bruxelles, 1953.
12. H. Cartan et J.-P. Serre, Un théorème de finitude concernant les variétés analytiques compactes, C. R. Acad. Sci. Paris **237** (1953), 128–130.
13. J.-P. Demailly, Complex Analytic and Differentiable Geometry, 2012, URL www-fourier.ujf-grenoble.fr/emailly/.
14. F. Docquier und H. Grauert, Levischen problem und Rungescher Satz für Teilgebiete Steinscher Mannigfaltigkeiten, Math. Ann. **140** (1960), 94–123.
15. P. Dolbeault, Sur la cohomologie des variétés analytiques complexes, C.R. Acad. Sci. Paris **236** (1953), 175–177.
16. J.E. Fornæss, A counterexample for the Levi problem for branched Riemann domains over C^n, Math. Ann. **234** (1978), 275–277.
17. J.E. Fornæss and B. Stensønes, Lectures on Counterexamples in Several Complex Variables, Math. Notes 33, Princeton University Press, Princeton N.J., 1987.
18. F. Forstnerič, Stein Manifolds and Holomorphic Mappings: the homotopy principle in complex analysis, Ergebnisse der Math. ihrer Grenzgebiete 56, Springer, Berlin, 2011.
19. K. Fritzsche and H. Grauert, From Holomorphic Functions to Complex Manifolds, Grad. Text Math. 213, Spinger, New York, 2002.

J. Noguchi, *Analytic Function Theory of Several Variables*,
DOI 10.1007/978-981-10-0291-5

20. R. Fujita, Domaines sans point critique intérieur sur l'espace projectif complexe, J. Math. Soc. Jpn. **15** (1963), 443–473.
21. H. Grauert, Charakterisierung der holomorph vollständigen komplexen Räume, Math. Ann. **129** (1955), 233–259.
22. H. Grauert, On Levi's problem and the imbedding of real-analytic manifolds, Ann. Math. **68** (1958), 460–472.
23. H. Grauert, Ein Theorem der analytischen Garbentheorie und die Moduleräume komplexer Strukturen, Publ. IHES **5** (1960), 233–292. Berichtigung: Publ. IHES **16** (1963), 131–132.
24. H. Grauert, Über Modifikationen und exzeptionelle analytische Mengen, Math. Ann. **146** (1962), 331–368.
25. H. Grauert, Bemerkenswerte pseudokonvexe Mannigfaltigkeiten, Math Z. **81** (1963), 377–391.
26. H. Grauert, Selected Papers, Springer, Berlin-New York, 1994.
27. H. Grauert and R. Remmert, Theorie der Steinschen Räume, Springer, 1977. Translated into English by Alan Huckleberry, Theory of Stein Spaces, Springer, 1979. Translated into Japanese by K. Miyajima, Stein Kukan Ron, Springer, Tokyo, 2009.
28. H. Grauert and R. Remmert, Coherent Analytic Sheaves, Springer, 1984.
29. R.C. Gunning and H. Rossi Analytic Functions of Several Complex Variables, Prentice-Hall, 1965.
30. H. Hironaka, Resolution of singularities of an algebraic variety over a field of characteristic zero. I, II, Ann. Math. (2) **79** (1964), 109–203; ibid. (2) **79** (1964) 205–326.
31. S. Hitotsumatsu, Analytic Function Theory of Several Variables, Baifukan, Tokyo, 1960.
32. L. Hörmander, L^2 estimates and existence theorems for the $\bar{\partial}$ operator, Acta Math. **113** (1965), 89–152
33. L. Hörmander, Introduction to Complex Analysis in Several Variables, First Edition 1966, Third Edition, North-Holland, 1989.
34. A. Kaneko, Introduction to Hyperfunctions (in Japanese), Univ. of Tokyo Press, Tokyo, 1996.
35. M. Kashiwara, Introduction to Algebraic Analysis (in Japanese), Iwanami-Shoten, Tokyo, 2008.
36. M. Kashiwara, T. Kawai and T. Kimura, Foundations of Algebraic Analysis, Princeton University Press, Princeton, N.J., 1986.
37. K. Kodaira, On Kähler varieties of restricted type, Ann. Math. **60** (1954), 28–48.
38. S. Lang, Algebra, Addison-Wesley Publ. Co., 1965.
39. E.E. Levi, Studii sui punti singolari essenziali delle funzioni analitche di due o più variabli complesse, Annali Mat. pur. appl. Ser. III, T. XVII, Milano (1910), 61–87.
40. S. Maeda, Functional Analysis (in Japanese), Morikita Publ. Co., Tokyo, 1974.
41. Y. Matsushima, Differential Manifolds, Dekker, New York, 1972; translated from Selected Series of Mathematics 5, Shokabo, Tokyo, 1965.
42. K. Miyajima et al. (Eds.), Complex Analysis in Several Variables—Memorial Conference of Kiyoshi Oka's Centennial Birthday, Kyoto/Nara 2001, Adv. Studies Pure Math. 42, Math. Soc. Jpn., Tokyo, 2004.
43. S. Morita, Introduction to Algebra (in Japanese), Selected Series of Mathematics 9, Shokabo, Tokyo, 1987.
44. S. Murakami, Manifolds (second edition, in Japanese), Kyoritsu Publ., Tokyo, 1989.
45. M. Nagata, Theory of Commutative Fields (in Japanese), Selected Series of Mathematics 6, Shokabo, 1967; Trans. Math. Mono. 125, Amer. Math. Soc. Providence, 1993.
46. R. Narasimhan, The Levi problem for complex spaces, II, Math. Ann. **146** (1962), 195–216.
47. R. Narasimhan, The Levi problem in the theory of functions of several complex variables, Proc. Internat. Congr. Mathematicians (Stockholm, 1962), pp. 385–388, Inst. Mittag-Leffler, Djursholm, 1963.
48. R. Narasimhan, Introduction to the Theory of Analytic Spaces, Lecture Notes in Math. 25, Springer, 1966.

49. T. Nishino, Function Theory in Several Complex Variables (in Japanese), The University of Tokyo Press, Tokyo, 1996; Translation into English by N. Levenberg and H. Yamaguchi, Amer. Math. Soc. Providence, R.I., 2001.
50. T. Nishino, Mathematics of Professor Kiyoshi Oka: Birth of "Idéaux de Domaines indéterminés" (in Japanese), Sugaku **49** (1997), pp. 144–157, Math. Soc. Jpn.; English translation by M. Naitou and T. Kizuka, Sugaku Expositions **12** No. 1, pp. 107–126, Amer. Math. Soc., 1999.
51. T. Nishino, Mathematics of Professor Oka—a landscape in his mind—(in Japanese), translated by T. Ohsawa, Complex Analysis in Several Variables—Memorial Conference of Kiyoshi Oka's Centennial Birthday, Kyoto/Nara 2001, Ed. K. Miyajima et al., Adv. Studies Pure Math. **42**, pp. 17–30, Math. Soc. Jpn., Tokyo, 2004.
52. J. Noguchi, Introduction to Complex Analysis, MMONO 168, Amer. Math. Soc. Rhode Island, 1997; translated from the original in Japanese, Shokabo, Tokyo, 1993.
53. J. Noguchi, Another direct proof of Oka's theorem (Oka IX), J. Math. Sci. Univ. Tokyo **19** (2012), 1–15.
54. J. Noguchi, A remark to a division algorithm in the proof of Oka's First Coherence Theorem, Internat. J. Math. **26** (4) (2015), DOI:10.1142/S0129167X15400054, 8 pp.
55. J. Noguchi, Inverse of Abelian integrals and ramified Riemann domains, to appear in Math. Ann.
56. J. Noguchi and T. Ochiai, Geometric Function Theory in Several Complex Variables, Math. Mono. 80, Amer. Math. Soc., 1990.
57. J. Noguchi and J. Winkelmann, Nevanlinna Theory in Several Complex Variables and Diophantine Approximation, Grundl. der Math. Wiss. Vol. 350, Springer, Tokyo-Heidelberg-New York-Dordrecht-London, 2014.
58. F. Norguet, Sur les domains d'holomorphie des fonctions uniformes de plusieurs variables complexes (Passage du local au global), Bull. Soc. Math. France **82** (1954), 137–159.
59. T. Ohsawa, Analysis of Several Complex Variables, Iwanami Shoten, Tokyo, 1998; translated by S. Gilbert Nakamura into Englsih, Transl. Math. Mono. vol. 211, Amer. Math. Soc., Providence, R.I., 2002.
60. T. Ohsawa, L^2 Approaches in Several Complex Variables, Springer Mono. Math., Springer, Tokyo, 2015.
61. K. Oka, Note sur les familles de fonctions multiformes etc., J. Sci. Hiroshima Univ. **4** (1934), 93–98 *[Rec. 20 jan 1934]*.
62. K. Oka, Sur les fonctions analytiques de plusieurs variables:
I—Domaines convexes par rapport aux fonctions rationnelles, J. Sci. Hiroshima Univ. Ser. A **6** (1936), 245–255 *[Rec. 1 mai 1936]*.
II—Domaines d'holomorphie, J. Sci. Hiroshima Univ. Ser. A **7** (1937), 115–130 *[Rec. 10 déc 1936]*.
III—Deuxième problème de Cousin, J. Sci. Hiroshima Univ. **9** (1939), 7–19 *[Rec. 20 jan 1938]*.
IV—Domaines d'holomorphie et domaines rationnellement convexes, Jpn. J. Math. **17** (1941), 517–521 *[Rec. 27 mar 1940]*.
V—L'intégrale de Cauchy, Jpn. J. Math. **17** (1941), 523–531 *[Rec. 27 mar 1940]*.
VI—Domaines pseudoconvexes. Tôhoku Math. J. **49** (1942(+43)), 15–52 *[Rec. 25 oct 1941]*.
VII—Sur quelques notions arithmétiques, Bull. Soc. Math. France **78** (1950), 1–27 *[Rec. 15 oct 1948]*.
VIII—Lemme fondamental, J. Math. Soc. Jpn. **3** (1951) No. 1, 204–214; No. 2, 259–278 *[Rec. 15 mar 1951]*.
IX—Domaines finis sans point critique intérieur, Jpn. J. Math. **23** (1953), 97–155 *[Rec. 20 oct 1953]*.
X—Une mode nouvelle engendrant les domaines pseudoconvexes, Jpn. J. Math. **32** (1962), 1–12 *[Rec. 20 sep 1962]*.

63. K. Oka, Sur les domaines pseudoconvexes, Proc. of the Imperial Academy, Tokyo (1941), 7–10 *[Comm. 13 jan 1941]*.
64. K. Oka, Note sur les fonctions analytiques de plusieurs variables, Kōdai Math. Sem. Rep. (1949), no. 5–6, 15–18 *[Rec. 19 déc 1949]*.
65. K. Oka, Sur les fonctions analytiques de plusieurs variables, Iwanami Shoten, Tokyo, 1961.
66. K. Oka, Posthumous Papers of Kiyoshi Oka, Eds. T. Nishino and A. Takeuchi, Kyoto, 1980–1983: URL http://www.lib.nara-wu.ac.jp/oka/.
67. K. Oka, Collected Works, Translated by R. Narasimhan, Ed. R. Remmert, Springer, Berlin-Heidelberg-New York-Tokyo, 1984.
68. K. Oka Digital Archives, Library of Nara Women's University, URL http://www.lib.nara-wu.ac.jp/oka/.
69. R. Remmert, Local Theory of Complex Spaces, Chap. I, Several Complex Variables VII, Ency. Math. Vol. 74, Springer, Berlin etc. 1994.
70. T. Sakai, Riemannian Geometry (in Japanese), Shokabo, Tokyo, 1992; translated into English by T. Sakai, Transl. Math. Mono. Vol. 149, Amer. Math. Soc., Providence, R.I., 1996.
71. L. Schwartz, Homomorphismes et applications complèment continues, C. R. l'Acad. Sci., Paris **236** (1953), 2472–2473.
72. J.-P. Serre, Faisceaux algébriques cohérents, Ann. Math. **61** No. 2 (1955), 197–278.
73. K. Stein, Topologische Bedingungen für die Existenz analytischer Funktionen komplexer Veränderichen zu vorgegebenen Nullstellenflächen, Math. Ann. **117** (1941), 727–757.
74. A. Takeuchi, Domaines pseudoconvexes infinis et la métrique riemannienne dans un espace projectif, J. Math. Soc. Jpn. **16** (1964), 159–181.
75. T. Tanisaki and R. Hotta, D-Modules and Algebraic Groups (in Japanese), Springer, Tokyo, 1995.
76. F. Trèves, Topological vector spaces, distributions and kernels, Pure appl. Mat. 25, Academic Press, New York, 1967.
77. T. Ueda, Modifications continues des variét'es de Stein, Publ. RIMS Kyoto Univ. **13** (1977), 681–686.
78. A. Weil, Introduction à l'Étude des Variétés Kähleriennes, Hermann, 1958.
79. T. Yamanaka, Theory of Topological Vector Spaces and Generalized Functions (in Japanese), Kyoritsu Publ. Co., Tokyo, 1966.

Index

J. Noguchi, *Analytic Function Theory of Several Variables*,
DOI 10.1007/978-981-10-0291-5

Symbols

J. Noguchi, *Analytic Function Theory of Several Variables*,
DOI 10.1007/978-981-10-0291-5

Added at galley-proof: A Simplified Proof of Cartan's Matrix Decomposition Lemma 4.2.2

In the arguments of Sect. 4.2 it was rather involved to determine a number of positive constants. Here, we give another simplified argument to determine the constants; we will not use exp $(\cdot)$ or log $(\cdot)$ of matrices.

In general, we denote the operator norm of a complex (p, p)-matrix A by

$$\|A\| = \max\{\|A\xi\|;\ \xi \in \mathbf{C}^p,\ \|\xi\| = 1\}.$$

We write $\|A\|_E = \sup\{\|A(z)\|;\ z \in E\}$ for a (p, p)-matrix valued function $A = A(z)$ in a subset $E \subset \mathbf{C}^n$. Let B be another (p, p)-matrix. Then the following holds:

(i) For elements a_{ij} of A, $\frac{1}{p}\|A\| \le \max_{i,j}\{|a_{ij}|\} \le \|A\|$.

(ii) $\|A + B\| \le \|A\| + \|B\|, \quad \|AB\| \le \|A\| \cdot \|B\|$.

(iii) For $A = A(z)$ $(z \in E)$ with $\|A\|_E \le \varepsilon(\text{constant}) < 1$, $\mathbf{1}_p - A(z)$ is invertible, and

$$(\mathbf{1}_p - A(z))^{-1} = \mathbf{1}_p + A(z) + A(z)^2 + \cdots .$$

Here, the right-hand side converges uniformly on E, and

$$\|(\mathbf{1}_p - A(z))^{-1}\| \le \frac{1}{1-\varepsilon}, \quad \|(\mathbf{1}_p - A(z))^{-1} - \mathbf{1}_p\| \le \frac{\|A(z)\|}{1-\varepsilon}.$$

In particular, if $\varepsilon = \frac{1}{2}$, $\|(\mathbf{1}_p - A(z))^{-1}\| \le 2$.

(iv) Suppose that for $k = 0, 1, \ldots$, positive numbers ε_k with $0 < \varepsilon_k < 1$ and (p, p)-matrix valued functions $A_k(z)$ $(z \in E)$ are given, so that $\|A_k\|_E \le \varepsilon_k$ and $\sum_{k=0}^{\infty} \varepsilon_k < \infty$. Then the following two infinite products

$$\lim_{k\to\infty} (\mathbf{1}_p - A_0(z)) \cdots (\mathbf{1}_p - A_k(z)), \quad \lim_{k\to\infty} (\mathbf{1}_p - A_k(z)) \cdots (\mathbf{1}_p - A_0(z))$$

converge uniformly on E.

The item (iv) follows from $\| \prod_{j=0}^{k} (\mathbf{1}_p - A_j)\|_E \le \prod_{j=0}^{k} (1 + \|A_j\|_E)$ and

$$\left\| \prod_{j=k}^{l} (\mathbf{1}_p - A_j) - \mathbf{1}_p \right\|_E \le \prod_{j=k}^{l} (1 + \|A_j\|_E) - 1 \le \exp\left(\sum_{j=k}^{l} \|A_j\|_E \right) - 1, \quad k < l.$$

These properties will be used without specific mentioning.

Proof of Lemma 4.2.2: Now, let A be the one given in Lemma 4.2.2. We use the notation, $\tilde{E}'_{(k)}, \tilde{E}''_{(k)}$, etc. for closed cubes defined in Sect. 4.2. We set $A = \mathbf{1}_p - B_1$. Applying (4.2.6) for this B_1, we have

(1) $$B_1 = B'_1 + B''_1,$$

(2) $$\|B'_1\|_{\tilde{E}'_{(2)}} \le \frac{1}{2\pi} \frac{2^2}{\delta} L \|B_1\|_{\tilde{E}'_{(1)} \cap \tilde{E}''_{(1)}}, \quad \|B''_1\|_{\tilde{E}''_{(2)}} \le \frac{1}{2\pi} \frac{2^2}{\delta} L \|B_1\|_{\tilde{E}'_{(1)} \cap \tilde{E}''_{(1)}}.$$

Assuming that $(\mathbf{1}_p - B_1')^{-1}$ and $(\mathbf{1}_p - B_1'')^{-1}$ exist, we put

$$M(B_1', B_1'') = (\mathbf{1}_p - B_1')^{-1}(\mathbf{1}_p - B_1' - B_1'')(\mathbf{1}_p - B_1'')^{-1} = \mathbf{1}_p - N(B_1', B_1'').$$

Then, $A = \mathbf{1}_p - B_1 = (\mathbf{1}_p - B_1')M(B_1', B_1'')(\mathbf{1}_p - B_1'')$. We repeat the process with replacing A by $M(B_1', B_1'')$:

$$A = (\mathbf{1}_p - B_1')(\mathbf{1}_p - B_2')M(B_2', B_2'')(\mathbf{1}_p - B_2'')(\mathbf{1}_p - B_1'').$$

The following estimate is the key for us to repeat this process with convergence.

Lemma 3 *Let P and Q be (p, p)-matrices with* $\max\{\|P\|, \|Q\|\} \le \frac{1}{2}$. *Then,*

$$\|N(P, Q)\| \le 2^2(\max\{\|P\|, \|Q\|\})^2.$$

Proof Noting that $(\mathbf{1}_p - Q)^{-1} = \mathbf{1}_p + Q(\mathbf{1}_p - Q)^{-1} = \mathbf{1}_p + Q + Q^2(\mathbf{1}_p - Q)^{-1}$, we have that

$$\begin{aligned}
M(P, Q) &= (\mathbf{1}_p - P)^{-1}(\mathbf{1}_p - P - Q)(\mathbf{1}_p - Q)^{-1} \\
&= (\mathbf{1}_p - (\mathbf{1}_p - P)^{-1}Q)(\mathbf{1}_p - Q)^{-1} \\
&= \mathbf{1}_p + Q + Q^2(\mathbf{1}_p - Q)^{-1} \\
&\quad - (\mathbf{1}_p + P(\mathbf{1}_p - P)^{-1})Q(\mathbf{1}_p + Q(\mathbf{1}_p - Q)^{-1}) \\
&= \mathbf{1}_p + Q + Q^2(\mathbf{1}_p - Q)^{-1} \\
&\quad - Q - Q^2(\mathbf{1}_p - Q)^{-1} - P(\mathbf{1}_p - P)^{-1}Q(\mathbf{1}_p - Q)^{-1} \\
&= \mathbf{1}_p - P(\mathbf{1}_p - P)^{-1}Q(\mathbf{1}_p - Q)^{-1}, \\
N(P, Q) &= P(\mathbf{1}_p - P)^{-1}Q(\mathbf{1}_p - Q)^{-1}.
\end{aligned}$$

It follows from the assumption that

$$\|N(P, Q)\| \le \|P\| \cdot 2 \cdot \|Q\| \cdot 2 = 2^2(\max\{\|P\|, \|Q\|\})^2. \qquad \triangle$$

Set

$$\varepsilon_1 = \max\left\{\|B_1'\|_{\tilde{E}'_{(2)}}, \|B_1''\|_{\tilde{E}''_{(2)}}\right\} \left(\le \frac{2L}{\pi\delta}\|B_1\|_{\tilde{E}'_{(1)} \cap \tilde{E}''_{(1)}}\right).$$

Taking $\delta > 0$ smaller if necessary, we may assume that $\frac{\pi\delta}{2^5 L} \le \frac{1}{2}$. Assume that $\|B_1\|_{\tilde{E}'_{(1)} \cap \tilde{E}''_{(1)}} \le \frac{\pi^2\delta^2}{2^6 L^2}$. Then,

$$\varepsilon_1 \le \frac{\pi\delta}{2^5 L} \le \frac{1}{2}. \tag{4}$$

Inductively, we assume that for $j = 1, \ldots, k$, (p, p)-matrix valued holomorphic functions

$$B_j'(z)\ (z \in \tilde{E}'_{(j+1)}),\ \ B_j''(z)\ (z \in \tilde{E}''_{(j+1)}),\ \text{and}\ \ \varepsilon_j := \max\left\{\|B_j'\|_{\tilde{E}'_{(j+1)}}, \|B_1''\|_{\tilde{E}''_{(j+1)}}\right\}$$

are determined, so that (the case of $j=1$ is due to (4))

$$(5) \qquad \varepsilon_j \leq \frac{\pi\delta}{2^{j+4}L}\left(\leq \frac{1}{2^j}\right), \quad 1 \leq j \leq k,$$

$$(6) \qquad A(z) = (\mathbf{1}_p - B'_1(z)) \cdots (\mathbf{1}_p - B'_k(z)) \cdot (\mathbf{1}_p - N(B'_k(z), B''_k(z))) \cdot (\mathbf{1}_p - B''_k(z)) \cdots (\mathbf{1}_p - B''_1(z)), \quad z \in \tilde{E}'_{(k+1)} \cap \tilde{E}''_{(k+1)}.$$

For $z \in \tilde{E}'_{(k+2)} \cap \tilde{E}''_{(k+2)}$, we set $B_{k+1}(z) = N(B'_k(z), B''_k(z))$, and by making use of $\gamma'_{(k+1)}, \gamma''_{(k+1)}$ defined in Sect. 4.2, we set

$$B'_{k+1}(z', z_n) = \frac{1}{2\pi i}\int_{\gamma'_{(k+1)}} \frac{B_{k+1}(z', \zeta)}{\zeta - z_n} d\zeta, \quad (z', z_n) \in \tilde{E}'_{(k+2)},$$

$$B''_{k+1}(z', z_n) = \frac{1}{2\pi i}\int_{\gamma''_{(k+1)}} \frac{B_{k+1}(z', \zeta)}{\zeta - z_n} d\zeta, \quad (z', z_n) \in \tilde{E}''_{(k+2)}.$$

Here, note that $|\zeta - z_n| \geq \frac{\delta}{2^{k+2}}$ in the above integrands; we hence infer from (5) and Lemma 3 that

$$\varepsilon_{k+1} \leq \frac{L}{2\pi}\frac{2^{k+2}}{\delta}\|N(B'_k, B''_k)\|_{\tilde{E}'_{(k+1)} \cap \tilde{E}''_{(k+1)}} \leq \frac{L}{2\pi}\frac{2^{k+2}}{\delta}2^2\varepsilon_k^2 \leq \frac{1}{2}\varepsilon_k \leq \frac{\pi\delta}{2^{k+5}L};$$

$$\mathbf{1}_p - B_{k+1} = (\mathbf{1}_p - B'_{k+1})(\mathbf{1}_p - N(B'_{k+1}, B''_{k+1}))(\mathbf{1}_p - B''_{k+1}) \text{ on } \tilde{E}'_{(k+2)} \cap \tilde{E}''_{(k+2)}.$$

Thus, (5) and (6) hold for "$k+1$".

Therefore we see by (5) that the following infinite products

$$A'(z) = \lim_{k\to\infty}(\mathbf{1}_p - B'_1(z)) \cdots (\mathbf{1}_p - B'_k(z)), \quad z \in \tilde{E}' := \bigcap_{k=1}^{\infty} \tilde{E}'_{(k)},$$

$$A''(z) = \lim_{k\to\infty}(\mathbf{1}_p - B''_k(z)) \cdots (\mathbf{1}_p - B''_1(z)), \quad z \in \tilde{E}'' := \bigcap_{k=1}^{\infty} \tilde{E}''_{(k)}$$

converge uniformly on each defined domain, and give rise to (p, p)-matrix valued holomorphic functions in their interiors.

For $z \in \tilde{E}' \cap \tilde{E}''$, we have that

$$\|N(B'_k(z), B''_k(z))\| \leq 2^2\varepsilon_k^2 \longrightarrow 0 \quad (k \to \infty).$$

Thus, from (6) we obtain $A(z) = A'(z)A''(z) = A'(z)(A''(z)^{-1})^{-1}$. □

Remark 6 (Estimate) In Lemma 4.2.2 there are positive constants η, C and closed cube neighborhood $\tilde{E}'$ (resp. $\tilde{E}''$) of E' (resp. E''), dependent only on E', E'' and U such that (i) $\tilde{E}' \cap \tilde{E}'' \subset U$; (ii) if $A = \mathbf{1}_p - B$ with $\|B\|_U \leq \eta$, then there are $A' = \mathbf{1}_p - B'$ and $A'' = \mathbf{1}_p - B''$ satisfying

$$\mathbf{1}_p - B(z) = (\mathbf{1}_p - B'(z))(\mathbf{1}_p - B''(z)), \quad z \in \tilde{E}' \cap \tilde{E}'',$$
$$\max\{\|B'\|_{\tilde{E}'}, \|B'''\|_{\tilde{E}''}\} \leq C\|B\|_U.$$

Printed in the United States
By Bookmasters